Understanding and Managing Emerald Ash Borer Impacts on Ash Forests

Understanding and Managing Emerald Ash Borer Impacts on Ash Forests

Special Issue Editors

Randall K. Kolka
Anthony W. D'Amato
Nam Jin Noh
Brian J. Palik
Thomas Pypker
Robert Slesak
Joseph W. Wagenbrenner

MDPI • Basel • Beijing • Wuhan • Barcelona • Belgrade

MDPI

Special Issue Editors

Randall K. Kolka	Anthony W. D'Amato	Nam Jin Noh
Northern Research Station	The University of Vermont	Michigan Technological University
USA	USA	USA

Brian J. Palik	Thomas Pypker	Robert Slesak
Northern Research Station	Thompson Rivers University	Minnesota Forest Resources Council
USA	Canada	USA

Joseph W. Wagenbrenner
USDA Forest Service, Pacific Southwest Research Station
USA

Editorial Office
MDPI
St. Alban-Anlage 66
Basel, Switzerland

This is a reprint of articles from the Special Issue published online in the open access journal *Forests* (ISSN 1999-4907) in 2018 (available at: http://www.mdpi.com/journal/forests/special_issues/EAB)

For citation purposes, cite each article independently as indicated on the article page online and as indicated below:

LastName, A.A.; LastName, B.B.; LastName, C.C. Article Title. *Journal Name* **Year**, *Article Number*, Page Range.

ISBN 978-3-03897-164-1 (Pbk)
ISBN 978-3-03897-165-8 (PDF)

Cover image courtesy of Leah Bauer's group.

Contents

About the Special Issue Editors

Randall K. Kolka, Dr., Research Soil Scientist, holds a B.S. degree in Soil Science from the University of Wisconsin-Stevens Point, and MS and PhD degrees in Soil Science from the University of Minnesota. Following his PhD in 1996, he was a post-doctoral Research Soil Scientist with the USDA Forest Service's Southern Research Station in South Carolina. In 1998 he became an Assistant Professor of Forest Hydrology in the Department of Forestry at the University of Kentucky. In 2002, he became Team Leader and Research Soil Scientist with the USDA Forest Service's Northern Research Station in Grand Rapids, MN. He currently leads a team of scientists, graduate students and post-docs conducting research on the cycling of water, carbon, nutrients, mercury and other pollutants at the plot to watershed scale in urban, agricultural, forested, wetland and aquatic ecosystems across the globe. He is an adjunct faculty member at seven universities and has published over 200 scientific articles.

Anthony W. D'Amato, Dr., is a Professor of Silviculture and Applied Forest Ecology and Director of the Forestry Program at the University of Vermont. His research focuses on long-term forest dynamics, disturbance effects on ecosystem structure and function, and silvicultural strategies for conferring adaptation potential within the context of global change. He received his B.S. in Forest Ecosystem Science from the University of Maine, M.S. in Forest Science from Oregon State University, and PhD in Forest Resources from University of Massachusetts. He was a faculty member for seven years at the University of Minnesota prior to joining the University of Vermont in January 2015.

Nam Jin Noh, Dr., is an ecosystem ecologist with broad expertise in climate change and forest management impacts on carbon and nutrient cycling. His research interests lie in understanding biogeochemical plant–soil feedback mechanisms in terrestrial ecosystems under climate change. He earned a PhD in Bioresource and Ecology from Korea University. He currently works at the Hawkesbury Institute for the Environment, Western Sydney University. He led the workshop on the Future of Ash Forests as a Postdoctoral Researcher at the School of Forest Resources and Environmental Science, Michigan Technological University. Previously he held research positions at River Basin Research Center, Gifu University, Japan.

Brian J. Palik, Dr., Research Ecologist, is science leader for applied forest ecology with the USDA Forest Service–Northern Research Station, in Grand Rapids MN. He has Ph.D. and M.S. degrees in forestry and ecology from Michigan State University and a B.S. from Alma College. He works broadly on questions related to the ecological sustainability of managed forests through use of operational-scale and long-term silviculture research.

Thomas Pypker, Dr., Associate Professor, is an ecohydrologist whose research focuses on how disturbance affects biophysical processes of managed ecosystems at scales ranging from the leaf to whole watersheds. The goal of his research is to provide insight into how management and climate change will affect the hydrology, stability and productivity of the Earth's ecosystems. Dr. Pypker earned a PhD in Forest Science from Oregon State University. He is currently faculty in the Department of Natural Resource Sciences at Thompson Rivers University in Kamloops, BC, Canada.

Robert Slesak, Dr., Director of Applied Research and Monitoring, directs the research program of the Minnesota Forest Resources Council in St. Paul MN. He has a Ph.D in Forest Soils from Oregon State University, and a M.S. and B.S. in ecosystem science and forestry from SUNY Environmental Science and Forestry. His research broadly focuses on sustainable forest management and the assessment of ecosystem functions related to hydrology and soils.

Joseph W. Wagenbrenner, Dr., is a hydrologist with broad research interests in disturbance and land use hydrology. He studies the effects of forest disturbances and management on runoff generation, soil erosion, and watershed processes and outputs. He earned a PhD in Biological and Agricultural Engineering from Washington State University. He currently works at the Pacific Southwest Research Station of the USDA Forest Service.

Preface to "Understanding and Managing Emerald Ash Borer Impacts on Ash Forests"

1. Introduction

Emerald ash borer (EAB) is rapidly spreading throughout Eastern North America, greatly impacting ecosystems where ash is a component tree. This rapid loss of ash trees has already resulted in ecological impacts on both terrestrial and aquatic ecosystems and is projected to be even more severe as EAB invades the ash-dominated wetlands of the western Great Lakes region. Due to the impact EAB is having on ash forests and the impending peril of ash-dominated wetlands as EAB spreads, we produced this Special Issue dedicated to "Understanding and Managing Emerald Ash Borer Impacts on Ash Forests". We are excited about the compilation of the 19 articles produced in this Special Issue that span the range of the entomology of the EAB, to the ecology of ash forests, to management recommendations for ash forests that are threatened by EAB. The idea for this Special Issue was spawned from a workshop on the Science and Management of Ash Forests After EAB held in Duluth, MN, on July 25–27, 2017. At that workshop we had 25 oral presentations and 21 poster presentations, and over 180 attendees that included representatives from county, state and federal agencies, universities, nonprofit organizations, Native American bands, and even private landowners. Of the 19 papers in this Special Issue, 14 were the result of presentations or ideas generated at the workshop.

The papers in this Special Issue can be grouped into five general categories: 1) entomology and biocontrol of EAB; 2) vegetation responses; 3) soil-related responses; 4) other ecosystem-related responses; and 5) adaptive forest management options and controlling the movement of EAB.

2. Entomology and Biocontrol of EAB

Biocontrol of EAB is an option to ameliorate EAB from infected ash ecosystems, however, our understanding of effective techniques is lacking. Using insecticides also has the potential to save individual trees. Finally, limitations from climate factors may be keeping EAB from spreading. In this Special Issue, three papers address the biocontrol of EAB, one assesses the use of insecticides on the genetic diversity of ash, and one paper looks at the cold hardiness of EAB.

2.1. Biocontrol of EAB

Duan et al. (2018) reviewed the premises, regulations, history, and challenges of developing a biological control program in North America to help combat EAB. The authors discussed the critical role of natural enemies in suppressing EAB populations in the pest's native range (Asia), as well as the effectiveness of some introduced Asiatic biocontrol agents in protecting regenerating ash against EAB in North America. Biological control of EAB began in southern Michigan in 2007 after the approval and release of three hymenopteran parasitoid species that specialize on EAB in Asia, with a fourth parasitoid species added in 2015. After reviewing the progress in EAB parasitoid releases and establishment in the U.S and Canada, the authors summarized the critical findings from a long-term field study in southern Michigan, which demonstrates the significant role that Tetrastichus planipennisi, an introduced larval parasitoid of EAB from China, plays in suppressing EAB densities in small ash trees and saplings in the aftermath of EAB invasion. Duan et al. (2018) concludes that parasitoids can provide significant EAB biocontrol services and enhance ash survival, but more work is needed to: "(1) evaluate the establishment and impact of biocontrol agents in different climate zones; (2) determine the combined effect of EAB biocontrol and host plant resistance or tolerance on the regeneration of North American ash species; and (3) expand foreign exploration for EAB natural enemies throughout Asia."

In associated research, Kashian et al. (2018) sampled parasitoid-release and non-release (control) plots in central and southeast Michigan to measure the health and recruitment of ash species from 2012–2015. Over this three-year period, they found a reduced mortality of the larger ash trees (> 4 cm diameter at breast height) and a generally greater diameter in the larger tree category in parasitoid-release plots compared to control plots. The results for ash saplings (2.5–4 cm diameter) were more variable for both mortality and diameter, and appeared to be location specific (Kashian et al. 2018). They concluded that "EAB biocontrol is likely to have a positive effect on ash populations, but that the study duration was not long enough to definitively deduce the long-term success of the biocontrol program in this region."

Margulies et al. (2018) conducted a similar study as Kashian et al. (2018) in southern Michigan, comparing vegetation responses in plots where parasitoid biocontrol agents were released vs. plots that had no biocontrol. They used hierarchical models to understand the important drivers of the re-establishment of ash saplings (< 10 cm diameter, but greater than 1 m in height) following EAB invasion and whether biocontrol has an effect on the establishment of invasive and weedy species. Margulies et al. (2018) found that the greater number of parasitoids released led to higher numbers of ash saplings. The sites with higher numbers of ash saplings also had lower densities of invasive woody vegetation. They conclude that the "protection of ash saplings by the biocontrol agent may help native recruitment during forest transition by supporting the growth of native hardwood seedlings over invasive and weedy species," and suggest that "research on the efficacy of EAB biocontrol should include all ash size classes and the community dynamics of co-occurring species."

2.2. Insecticide Control of EAB and Relationships with Genetic Diversity

The use of insecticides to treat individual ash trees has been very successful, but is expensive, up to about \$150/tree every three years (Flower et al. 2018). In some cases this cost is feasible in urban environments, but to preserve native natural stand genetic diversity we need to consider the optimal treatment densities in the wild. Flower et al. (2018) established 27 plots across 17 populations of white ash in the Allegheny National Forest in Pennsylvania and measured genetic diversity across a gradient in the percentage of ash trees treated with insecticide (between 10%–95% of ash trees in the individual stands were treated). They found that genetic diversity was similar in treated and untreated trees, with little evidence of genetic differentiation or inbreeding, indicating that insecticide treatments were conserving local genetic diversity. By conducting a series of simulations, they indicated that the best practice to maintain genetic diversity in white ash was to treat more populations rather than more trees within a population (Flower et al. 2018). Their results suggest that forest managers can use genetic screening to "select highly diverse and unique populations to maximize diversity and reduce expenditures (by up to 21%) to help practitioners develop cost-effective strategies to conserve genetic diversity."

2.3. Cold Hardiness of EAB

One fortunate obstacle that may be keeping EAB at bay is the inability of larvae to survive in extremely cold climates. Moreover, the species of ash might influence EAB's cold hardiness. Although some research has been carried out into EAB cold hardiness in green ash, little has been studied in black ash. Christianson and Venette (2018) compared the cold hardiness of EAB larvae in both ash species over three years by using artificially- and naturally-infested ash trees that were cut into logs. The larvae were extracted from the logs and cooled to one of five temperature treatments as low as −40 °C. They found few differences in cold hardiness between EAB larvae extracted from black ash vs. green ash logs. The ash species did not consistently affect the EAB mortality rates when cooled. Independent of the ash species, once temperatures got to −35 °C, EAB mortality was high (Christianson and Venette 2018). They conclude that regions that reach these winter temperatures "may provide thermal refugia that are vital to the local persistence of native ash stands."

3. Vegetation Responses to EAB

Understanding the vegetation response once EAB has infected and removed ash from the canopy is important for the future management of those stands. Depending on the areal density of the ash, removing canopy trees changes the water and nutrient balance and light environment, which cumulatively affects how the sub-canopy vegetation will respond. In the Special Issue there are five papers that address vegetation responses following simulated or actual EAB invasions. Two papers measure the movement of water up the stem of ash (i.e., sap flux), one paper measures regeneration and other seedling responses, and one paper assesses the effect on neighboring trees and nutrient dynamics.

3.1. Sap Flux of Ash Following EAB Invasion

One of the important drivers that kills ash following EAB invasion is the tree's inability to uptake water as a result of the phloem and xylem being destroyed by EAB. Flower et al. (2018) measured sap flux, leaf gas exchange and morphology, isotopes, and carbohydrates on mature green ash in Ohio. Their data indicate that EAB damage can decrease tree water use by as much as 80% and decrease the leaf area and leaf mass. The authors concluded that altered leaf gas exchange, increased water stress and reduced photosynthetic rates indicate that EAB damages the xylem in addition to the phloem (Flower et al. 2018).

In an effort to understand the influence of black ash on the hydrologic cycle in black-ash-dominated wetlands in the Upper Peninsula of Michigan, Shannon et al. (2018) measured sap flux rates and the vapor pressure deficit at various water table levels of three common species found in those wetlands. In comparing red maple and yellow birch to black ash, black ash had higher sap flux rates at all water table levels than the non-ash species. Black ash showed a significant increase in sap flux as the water tables decreased, whereas the non-ash species did not. Black ash, and to a lesser degree, red maple, had greater response to increasing vapor pressure deficits as the water tables decreased. Overall, the data from Shannon et al. (2018) indicate that black ash appears to be the most adapted to the wetland conditions and they conclude that "understanding how a replacement species will respond to the expected altered hydrologic regimes of black ash wetlands following EAB infestation will improve species selection."

3.2. Regeneration and Seedling Responses to EAB Invasions

Following EAB invasion, Dolan and Kilgore (2018) measured the impacts of ash mortality on the recruitment of woody and non-woody vegetation in 14 plots in Ohio and Pennsylvania. Their plot network had varying percentages of ash in the overall basal area, but generally they were dominated by other species. Using change in the relative basal of ash, they found that changes in canopy cover were not correlated with the loss of ash and only the density of shade-tolerant shrubs increased with increasing ash mortality. The lack of change in canopy cover underscores the low abundance of ash in these stands as other species were likely able to take advantage to expand their canopies. They also found that native and non-native shrub species increased in sites where they were present before EAB (Dolan and Kilgore 2018). The authors conclude that the "shifts in understory vegetation indicate that ash mortality enhances the rate of succession to shade-tolerant species."

In addition to regeneration following EAB invasion, the planting of seedlings may facilitate future canopy cover. Bolton et al. (2018) used simulated short (girdled ash) and long-term (ash cut and left in place) EAB invasions to assess planted seedling responses in the Ottawa National Forest in the Upper Peninsula of Michigan. They also planted seedlings in a municipal forest near Superior, Wisconsin. Silver maple, red maple, American elm and northern white cedar may be a good option for planting, as they experienced significantly better growth and survival. In contrast, tamarack, black spruce, balsam fir and yellow birch did not survive well. The microsite was also an important factor in survival, with those seedlings planted on hummocks having much higher survival than those on hollows (Bolton et al. 2018). They conclude that "regional landowners and forest managers can use these results to help mitigate the canopy and structure losses from EAB and maintain forest cover and hydrologic function in black ash-dominated wetlands after infestation."

3.3. EAB Effects on Neighboring Trees and Nutrient Dynamics

Following the death of ash trees from EAB, nearby trees maybe susceptible to similar fates because of their nearness to other infected trees. Kappler et al. (2018) examined tree health and soil nutrients following the death of >90% of green ash in a floodplain environment in Ohio. Although they tested a suite of A-horizon nutrients, comparing areas without trees to areas with remaining ash, only sulfur and phosphorus varied, with both being greater in the ash-occupied areas (Kappler et al. 2018). As a result, they were not able to make a direct connection between the health of ash and soil nutrients. The remaining ash trees were healthiest as the number of neighboring dead ash declined. The ash trees were healthiest if there was a minimum of 6 m to the nearest ash neighbor. No other neighboring tree species had any impacts on ash health. The authors conclude that their research, "highlights scale-dependent neighborhood composition drivers of tree susceptibility to pests and suggests that drivers during initial infestation differ from drivers in aftermath forests."

4. Other Ash Ecosystem Responses Following Invasion of EAB

Other than plant responses, EAB invasions have been shown to have impacts on soils, fauna, greenhouse gas fluxes and other ecosystem functions. During the removal of an important canopy species, cascading effects can be seen through the ecosystem, starting with increases in dead wood inputs and vegetation changes that can lead to changes in soil biogeochemistry. Concurrently, those changes in inputs and vegetation can lead to changes in soil fauna that drive various ecosystem functions, such as greenhouse emissions. In this Special Issue we have seven papers that address ecosystem responses following EAB invasions, two of which are review papers, one directly measures changes in coarse wood debris inputs, two that examine the effects of soil fauna and two that assess greenhouse gas fluxes.

4.1. Review of Ecosystem Impacts Following EAB Invasion

Two review papers assessed ecosystem level impacts following EAB invasion but differed in their focus. Kolka et al. (2018) took a more general approach, reviewing the literature across the geography of black ash, focusing on black ash wetlands and what the future holds for these threatened ecosystems and the information gaps that need to be studied. Using two companion studies in northern Minnesota and the upper peninsula of Michigan that simulated short- (girdled) and long-term (ash cut) invasions, as well as the literature, Kolka et al. (2018) indicates that changes in hydrology could have important impacts on ecosystem functions. Higher water tables resulting from the removal of black ash canopies will impact vegetation and faunal communities, nutrient and carbon cycling, as well as the balance between plant production and decomposition. Although much progress has been made in the past 10 years developing an understanding of EAB impacts on black ash wetlands (much of which is included in this Special Issue), Kolka et al. (2018) indicate that assessing the impacts of the loss of black ash on biota is the strongest outstanding need, followed by how EAB infestations will alter physical and ecological processes across the range of black ash systems. They conclude that "research to address these fundamental needs and existing mitigation strategies is currently underway and should provide critical information for further refinement of mitigation and management strategies."

Klooster et al. (2018) focused their review on the ecological impacts at the "epicenter" of where EAB was first introduced in southeast Michigan. Although first detected in 2002, dendrochronological studies indicate that EAB had been infecting trees since the 1990s. By 2009, the mortality of the ash species in the region had exceeded 99%, with no seed production, soil seed bank, or new recruitment for future generations. They found no difference in mortality across gradients of ash density, ash importance, or community composition. Basically, if an ash was in the landscape, it would be infested. Trees tended to die over a period of about five years, resulting in widespread gaps in the canopy. As a result, these ecosystems saw large increases in coarse woody debris and increased growth rates of non-native and invasive plants (Klooster et al. 2018). Changes in dead wood inputs and vegetation communities are leading to the local extirpation of specialist herbivores of ash, changes in the diversity and abundance of soil invertebrates, and the behavior and abundance of certain bird species. The authors conclude that "these and other impacts on forest ecosystems are likely to be experienced elsewhere as EAB continues to spread."

4.2. Changes in Inputs of Coarse Woody Debris Following EAB Invasion

One of the more important ecological impacts resulting from EAB invasion is the relatively short-term influx of large quantities of coarse woody debris from newly dead or dying trees. Coarse woody debris has both short- and long-lasting impacts on nutrient and carbon cycles, as well as providing habitat for both flora and fauna. Perry et al. (2018) assessed coarse woody debris dynamics at the "epicenter" of the EAB invasion across the same areas as the Klooster et al. (2018) review above. Over the period of 2008 to 2012, inputs of coarse woody debris increased by 53%, with most of ash snapping at the bole. By 2012, only 31% of the ash snags had fallen, with the expectation that continued large inputs of coarse woody debris will continue to occur in the future. The coarse woody debris decay class also increased from 2008–2012, indicating greater decomposition over time despite high new inputs into the coarse woody debris pool. The authors conclude that "as the range of EAB expands, similar patterns of down coarse wood debris dynamics are expected in response to extensive ash mortality."

4.3. EAB Invasion Impacts on Soil Fauna

Although very different in their ecological roles, soil bacteria and ground-dwelling beetles are important for various ecosystem functions. Soil bacteria drive decomposition and carbon and nutrient cycles, as well the emission of greenhouse gases. Beetles are a good indicator of ecosystem disturbance, because some species have very specific fielding guilds, which are affected by changes in vegetation, light regimes and nutrient cycles.

Ricketts et al. (2018) compared communities of soil bacteria on ash plots and non-ash plots in Ohio. Non-ash plots were areas of forest composed of other tree species, adjacent to ash plots, and with a similar landscape position, but which had not been infected by EAB due to the lack of ash trees in the plots. They used DNA sequencing to compare soil bacterial communities in the ash and non-ash plots and found that they were different, suggesting a potential association between ash trees and their belowground bacterial communities. The authors' findings revealed that Acidobacteria abundance was driving this difference and was greater in the non-ash plots. Acidobacteria contribute to the metabolism of sugars leading to glycosis, but are not as involved in metabolic transformations more central to the TCA cycle, which has important implications for the carbon cycle. Although they found no difference in the bacterial functional potential which controls nitrogen cycles, they did find that both the phosphorus and sulfur metabolic potential was higher in the non-ash plots. The authors conclude that the "ash-soil microbiome association implies that EAB-induced ash decline may promote belowground successional shifts, altering carbon and nutrient cycling and changing soil properties beyond the effects of litter additions caused by ash mortality."

Savage and Rieske (2018) assessed beetle populations across an EAB invasion front in northern Kentucky. They found that native predatory beetles are being recruited to forests invaded by EAB, suggesting a possible response to an increased prey base. Similarly, herbivores, fungivores, and parasitic beetles were more abundant where ash decline was the greatest, which is likely indicative of changes in the plant and fungal communities. The authors conclude that "ash forests are changing, and a deeper understanding of how arthropod communities and trophic guilds are responding will contribute to more proficient monitoring and protection."

4.4. EAB Invasion Impacts on Soil Gas Fluxes

A poorly understood aspect of the ecological impacts is the effect EAB and subsequent ash decline has on soil carbon, especially those related to the balance of greenhouse gas fluxes. The two studies in this Special Issue cross a gradient of soil moisture, from sandy upland soils to wetlands, which lead to somewhat different results.

Hatala Matthes et al. (2018) measured soil respiration and methane production near white ash trees prior to, during, and following EAB invasion in New Hampshire. The site was characterized by sandy upland soils with a mixed canopy with only about 30% ash. Although they developed good relationships with the drivers of both carbon dioxide and methane fluxes (e.g. temperature, soil moisture, and soil sand fraction), no effect of ash decline was seen on gas fluxes or soil microclimate. The authors conclude that "our results indicate that short-term changes in soil carbon flux following insect disturbances may be minimal, particularly in forests with well-drained soils and a mixed-species canopy."

Van Grinsven et al. (2018) also measured carbon dioxide and methane fluxes from black ash wetland soils in the Upper Peninsula of Michigan following simulated short- (girdled) and long-term (ash-cut) EAB invasion, including an undisturbed control. Similar to Hatala Matthes et al. (2018), Van Grinsven et al. (2018) found that soil temperature and soil water conditions (i.e., water table levels) were strong predictors of carbon dioxide and methane fluxes. However, Van Grinsven et al. (2018) did see a response to disturbance, as carbon dioxide fluxes were greater in both the girdled and ash cut sites when compared to the undisturbed control. Methane fluxes were variable, with the ash cut sites having higher fluxes than the girdled sites, but neither was different from the undisturbed control. The authors conclude that "black ash wetland soil carbon mineralization processes were sensitive, but also showed resilience to EAB-induced impacts" and that "the strong connection between depressional black ash wetland study sites and groundwater likely buffered the magnitude of disturbance-related impact on water tables and carbon cycling."

5. Adaptive Forest Management Options and Controlling the Movement of EAB

EAB will continue to spread in North America and kill ash species. However, we can affect how fast it spreads and minimize the ecological and economic damage it does to forests dominated by ash. As EAB continues to impact forests across the US, we really have two options. First we can "slow the spread" which is the motto that has been taken up by many states. Second, either before or after detection, we can manage our forests in light of EAB in an attempt to increase resilience to the impacts of this insect on core ecosystem functions. The final two papers addressed these two areas where we have opportunities to manage the damage EAB does to the landscape, with the first focused on approaches to minimizing spread and the second serving as a review of experience to date with adaptive strategies addressing the threat posed by EAB to black ash wetlands.

Diss-Torrance et al. (2018) used surveys over the period of 2006–2015 to assess how the regulation of firewood allowed into state parks and associated education on the risk of spread of invasive species (e.g. EAB) influenced the firewood transport behavior of campers in Wisconsin. They found that awareness of EAB and the risk of its spread in firewood increased dramatically in the first two years of firewood regulation and education and then stayed steady at $\approx 95\%$. Compliance with firewood regulation on state campgrounds increased for the first four years then remained steady at $\approx 90\%$ compliant. When supplying wood for use on their own property, campers reduced the distance from which the wood was obtained from an average of 55 to 22 miles. They conclude that "regulation and persuasion based on motivational principles can lead to changes in behavior that extend beyond specific situations where regulation takes place, and that public properties can provide a venue for encouraging other types of environmentally responsible behavior."

Building on the findings of extensive research and field trials in black ash wetlands in the western Lake States, D'Amato et al. (2018) presented a review of the current understanding regarding the ecological context of black ash wetlands and the most effective strategies to date for increasing the resilience of these areas to EAB impacts. Although a dire message, they found that, because of its unique roll with regards to hydrology regulation and vegetation dynamics, there is no replacement for black ash in these wetland ecosystems. Nonetheless, by increasing the representation of non-ash species in these areas, there is the potential to maintain these ecosystems as forests and prevent retrogression into marshes as water tables rise following the death of black ash. Currently, the best approach is to develop adaptive silvicultural strategies such as planting in the understory and regeneration harvests (i.e., group selection and shelterwood) that maintain cover while seedlings can be established. The authors conclude that "regardless of treatment or site evaluated, all adaptive management efforts should include retention of mature, seed-bearing black ash to maintain its unique ecological functions prior to EAB arrival and provide opportunities for natural resistance and reestablishment after invasion." There is some hope for the future.

6. Conclusion

The introduction of EAB to North America in the early 2000s has irreversibly changed many areas where ash once formed a significant component of urban and rural forest systems. The continued spread of this insect across much of the US and Canada suggests that many new regions will face similar challenges associated with EAB to those observed across portions of the Great Lakes and northeastern US/Canada over the past two decades. The body of work presented in this Special Issue provides an invaluable knowledge base with which to gauge appropriate integrated pest management responses to new outbreaks, anticipate ecological impacts, and inform holistic, adaptive silvicultural strategies to increase ecosystem resilience to EAB. Although the introduction of EAB has been truly devastating for ash forests, a potential positive aspect is the broad body of research it has motivated to better understand what we may be losing and the appropriate strategies for sustaining these forests and associated functions in the future.

Randall K. Kolka, Brian J. Palik, Anthony W. D'Amato , Nam Jin Noh , Thomas Pypker,
Robert Slesak , Joseph W. Wagenbrenner
Special Issue Editors

forests

MDPI

Article

Buying Time: Preliminary Assessment of Biocontrol in the Recovery of Native Forest Vegetation in the Aftermath of the Invasive Emerald Ash Borer

Elan Margulies [1], Leah Bauer [2] and Inés Ibáñez [1,*]

[1] School for Environment and Sustainability, University of Michigan, Ann Arbor, MI 48109, USA;
     elmar@umich.edu
[2] United States Department of Agriculture (USDA) Forest Service, Northern Research Station,
     Lansing, MI 48910, USA; lbauer@fs.fed.us
*   Correspondence: iibanez@umich.edu; Tel.: +1-734-615-8817

Received: 27 August 2017; Accepted: 25 September 2017; Published: 28 September 2017

Abstract: Introduced forest pests have become one of the major threats to forests, and biological control is one of the few environmentally acceptable management practices. Assessing the impacts of a biocontrol program includes evaluating the establishment of biocontrol agents, the control of target pest, the impact on the affected organism, and the indirect impacts that the biocontrol agent may have on the whole community. We assessed the recovery of forest vegetation following the mortality of ash trees caused by the invasive emerald ash borer (EAB) pest in forest stands where biocontrol agents were released or not. We used a multilevel framework to evaluate potential indirect effects of the biocontrol agents on native forest seedlings. Our results showed a higher number of ash saplings where increasing numbers of the dominant EAB biocontrol agent were released, while the number of invasive and weedy saplings was negatively associated with the number of ash saplings, and the density of native seedlings was negatively associated with invasive and weedy saplings. The protection of ash saplings by the biocontrol agent may help native recruitment during forest transition by supporting the growth of native hardwood seedlings over invasive and weedy species. These results show that research on the efficacy of EAB biocontrol should include all ash size classes and the community dynamics of co-occurring species.

Keywords: *Agrilus planipennis*; *Fraxinus* spp.; gap dynamics; invasive species; *Oobius agrili*; southeastern Michigan; *Spathius agrili*; temperate forests; *Tetrastichus planipennisi*

---

1. Introduction

Some of the major challenges facing North American forests are invasions of non-native pests [1–3]. In recent decades, the increase in new species introductions, associated primarily with increasing global trade, has exacerbated the frequency and impact of these invasive pest outbreaks [4]. Outbreaks decimate forests and, in some instances, result in the local elimination of tree species [5,6] substantially changing forest ecosystems [1,7]. Among North American forests, the northeastern region is being particularly affected by these introductions [8,9], and one of the most recent pests arriving to this area is the emerald ash borer, *Agrilus planipennis* Fairmaire (Coleoptera: Buprestidae) [10].

The emerald ash borer (EAB) is an invasive wood-boring beetle from Asia that threatens North American ash species (*Fraxinus* spp.) [11,12]. Emerald ash borer was first discovered near Detroit, Michigan in 2002 [13,14]. Despite early eradication programs and ongoing quarantines by U.S. and Canadian regulatory agencies, EAB continues to spread and is now known in 30 states, Washington D.C. and two provinces [15]. While Asian species of ash are relatively resistant to EAB [16–18], most North American ash species show little resistance to this pest and most overstory ash trees die within three to six years of initial infestation [19–21]. This is relevant because ash species are widespread in North American eastern forests amounting to ~2.5% of the above-ground biomass [22], and ash trees have been widely planted in urban settings [23]. The potential cost of EAB to non-urban forests was estimated at more than $282 billion [24], and the undiscounted value of ash trees in the urban forests in the United States was estimated at $20–60 billion [25], making EAB, among those quantified, the most economically devastating insect pest in North American history [26].

Soon after EAB was discovered in North America, researchers began studying the role that natural enemies play in regulating EAB densities. Results from early studies revealed a low diversity and prevalence of parasitoids attacking EAB compared to those attacking native *Agrilus* species [27,28]. Furthermore, in regions of Asia where EAB is native, several specialized hymenopteran parasitoids that co-evolved with EAB, suppress EAB densities below a tolerance threshold for survival of native and some exotic ash species [16,29,30]. Several of these parasitoids became the basis of a biological control program aimed at reducing the density of EAB populations in North America [31].

Following extensive decline and mortality among the mature overstory ash trees, in 2007, EAB biocontrol began in southern Michigan. Three EAB parasitoid species from China were introduced: the egg parasitoid *Oobius agrili* (*O. agrili*) Zhang and Huang (Hymenoptera: Encyrtidae), the larval endoparasitoid *Tetrastichus planipennisi* (*T. planipennisi*) Yang (Hymenoptera: Eulophidae), and the larval ectoparasitoid *Spathius agrili* (*S. agrili*) Yang (Hymenoptera: Braconidae) [31,32]. To date, the establishment of both *T. planipennisi* and *O. agrili* have been confirmed at several release sites in Michigan and several other states [33]. Parasitism rates are not yet as high as in China [16,34], and it is too early to know the full extent to which these two biocontrol agents can protect the surviving ash saplings and trees as they mature into large-diameter size classes [31,35].

After the death of mature overstory trees, the ash seedling, sapling, and basal sprout bank became an important transitional resource for forest recovery [36,37]. In particular, seedlings of white ash, *Fraxinus Americana* (*F. Americana*), a common species in many North American eastern forests [12], can tolerate very low light conditions [38], allowing for the establishment of a robust seedling bank of up to 10,000–20,000 seedlings per hectare pre-EAB [39]. The high mortality rates due to EAB for mature ash is not paralleled among seedlings and saplings (<2.5 cm diameter at breast height, DBH) [39], as these size classes are too small to support EAB larval feeding. Researchers working in southern Michigan, where EAB biocontrol began, found that *T. planipennisi* had higher parasitism in thin-barked, small-diameter ash trees [40], which is likely correlated with its relatively short ovipositor [41]. Thus, when these seedlings and saplings grow in response to the death of mature trees, they are temporally protected by this biocontrol agent [37].

The EAB invasion also affects the entire plant community, as gaps and other disturbances caused by death of canopy trees set up the process of succession. Forest succession will then reflect the composition of the advance regeneration layer and of the seeds available for germination [42–44]. In forest ecosystems, most seeds originate locally [45,46] and their abundances are correlated with the basal area of nearby trees [47], while off-site propagules often come from the neighboring landscape [48]. If the surrounding areas are largely intact, native species will account for the majority of the seeds reaching a site [7,49]. However, as the surrounding landscape becomes more human-altered, weedy and invasive propagules could constitute a large proportion of those seeds [50,51].

Besides propagule availability, successful establishment of invasive and weedy plant species in a new location often depends on the higher level of resources, e.g., light [50,52,53]. Moreover, disturbance is frequently necessary for these plant species to penetrate interior forest communities [49,54,55], and it is often in disturbed habitats when invasive plants can outcompete native species [56,57]. In particular, researchers have observed that forests with simulated EAB damage were more susceptible to invasive plants [58]. And, early EAB quarantine efforts that cut mature ash trees, which caused the creation of forest gaps and higher light levels, showed an increased in the likelihood of plant invasions [59]. However, if an abundant ash seedling bank exists and it rapidly responds to the canopy opening [38], the resulting ash sapling layer, if protected from EAB, will then shade the stand and potentially curtail the success of the invasion (Figure 1a).

In our study, we investigated the impacts of recently introduced EAB biocontrol agents on ash sapling densities and forest vegetation, native and introduced, in the vicinity of the EAB-invasion epicenter in southeast Michigan, USA [60]. We collected data on the forest structure and composition of stands in this region and analyzed the data as a function of biocontrol release levels (Figure 1b). To understand the relationship between the release of biocontrol and native tree seedling populations, we investigated three dynamics: (i) does parasitoid release affect ash sapling density? (ii) What are the variable driving the establishment of invasive and weedy saplings in this system? And, (iii) is the density of invasive and weedy species associated with native seedling density? Our expectation was that with the protection from the biocontrol agents, the ash sapling layer would prevent the colonization of EAB-disturbed areas by invasive and weedy species, thereby buying time for other native, and more shade tolerant, species to establish and recruit (Figure 1a).

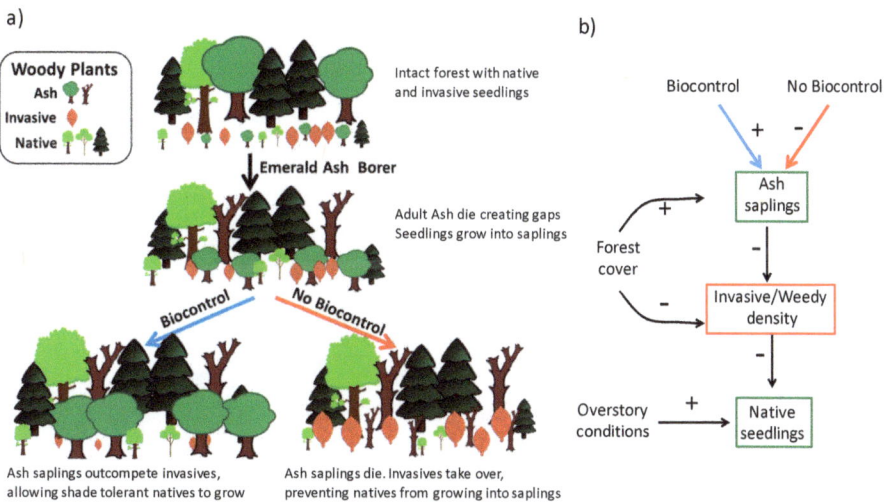

Figure 1. (**a**) Visual representation of the forest transition with and without biocontrol. This graphic outlines the hypothesis that as gaps are created from the loss of mature ash trees, EAB biocontrol agents, mainly *T. planipennisi* an introduced EAB natural enemy, can protect ash saplings, buying time for native seedlings to grow and fill gaps (left pathway). Without the influence of the EAB biocontrol program, ash do not survive long enough to allow the native community to recruit, resulting in invasive and weedy species taking over (right pathway); (**b**) Graphical representation of analysis testing the hypothesis, with positive and negative signs indicating our original expectations.

2. Materials and Methods

2.1. Study Sites

In the summer of 2014, we sampled forest composition at sites of varying distance from parasitoid release locations in southeastern Michigan, USA. A total of 21 sites were selected and surveyed (Figure 2). The sites were classified as a release site if parasitoids were released <1 km away (Supplementary Material 1). Seven release sites were chosen at random from the 22 parasitoid release locations within the counties of Ingham, Jackson, Livingston, Oakland and Washtenaw in 2014 [61]. The parasitoids were released in wooded stands larger than 16 ha, that were >100 m from a road and that would not be harvested or developed for at least five years after release. The 14 control sites were selected in the general area of the release plots to reflect similar vegetation types. Control sites were between 4.1 km and 20.5 km to the nearest release location (Figure 2, Supplementary Material 1). The climate in the area is characterized by cold winters (average minimum temperature in January is −8.8 °C) and warm summers (average maximum temperature in July is 28.8 °C) with a total annual precipitation of 953 mm evenly distributed along the year. Most soils of the region derive from end-moraine ridges and ground moraines, and from alluvial deposits along river valleys.

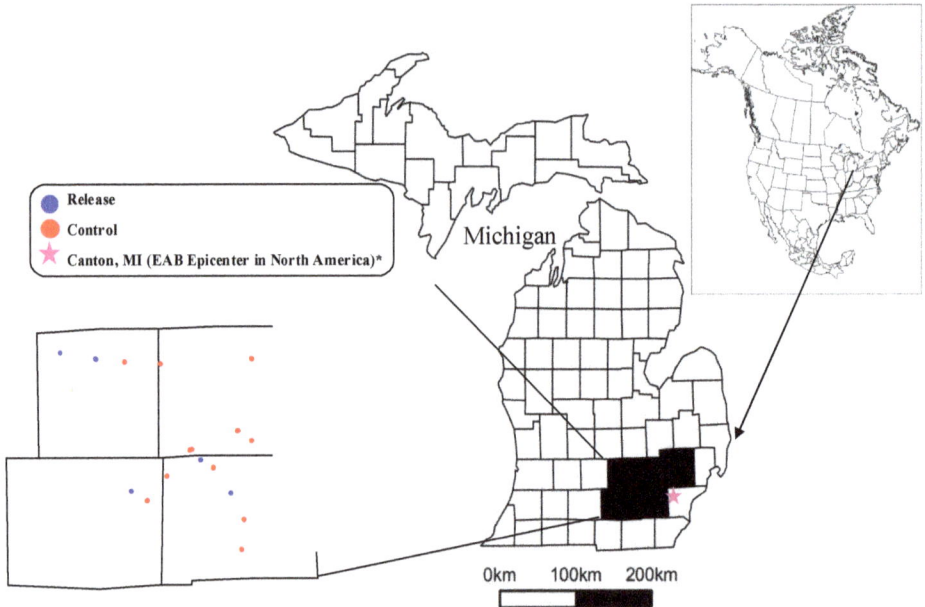

Figure 2. Map of Michigan showing the location of our 21 vegetative study sites (Lambert projection). Insert shows the five Michigan counties where the release and control vegetation plots were located (colored dots in the map).

2.2. Biocontrol Releases

The EAB parasitoid release data was obtained at mapbiocontrol.org, a geospatial framework for biocontrol information [59]. At the seven biocontrol sites, there were 44 discrete release events with a sum of 14,065 individual releases that took place between the years of 2007 and 2012 (Supplementary Material 2). At the release sites, the relative release proportions for *O. agrili*, *T. planipennisi*, and *S. agrili* were 19%, 68% and 13% respectively. Both *O. agrili* and *T. planipennisi* were confirmed established at one of the seven parasitoid-release plots, whereas establishment of *T. planipennisi* was confirmed at two other release plots; establishment of *S. agrili* has not been confirmed [61–63].

2.3. Vegetation Sampling

To investigate the succession process at both the control and release study sites, two 20 × 20 m plots were set up at each site. Plots were centered around a dead standing canopy tree, an ash tree if possible. All living woody species >10 cm DBH were classified as trees and identified to species level in the 400 m² plot. We measured the DBH of all the trees in the plot to calculate the plot's basal area. We established four 2 × 10 m sampling transects, totaling 80 m² per plot where native, invasive and weedy saplings (>1 m in height and <10 cm DBH) were counted. We also set up four 1 × 10 m groundcover (all vegetation <1 m in height) and seedling transects, totaling 40 m² per plot. Within each 1 m², we quantified percent groundcover by invasive and weedy species. Plants were characterized as weedy species if they are native, not typically growing in closed canopy forests but rather in more open, higher-light environments. Plants were characterized as invasive if they are not native to North America. The seedlings of woody plants (<1 m in height) were counted and identified to species within the same groundcover transects. All plot measures were estimated per m^2 unit area, and plot-level averages and standard deviations were used in the analyses. See Supplementary Materials 3 and 4 for data summaries.

2.4. Environmental and Land Cover Data

To account for resource availability, we measured the light and moisture levels at each plot when vegetation was surveyed (Supplementary Material 5). Photosynthetically active radiation (PAR) was measured as an indicator of light availability in every meter radiating along the cardinal axes from the central dead tree. This was repeated 3×, totaling 120 readings per plot using a LightScout Quantum Light 6 Sensor Bar and the LightScout Light Sensor Reader from Spectrum Technologies, Plainfield, IL, USA. Volumetric water content (VMC) was used as an indicator of soil moisture, it was measured every meter radiating along the cardinal axes from the central dead tree, with a total of 40 moisture readings per plot. VMC was measured in the top 15 cm using Fieldscout-TDR 300 Soil Moisture Meter from Spectrum Technologies, Plainfield, IL, USA. To determine the percent forested area surrounding each plot, we used available land cover data from 2002, using ArcGIS 10.3 (ESRI, Redlands, CA, USA) we estimated the percent of forested land within 1 km of the study sites [64]. Total land area was calculated by subtracting the area covered by water from the total area. See Supplementary Material 5 for light, soil moisture and forest cover data.

2.5. Statistical Analysis

To evaluate the relationship between native seedlings and the release of parasitoids, we first carried out extensive exploratory data analysis and then developed a multilevel, or hierarchical, model where estimates from a submodel were used as predictors in subsequent models (Figure 1b). First, parasitoid release information (number of released *T. planipennisi*, as this was the most successful parasitoid establishing and spreading [37,62]) was used to analyze ash sapling density, then invasive and weedy species density was analyzed as a function of the estimated ash sapling densities, and we finished by using estimates of invasive and weedy species densities to analyze the native tree seedling data (Figure 1b). This multilevel approach allowed sharing of information across the data sets [65], potentially better informing the dynamics taking place in these plots. Following this hierarchical approach, we ran variants of this basic model that included some additional explanatory variables (e.g., forest cover around the plots, basal area in the plots, ground cover, soil moisture, light, total releases, distance to release), we describe below the model best supported by the data (based on goodness of fit and deviance information criterion, DIC; [66]).

We first estimated the abundance of ash saplings, *AshSaplings*, as a function of the percent of forest cover (*Forest cover*) around the plots within 1 km radius. We used forest cover as a proxy for source of propagules determining the strength of the seedling bank growing into saplings. We also estimated ash sapling density as a function of the number of parasitoids released (*T. planipennisi*). Because these two variables were correlated, *r*: 0.66, we orthogonalized the number of released parasitoids with respect to forest cover, and used the residuals ($\varepsilon_{release}$) in the analysis. This approach allowed us a better assessment of the independent effect of the biocontrol treatment on ash sapling density once the strength of the source of propagules, the major driver of sapling density, was accounted for. The likelihood for the average density of ash saplings in plot *i*, was:

$$AshSaplings_i \sim \text{Poisson}(Ash_i) \tag{1}$$

and process model:

$$\ln(Ash_i) = \alpha_1 + \alpha_2 Forest\ Cover_i + \alpha_3 \varepsilon_{release\ i} \tag{2}$$

The density of invasive and weedy saplings, *InvWeedyS*, was analyzed as a function of the estimated density of ash saplings (*Ash*), the main native competitor after disturbance, and of the percentage of forest cover within 1 km, used here again as a proxy for sources of propagules [67,68], but in this case assuming that areas with higher forest cover are likely to have fewer invasive and weedy species, likelihood:

$$InvWeedyS_i \sim \text{Poisson}(IW_i) \tag{3}$$

and process model:

$$\ln(IW_i) = \beta_1 + \beta_2 Ash_i + \beta_3 Forest\ Cover_i \tag{4}$$

The average density of native woody seedlings, *NativeSeedlings*$_i$, was then analyzed as a function of the basal area of the stand (*BA* in units of m^2/m^2) to reflect sources of seeds [69], and of the estimated density of invasive and weedy saplings (*IW*) that could be competing with the native vegetation, likelihood:

$$NativeSeedlings_i \sim \text{Normal}\ (Native_i,\ NSvar_i) \tag{5}$$

and process model:

$$Native_i = \gamma_1 + \gamma_2 BA_i + \alpha\gamma_3 IW_i \tag{6}$$

The variances associated with each plot, *NSvar*$_i$, were estimates from our data. Due to the multilevel structure of the model we followed a Bayesian approach in the estimation of the parameters [70]. Parameters were estimated from non-informative distributions, α^*, β^*, γ^* ~Normal (0, 10,000). The analysis was run in OpenBugs [71] (see Supplementary Material 6 for code), and three chains were run simultaneously to assess convergence. Parameter posterior means, variances and 95% credible intervals, were calculated after convergence, thinning every 100th iteration. Parameters associated with the covariates were considered statistically significant if the 95% credible interval (CI) around their means did not overlap with zero.

3. Results

In total, we surveyed 41 plots at 21 sites, which included 688 trees, 3,826 saplings, 19,583 seedlings, and 12,961 distinct recordings for groundcover (see Supplementary Material 3 for detailed data of each plot). The most common tree species was *Prunus serotina*, found in 58% of the plots, followed by *Ulmus americana* and *F. americana* (this last one included dead stems) growing in 44% of the sampled plots. Other common species were *Acer rubrum*, 48%, *Quercus rubra*, 36.5%, *A. saccharum* 30% and *Q. alba* 26.8% (Supplementary Material 3). Among the invasive and weedy saplings the most common species were *Elaeagnus umbellata* found in 44% of the plots, *Rubus* spp. in 31.7% of the plots, *Rosa multiflora* and *Lonicera maackii* in 26%, and *L. tatarica* in 24.4% of the plots (Supplementary Material 4). Of the additional variables included only forest cover around the plots (ash and invasive and weedy species submodels) and basal area in the plot (native seedlings submodel) improved the fit of the models, other variables (ground cover, soil moisture, light, total releases, distance to release) when included did not improve the fit of the model and we opted to exclude them in our final analysis. The hierarchical structure of the analysis greatly improved the fit of the model (see Supplementary Material 7 for alternative model comparisons). All parameter estimates from the analysis are reported in Table 1.

3.1. Results from the Ash Sapling Submodel

Increase in percent forest cover around the plots was associated with a higher number of ash saplings (α_2 parameter was positive and statistically significant; Table 1), this variable had the strongest impact on ash sapling densities (Figure 3a). The number of released parasitoids (*T. planipennisi*) was also statistically significant and positively associated with higher densities of ash saplings (α_3 parameter; Table 1, Figure 3).

Table 1. Posterior means, standard deviations (SD) and 95% credible intervals (CI) for all the parameters included in the analysis. Coefficients associated with the explanatory variables that were statistically significant (95% CI did not include zero) are shown in bold.

Parameter		Mean ± SD	95% CI	
Ash saplings submodel:				
α_1	intercept	-0.382 ± 0.002	-0.385	-0.377
α_2	forest cover	**0.406 ± 0.003**	**0.398**	**0.412**
α_3	number of *T. planipennesis* released	**0.001736 ± 0.0000019**	**0.001734**	**0.001737**
Invasive and weedy species submodel:				
β_1	intercept	0.8999 ± 0.00015	0.8996	0.9002
β_2	ash saplings	**-1.506 ± 0.005**	**-1.512**	**-1.492**
β_3	forest cover	**0.2737 ± 0.0015**	**0.2715**	**0.2753**
Native seedlings submodel:				
γ_1	intercept	1.97 ± 0.19	-2.31	-1.64
γ_2	basal area	**3830 ± 3.62**	**3822**	**3835**
γ_3	invasive and weedy saplings	**-4.95 ± 0.05**	**-5.04**	**-4.86**

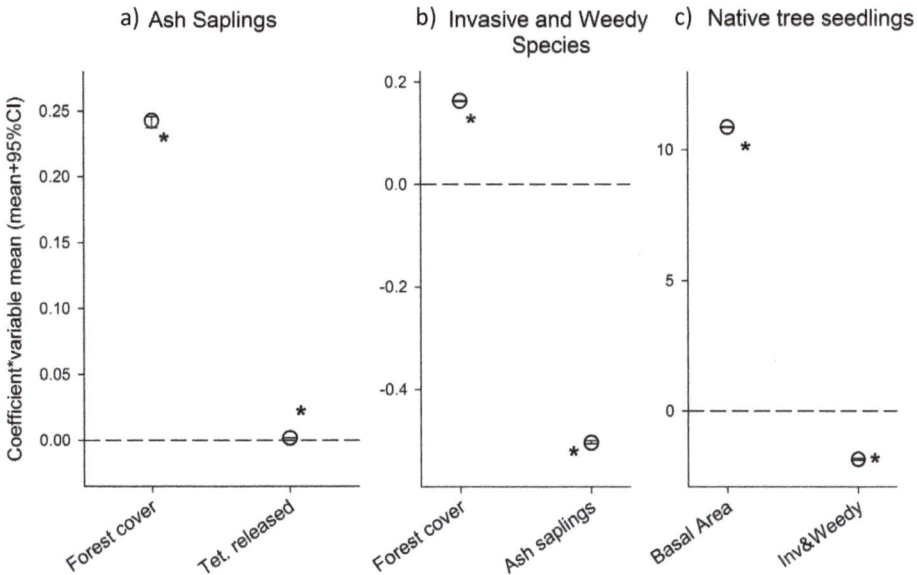

Figure 3. Posterior parameter means (+95% CI) of each of the parameters included in the analyses for each submodel, ash saplings (**a**); invasive and weedy species (**b**); and native tree seedlings (**c**). Parameters have been standardized (i.e., multiplied by the covariate mean, except for *T. planipennisi* releases where we used the orthogonalized variable which is centered around zero) to assess their influence. Coefficients that were statistically significant (95% CI did not overlap with zero) are indicated by an asterisk. Note: symbols are larger than the 95% CIs.

3.2. Results from the Invasive and Weedy Species Submodel

The abundance of invasive and weedy species was negatively associated with increasing number of ash saplings, which had the strongest effect, and was positively associated with increasing forest cover around the plots (parameters β_2 and β_3 were statistically significant; Table 1, Figure 3b).

3.3. Results from the Native Seedlings Submodel

Increase in plot basal area was significantly associated with a higher number of native seedlings (parameter γ_2; Table 1, Figure 3) and had the largest effect. The density of invasive and weedy species was associated with lower number of native seedlings, and the effect was statistically significant (parameter γ_3; Table 1, Figure 3c).

4. Discussion

Most biocontrol evaluation research focuses on the establishment and spread of introduced biocontrol agents, their impacts on the target pest or invasive species, and their potential interactions with non-target organisms and other natural enemies [72–74]. However, it may take several years before the impacts or recovery of affected species and communities can be assessed. In the case of trees, which have a long-life cycle, a comparatively long lag time may be needed before the impacts of biocontrol can be realized. In our system, the biocontrol program for EAB is still in the early phases (for a review see [31]). Nevertheless, researchers have found that one species of introduced parasitoid, *T. planipennisi*, is now the dominant natural enemy of EAB larvae in young ash trees and saplings, which are growing in large numbers in forest gaps after EAB decimated the overstory ash trees [37,62,75]. To evaluate the delayed impacts of EAB biocontrol on the entire woody community, we carried out an analysis that linked biocontrol with tree recruitment dynamics taking place since death of overstory ash trees. Our analyses showed a positive association between the release numbers of the parasitoid, *T. planipennisi*, and the density of ash saplings, the EAB tree host. We also documented negative associations between ash saplings and invasive plant species, and between invasive species and native seedlings. Our results indicate that even if *T. planipennisi* does not protect larger ash trees from EAB, it can still protect ash saplings which are likely out-shading invasive species, and thus, buying time for slower growing native species to recruit into these sites. This illustrates that there is a secondary positive community-level benefit of EAB biocontrol in these forests.

4.1. Do Biocontrol Agents Affect Ash Sapling Density?

Emerald ash borer biocontrol releases began in Michigan in 2007 and >14,000 parasitoids were released at our seven release plots during a six-year period [61] (Supplementary Material 2). Although parasitoid prevalence in some ash trees is as high as 35% for *O. agrili* and >90% for *T. planipennisi* at some study sites, the mature ash trees still experienced high mortality as they were already dead or dying when parasitoid releases began [31,37]. In areas no treated with biocontrol, up to 19% of the stems in the regenerating sapling layer can also be infected [76]. However, younger, thin barked ash trees and saplings growing at these release sites seem to be protected by the dominant biocontrol agent *T. planipennisi*, a small parasitoid with a short ovipositor that successfully parasitizes EAB larvae in ash trees <10 cm DBH [37,40,77,78]. Our results illustrate that this is likely the case at our study sites where a higher density of ash saplings was associated with higher release numbers of *T. planipennisi* (Figure 3).

This relationship was maintained after we controlled for the percent of forest cover around the plots, our proxy for the source of pre-EAB ash propagules that would subsequently grow into the seedling layer. And, as we sampled in areas with a canopy opening, we were able to document the seedling transition into a sapling layer in response to this disturbance. It would have been at this stage that the biocontrol agent, *T. planipennisi*, became most effective in protecting ash, as saplings have relatively thin barks. Previous work in this system have shown biocontrol can reduce EAB infestation in saplings (2.5–8 cm DBH) by >50% [37], ensuring a healthy sapling layer. Moreover, ash can reproduce at small sizes (8 cm DBH; [38]) and produce large number of seeds during mast years [36]. This could ensure that under the influence of the biocontrol agents ash populations will not entirely disappear because of EAB.

4.2. What Are the Variable Driving the Establishment of Invasive and Weedy Saplings in This System?

The sites in our study had experienced overstory tree dieback and were surrounded by agricultural, developed and other forested areas. Therefore, the likelihood of invasive plant species rapidly colonizing an area after a disturbance was relatively high. Still, one of the woody species that can rapidly take over after an opening in the forest canopy is *F. americana*, white ash. White ash seedling and basal sprouts densities have been observed as high as 20,000 per hectare [36,39]. Thus, even if adult trees succumb to the EAB, seedlings can rapidly grow into the sapling layer and, if protected by biocontrol, shade the ground vegetation. Our results revealed that this is likely the case at our study sites, where we found a negative association between ash saplings and the abundance of invasive and weedy species, unveiling an indirect beneficial effect of EAB biocontrol on the entire community.

Unexpectedly, we also documented a positive association of forest cover around the study sites and the incidence of invasive and weedy species. However, when we included other variables, e.g., light, we did not find an association with density of invasive and weedy species [79]. We had hypothesized that higher forest percent cover around our sites, which ranged from 25 to 90%, would be linked to a decrease in invasive species propagules. Invasive plant species are not common in forest interiors [7,80,81], and under natural disturbances the native vegetation, including both fast and slow growing species, can recover [82–84]. However, our results illustrate an opposite trend emphasizing the importance of assessing the risk of plant invasion not only at the site, habitat characteristics level, but also within the context of the historical landscape [85]. In our study area, forests have been under considerable human influence for almost two centuries, being highly fragmented and having a large edge to area ratio [86] where invasive and weedy species thrive, i.e., showing fast growing and reproductive rates [85]. As a result, the availability of propagules from introduced species is likely to be widespread in the region.

4.3. Is the Density of Invasive and Weedy Species Associated with Native Seedling Density?

One of the indirect effects a pest outbreak may have is the creation of optimal conditions for the establishment of harmful species. The conditions created by closed canopy forests and the lack of propagules are thought to buffer mature native forests from invasive plants [87,88]. It is mostly after disturbance events that invasive species are able to establish populations large enough to negatively affect the native community [54,89,90]. Our study supports this trend, as we observed a negative association between invasive and weedy saplings and native seedlings. This is due to some of the traits that prevail among the observed invasive and weedy species, i.e., fast growth rates and/or prolific seed production when resources are plentiful, which confer greater competitive ability to these species over natives, mostly non-ash seedlings in our sites, in disturbed forest areas where light is not limiting [91,92].

The understory vegetation response to disturbance mostly follows the direct regeneration hypothesis (DRH), which posits that tree communities will regenerate from existing seedlings to pre-disturbance levels within decades [93]. The resiliency of the DRH is based on the regeneration capacity of trees, which is proportional to basal area [69]. Since the degree of disturbance was similar among sites (one dead canopy tree), our analysis supports this hypothesis, we found a very strong effect of basal area on the density of seedlings from woody species (Figure 3). However, under highly modified contemporary landscapes, the availability of propagules from introduced harmful species increases with the level of development and roads around remnant vegetation patches [85]. Thus, after a disturbance event, a site with a high basal area may still be threatened by the establishment of invasive plant species.

5. Conclusions

In summary, this study provides a linkage between post disturbance vegetation recovery and an invasive insect biocontrol program. We found that biocontrol protection on ash saplings has secondary effects that protect native seedling recovery from invasive and weedy saplings. Our results also indicate that after a disturbance event, biocontrol efforts can help forest recovery by buying time for the native community recruitment by reducing the threat of invasive and weedy species. We found that the efficacy of the ash biocontrol program could not be solely evaluated using the effect it had on parasitism levels or protecting mature ash trees, but should also include other ash size classes and the community dynamics of the adjacent species.

Supplementary Materials: The following are available online at www.mdpi.com/1999-4907/8/10/369/s1, Supplementary Material 1: Information on vegetation study sites. Supplementary Material 2: Parasitoid release information. Supplementary Material 3: Vegetation information. Supplementary Material 4: List of invasive and weedy species. Supplementary Material 5: Environmental data. Supplementary Material 6: Model code. Supplementary Material 7: Model comparisons.

Acknowledgments: This work is/was supported by the USDA National Institute of Food and Agriculture, McIntire Stennis project [2012-32100-06099]. We also appreciate the ongoing support provided for this research by USDA Forest Service, Northern Research Station, Lansing, MI, USA; USDA Agricultural Research Service, Beneficial Insects Introduction Research Unit, Newark, DE, USA; USDA Animal and Plant Health Inspection Service, Brighton, MI, USA and Buzzards Bay, MA, USA; and the Department of Entomology, Michigan State University, East Lansing, MI, USA.

Author Contributions: E.M., L.B. and I.I. conceived and designed the study; E.M. collected the data; I.I. analyzed the data; E.M., L.B. and I.I. wrote the paper.

Conflicts of Interest: The authors declare no conflict of interest. The founding sponsors had no role in the design of the study; in the collection, analyses, or interpretation of data; in the writing of the manuscript, and in the decision to publish the results.

References

1. Liebhold, A.M.; Macdonald, W.L.; Bergdahl, D.; Mastro, V.C. Invasion by exotic pests: A threat to forest ecosystems. *For. Sci. Monogr.* **1995**, *30*, 1–49.

2. Lovett, G.M.; Weiss, M.; Liebhold, A.M.; Holmes, T.P.; Leung, B.; Lambert, K.F.; Orwig, D.A.; Campbell, F.T.; Rosenthal, J.; Mccullough, D.G.; et al. Nonnative forest insects and pathogens in the United States: Impacts and policy options. *Ecol. Appl.* **2016**, *26*, 1437–1455. [CrossRef] [PubMed]

3. Flower, C.E.; Gonzalez-Meler, M.A. Responses of Temperate Forest Productivity to Insect and Pathogen Disturbances. *Annu. Rev. Plant Biol.* **2015**, *66*, 547–569. [CrossRef] [PubMed]

4. Aukema, J.E.; McCullough, D.G.; Von Holle, B.; Liebhold, A.M.; Britton, K.; Frankel, S.J. Historical Accumulation of Nonindigenous Forest Pests in the Continental United States. *BioScience* **2010**, *60*, 886–897. [CrossRef]

5. Busby, P.E.; Canham, C.D. An exotic insect and pathogen disease complex reduces aboveground tree biomass in temperate forests of eastern North America. *Can. J. For. Res.* **2010**, *41*, 401–411. [CrossRef]

6. Van de Gevel, S.L.; Hart, J.L.; Spond, M.D.; White, P.B.; Sutton, M.N.; Grissino-Mayer, H.D. American chestnut *Castanea dentata* to northern red oak *Quercus rubra*: Forest dynamics of an old-growth forest in the Blue Ridge Mountains; USA. *Can. J. Bot.* **2012**, *90*, 1263–1276. [CrossRef]

7. Mosher, E.S.; Silander, J.A.; Latimer, A.M. The role of land-use history in major invasions by woody plant species in the northeastern North American landscape. *Biol. Invasions* **2009**, *11*, 2317–2328. [CrossRef]

8. Lovett, G.M.; Canham, C.D.; Arthur, M.A.; Weathers, K.C.; Fitzhugh, R.D. 2006 Forest Ecosystem Responses to Exotic Pests and Pathogens in Eastern North America. *BioScience* **2006**, *56*, 395–405. [CrossRef]

9. Liebhold, A.M.; Mccullough, D.G.; Blackburn, L.M.; Frankel, S.J.; Von Holle, B.; Aukema, J.E. A highly aggregated geographical distribution of forest pest invasions in the USA. *Divers. Distrib.* **2013**, *19*, 1208–1216. [CrossRef]

10. Morin, R.S.; Liebhold, A.M.; Pugh, S.A.; Crocker, S.J. Regional assessment of emerald ash borer, *Agrilus planipennis*, impacts in forests of the Eastern United States. *Biol. Invasions* **2017**, *19*, 703–711. [CrossRef]

11. Cappaert, D.L.; McCullough, D.G.; Poland, T.M.; Siegert, N.W. Emerald ash borer in North America: A research and regulatory challenge. *Am. Entomol.* **2005**, *51*, 152–165. [CrossRef]

12. Rebek, E.J.; Herms, D.A.; Smitley, D.R. Interspecific Variation in Resistance to Emerald Ash Borer (Coleoptera: Buprestidae) Among North American and Asian Ash (*Fraxinus* spp.). *Environ. Entomol.* **2008**, *37*, 242–246. [CrossRef]

13. Haack, R.A.; Jendek, E.; Liu, H.; Marchant, K.R.; Petrice, T.R.; Poland, T.M.; Ye, H. The emerald ash borer: A new exotic pest in North America. *Newsl. Mich. Entomol. Soc.* **2002**, *47*, 1–5.

14. Poland, T.M.; McCullough, D.G. Emerald ash borer: Invasion of the urban forest and the threat to North America's ash resource. *J. For.* **2006**, *104*, 118–124.

15. USDA-APHIS. Emerald Ash Borer. Available online: https://www.aphis.usda.gov/plant_health/plant_pest_info/emerald_ash_b/downloads/MultiState.pdf (accessed on 10 July 2017).

16. Liu, H.; Bauer, L.S.; Gao, R.; Zhao, T.; Petrice, T.R.; Haack, R.A. Exploratory survey for the emerald ash borer; *Agrilus planipennis* Coleoptera: Buprestidae; and its natural enemies in China. *Gt. Lakes Entomol.* **2003**, *36*, 191–204.

17. Eyles, A.; Jones, W.; Riedl, K.; Cipollini, D.; Schwartz, S.; Chan, K.; Herms, D.A.; Bonello, P. Comparative phloem chemistry of Manchurian *Fraxinus mandshurica* and two North American ash species *Fraxinus americana* and *Fraxinus pennsylvanica*. *J. Chem. Ecol.* **2007**, *33*, 1430–1448. [CrossRef] [PubMed]

18. Villari, C.; Herms, D.A.; Whitehill, J.G.A.; Cipollini, D.; Bonello, P. Progress and gaps in understanding mechanisms of ash tree resistance to emerald ash borer; a model for wood-boring insects that kill angiosperms. *New Phytol.* **2016**, *209*, 63–79. [CrossRef] [PubMed]

19. Anulewicz, A.C.; McCullough, D.G.; Cappaert, D.L.; Poland, T.M. Host range of the emerald ash borer *Agrilus planipennis* Fairmaire Coleoptera: Buprestidae in North America: Results of multiple-choice field experiments. *Environ. Entomol.* **2008**, *37*, 230–241. [CrossRef]

20. Gandhi, K.J.K.; Herms, D.A. Potential biodiversity loss due to impending devastation of the North American genus *Fraxinus* by the exotic emerald ash borer. *Biol. Invasions* **2010**, *12*, 1839–1846. [CrossRef]

21. Spei, B.A.; Kashian, D.M. Potential for persistence of blue ash in the presence of emerald ash borer in southeastern Michigan. *For. Ecol. Manag.* **2017**, *392*, 137–143. [CrossRef]

22. Flower, C.E.; Knight, K.S.; Gonzalez-Meler, M.A. Impacts of the emerald ash borer *Agrilus planipennis* Fairmaire induced ash *Fraxinus* spp. mortality on forest carbon cycling and successional dynamics in the eastern United States. *Biol. Invasions* **2013**, *15*, 931–944. [CrossRef]

23. Sadof, C.S.; Hughes, G.P.; Witte, A.R.; Peterson, D.J.; Ginzel, M.D. Tools for staging and managing emerald ash borer in the urban forest. *Arboric. Urban For.* **2007**, *43*, 15–26.

24. Nowak, D.; Crane, D.; Stevens, J.; Walton, J. *Potential Damage from Emerald Ash Borer*; United States Department of Agriculture; Forest Service; Northern Research Station: Syracuse, NY, USA, 2003; pp. 1–5. Available online: https://www.nrs.fs.fed.us/disturbance/invasive_species/eab/local-resources/downloads/EAB_potential.pdf (accessed on 26 September 2017).

25. Federal Register 2003. Emerald Ash Borer; Quarantine and Regulations. 7 CFR Part 301 [docket Number 02-125-1]. Available online: https://www.federalregister.gov/articles/2003/10/14/03-25881/emerald-ash-borer-quarantine-and-regulations (accessed on 10 May 2017).

26. Herms, D.A.; McCullough, D.G. Emerald ash borer invasion of North America: History, biology, ecology, impacts, and management. *Annu. Rev. Entomol.* **2014**, *59*, 13–30. [CrossRef] [PubMed]

27. Taylor, P.B.; Duan, J.J.; Fuester, R.W.; Hoddle, M.; Van Driesche, R.G. Parasitoid Guilds of *Agrilus* Woodborers Coleoptera: Buprestidae: Their Diversity and Potential for Use in Biological Control. *Psyche* **2012**, *2012*, 1–10. [CrossRef]

28. Bauer, L.S.; Duan, J.J.; Gould, J.R. Emerald ash borer *Agrilus planipennis* Fairmaire Coleoptera: Buprestidae. In *The Use of Classical Biological Control to Preserve Forests in North America*; van Driesche, R., Reardon, R., Eds.; United States Department of Agriculture; Forest Service; Forest Health and Technology Enterprise Team; FHTET-2013-2: Morgantown, WV, USA, 2014; pp. 189–209.

29. Duan, J.J.; Yurchenko, G.; Fuester, R. Occurrence of emerald ash borer Coleoptera: Buprestidae and biotic factors affecting its immature stages in the Russian Far East. *Environ. Entomol.* **2012**, *41*, 245–254. [CrossRef] [PubMed]

30. Wang, X.Y.; Jennings, D.J.; Duan, J.J. Trade-offs in parasitism efficiency and brood size mediate parasitoid coexistence; with implications for biological control of the invasive emerald ash borer. *J. Appl. Ecol.* **2015**, *52*, 1255–1263. [CrossRef]

31. Bauer, L.S.; Duan, J.J.; Gould, J.R.; Van Driesche, R.G. Progress in the classical biological control of *Agrilus planipennis* Fairmaire Coleoptera: Buprestidae in North America. *Can. Entomol.* **2015**, *147*, 300–317. [CrossRef]

32. Federal Register. Availability of an Environmental Assessment for the Proposed Release of Three Parasitoids for the Biological Control of the Emerald Ash Borer Agrilus planipennis in the Continental United States. *Fed. Regist.* **2007**, *72*, 28947–28948, [docket number APHIS-2007-006]. Available online: http://www.regulations.gov/#!documentDetail;D=APHIS-2007-0060-0043 (accessed on 10 May 2017).

33. Bauer, L.; Jennings, D.; Duan, J.; van Driesche, R.; Gould, J.; Kashian, D.; Miller, D.; Petrice, T.; Morris, E.; Poland, T. Recent Progress in Biological Control of Emerald Ash Borer. In Proceedings of the 27th USDA Interagency Research Forum on Invasive Species, Annapolis, MD, USA, 12–15 January 2016; pp. 22–25.

34. Liu, H.; Bauer, L.S.; Miller, D.L.; Zhao, T.; Gao, R.; Song, L.; Luan, Q.; Jin, R.; Gao, C. Seasonal abundance of *Agrilus planipennis* Coleoptera: Buprestidae and its natural enemies *Oobius agrili* Hymenoptera: Encyrtidae and *Tetrastichus planipennisi* Hymenoptera: Eulophidae in China. *Biol. Control* **2007**, *42*, 61–71. [CrossRef]

35. Knight, K.S.; Brown, J.P.; Long, R.P. Factors affecting the survival of ash *Fraxinus* spp. trees infested by emerald ash borer *Agrilus planipennis*. *Biol. Invasions* **2013**, *15*, 371–383. [CrossRef]

36. Kashian, D.M. Sprouting and seed production may promote persistence of green ash in the presence of the emerald ash borer. *Ecosphere* **2016**, *7*, 1–15. [CrossRef]

37. Duan, J.J.; Bauer, L.S.; Van Driesche, R.G. Emerald ash borer biocontrol in ash saplings: The potential for early stage recovery of North American ash trees. *For. Ecol. Manag.* **2017**, *394*, 64–72. [CrossRef]

38. Schlesinger, R.C. *Fraxinus americana* L.: White ash. In *Silvics of North America: Hardwoods; Agriculture Handbook 654*; Burns, R.M., Honkala, B.H., Eds.; USDA Forest Service: Washington, DC, USA, 1990; pp. 333–338.

39. Kashian, D.M.; Witter, J.A. Assessing the potential for ash canopy tree replacement via current regeneration following emerald ash borer-caused mortality on southeastern Michigan landscapes. *For. Ecol. Manag.* **2011**, *261*, 480–488. [CrossRef]

40. Abell, K.J.; Duan, J.J.; Bauer, L.; Lelito, J.P.; Van Driesche, R.G. The effect of bark thickness on host partitioning between *Tetrastichus planipennisi* Hymen: Eulophidae and *Atanycolus* spp. Hymen: Braconidae; two parasitoids of emerald ash borer Coleop: Buprestidae. *Biol. Control* **2012**, *63*, 320–325. [CrossRef]

41. Duan, J.J.; Oppel, C.B. Critical rearing parameters of *Tetrastichus planipennisi* Hymenoptera: Eulophidae as affected by host-plant substrate and host-parasitoid group structure. *J. Econ. Entomol.* **2012**, *105*, 792–801. [CrossRef] [PubMed]

42. Costilow, K.C.; Knight, K.S.; Flower, C.E. Disturbance severity and canopy position control the radial growth response of maple trees (*Acer* spp.) in forests of northwest Ohio impacted by emerald ash borer (*Agrilus planipennis*). *Ann. For. Sci.* **2017**, *74*, 1–10. [CrossRef]

43. Rejmanek, M. Invasibility of Plant Communities. In *Biological Invasions: A Global Perspective*; Drake, J.A., Mooney, H., di Castri, F., Groves, R., Kruger, F., Rejmánek, M., Williamson, M., Eds.; Wiley: Chichester, UK, 1989; pp. 369–388.

44. Baraloto, C.; Goldberg, D.E.; Bonal, D. Performance Trade-offs among Tropical Tree Seedlings in Contrasting Microhabitats. *Ecology* **2005**, *86*, 2461–2472. [CrossRef]

45. Muller-Landau, H.C.; Wright, S.J.; Calderón, O.; Condit, R.; Hubbell, S.P. Interspecific variation in primary seed dispersal in a tropical forest. *J. Ecol.* **2008**, *96*, 653–667. [CrossRef]

46. Zhao, F.; Qi, L.; Fang, L.; Yang, J. Influencing factors of seed long-distance dispersal on a fragmented forest landscape on Changbai Mountains; China. *Chin. Geogr. Sci.* **2016**, *26*, 68–77. [CrossRef]

47. Greene, D.F.; Johnson, E.A. Modelling the temporal variation in the seed production of North American trees. *Can. J. For. Res.* **2004**, *34*, 65–75. [CrossRef]

48. Jasper, J.M.; Bleher, B.; Bohing-Gaese, K.; Chira, R.; Farwig, N. Fragmentation and local disturbance of forests reduce frugivore diversity and fruit removal in *Ficus thonningii* trees. *Basic Appl. Ecol.* **2008**, *9*, 663–672.

49. Lundgren, M.R.; Small, C.J.; Dreyer, G.D. Influence of Land Use and Site Characteristics on Invasive Plant Abundance in the Quinebaug Highlands of Southern New England. *Northeast. Nat.* **2004**, *11*, 313–332. [CrossRef]

50. With, K.A. The landscape ecology of invasive species. *Conserv. Biol.* **2002**, *16*, 1192–1203. [CrossRef]

51. González-Moreno, P.; Diez, J.M.; Ibáñez, I.; Font, X.; Vilà, M. Plant invasions are context-dependent: Multiscale effects of climate; human activity and habitat. *Divers. Distrib.* **2014**, *20*, 720–731. [CrossRef]

52. Stachowicz, J.J.; Tilman, D. Species Invasion and the Relationships between Species Diversity, Community Saturation and Ecosystem Functioning. In *Species Invasions: Insights into Ecology, Evolution, and Biogeography*; Sax, D.F., Ed.; Sinauer Associates Inc.: Massachusetts, MA, USA, 2005; pp. 41–64.

53. Huston, M.A. Management strategies for plant invasions: Manipulating productivity; disturbance; and competition. *Divers. Distrib.* **2004**, *10*, 167–178. [CrossRef]
54. Pavlovic, N.B.; Leicht-Young, S.A. Are temperate mature forests buffered from invasive lianas? *J. Torrey Bot. Soc.* **2011**, *138*, 85–92. [CrossRef]
55. Simberloff, D.; Souza, L.; Nuñez, M.A.; Barrios-Garcia, M.N.; Bunn, W. The natives are restless, but not often and mostly when disturbed. *Ecology* **2012**, *93*, 598–607. [CrossRef] [PubMed]
56. Von Holle, B.; Delcourt, H.R.; Simberloff, D. Theimportance of biological inertia in pant community resistance to invasion. *J. Veg. Sci.* **2003**, *14*, 425–432. [CrossRef]
57. Brym, Z.T.; Allen, D.; Ibáñez, I. Community control on growth and survival of an exotic shrub. *Biol. Invasions* **2014**, *16*, 2529–2541. [CrossRef]
58. Davis, J.C.; Shannon, J.P.; Bolton, N.W.; Kolka, R.K.; Pypker, T.G. Vegetation responses to simulated emerald ash borer infestation in *Fraxinus nigra* dominated wetlands of Upper Michigan; USA. *Can. J. For. Res.* **2017**, *47*, 319–330. [CrossRef]
59. Hausman, C.E.; Jaeger, J.F.; Rocha, O.J. Impacts of the emerald ash borer EAB eradication and tree mortality: Potential for a secondary spread of invasive plant species. *Biol. Invasions* **2010**, *12*, 2013–2023. [CrossRef]
60. Siegert, N.W.; McCullough, D.G.; Liebhold, A.M.; Telewski, F.W. Dendrochronological reconstruction of the epicentre and early spread of emerald ash borer in North America. *Divers. Distrib.* **2014**, *20*, 847–858. [CrossRef]
61. Mapbiocontrol Agent Release Tracking and Data Management for Federal; State; and Researchers Releasing Three Biocontrol Agents Released Against Emerald Ash Borer. 2017. Available online: www.mapbiocontrol. org (accessed on 14 August 2014).
62. Duan, J.J.; Bauer, L.S.; Abell, K.J.; Lelito, J.P.; Van Driesche, R.G. Establishment and abundance of *Tetrastichus planipennisi* (Hymenoptera: Eulophidae) in Michigan: Potential for success in classical biocontrol of the invasive emerald ash borer (Coleoptera: Buprestidae) establishment and abundance of *Tetrastichus planipennisi*. *J. Econ. Entomol.* **2013**, *106*, 1145–1154. [PubMed]
63. Abell, K.J.; Bauer, L.S.; Duan, J.J.; Van Driesche, R.G. Long-term monitoring of the introduced emerald ash borer Coleoptera: Buprestidae egg parasitoid; *Oobius agrili* Hymenoptera: Encyrtidae, in Michigan, USA and evaluation of a newly developed monitoring technique. *Biol. Control* **2014**, *79*, 36–42. [CrossRef]
64. Michigan Geographic Data Library. 2014. Available online: http://www.mcgi.state.mi.us/mgdl/?rel=ext& action=sext (accessed on 20 May 2014).
65. Clark, J.S. Why environmental scientists are becoming Bayesians. *Ecol. Lett.* **2005**, *8*, 2–14. [CrossRef]
66. Spiegelhalter, D.J.; Best, N.G.; Carlin, B.P.; van Der Linde, A. Bayesian Measures of Model Complexity and Fit. *J. R. Stat. Soc.* **2002**, *64*, 583–639. [CrossRef]
67. Chytrý, M.; Jarošík, V.; Pyšek, P.; Hájek, O.; Knollová, I.; Tichy, L.; Danihelka, J. 2008 Separating habitat invasibility by alien plants from the actual level of invasion. *Ecology* **2008**, *89*, 1541–1553. [CrossRef] [PubMed]
68. González-Moreno, P.; Pino, J.; Carreras, D.; Basnou, C.; Fernández-Rebollar, I.; Vilà, M. Quantifying the landscape influence on plant invasions in Mediterranean coastal habitats. *Landsc. Ecol.* **2003**, *28*, 891–903. [CrossRef]
69. Ilisson, T.; Chen, H.Y.H. The direct regeneration hypothesis in northern forests. *J. Veg. Sci.* **2009**, *20*, 735–744. [CrossRef]
70. Gelman, A.; Hill, J. *Data Analysis Using Regression and Multilevel/Hierarchical Models*; Cambridge University Press: New York, NY, USA, 2007.
71. Thomas, A.; O'Hara, B.; Ligges, U.; Sturtz, S. Making BUGS open. *R News* **2006**, *6*, 12–17.
72. Meyer, J.Y.; Fourdrigniez, M. Conservation benefits of biological control: The recovery of a threatened plant subsequent to the introduction of a pathogen to contain an invasive tree species. *Biol. Conserv.* **2011**, *144*, 106–113. [CrossRef]
73. Denslow, J.S.; D'Antonio, C.M.D. After biocontrol: Assessing indirect effects of insect releases. *Biol. Control* **2005**, *35*, 307–318. [CrossRef]
74. Barton, J.; Fowler, S.V.; Gianotti, A.F.; Winks, C.J.; Beurs, M.D.; Arnold, G.C.; Forrester, G. Successful biological control of mist flower *Ageratina riparia* in New Zealand: Agent establishment; impact and benefits to the native flora. *Biol. Control* **2007**, *40*, 370–385. [CrossRef]

75. Duan, J.J.; Bauer, L.S.; Abell, K.J.; Ulyshen, M.D.; Van Driesche, R.G. Population dynamics of an invasive forest insect and associated natural enemies in the aftermath of invasion: Implications for biological control. *J. Appl. Ecol.* **2015**, *52*, 1246–1254. [CrossRef]

76. Aubin, I.; Cardou, F.; Ryall, K.; Kreutzweiser, D.; Scarr, T. Ash regeneration capacity after emerald ash borer (EAB) outbreaks: Some early results. *For. Chron.* **2015**, *91*, 291–298. [CrossRef]

77. Wang, X.; Yang, Z.; Wu, H.; Liu, S.; Wang, H.; Bai, L. Parasitism and reproductive biology of *Spathius agrili* Yang Hymenoptera: Braconidae. *Acta Entomol. Sin.* **2007**, *50*, 920–926.

78. Duan, J.J.; Bauer, L.S.; Hansen, J.A.; Abell, K.J.; Van Driesche, R.G. An improved method for monitoring parasitism and establishment of *Oobius agrili* (Hymenoptera: Encyrtidae), an egg parasitoid introduced for biological control of the emerald ash borer (Coleoptera: Buprestidae) in North America. *Biol. Control* **2012**, *60*, 255–261. [CrossRef]

79. Martin, P.H.; Canham, C.D.; Marks, P.L. Why forests appear resistant to exotic plant invasions: Intentional introductions, stand dynamics, and the role of shade tolerance. *Front. Ecol. Envirion.* **2009**, *7*, 142–179. [CrossRef]

80. Yates, E.D.; Levia, D.F.; Williams, C.L. Recruitment of three non-native invasive plants into a fragmented forest in southern Illinois. *For. Ecol. Manag.* **2004**, *190*, 119–130. [CrossRef]

81. Flory, S.L.; Clay, K. Effects of roads and forest successional age on experimental plant invasions. *Biol. Conserv.* **2009**, *142*, 2531–2537. [CrossRef]

82. Kuuluvainen, T. Gap disturbance, ground microtopography; and the regeneration dynamics of boreal coniferous forests in Finland: A review. *Ann. Zool. Fenn.* **1994**, *31*, 35–51.

83. Kueffer, C.; Schumacher, E.; Dietz, H.; Fleischmann, K.; Edwards, P.J. Managing successional trajectories in alien-dominated; novel ecosystems by facilitating seedling regeneration: A case study. *Biol. Conserv.* **2010**, *143*, 1792–1802. [CrossRef]

84. Thompson, J.R.; Carpenter, D.N.; Cogbill, C.V.; Foster, D.R. Four centuries of change in northeastern United States forests. *PLoS ONE* **2013**, *8*, e72540. [CrossRef] [PubMed]

85. Vilà, M.; Ibáñez, I. Plant invasions in the landscape. *Landsc. Ecol.* **2011**, *26*, 461–472. [CrossRef]

86. Dickmann, D.L.; Leefers, L.A. *The Forests of Michigan*; University of Michigan Press: Ann Arbor, MI, USA, 2003.

87. Hutchinson, T.F.; Vankat, J.L. Society for Conservation Biology Invasibility and Effects of Amur Honeysuckle in Southwestern Ohio Forests. *Conserv. Biol.* **1997**, *11*, 1117–1124.

88. Ohlemüller, R.; Walker, S.; Wilson, J.B.; Memmott, J. Local vs. regional factors as determinants of the invasibility of indigenous forest fragments by alien plant species. *Oikos* **2006**, *112*, 493–501. [CrossRef]

89. Lookwood, J.L.; Hoopes, M.F.; Marchetti, M.P. *Invasion Ecology*; Wiley-Blackwell: Oxford, UK, 2007.

90. Ruckli, R.; Rusterholz, H.P.; Baur, B. Invasion of an annual exotic plant into deciduous forests suppresses arbuscular mycorrhiza symbiosis and reduces performance of sycamore maple saplings. *For. Ecol. Manag.* **2014**, *318*, 285–293. [CrossRef]

91. Levine, J.M.; Vilà, M.; D'Antonio, C.M.; Dukes, J.S.; Grigulis, K.; Lavorel, S. Mechanisms underlying the impacts of exotic plant invasions. *Proc. R. Soc. Lond. Biol. Sci.* **2003**, *270*, 775–781. [CrossRef] [PubMed]

92. Blumenthal, D.M.; Hufbauer, R.A. Increased plant size in exotic populations: A common-garden test with 14 invasive species. *Ecology* **2007**, *88*, 2758–2765. [CrossRef] [PubMed]

93. Yih, K.; Boucher, D.H.; Vandermeer, J.H.; Zamora, N. Recovery of the rain forest of southeastern Nicaragua after destruction by Hurricane Joan. *Biotropica* **1991**, *23*, 106–113. [CrossRef]

forests

MDPI

Article

Tree Stress and Mortality from Emerald Ash Borer Does Not Systematically Alter Short-Term Soil Carbon Flux in a Mixed Northeastern U.S. Forest

Jaclyn Hatala Matthes [1,*], Ashley K. Lang [2], Fiona V. Jevon [2] and Sarah J. Russell [1]

[1] Department of Biological Sciences, Wellesley College, 106 Central St., Wellesley, MA 02481, USA; srussell@wellesley.edu

[2] Department of Biological Sciences, Dartmouth College, 78 College St., Hanover, NH 03755, USA; ashley.k.lang.gr@dartmouth.edu (A.K.L.); fiona.v.jevon.gr@dartmouth.edu (F.V.J.)

* Correspondence: jmatthes@wellesley.edu; Tel.: +1-781-283-3159

Received: 22 December 2017; Accepted: 13 January 2018; Published: 16 January 2018

Abstract: Invasive insect pests are a common disturbance in temperate forests, but their effects on belowground processes in these ecosystems are poorly understood. This study examined how aboveground disturbance might impact short-term soil carbon flux in a forest impacted by emerald ash borer (*Agrilus planipennis* Fairmaire) in central New Hampshire, USA. We anticipated changes to soil moisture and temperature resulting from tree mortality caused by emerald ash borer, with subsequent effects on rates of soil respiration and methane oxidation. We measured carbon dioxide emissions and methane uptake beneath trees before, during, and after infestation by emerald ash borer. In our study, emerald ash borer damage to nearby trees did not alter soil microclimate nor soil carbon fluxes. While surprising, the lack of change in soil microclimate conditions may have been a result of the sandy, well-drained soil in our study area and the diffuse spatial distribution of canopy ash trees and subsequent canopy light gaps after tree mortality. Overall, our results indicate that short-term changes in soil carbon flux following insect disturbances may be minimal, particularly in forests with well-drained soils and a mixed-species canopy.

Keywords: emerald ash borer; forest disturbance; *Fraxinus*; soil respiration; methane oxidation

1. Introduction

Invasive forest insects can create biotic disturbances that critically alter ecosystem processes and ecosystem services [1–3]. However, the impacts of invasive insects on the ecosystem carbon cycle remain uncertain, particularly for indirect impacts such as changes to soil carbon cycling [4,5]. Many invasive insects cause tree stress and mortality through wood-boring, phloem-feeding, or defoliation, which potentially changes plant carbon and nutrient allocation strategies and subsequent plant-soil feedbacks [6,7]. This study examined the short-term, two-year impact of the invasive emerald ash borer (EAB; *Agrilus planipennis* Fairmaire) on soil carbon dioxide (CO_2) and methane (CH_4) fluxes shortly following the first detection of EAB in a mixed deciduous forest in New Hampshire, USA.

EAB is an invasive wood-boring insect that kills trees of the genus *Fraxinus* by effectively girdling stem xylem tissue with larval feeding tunnels [8]. EAB was first identified in the state of New Hampshire, USA, in 2013. Mortality in *Fraxinus* trees infested with EAB is rapid and trees generally die 1–2 years after EAB activity is detected [9–11]. Initial stages of infestation are characterized by a visible thinning of the canopy [12]. Currently there are several management strategies under consideration, including preemptive and salvage logging, insecticide, and biological control; however, none of these have been particularly effective in slowing the spread of EAB [13]. Although losses to aboveground biomass due to EAB-induced mortality have been documented in previous studies [14], impacts to the soil carbon processes beneath *Fraxinus* are less well understood. Given that forest soil carbon stocks are approximately twice the size of aboveground carbon stocks, it is critical that we understand how invasive pests impact this globally important carbon pool [15].

Previous studies have examined the complex feedbacks between biotic disturbances and soil carbon flux in other systems, which might provide a useful analog to EAB impacts for short-term changes in soil respiration. Defoliation studies in grasslands have found that moderate clipping or defoliation can increase [16], decrease [17,18], or have no significant effect [19,20] on soil CO_2 flux within a year of vegetation removal. In forest ecosystems, the effects of biotic disturbances are similarly variable, with evidence for higher [21,22], lower [23,24], or no net change [25] in soil CO_2 flux following disturbance. Higher soil respiration following defoliation may result from stress-induced root carbon exudation [26], higher soil temperature caused by canopy gaps, and higher soil moisture caused by reduced transpiration under defoliated trees [27]. However, insect disturbances that are severe enough to cause host mortality or significant loss of active tissues may also dampen soil carbon emissions by reducing root respiration [28]. All together, these factors suggest a complex set of mechanisms that could enhance, reduce, or produce no net change in soil CO_2 efflux. For many insect disturbances, these mechanisms are likely to change with time and infestation severity [29].

In addition to the effects of insect disturbance on soil CO_2 flux, these disturbances may alter rates of CH_4 oxidation in forest soils. Globally, methanotrophic bacteria in upland forest soils oxidize approximately 28 teragrams of CH_4 each year to CO_2 [30]. This CH_4 uptake reduces the radiative forcing of greenhouse gases in the atmosphere, as CH_4 has a stronger radiative forcing capacity than CO_2, and is an important component of the global CH_4 budget [31,32]. Rates of CH_4 uptake in soil may depend on several chemical and biological factors, including soil temperature, pH, texture, moisture, nutrient content, and the dynamics of the soil microbial community [33–37]. Previous studies that manipulated precipitation in forests have found lower CH_4 oxidation rates as soil moisture increased, likely due to a corresponding decrease in diffusivity [38,39]. However, to our knowledge, no previous studies have assessed potential changes in CH_4 oxidation in forest soils beneath trees experiencing biotic disturbance. Biotic disturbances may have important indirect effects on this component of the forest carbon cycle, particularly through changes in soil microclimate that follow tree mortality and tissue loss.

The two primary goals of this study were: (1) to identify patterns in soil CO_2 and CH_4 fluxes that related to EAB impact; and (2) to suggest mechanisms for changes in these fluxes under EAB infestation. Because different environmental parameters control soil CO_2 efflux and CH_4 uptake, we formulated independent sets of alternative hypotheses for each process. Our null expectation was that EAB disturbance does not produce a net change in CO_2 nor CH_4 flux, either because stress and mortality produce no effect on soil CO_2 or CH_4 flux or because multiple concurrent changes to the soil environment counteract each other such that no net effect is produced. In this study, we assessed two alternate hypotheses for the potential effects of EAB impact on either soil CO_2 flux or CH_4 uptake (Figure 1):

- CO_2 flux, H_1: Soils beneath EAB-impacted *Fraxinus americana* L. (*F. americana*) will have lower rates of CO_2 flux due to lower active root metabolism and exudation of labile carbon substrates compared to soil beneath visibly healthy trees.
- CO_2 flux, H_2: Soils beneath EAB-impacted *F. americana* will have higher rates of CO_2 flux due to an increase in dead root biomass available for decomposition, as well as an increase in soil temperature and moisture, as infestation results in large canopy gaps and reduced evapotranspiration.
- CH_4 uptake, H_1: Soils beneath EAB-impacted *F. americana* will have lower rates of CH_4 uptake due to an increase in soil moisture from decreased evapotranspiration.
- CH_4 uptake, H_2: Soils beneath EAB-impacted *F. americana* will have higher rates of CH_4 uptake due to an increase in soil temperature, as canopy gaps allow more sunlight to reach the forest floor.

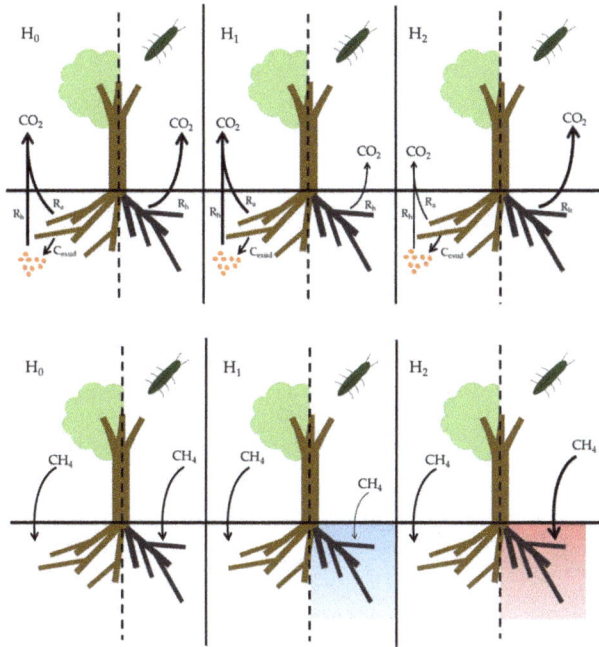

Figure 1. Null hypothesis (H_0) and two alternate hypotheses (H_1 and H_2) for the effect of tree stress induced by the emerald ash borer (EAB) on soil CO_2 flux and CH_4 uptake. H_0 poses that there is no net change on soil flux, including autotrophic (R_a) and heterotrophic (R_h) respiration, since processes that differ beneath live and dead *Fraxinus americana* L. might balance each other. Other alternate hypotheses suggest that either live or dead *F. americana* might have higher soil flux due to biological and physical changes to the soil beneath impacted trees.

2. Materials and Methods

2.1. Study Site

Our study site was located in Canterbury, NH, USA, in a mixed forest comprised of approximately 30% *Fraxinus americana* L., and also *Acer saccharum* Marsh., *Tsuga canadensis* L., and *Acer rubrum* L. The 1981–2010 climatological mean annual temperature at the nearest long-term weather station was 7.96 ± 0.58 C (mean ± standard deviation) and the annual precipitation was 1020 ± 178 mm (Concord, NH weather station code GHCND:USW00014745, 21 km from the study site [40]). For climate variables during our study period, we used total monthly precipitation data and mean monthly temperature data from the same station as the 30-year climatology.

2.2. Study Design

At our study site, we classified *F. americana* trees into three categories based on visual canopy cover estimation: healthy, impacted, or dead. Trees were considered healthy if their canopy was fully leafed-out in midsummer, with no bare branches or visible signs of stress (i.e., epicormic branching). Trees were classified as impacted if their canopy showed evidence of thinning (Figure 2b) or epicormic branching. Trees designated as dead were examined for signs of EAB infestation (D-shaped exit holes; Figure 2a) and only trees killed by EAB were included in the study. On 28 April 2016, sixteen *F. americana* trees were selected across a range of classes impacted by EAB: three healthy, six impacted, and seven recently dead, and we measured tree diameter at 1.3 m from ground level (DBH, cm).

Trees were selected for our study within two separate tracts of about 100 m^2 area less than 1 km from each other, where *F. americana* were spaced about 5–20 m with other species typically occurring between *F. americana* individuals (Figure 2d,e). Eight additional *F. americana* trees, five healthy and three impacted, were added to the study on 8 August 2016 to achieve higher replication ($n = 24$ from this date onward), and we measured DBH for these additional replicates. On 16 June 2017, the impact classes of individual trees were re-evaluated.

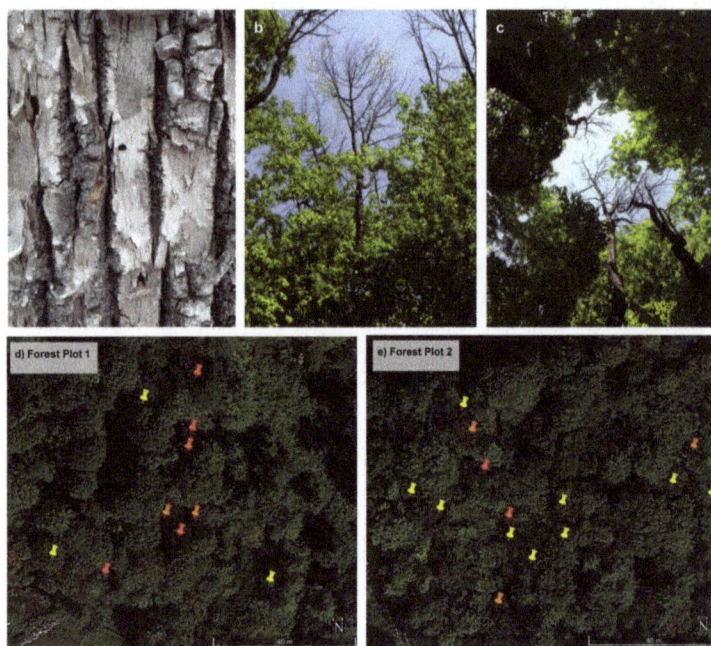

Figure 2. Evidence of emerald ash borer (EAB) impact and relative locations of study trees: (**a**) D-shaped exit hole in trunk from an emerging adult beetle; (**b**) thinning canopy of an impacted tree; (**c**) bare canopy and conspicuous light gap surrounding EAB-killed trees; (**d**) study trees from Plot 1; and (**e**) Plot 2 on a 11 September 2017 true-color satellite image, where red indicates trees that were dead from the start of the study, orange indicates trees that died during the study, and yellow indicates trees that were stressed by the end of the study.

On the same dates listed above when trees were selected (either 16 June or 8 August 2016), we installed 10.16 cm diameter polyvinyl chloride (PVC) soil flux collars that remained in place for the duration of the study, following standard soil flux measurement protocol [41]. Respiration collars were installed 1.5 m from the base of the tree in a randomly chosen direction, to a depth of approximately 7 cm to capture aggregate changes in both microbial and root respiration. We waited one month to measure CO_2 and CH_4 fluxes at the soil collars to allow for belowground acclimation to the installation disturbance. We measured the height of each collar at the four polar coordinates and took the mean of those four measurements as the collar height in cm. To validate soil flux models at different spots beneath the same trees, we installed a second collar at half the trees within the study ($n = 12$). The second collars were installed on four healthy, four impacted, and four dead trees, as evaluated in 2016.

We mapped the location of all study trees with a GPS unit (Garmin, Olathe, KS, USA). We plotted our study tree coordinates on a ~2 m resolution WorldView satellite image collected on 11 September 2017 in Google Earth (Google Earth Pro, v. 7.3.0.3832, Mountain View, CA, USA) to visualize the mosaic composition of the forest at our study site (Figure 2). By visually locating the approximate canopy extent of trees that were either dead from the start of the study or died during the study, we used the area calculator in Google Earth on the 11 September 2017 satellite image to roughly estimate the canopy gap size created by individual ash mortality.

2.3. Chamber Measurements and Flux Calculations

Soil CO_2 and CH_4 fluxes were measured at each collar during twelve sampling events: once per month from June to November 2016 (six sampling dates) and May to October 2017 (six sampling dates). Measurements were collected across all collars on each sampling date between approximately 09:00–12:00. Previous studies have demonstrated that instantaneous soil carbon flux in the late-morning is a good approximation of the daily mean soil CO_2 flux at a northeastern U.S. temperate forest site [42], thus avoiding potential biases in the effect of measurement time on soil flux [43]. During each chamber measurement, soil temperature was measured with a digital thermometer inserted 10 cm below the soil surface adjacent to the chamber. Integrated soil moisture was measured with a time domain reflectometer. We used a Decagon G3 Soil Moisture and Conductivity Sensor (Decagon Devices, Pullman, WA, USA, ± 0.03 m^3 m^{-3} accuracy) to measure soil moisture across the top 10 cm of soil on 2016 sampling dates and on 19 May 2017, and integrated soil moisture and conductivity across the top 12 cm of soil was measured with an H2 HydroSense II (Campbell Scientific, Logan, UT, USA, ± 0.03 m^3 m^{-3} accuracy) on the remaining five 2017 sampling dates.

During each soil CO_2 and CH_4 flux sampling event, we connected a portable greenhouse gas analyzer to the PVC respiration collar for a period of two minutes. On 2016 sampling dates, a Los Gatos Research (LGR) Ultraportable Greenhouse Gas Analyzer (Los Gatos Research, Los Gatos, CA, USA) cavity ring-down spectrometer was used to measure the CO_2, CH_4, and H_2O concentration inside the PVC chamber at 0.2 Hz temporal resolution (0.3 ppm CO_2 precision and 2 ppb CH_4 precision). On 2017 sampling dates, a Picarro GasScouter G4301 Gas Concentration Analyzer (Picarro Inc., Santa Clara, CA, USA) cavity ring-down spectrometer was used to measure the CO_2, CH_4, and H_2O concentration inside the PVC chamber at 1 Hz temporal resolution (0.4 ppm CO_2 precision and 3 ppb CH_4 precision). We used values of dry CO_2 and CH_4 concentration data (in ppm) that were corrected by the internal software for each greenhouse gas analyzer for CO_2 and CH_4 flux calculations.

We calculated soil CO_2 and CH_4 flux by fitting a linear regression model to the concentration data collected in the center 70 s of the two-minute measurement period to reduce potential errors associated with securing and removing the chamber top, which is a reasonable measurement interval for high-resolution soil flux chamber measurements [44]. We estimated the flux as the slope of a linear fit between concentration and time during each chamber measurement and converted units from ppm (provided by the analyzer) to the mass flux in μmol m^{-2} s^{-1} using the measured chamber height and surface area, and temperature and pressure at the time of measurement.

2.4. Soil Characteristics

After the last sampling event at the study site, we harvested the soil within each replicate collar to 7 cm depth to measure root biomass, soil texture, and soil pH. We sieved soils to 1 mm, separating roots from particulates, dried roots from each soil collar at 45 C for 72 h, and weighed dried roots within each replicate. We measured soil pH with an electrode in a 2:1 deionized water-to-soil suspension (Thermo Scientific, Waltham, MA, USA). We measured soil texture for a well-mixed 2 mL subsample of the remaining mineral soil fraction (<1 mm) within each soil collar. To measure soil texture, we dispersed each 2 mL subsample in 10 mL sodium hexametaphosphate and then placed each subsample on a shaking table for 24 h prior to measurement on a laser diffraction particle size analyzer (LS 13-320, Beckman Coulter, Brea, CA, USA). Output from the particle size analyzer was classified into volume fraction of clay, silt, and sand for each sample according to United States Department of Agriculture (USDA) particle size classes [45]. These soil variables—root biomass, pH, and fraction sand content—were used as covariates in the mixed models.

2.5. Statistical Methods

We nested all of our statistical analyses within each tree in the study ($n = 24$). For trees that had two replicate collars, we randomly selected one collar to be in the model fit dataset (training data) and the second replicate collar to be in a model validation dataset. To test our hypotheses that the level of EAB impact altered soil moisture and temperature (Figure 1), we created two mixed effects models with either soil moisture or temperature data as the response variable. For the soil moisture model, we fit two linear fixed effects for total monthly rainfall and the fraction of soil sand content and a fixed effect for EAB impact status (healthy, impacted, or dead). Additionally, we fit random effects for each replicate tree to reflect the non-independence of the repeated measurements and a second set of random effects for sampling month. For the temperature model, we fit a fixed effect for EAB impact status, and random effects for replicate tree and for month sampled.

To test our hypotheses that EAB presence influenced soil CO_2 and CH_4 flux, we constructed two additional sets of Bayesian linear mixed models with either soil CO_2 or CH_4 flux as the response variable. In both models, we fit separate random effects for measurement month and replicate (i.e., each tree) to reflect our repeated measures design. Our fixed effects for the CO_2 flux model were: (1) soil temperature and (2) soil moisture at the time of flux measurement; (3) monthly total rainfall; (4) DBH of the nearest *F. americana* tree; (5) root biomass of a 7-cm-deep core collected within the chamber at the conclusion of the experiment; (6) fraction of sand content; and (7) soil pH, with each measured from the same core.

Our fixed effects for the CH_4 uptake model were: (1) soil temperature and (2) soil moisture at the time of flux measurement; (3) monthly total rainfall; (4) fraction of soil sand content measured from the same core; and (5) soil pH. The CO_2 flux and CH_4 uptake models are each represented as Equation (1):

$$y_i = M_{t,i} + T_{j,i} + \beta_k X_k + \varepsilon_i \tag{1}$$

where y_i is each CO_2 flux or CH_4 flux measurement and $i = 380$ chamber flux measurements, M_t is a random month effect fit across all measurements within $t = 7$ study months, T_j is a random tree replicate effect fit across all measurements within $j = 24$ study trees, β_k is a set of linear regression coefficients for the X_k fixed effects variables, and ε_i is the residual variance. In our CO_2 flux model, $k = 7$ and $k = 5$ in the CH_4 uptake model, where X_k in each case represents the fixed effects described above. The random effects M_t and T_j were each modeled as normally distributed with means of zero and respective variances σ_M and σ_T, with uninformative priors on the variance ($\sigma_M = \sigma_T = 10^6$). We included a monthly random effect in addition to the fixed effects of soil temperature and moisture to account for potential seasonal changes in tree physiology (e.g., root carbon exudation, root elongation), which may have affected the soil carbon flux measurements.

We modeled CO_2 flux as $y_i \sim \text{lognormal}(\mu_i, \sigma_e)$, where y_i is each flux measurement, μ_i is the process mean fit as the sum of the modeled fixed and random effects, and σ_e is the measurement and process variance. Since all measured values of CH_4 flux in this study were negative, we took the absolute value of CH_4 flux (to conceptually represent the CH_4 uptake flux) and modeled the absolute value of CH_4 flux as $y_i \sim N(\mu_i, \sigma_e)$, where σ_e represents the measurement variance and i is each observation. Fitting the model with the absolute value of the CH_4 flux allowed for more congruent comparison between changes in CO_2 flux and CH_4 uptake.

We fit the mixed effects models in a Bayesian statistical framework with the RStan package [46]. All models were fit with standard Bayesian Markov chain Monte Carlo (MCMC) sampling techniques, with 100,000 iterations and four MCMC chains per model. For all fitted parameters, effective sample size exceeded 80,000 and R_hat = 1, indicating MCMC chain convergence and reliable estimates of posterior probability density [47]. This analysis was conducted entirely in R [48]. The code and raw data to fully reproduce the analysis within this paper is available at [49] and additionally as the EAB_soilflux.zip file included as Supplementary Materials. The code for this analysis relied on several R packages to facilitate data cleaning, processing, and visualization [50–55].

3. Results

3.1. F. americana Decline over Time

F. americana quickly declined during the study period. From May 2016 through October 2017, there was rapid progression of stress and mortality within *F. americana* as determined by visual survey methods: all trees that were healthy in 2016 had been impacted by 2017, leaving no trees in the "healthy" category for 2017 at this site (Figure 3). The mean canopy gap size created by individual ash mortality was 28 m^2 (full range of estimated trees 8.5–48 m^2), which, if assuming that the gap is a circle, translated to a mean gap diameter of 2.9 m (full range 1.7–3.9 m). Despite widespread ash decline, the non-dominance of ash at this site meant that much of the forest remained intact by the end of the study (Figure 2d,e).

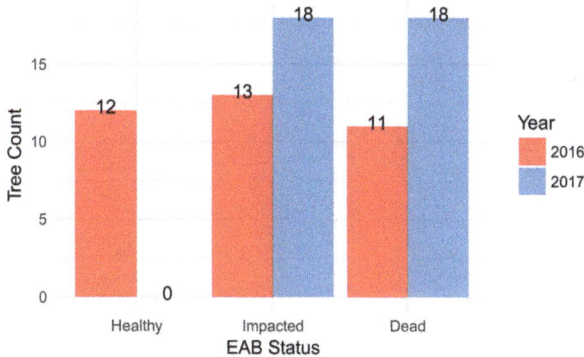

Figure 3. Emerald Ash Borer (EAB) impact classes of *F. americana* in July 2016 and 2017 from visual surveys. EAB impacts, estimated by visual surveys well after leaf-out, rapidly progressed within this forest stand from 2016 to 2017, where no *F. americana* remained in the Healthy class in 2017.

3.2. Soil Characteristics and Soil Microclimate Trends

Soil pH was between 4.05 and 5.43 at this site and sand content was high, ranging from 0.92 to 0.99 volumetric particle fraction (Table 1). Air temperature during the growing season was significantly higher than the upper quartile of the climatological monthly means for July, August, and September 2016 and September and October 2017 (Figure 4a). May 2017 was below the lower quartile for the climatological mean monthly temperature.

Table 1. Fixed regression parameters were estimated separately for CO_2 flux and CH_4 uptake Bayesian linear mixed models. Bolded regression parameters reflect posterior estimates with more than 90% of the posterior probability density different from zero, and italicized parameter estimates reflect posterior distributions where >50% of the probability density bridges zero. DBH: diameter at breast height (1.3 m from ground level).

Parameter	Variable Range	CO_2 Param Median	CO_2 95% Interval	CH_4 Param Median	CH_4 95% Interval
Intercept	n/a	−7.65	−15.64–0.53	−4.25	−21.0–12.2
Soil Temperature	7–20 °C	**0.06**	**0.03–0.08**	*0.03*	*−0.03–0.08*
Soil Moisture	0.009–0.530 m^3 m^{-3}	0.21	−0.35–0.76	**−6.00**	**−7.99−−3.91**
Monthly Rainfall	5.64–19.9 cm	**−0.02**	**−0.03–0.00**	−0.01	−0.06–0.03
Fraction Sand	0.92–0.99	**7.74**	**0.60–14.6**	12.63	−1.45–27.11
pH	4.05–5.43	*0.09*	*−0.34–0.55*	**−0.83**	**−1.73–0.08**
DBH	24.7–66.5 cm	*0.01*	*−0.01–0.01*	n/a	n/a
Root Biomass	0.67–7.55 g	*−0.06*	*−0.14–0.03*	n/a	n/a

Strong differences existed in monthly precipitation between 2016 and 2017 in May and June, where precipitation in 2016 was about half that of 2017 in each of those two months (Figure 4b). Although monthly total precipitation had high variability in those months, instantaneous soil moisture measurements collected at the time of soil flux measurement were unrelated to total monthly precipitation (Figure S1), likely due to the high sand content of soils at this site leading to strong drainage capacity (Table 1). Given that monthly precipitation and soil moisture were not correlated, we included both variables as fixed effects within the mixed models, since conceptually they might represent different processes. For example, soil moisture inhibits diffusion locally and total monthly precipitation could enhance tree productivity and root carbon exudation on longer timescales, impacting soil flux through different mechanisms.

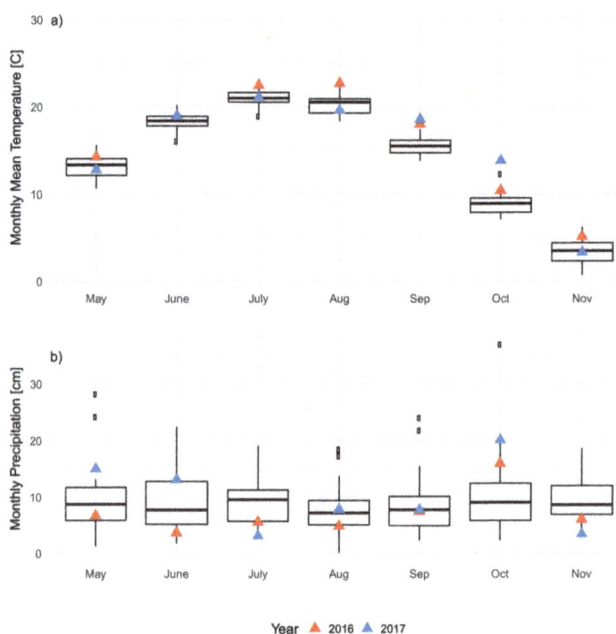

Figure 4. Monthly Mean Temperature and Total Precipitation in 2016 and 2017: (**a**) Monthly mean temperature was significantly higher than the 75% quartile range of the 30-year climatology for July, August, September, and November in 2016, and for September and October in 2017. (**b**) Monthly total precipitation was much higher in 2017 than in 2016 for May, June, and October, and the values for 2017 in those months were also higher than the 75% quartile of the 30-year climatology for total precipitation (box plot).

3.3. EAB Impact on Soil Microclimate and Gas Flux

We evaluated the mixed effects models fit to the repeated soil moisture and soil temperature measurements to determine whether soil microclimate was impacted by EAB infestation. In the model fit to measured soil moisture, the posterior probability density for the coefficient for total monthly rainfall was not significantly different from zero, but the coefficient for fraction soil sand content was negative (median = −1.69, 95% interval = −2.63–−0.79; Table S1). The random effects fit to EAB impact status (healthy, impacted, or dead) were not significantly different from zero (Table S1). For the model fit to the repeated soil temperature measurements, the random effects fit to EAB impact status were also not significantly different from each other (Table S1). For both the soil temperature and soil moisture models, the parameter estimates for the EAB status effects led us to conclude that EAB impact was not a reliable predictor of soil microclimate at this site.

The linear regression flux models for CO_2 flux and CH_4 uptake that were measured at individual collars had a mean standard error of 0.028 µmol m^{-2} s^{-1} for CO_2 flux (0.026 µmol m^{-2} s^{-1} for the LGR analyzer in 2016 and 0.030 µmol m^{-2} s^{-1} for the Picarro analyzer in 2017) and 0.042 nmol m^{-2} s^{-1} for CH_4 flux (0.035 nmol m^{-2} s^{-1} for the LGR analyzer in 2016 and 0.047 nmol m^{-2} s^{-1} for the Picarro analyzer in 2017). Since these errors represent less than one percent of the total measured flux magnitude, we did not explicitly incorporate the error from this flux calculation in subsequent analysis. In both 2016 and 2017, CO_2 flux followed a seasonal pattern with the peak annual flux in August 2016 and July 2017 (Figure 5a). CH_4 flux also had a strong seasonal pattern, with peak CH_4 uptake lagging the peak in CO_2 flux, in September 2016 and August 2017. Peak CH_4 uptake during mid-summer was much higher in 2016 compared to 2017, and fluxes for 2016 and 2017 were significantly different in July, August, and September (Figure 5b; paired *t*-test, $p < 0.05$).

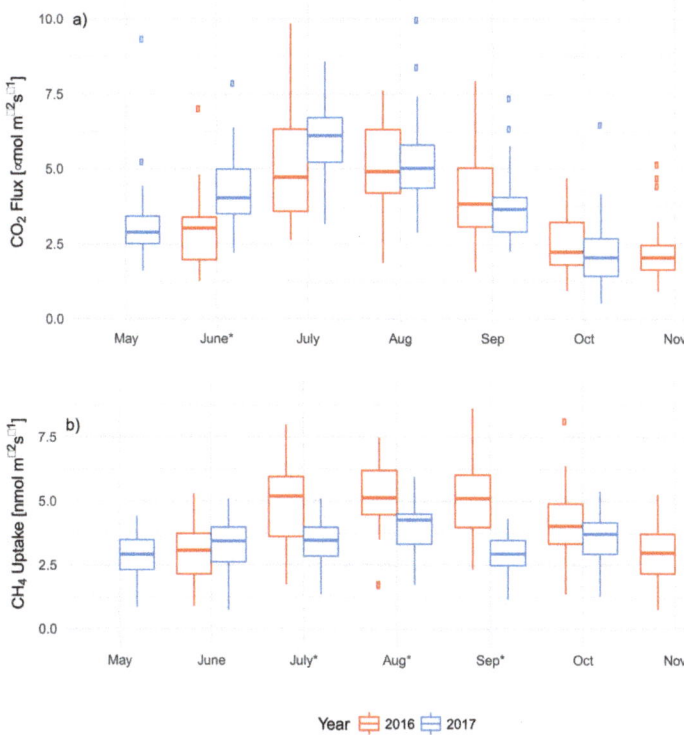

Figure 5. Seasonal patterns in CO_2 flux and CH_4 uptake: (**a**) monthly soil CO_2 flux and (**b**) monthly soil CH_4 uptake aggregated across all replicates both followed a clear seasonal pattern, with the highest efflux in mid-summer. For CO_2 flux, the measurements collected in 2016 and 2017 were significantly different only during June and July (paired *t*-test, $p < 0.05$) and the CH_4 uptakes in 2016 and 2017 were significantly different during July, August, and September (paired *t*-test, $p < 0.05$).

To evaluate potential changes in soil flux with *F. americana* decline, we used the three EAB impact categories (healthy, impacted, and dead) evaluated in 2016 and 2017 to assign two groups that described the trajectories of ash decline: trees that experienced EAB stress and mortality during the experiment (Impacted) and trees that were already dead at the start of the experiment (Dead).

These two transitional categories allowed us to compare changes in soil CO_2 flux and CH_4 uptake as *F. americana* declined. Patterns in the CO_2 flux and CH_4 uptake difference between 2016 and 2017 organized by these two classes were not significantly different, and did not exhibit any discernible trends across the sampling dates (Figure 6). It is important to note that this visual summary of the mean patterns in CO_2 flux and CH_4 uptake by category does not incorporate the covariate effects that also varied between the two years, which are additionally important drivers of CO_2 flux and CH_4 uptake.

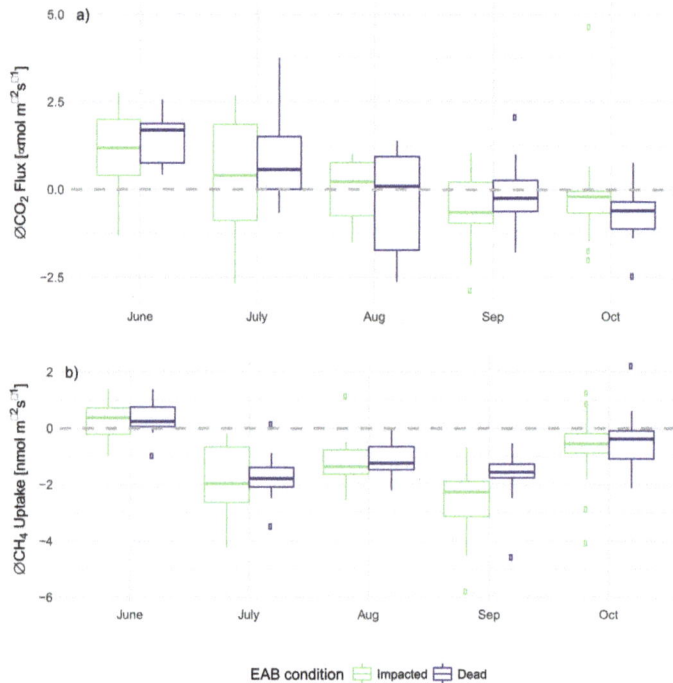

Figure 6. Monthly CO_2 flux and CH_4 uptake differences in 2016 and 2017 organized by trees that were stressed by emerald ash borer (EAB) during the experiment (Impacted) or already dead at the start of the experiment (Dead): (**a**) monthly soil CO_2 flux 2016–2017 differences and (**b**) monthly soil CH_4 uptake 2016–2017 differences had no consistent trends. Whether a tree was impacted by EAB during the experiment or dead from the start of the experiment did not have a significant impact on soil CO_2 flux or CH_4 uptake differences between 2016 and 2017.

3.4. Modeled Predictors of CO_2 Flux and CH_4 Uptake

In the mixed effects model fit to CO_2 flux data, the modeled regression coefficient for temperature was positive (Table 1). The regression coefficient for soil moisture was also positive, but the coefficient for monthly rainfall was negative. Fraction sand content had a large, but uncertain, positive effect on soil CO_2 flux. The posterior distributions of the regression coefficients for DBH of the nearest tree, root biomass, and soil pH were all not significantly different from zero for the CO_2 flux model.

In the mixed effects model fit to the CH_4 uptake data, the regression coefficient for soil temperature was not significantly different from zero (Table 1). The coefficients for soil moisture and monthly rainfall were both negative (although only the soil moisture coefficient is significantly different from zero), in contrast to the CO_2 flux model where the moisture coefficient was positive and the rainfall coefficient was negative. In the CH_4 uptake model, the coefficient for the fraction sand content was positive but had large uncertainty bounds, as was the case for fraction sand regression coefficient in the CO_2 flux model. The coefficient for soil pH was negative for the CH_4 uptake model, in contrast to the CO_2 flux model where the pH coefficient was not significantly different from zero (Table 1).

Random month effects fit for both the CO_2 flux and CH_4 uptake models broadly paralleled trends apparent within Figure 5, with the largest positive effects in August and September for both CO_2 flux and CH_4 uptake, but with more variability in the order of the remainder of the months (Table S2). Because the random month effects are fit together with soil temperature, moisture, and rainfall, these effects represent patterns that move beyond the relationship with these other fixed factors, and could yield insight into which types of phenological signals might be important for soil processes. Random tree replicate effects were significant, but did not follow patterns that were associated with EAB impact class (Figure S2). Distribution statistics for all estimated parameters in both the CO_2 flux and CH_4 uptake models are provided in Table S2. We used the soil flux measurements that were replicated at half the trees ($n = 12$) and found that modeled CO_2 flux and CH_4 uptake generally corresponded (Figure S3). However, there remained large variability in the correspondence between measured and modeled soil flux (Figure S3). We caution that this is not necessarily a reliable predictive model, likely due to large variability among soil flux collars, even from within the same tree.

4. Discussion

4.1. Impact of EAB on Microclimate

We hypothesized that EAB-induced tree mortality would influence soil microclimate and subsequently alter CO_2 flux and CH_4 uptake in soil beneath affected trees. In our system, tree mortality and stress by EAB did not change soil microclimate conditions in a systematic direction. In previous studies of forest pests, outbreaks are often associated with increases in soil moisture, as transpiration of infested trees declines [3,24,56]. However, the fact that soils at our study site had high sand content in combination with the decoupling of monthly precipitation and soil moisture (Figure S1) indicated that our site was well drained, so that changes to plant transpiration from EAB damage might not be strong enough to influence overall soil moisture. Similarly, while previous studies indicate that forest insect disturbance may increase soil temperature [27], this was not the case in our study. We suspect this is due to the small relative size of the canopy gaps created by dead and dying *F. americana* (see Figure 2c). Indeed, previous studies on the effects of canopy gaps on soil temperature and moisture have found mixed results, but generally see greater impact with larger gaps [57].

4.2. Impact of EAB on CO_2 Flux

Our findings agreed with previous studies that soil temperature and moisture overwhelmingly control soil CO_2 flux [58]. As we did not find a discernible effect of EAB infestation on these microclimate factors, we are therefore unsurprised that CO_2 flux also does not appear to be influenced by EAB impact. Another possibility is that our sample size was not adequate to detect a possible effect of EAB on soil flux. In studies that did find an effect of biotic disturbances on soil CO_2 flux [24], the impacted tree species was highly monodominant. In this study, the impacted species composed approximately one-third of the canopy. This difference is particularly important when considering the mechanisms suggested by our models: given that the variables that significantly influence soil CO_2 flux included climate and microclimate variables, but did not include root biomass, we would not expect that a diffuse infestation would immediately influence CO_2 flux if the infestation does not affect the soil microclimate. In our system, canopy gaps beneath affected trees were relatively small and reductions in transpiration by beetle-killed trees may have been balanced by increased water use of nearby trees. Thus, non-ash tree species unaffected by EAB may have buffered the effects of ash mortality on soil microclimate, and thus on soil CO_2 flux.

4.3. Impact of EAB on CH$_4$ Uptake

To our knowledge, this was the first study to assess the potential effects of tree mortality due to insect infestation on soil CH$_4$ uptake. In our mixed effects model, CH$_4$ uptake was strongly inhibited by soil moisture, monthly precipitation, and greatly enhanced by the fraction of sand content in the soil. This agrees with many previous studies that suggest that physical limits to atmosphere-gas diffusion might play a more important role than other soil carbon cycling components of CH$_4$ uptake [35–37,59]. While our data did not indicate a relationship between EAB-induced mortality and changes in soil moisture, this may be due to the mixed composition of our study site, and potential effects might be stronger in stands dominated by *Fraxinus* spp., for example, in riparian areas dominated by *Fraxinus nigra* Marsh. and/or *Fraxinus pennsylvanica* Marsh. The lack of *F. americana* dominance likely also dampened a potential temperature response feedback for CH$_4$ uptake at our site, and areas with more complete canopy mortality will likely experience much higher soil temperatures following EAB infestation.

Our findings suggested that the effect of insect disturbances on carbon dynamics in forests may depend on both edaphic and vegetative properties, and an understanding of these characteristics will complement future studies aiming to connect insect disturbances with ecosystem carbon storage. Sites where soils are poorly drained might experience a stronger effect of EAB infestation on soil carbon flux; for example, in stands of *Fraxinus* species that grow in riparian areas. Pre- and post-disturbance forest compositions are also likely to affect the magnitude of insect invasions on ecosystem carbon dynamics. In northern temperate forests, *F. americana* represents a small component of standing tree biomass, and the physiological response of co-dominant tree species in stands affected by EAB may reduce the overall effect of ash mortality on soil carbon fluxes. In less diverse systems where ash represents a greater fraction of forest biomass, the effects of EAB on the forest carbon balance are stronger [14,60]. However, the influence of EAB on forest carbon balance extends far beyond the period of active infestation. Our results support the idea that short-term impacts of insect disturbance on temperate forests are likely dwarfed by the effects of longer-term compositional changes in the tree community and transfer of *F. americana* snags into the soil pool, which are known to play a critical role in future carbon dynamics of disturbed forests [56,61]. If so, the impact of EAB on forest carbon dynamics will vary widely on a regional and local scale as different species replace ash in affected stands. A predictive framework for these compositional changes will be critical for understanding how EAB may alter forest carbon stocks into the future.

Our results may be relevant when considering management options for forests impacted by EAB. Early management strategies in the U.S. included attempts to use preemptive and salvage logging as a control measure in locations with confirmed EAB infestation [8,62]. While most of these initial attempts were abandoned as regional control strategies, salvage logging, both before and after EAB detection, remains a management strategy in some places [8,62]. Unlike other management techniques, such as insecticide or biological control, preemptive and salvage logging results in the immediate formation of canopy gaps at the site of the removed *Fraxinus* trees. Our results suggest that changes in microclimate, which is likely linked to the size and distribution of canopy gaps [57], are the critical factor in determining soil carbon flux responses to tree mortality at this scale. Therefore, this management strategy in particular may influence soil carbon fluxes if canopy gaps created by tree removal are large enough to change soil moisture or temperature.

5. Conclusions

As biotic disturbances become more common in temperate forests [63], there is a need to understand the consequences of these events for ecosystem carbon dynamics. Our results demonstrated that the short-term effects of EAB-induced ash mortality on soil carbon dynamics may be minimal in some ecosystems. In our study, EAB infestation did not lead to changes in soil microclimate conditions, possibly owing to coarse soil texture and sufficient drainage, and to the diffuse nature of the infested trees in this mixed forest. EAB-induced tree mortality created no change in CO$_2$ efflux

Forests **2018**, *9*, 37

or CH$_4$ uptake from soil beneath affected trees. In ecosystems with poorly drained soils or higher proportions of *Fraxinus* in the canopy, changes to soil moisture or temperature following widespread *Fraxinus* mortality may be sufficient to alter these fluxes. Our findings emphasized that soil properties may mediate the effect of tree-killing insects on ecosystem carbon fluxes and should be considered when assessing how invasive pests may alter forest carbon balance following disturbance.

Supplementary Materials: The following are available online at www.mdpi.com/1999-4907/9/1/37/s1, Figure S1: Soil Moisture and Monthly Precipitation, Figure S2: Posterior distributions of random tree effects; Figure S3: Model-data correspondence for replicate tree validation data; Table S1: model posterior distribution statistics for soil microclimate models, Table S2: model posterior distribution statistics for soil CO$_2$ flux and CH$_4$ uptake models.

Acknowledgments: This research was supported by NSF-1638406 and the Wellesley College Provost's Office. Many thanks to Jim, Susan, and Holly Snyder of Tamarack Farm for access to this study site and for their support and encouragement to conduct this experiment. Emma Conrad-Rooney, Erica Huang, and Lara Jones of Wellesley College all provided laboratory assistance in processing soils from this study. We also appreciate assistance and advice from Dan Brabander, Alden Griffith, and Katrin Monecke in the measurement of soil moisture, soil texture, and soil pH.

Author Contributions: J.H.M., A.K.L., and F.V.J. conceived and designed the experiments; J.H.M., A.K.L., F.V.J., and S.J.R. performed the experiments; J.H.M., A.K.L., and F.V.J. analyzed the data; J.H.M., A.K.L., and F.V.J. wrote the paper.

Conflicts of Interest: The authors declare no conflict of interest.

References

1. Lovett, G.M.; Weiss, M.; Liebhold, A.M.; Holmes, T.P.; Leung, B.; Lambert, K.F.; Orwig, D.A.; Campbell, F.T.; Rosenthal, J.; McCullough, D.G.; et al. Nonnative forest insects and pathogens in the United States: Impacts and policy options. *Ecol. Appl.* **2016**, *26*, 1437–1455. [CrossRef] [PubMed]
2. Boyd, I.L.; Freer-Smith, P.H.; Gilligan, C.A.; Godfray, H.C.J. The consequence of tree pests and diseases for ecosystem services. *Science* **2013**, *342*. [CrossRef] [PubMed]
3. Clark, K.L.; Skowronski, N.; Hom, J. Invasive insects impact forest carbon dynamics. *Glob. Chang. Biol.* **2010**, *16*, 88–101. [CrossRef]
4. Hicke, J.A.; Allen, C.D.; Desai, A.R.; Dietze, M.C.; Hall, R.J.; Hogg, E.H.; Kashian, D.M.; Moore, D.; Raffa, K.F.; Sturrock, R.N.; et al. Effects of biotic disturbances on forest carbon cycling in the United States and Canada. *Glob. Chang. Biol.* **2012**, *18*, 7–34. [CrossRef]
5. Peltzer, D.A.; Allen, R.B.; Lovett, G.M.; Whitehead, D.; Wardle, D.A. Effects of biological invasions on forest carbon sequestration. *Glob. Chang. Biol.* **2010**, *16*, 732–746. [CrossRef]
6. Schultz, J.C.; Appel, H.M.; Ferrieri, A.P.; Arnold, T.M. Flexible resource allocation during plant defense responses. *Front. Plant Sci.* **2013**, *4*, 324. [CrossRef] [PubMed]
7. Kozlowski, T. Tree physiology and forest pests. *J. For.* **1969**, *67*, 118–123.
8. Herms, D.A.; Mc Cullough, D.G. Emerald ash borer invasion of North America: History, biology, ecology, impacts, and management. *Annu. Rev. Entomol.* **2014**, *59*, 13–30. [CrossRef] [PubMed]
9. Knight, K.; Robert, P.; Rebbeck, J. How fast will trees die? A transition matrix model of ash decline in forest stands infested by emerald ash borer. In Proceedings of the Emerald Ash Borer Research and Development Meeting, Pittsburgh, PA, USA, 23–24 October 2008; pp. 28–29.
10. McCullough, D.G.; Katovich, S.A. *Pest Alert: Emerald Ash Borer*; Canadian Food Inspection Agency: Ottawa, ON, USA, 2004; pp. 4–5.
11. Klooster, W.S.; Herms, D.A.; Knight, K.S.; Herms, C.P.; McCullough, D.G.; Smith, A.; Gandhi, K.J.K.; Cardina, J. Ash (*Fraxinus* spp.) mortality, regeneration, and seed bank dynamics in mixed hardwood forests following invasion by emerald ash borer (*Agrilus planipennis*). *Biol. Invasions* **2014**, *16*, 859–873. [CrossRef]
12. Flower, C.E.; Knight, K.S.; Rebbeck, J.; Gonzalez-Meler, M.A. The relationship between the emerald ash borer (*Agrilus planipennis*) and ash (*Fraxinus* spp.) tree decline: Using visual canopy condition assessments and leaf isotope measurements to assess pest damage. *For. Ecol. Manag.* **2013**, *303*, 143–147. [CrossRef]
13. Vannatta, A.R.; Hauer, R.H.; Schuettpelz, N.M. Economic Analysis of Emerald Ash Borer (Coleoptera: Buprestidae) Management Options. *J. Econ. Entomol.* **2012**, *105*, 196–206. [CrossRef] [PubMed]

14. Flower, C.E.; Knight, K.S.; Gonzalez-Meler, M.A. Impacts of the emerald ash borer (*Agrilus planipennis* Fairmaire) induced ash (*Fraxinus* spp.) mortality on forest carbon cycling and successional dynamics in the eastern United States. *Biol. Invasions* **2013**, *15*, 931–944. [CrossRef]

15. Scharlemann, J.P.W.; Tanner, E.V.J.; Hiederer, R.; Kapos, V. Global soil carbon: Understanding and managing the largest terrestrial carbon pool. *Carbon Manag.* **2014**, *5*, 81–91. [CrossRef]

16. Hamilton, E.W.; Frank, D.A.; Hinchey, P.M.; Murray, T.R. Defoliation induces root exudation and triggers positive rhizospheric feedbacks in a temperate grassland. *Soil Biol. Biochem.* **2008**, *40*, 2865–2873. [CrossRef]

17. Guitian, R.; Bardgett, R.D. Plant and soil microbial responses to defoliation in temperate semi-natural grassland. *Plant Soil* **2000**, *220*, 271. [CrossRef]

18. Craine, J.M.; Wedin, D.A.; Stuart Chapin, F. Predominance of ecophysiological controls on soil CO_2 flux in a Minnesota grassland. *Plant Soil* **1999**, *207*, 77–86. [CrossRef]

19. Uhlířová, E.; Šimek, M.; Šantrůčková, H. Microbial transformation of organic matter in soils of montane grasslands under different management. *Appl. Soil Ecol.* **2005**, *28*, 225–235. [CrossRef]

20. Gavrichkova, O.; Moscatelli, M.C.; Kuzyakov, Y.; Grego, S.; Valentini, R. Influence of defoliation on CO_2 efflux from soil and microbial activity in a Mediterranean grassland. *Agric. Ecosyst. Environ.* **2010**, *136*, 87–96. [CrossRef]

21. Frost, C.J.; Hunter, M.D. Insect canopy herbivory and frass deposition affect soil nutrient dynamics and export in oak mesocosms. *Ecology* **2004**, *85*, 3335–3347. [CrossRef]

22. Ruess, R.W.; Hendrick, R.L.; Bryant, J.P. Regulation of fine root dynamics by mammalian browsers in early successional Alaskan taiga forests. *Ecology* **1998**, *79*, 2706–2720. [CrossRef]

23. Moore, D.J.P.; Trahan, N.A.; Wilkes, P.; Quaife, T.; Stephens, B.B.; Elder, K.; Desai, A.R.; Negron, J.; Monson, R.K. Persistent reduced ecosystem respiration after insect disturbance in high elevation forests. *Ecol. Lett.* **2013**, *16*, 731–737. [CrossRef] [PubMed]

24. Nuckolls, A.E.; Wurzburger, N.; Ford, C.R.; Hendrick, R.L.; Vose, J.M.; Kloeppel, B.D. Hemlock Declines Rapidly with Hemlock Woolly Adelgid Infestation: Impacts on the Carbon Cycle of Southern Appalachian Forests. *Ecosystems* **2009**, *12*, 179–190. [CrossRef]

25. Morehouse, K.; Johns, T.; Kaye, J.; Kaye, M. Carbon and nitrogen cycling immediately following bark beetle outbreaks in southwestern ponderosa pine forests. *For. Ecol. Manag.* **2008**, *255*, 2698–2708. [CrossRef]

26. Hamilton, E.W.; Frank, D.A. Can plants stimulate soil microbes and their own nutrient supply? Evidence from a grazing tolerant grass. *Ecology* **2001**, *82*, 2397–2402. [CrossRef]

27. Classen, A.T.; Hart, S.C.; Whitman, T.G.; Cobb, N.S.; Koch, G.W. Insect Infestations Linked to Shifts in Microclimate: Important Climate Change Implications. *Soil Sci. Soc. Am. J.* **2006**, *69*, 2049–2057. [CrossRef]

28. Edburg, S.L.; Hicke, J.A.; Lawrence, D.M.; Thornton, P.E. Simulating coupled carbon and nitrogen dynamics following mountain pine beetle outbreaks in the western United States. *J. Geophys. Res.* **2011**, *116*, G04033. [CrossRef]

29. Ehrenfeld, J.G. Ecosystem Consequences of Biological Invasions. *Annu. Rev. Ecol. Evol. Syst.* **2010**, *41*, 59–80. [CrossRef]

30. Curry, C.L. Modeling the soil consumption of atmospheric methane at the global scale. *Glob. Biogeochem. Cycles* **2007**, *21*, GB4012. [CrossRef]

31. Le Mer, J.; Roger, P. Production, oxidation, emission and consumption of methane by soils: A review. *Eur. J. Soil Biol.* **2001**, *37*, 25–50. [CrossRef]

32. Dutaur, L.; Verchot, L.V. A global inventory of the soil CH4 sink. *Glob. Biogeochem. Cycles* **2007**, *21*, GB4013. [CrossRef]

33. Serrano-Silva, N.; Sarria-Guzmán, Y.; Dendooven, L.; Luna-Guido, M. Methanogenesis and Methanotrophy in Soil: A Review. *Pedosphere* **2014**, *24*, 291–307. [CrossRef]

34. Mancinelli, R.L. The regulation of methane oxidation in soil. *Annu. Rev. Microbiol.* **1995**, *49*, 581–605. [CrossRef] [PubMed]

35. Adamsen, A.P.S.; King, G.M. Methane consumption in temperate and subarctic forest soils: Rates, vertical zonation, and responses to water and nitrogen. *Appl. Environ. Microbiol.* **1993**, *59*, 485–490. [PubMed]

36. Striegl, R.G. Diffusional limits to the consumption of atmospheric methane by soils. *Chemosphere* **1993**, *26*, 715–720. [CrossRef]

37. Castro, M.S.; Steudler, P.; Melillo, J.M. Factors controlling atmospheric methane consumption by temperate forest soils. *Glob. Biogeochem. Cycles* **1995**, *9*, 1–10. [CrossRef]

38. Borken, W.; Davidson, E.A.; Savage, K.; Sundquist, E.T.; Steudler, P. Effect of summer throughfall exclusion, summer drought, and winter snow cover on methane fluxes in a temperate forest soil. *Soil Biol. Biochem.* **2006**, *38*, 1388–1395. [CrossRef]

39. Billings, S.A.; Richter, D.D.; Yarie, J. Sensitivity of soil methane fluxes to reduced precipitation in boreal forest soils. *Soil Biol. Biochem.* **2000**, *32*, 1431–1441. [CrossRef]

40. National Oceanic and Atmospheric Administration. National Centers for Environmental Information Climate Data Online. Available online: https://www.ncdc.noaa.gov/cdo-web/ (accessed on 14 December 2017).

41. Davidson, E.A.; Savage, K.; Verchot, L.V.; Navarro, R. Minimizing artifacts and biases in chamber-based measurements of soil respiration. *Agric. For. Meteorol.* **2002**, *113*, 21–37. [CrossRef]

42. Davidson, E.A.; Belk, E.; Boone, R.D. Soil water content and temperature as independent or confounded factors controlling soil respiration in a temperate mixed hardwood forest. *Glob. Chang. Biol.* **1998**, *4*, 217–227. [CrossRef]

43. Cueva, A.; Bullock, S.H.; López-Reyes, E.; Vargas, R. Potential bias of daily soil CO_2 efflux estimates due to sampling time. *Sci. Rep.* **2017**, *7*, 11925. [CrossRef] [PubMed]

44. Pirk, N.; Mastepanov, M.; Parmentier, F.-J.W.; Lund, M.; Crill, P.; Christensen, T.R. Calculations of automatic chamber flux measurements of methane and carbon dioxide using short time series of concentrations. *Biogeosciences* **2016**, *13*, 903–912. [CrossRef]

45. United States Department of Agriculture. *Soil Mechanics Level 1, Module 3: USDA Textural Soil Classification*; United States Department of Agriculture: Washington, DC, USA, 1987.

46. Stan Development Team. *RStan: The R Interface to Stan*; Stan Development Team: New York, NY, USA, 2017.

47. Brooks, S.P.; Gelman, A. Interface Foundation of America General Methods for Monitoring Convergence of Iterative Simulations General Methods for Monitoring Convergence of Iterative Simulations. *J. Comput. Graph. Stat.* **1998**, *7*, 434–455. [CrossRef]

48. R Core Team. *R: A Language and Environment for Statistical Computing*; R Core Team: Vienna, Austria, 2017.

49. EAB_soilflux GitHub Code Respository. Available online: https://github.com/jhmatthes/EAB_soilflux (accessed on 12 January 2018).

50. Wickham, H. *Tidyverse: Easily Install and Load "Tidyverse" Packages*, version 1.2.1; R Core Team: Vienna, Austria, 2017.

51. Wickham, H. *Forcats: Tools for Working with Categorical Variables (Factors)*, version 0.2.0; R Core Team: Vienna, Austria, 2017.

52. Wickham, H. *Stringr: Simple, Consistent Wrappers for Common String Operations*, version 1.2.0; R Core Team: Vienna, Austria, 2017.

53. Grolemund, G.; Wickham, H. Dates and Times Made Easy with lubridate. *J. Stat. Softw.* **2011**, *40*, 1–25. [CrossRef]

54. Wilke, C.O. *ggridges: Ridgeline Plots in "ggplot2"*, version 0.4.1; R Core Team: Vienna, Austria, 2017.

55. Auguie, B. *gridExtra: Miscellaneous Functions for "Grid" Graphics*, version 2.3; R Core Team: Vienna, Austria, 2017.

56. Reed, D.E.; Ewers, B.E.; Pendall, E. Impact of mountain pine beetle induced mortality on forest carbon and water fluxes. *Environ. Res. Lett.* **2014**, *9*. [CrossRef]

57. Muscolo, A.; Bagnato, S.; Sidari, M.; Mercurio, R. A review of the roles of forest canopy gaps. *J. For. Res.* **2014**, *25*, 725–736. [CrossRef]

58. Raich, J.W.; Tufekciogul, A. Vegetation and soil respiration: Correlations and controls. *Biogeochemistry* **2000**, *48*, 71–90. [CrossRef]

59. Blankinship, J.C.; Brown, J.R.; Dijkstra, P.; Allwright, M.C.; Hungate, B.A. Response of Terrestrial CH4 Uptake to Interactive Changes in Precipitation and Temperature Along a Climatic Gradient. *Ecosystems* **2010**, *13*, 1157–1170. [CrossRef]

60. Flower, C.E.; Gonzalez-Meler, M.A. Responses of temperate forest productivity to insect and pathogen disturbances. *Annu. Rev. Plant Biol.* **2015**, *66*, 547–569. [CrossRef] [PubMed]

61. Crowley, K.F.; Lovett, G.M.; Arthur, M.A.; Weathers, K.C. Long-term effects of pest-induced tree species change on carbon and nitrogen cycling in northeastern U.S. forests: A modeling analysis. *For. Ecol. Manag.* **2016**, *372*, 269–290. [CrossRef]

62. Van Driesche, R.G.; Reardon, R.C. *Biology and Control of Emerald Ash Borer*; USDA: Washington, DC, USA, 2015; p. 171.

63. Sturrock, R.N.; Frankel, S.J.; Brown, A.V.; Hennon, P.E.; Kliejunas, J.T.; Lewis, K.J.; Worrall, J.J.; Woods, A.J. Climate change and forest diseases. *Plant Pathol.* **2011**, *60*, 133–149. [CrossRef]

![forests logo] *forests*

MDPI

Article

Coleopteran Communities Associated with Forests Invaded by Emerald Ash Borer

Matthew B. Savage and Lynne K. Rieske *

Department of Entomology, University of Kentucky, S-225 Ag North, Lexington, KY 40546, USA;
matthew.savage@uky.edu
* Correspondence: Lrieske@uky.edu; Tel.: +1-859-257-1167

Received: 19 December 2017; Accepted: 25 January 2018; Published: 30 January 2018

Abstract: Extensive ash mortality caused by the non-native emerald ash borer alters canopy structure and creates inputs of coarse woody debris as dead and dying ash fall to the forest floor; this affects habitat heterogeneity; resource availability; and exposure to predation and parasitism. As EAB-induced (emerald ash borer-induced) disturbance progresses the native arthropod associates of these forests may be irreversibly altered through loss of habitat; changing abiotic conditions and altered trophic interactions. We documented coleopteran communities associated with EAB-disturbed forests in a one-year study to evaluate the nature of these changes. Arthropods were collected via ethanol-baited traps on five sites with varying levels of EAB-induced ash mortality from May to September; captured beetles were identified to the family level and assigned to feeding guilds (herbivore; fungivore; xylophage; saprophage; predator; or parasite). Over 11,700 Coleoptera were identified in 57 families. In spite of their abundance; herbivores comprised a relatively small portion of coleopteran family richness (8 of 57 families). Conversely, coleopteran fungivore richness was high (23 families), and fungivore abundance was low. Herbivores and fungivores were more abundant at sites where ash decline was most evident. The predatory Trogossitidae and Cleridae were positively correlated with ash decline, suggesting a positive numerical response to the increased prey base associated with EAB invasion. Ash forests are changing, and a deeper understanding of arthropod community responses will facilitate restoration.

Keywords: *Fraxinus*; invasive species; trophic guild; natural enemies

1. Introduction

Invasions by non-native invasive species pose significant threats to forest ecosystem function [1] and native biodiversity [2,3], and have widespread economic impacts [4,5]. Ash (*Fraxinus* spp.) are a consistent component of hardwood forests throughout the United States [6,7]; their prevalence and persistence is threatened by the emerald ash borer (EAB, *Agrilus planipennis* Fairmaire, Coleoptera: Buprestidae). Since its discovery near Detroit, MI in 2002 [8], EAB has spread rapidly through much of the eastern contiguous United States and southeastern Canada [9] inflicting extensive ash mortality in invaded regions. Larvae feed on phloem beneath the bark, forming serpentine galleries and destroying the vascular tissue, disrupting translocation of water and nutrients to the canopy, ultimately girdling the tree [10,11]. The majority of EAB-induced ash mortality (>85%) occurs within 3–5 years of the initial invasion [12,13]. All North American *Fraxinus* species are susceptible to attack and EAB readily colonizes healthy trees [10].

The direct effects of EAB invasion include altered forest structure due to rapid ash mortality, with subsequent alterations in ash-associated communities [14–18]. The indirect effects of rapid and broad scale tree mortality include increased gap formation which alters light penetration to the forest floor, accumulation of coarse woody debris, and qualitative and quantitative alterations in litter inputs causing shifting temperature and moisture regimes on the forest floor [19,20]. Such changes associated with EAB-induced ash mortality is affecting arthropod community associates of these invaded forests. In particular, changes at the soil-surface interface via increased leaf litter and coarse woody debris inputs can influence the abundance and distribution of soil biota [21,22]. Coleopterans, in particular the Carabidae, are well documented indicators of disturbance [23,24], and have been shown to respond to EAB-induced changes [25,26].

We sought to gain a broader understanding of the extent to which EAB-induced ash mortality might affect arthropod community associates, and focus here on aerial Coleoptera. We evaluated the extent to which coleopteran abundance and richness are affected by widespread changes in forest structure associated with the EAB invasion, and further considered these changes in relation to trophic guilds. We hypothesized that EAB-induced changes in forest composition and structure will lead to guild-specific changes in coleopteran communities. Specifically, we expected that xylophage, saprophage, and fungivore abundance and overall richness would increase in response to increases in habitat caused by inputs due to EAB disturbance.

2. Materials and Methods

2.1. Study Sites

Five study sites were established in mixed mesophytic forests in north-central Kentucky along the forefront of the expanding EAB invasion [27], in Anderson, Fayette, Henry, Shelby, and Spencer counties. Ash thrive on the moist and fertile soils that predominate in this region [28,29], and were historically a significant component of these forests [25]. At the onset of the study EAB was present at the Anderson, Henry, and Shelby sites (initially reported in November 2011, October 2009, and May 2009, respectively). EAB was first detected at the Fayette and Spencer sites in 2014, but there were little to no signs of EAB-induced stress.

At each site, 0.04 ha circular whole plots, situated ≥50 m apart, were placed in contiguous forests in blocks of three, with three blocks at each site, for a total of 45 plots across all five sites [27]. Ash canopy dieback was visually assessed by a single observer and each tree assigned a crown dieback value from 0% (healthy) to 100% (dead). When split or sloughing bark, larval galleries, or adult exit holes were evident, dieback was attributed to EAB. Our sites represented the full spectrum of forest disturbance associated with the EAB invasion, including pre-invasion at Fayette and Spencer (newly detected; <17% ash canopy dieback, <2% ash mortality), peak invasion at Shelby (EAB populations high; 25–30% ash canopy dieback, ~10% ash mortality), and post-invasion forests at Henry and Anderson (EAB populations low; >55% canopy dieback, 19–50% ash mortality) [27].

2.2. Arthropod Monitoring

Native coleopteran communities in the sub-canopy strata were monitored using 12-unit Lindgren multi-funnel traps (one per plot, $N = 45$) from 20 May to 12 September 2014. Traps were suspended over an ash branch (~4 m) and fitted with two 50 mL vials of 70% ethanol, a commonly used lure for xylophagous insects [30–32], hung from the funnel edge, and with a dichlorvos strip (2×5 cm^2) (American Vanguard Corporation Chemical Corp., Los Angeles, CA, USA) placed in each trap bottom. Traps were monitored every 7–14 days; contents were removed and stored in 70% EtOH in resealable plastic bags, and lures were replenished. In the laboratory samples were sorted to order [33]; Coleoptera were sorted and identified to family using available keys [33–36], counted, and assigned to trophic guilds based on larval feeding habits, including herbivore, saprophage, fungivore, xylophage, predator, or parasite [37]. We used family-level identifications, which are deemed taxonomically sufficient when undertaking a study of this nature [38,39]. This approach provides a good estimate of invertebrate populations within a given community when using a given sampling method, and has been utilized in a number of invertebrate studies [39–43]. Ordinally the Coleoptera are trophically diverse, but more or less trophically uniform within families [37,44], which allows classifying families into feeding guilds that exploit resources in a similar manner [45]. In our study, the carrion feeders, including the Silphidae, some Staphylinidae (e.g., *Aleochara* spp.), some Histeridae, some Nitidulidae (e.g., *Nitidulia* spp.), and some Leiodidae, were responding to the decaying trap contents rather than the ethanol lure, which resulted in excessive fluctuations in abundance, and so were excluded from additional analyses.

2.3. Analysis

We used assessments of ash mortality from Davidson and Rieske [27] and also evaluated ash canopies for decline, ranging from low (Fayette) to high (Henry), to assess the influence of ash decline on coleopteran abundance and richness. The abundance of aerial coleopterans was evaluated with funnel traps (total no. trapped). Richness (total no. families captured) and evenness (E_{var}) [46] was derived by site. Diversity indices were not derived because of data gaps caused by intermittent difficulties in accessing monitoring sites. Data were tested for normality (PROC UNIVARIATE) and transformed when necessary. Significance was determined at $\alpha = 0.05$ unless stated otherwise. All analyses were performed using SAS (v9.3, SAS Campus Drive Cary, NC, USA) [47].

Overall coleopteran abundance and cumulative richness by site were analyzed using a repeated measure mixed linear model (PROC MIXED), with sample interval as the repeated measure and individual plots (traps) as subjects. The difference of least squares (Least Squares Means) was used to separate means for these population parameters. Coleopteran feeding guild abundance and richness summed over the 16-week sampling period were analyzed using a generalized linear mixed model (PROC GLM) to compare guild × site interactions. Feeding guild abundance was transformed using a square root transformation for total counts and arcsine transformation for percent abundance. Feeding guild abundance (absolute and percent) was compared across all sites where the difference of least squares was used to separate means and post-hoc analysis was performed using pairwise T-Comparisons if differences arose. A chi squared analysis was used to determine differences in trophic guild abundance across sites. Correlations between the predator guild and ash canopy decline were evaluated (PROC CORR).

3. Results

3.1. Study Sites

Across our study sites, ash composition ranged from 12–26% for stems >2.5 cm diameter. EAB-induced ash mortality ranged from 0–50% and was highly correlated with EAB abundance [27]. Ash canopy dieback ranged from a low of ~7% at our least-disturbed, most recently invaded site to a high of 74% at our most degraded site (Table 1).

Table 1. Ash canopy dieback and coleopteran abundance at five sites in north-central Kentucky used to evaluate the colopteran community associated emerald ash borer-induced ash decline. Means followed by the same letter do not differ ($\alpha = 0.05$).

Site	Location (Lat., Long.)	Sample Intervals (Site Visits)	*Fraxinus* spp. Canopy Dieback (Mean % (s.e.))	Coleoptera Abundance [1]	Coleoptera Evenness [2]
Henry	38.56572, 85.14665	14	73.9 (4.6) a	1.30 (0.07) b	0.27 (0.01) a
Anderson	32.00857, 84.95980	13	56.9 (1.9) b	1.53 (0.07) ab	0.23 (0.01) a
Shelby	38.27980, 85.36258	16	27.4 (3.9) c	1.64 (0.06) a	0.27 (0.01) a
Spencer	38.02163, 85.27577	6	16.2 (2.8) cd	1.49 (0.11) ab	0.11 (0.01) b
Fayette	37.89653, 84.39270	10	7.4 (3.0) d	1.27 (0.08) b	0.17 (0.02) ab
			$F_{3,350} = 58.6; p < 0.001$	$F_{4,527} = 2.1; p < 0.02$	$F_{4,14} = 35.0; p < 0.01$

[1] Number of individuals per day (LS-means ± s.e.) captured in ethanol-baited funnel traps. Means separation on transformed data. [2] Evenness index: $E_{var} = 1 - \frac{2}{\pi} arctan\{\sum_{s=1}^{S}(\ln(x_s) - \sum_{t=1}^{s} \ln(x_t)/S)^2/S\}$.

3.2. Arthropods

Funnel traps yielded 16,455 arthropods, including 11,786 coleopterans (>71%) representing 57 families, excluding carrion feeders (Table A1). Elateridae was the most abundant family (Figure 1), with 16% of the total, followed by the Curculionidae and Staphylinidae (13 and 12%, respectively); these three families comprised nearly 41% of the coleopterans captured. The next most abundant families were the Ptilodactylidae (9%), the Latridiidae (9%), and the Histeridae (8%); collectively they comprised almost 27% of the total coleopterans.

Relative Family Abundance

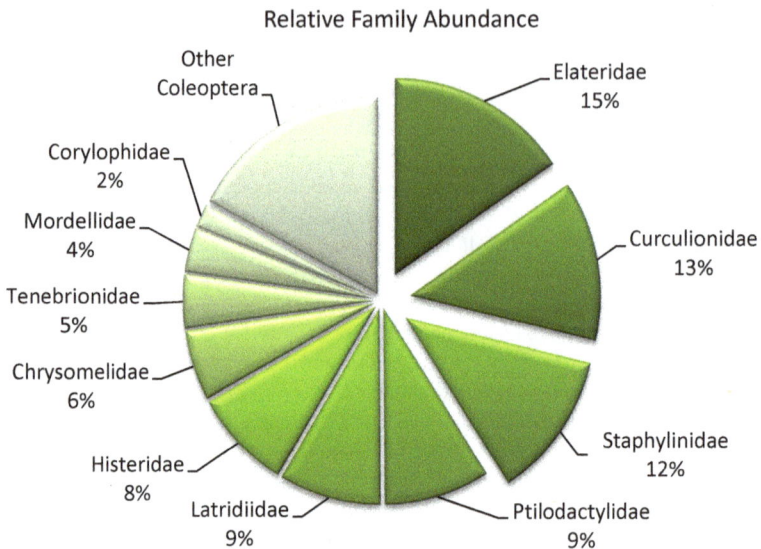

Figure 1. Relative abundance of the 10 numerically dominant coleopteran families found in forests associated with emerald ash borer-induced ash mortality.

Coleopteran abundance (Table 1), but not cumulative family richness (Figure 2) differed significantly among study sites; both tended to be lowest in pre-(Fayette) and post-disturbed (Henry) sites, and greatest at the site typifying peak invasion (Shelby). Henry, Anderson, and Shelby had the highest coleopteran evenness, and Spencer had the lowest (Table 1).

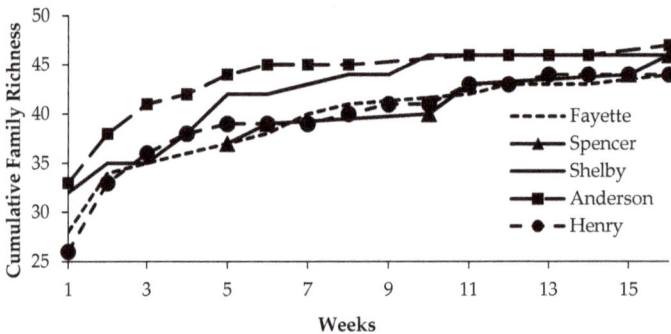

Figure 2. Cumulative coleopteran family richness at five forested sites in north central Kentucky varying in levels of EAB-induced (Emerald Ash Borer-induced) disturbance.

Coleopteran abundance was greatest among herbivores (4207 individuals, 36%) (Table 2), comprised primarily of the Elateridae, which feed on flowers, nectar, pollen, and rotting fruit [35].

Table 2. Relative abundance and richness of Coleopteran feeding guilds sampled from five sites affected by emerald ash borer ash decline.

Trophic Guild	Coleopteran Family-Level	
	Abundance (%)	Richness (%)
Herbivore	36	14
Fungivore	17	40
Predator	26	19
Xylophage	10	12
Saprophage	10	10
Parasite	<1	5
Unidentified	<1	–
Total	100	100

However, in spite of their abundance, herbivores comprised only 14% of total family richness (8 families). In contrast, coleopteran fungivore richness was highest at 40% (23 families), in spite of the fact that abundance was relatively low (2082 individuals, 17%) (Table 2; Figure 3).

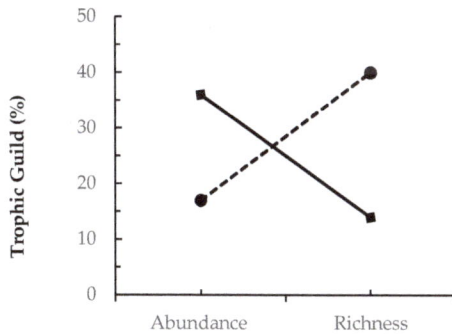

Figure 3. Relationship between relative herbivore (◆—◆) and fungivore (●—●) abundance and richness among coleopterans sampled from EAB-impacted (Emerald Ash Borer-impacted) forests.

Fungivores were dominated by the Latridiidae (Figure 1), which typically feed on the reproductive structures of fungi and are commonly found in plant debris [35]. Predators comprised 26% of the total (3050 individuals) and 19% of the coleopteran family richness (11 families) (Table 2). Predators consisted primarily of the Staphylinidae (Figure 1), which are generalists, and the Histeridae, which has one subfamily associated with bark beetle (Curculionidae) galleries, and another subfamily that feeds principally on fly and beetle larvae associated with dung [35]. Saprophages and xylophages made up ~10% of the abundance, and similarly 10 and 12% of coleopteran family richness, and consisted primarily of Ptilodactylidae and Scolytines (Figure 1; Table 2).

Trophic guild abundance across sites varied (x^2 = 1045.6; df = 20; $p < 0.0001$) (Figure 4). Herbivore and saprophage abundance was greatest at Shelby (Figure 4a,b), which represented the greatest EAB activity, reflected in EAB intercept trap catch [27], among the five sites. Fungivore abundance (Figure 4c) was positively correlated with ash decline and was greatest at Shelby, Anderson, and Henry, where disturbance caused by the EAB invasion was more advanced, and lowest at Fayette and Spencer, where EAB-related disturbance was minimal. Xylophages were most abundant at Spencer (Figure 4d), again representing relatively early stages of EAB invasion. Predator abundance (Figure 4e) was lowest at Fayette and Henry, representing both pre- and post-EAB invasion and highest at the sites where the invasion is nearer its peak. Parasite abundance (Figure 4f) was similar across all sites.

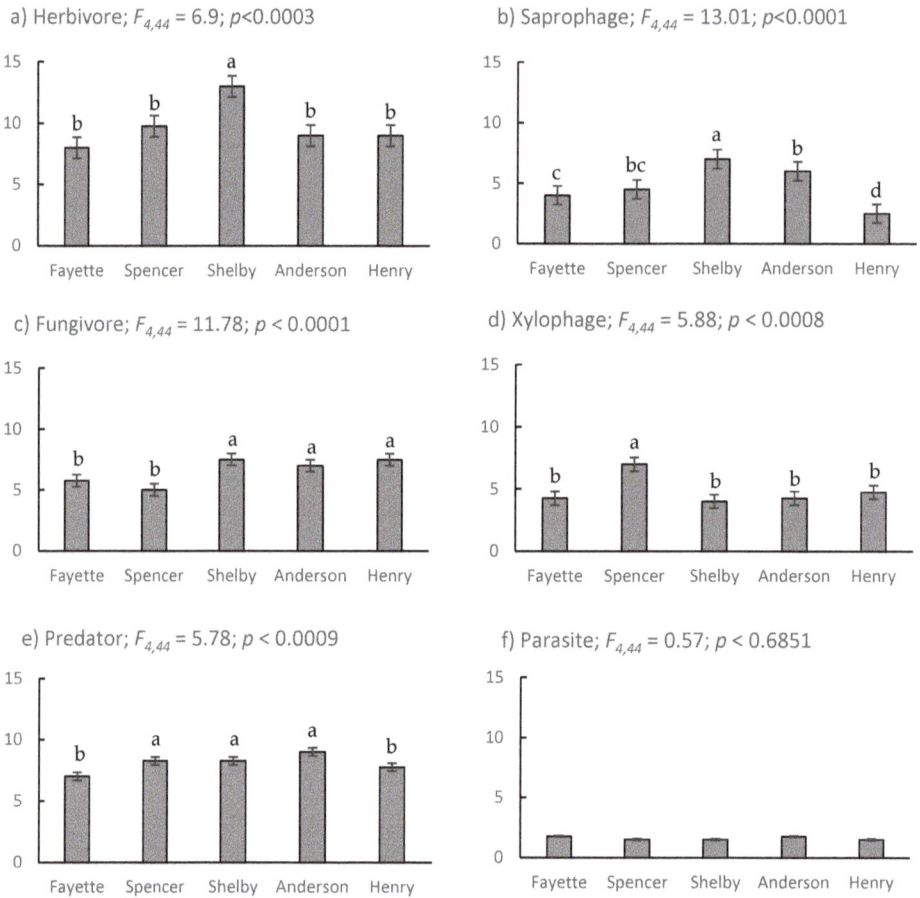

Figure 4. Coleopteran feeding guild abundance at five forested sites in north central Kentucky, including (**a**) herbivores, (**b**) saprophages, (**c**) fungivores, (**d**) xylophages, (**e**) predators, and (**f**) parasitoids. Means followed by the same letter do not differ ($\alpha = 0.05$).

Among the predators, Trogossitidae and Cleridae abundance (7% and 5% of total predator abundance, respectively) were positively correlated ($\alpha < 0.1$) with ash decline (Figure 5).

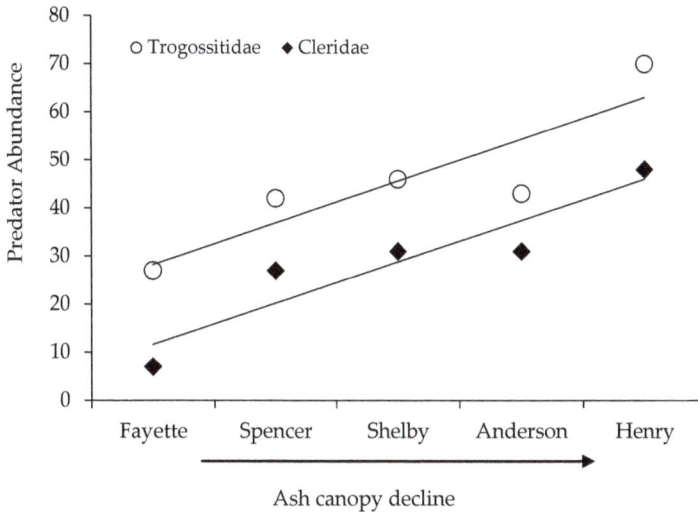

Figure 5. Relationship between *Fraxinus* canopy decline and abundance of two coleopteran predators: the Trogossitidae (Trogossitid abundance = 8.7 × canopy decline (%) + 19.5, R^2 = 0.71, p = 0.07) and the Cleridae (Clerid abundance = 8.6 × canopy decline (%) + 3, R^2 = 0.74, p = 0.06).

4. Discussion

Changes in structure, composition, and succession [14–18], alterations in light availability to the forest floor [19,20], and inputs of coarse woody debris associated with forest disturbance create and eliminate habitats [14–17,21,22,25], and affect resource availability and trophic relationships. Ash-dominated forests in the wake of the EAB invasion are expecting a loss of overall arthropod richness and are facing cascading ecological impacts and altered ecosystem processes. Of the 282 native arthropods associated with ash, 43 are monophagous; nine of these monophages are coleopterans [24]. Undoubtedly, these ash specialists will be negatively affected and may experience localized extirpation. We found discernible differences in aerial coleopteran communities associated with EAB-disturbed forests. The increase in abundance and cumulative richness of coleopteran associates where EAB activity density was at its greatest may be attributable to increases in habitat availability due to newly forming snags and coarse woody debris, and to volatile emissions from dying trees [30,31,48,49]. These resources are relatively transient, and as they decline we can anticipate a corresponding decline in coleopteran taxa reliant on their persistence. Contrary to expectations, coleopteran family evenness was highest in sites more heavily disturbed by EAB, where these transient resources would be most plentiful.

We found coleopteran abundance to be greatest among herbivores, comprised primarily of the Elateridae, which feed on flowers, nectar, pollen, and rotting fruit [35]. The Latridiidae, which feed on fungal reproductive structures and are common in plant debris [35], were the dominant fungivore family. The increase in fungivore richness that we observed may be a response to an increase in availability of these types of resources as ash decline progresses. Predators comprised 26% of the total trap catch, and consisted primarily of the Staphylinidae, which are generalist predators, and the Histeridae, which has one subfamily most commonly associated with bark beetle (Curculionidae) galleries [35,36]. Among the predators, Trogossitidae and Cleridae were positively correlated with ash decline. Trogossitids consisted of 228 individuals of primarily *Tenebroides* sp.; these are bark-gnawing beetles found beneath the bark of dead trees and are associated with wood-boring beetles [35]. Clerids (144 individuals) consisted primarily of *Enoclerus* sp.; these checkered beetles are associated with dead wood, and are often found predating larval Curculionidae, Cerambycidae, and Buprestidae [35].

Interestingly, the parastic Passandridae, comprised entirely of *Catogenus rufus* (Fabricius), were consistently present in low numbers, regardless of the extent of forest disturbance (Figure 4f). *Catogenus rufus* has been found as both larvae and adults in EAB galleries from dead ash in EAB-invaded forests [27]; it was present in low numbers across sites and appeared unaffected by the stage of the EAB invasion or by the corresponding decline in ash canopies. Its presence at all sites in similar abundance suggests that it utilizes a variety of wood-boring hosts and is not demonstrating a numerical response to the EAB invasion in these forests, though it may still be utilizing EAB as a resource. *Tenebroides*, *Enoclerus* and *Catogenus* spp. have been documented in association with EAB larvae and pupae near the epicenter of the EAB invasion in North America [50], suggesting that they may be playing a role in population dynamics of this aggressive invader.

Our use of family level identifications in the evaluation of aerial coleopteran communities could be viewed as a limitation of this study. However, "taxonomic sufficiency" (sensu Ellis 1985 [38]) recognizes that, within a community, changes at the species level are often reflected at coarser taxonomic levels. The use of coarser taxonomic identifications reduces the inputs associated with large scale community level studies [40–42,51,52]. Family level richness is a good predictor of species richness for a variety of taxa, including butterflies [51]. Family level identifications of benthic fauna are appropriate for calculating stream quality indices [53–56] and multivariate analyses of community data [54], and can reliably detect moderate ecosystem impacts [57]. Identifications beyond the family level may not yield much more information and may not be cost effective [58]. Targeting coarser taxonomic resolution, rather than insisting upon species level identification for woody plant surveys, significantly reduces costs of field work [52]. Clearly, accepting coarser taxonomic sufficiency provides us with an effective approach to conduct rapid studies on ephemeral systems such as ours, as well as larger landscape scale studies over longer periods of time to answer broad questions regarding arthropod community responses to change. However, findings must be treated with caution, as not all members within a family are trophically equivalent (e.g., Formicidae) [58,59], leading to potentially misleading conclusions [60].

We compare the assemblage of aerial coleopterans in forest plots with no apparent EAB to the assemblage associated with the projected post-EAB community [17], which allows projections about long-term effects of ash loss on aerial coleopterans. Our comparative approach does not describe unforeseen alterations in successional trajectories independent of the EAB invasion, nor does it compare the ecological histories of the communities within these distinct forests, but it does provide a means of estimating potential long-term changes in arthropod community structure as a result of EAB-induced ash mortality (see [42,43]).

5. Conclusions

Endemic aerial coleopterans are readily utilizing the influx of resources provided by the EAB invasion. Collectively, our data suggest that native predators and parasites are being recruited to forests impacted by EAB, and that these native natural enemies may be a viable component of post-invasion EAB population dynamics in eastern North American forests.

Unprotected ash are devastated by the emerald ash borer. Following depletion of the ash resources, EAB populations sharply decline [61], greatly reducing the pest pressure on regenerating seedlings and saplings. The decline in pest pressure increases the chance of continued survival of young ash in North American forests [62], providing essential resources for ash specialists.

Ash forests are changing, and a deeper understanding of how arthropod communities and trophic guilds are responding will contribute to more proficient monitoring and protection.

Acknowledgments: The authors thank Bill Davidson, Eric Chapman, Ignazio Graziosi, Abe Nielsen, and Chris Strohm for assistance with field and laboratory work, and Sarah Witt and Edward Rouales for help with statistical analyses. Mary Arthur and Lee Townsend provided comments on an earlier version of this manuscript. We also thank Lee Crawfort, Bonnie Cecil, Taylorsville Lake State Park, Shelby County Parks and Recreation, and Lexington-Fayette Urban County Government for providing access to land for this project. This is publication number 17-08-120 of the Kentucky Agricultural Experiment Station paper and is published with the approval of the Director. This work is supported by the Kentucky Division of Forestry and the USDA Forest Service through a Landscape Scale Restoration Grant, and by McIntire Stennis Funds under 2351197000.

Author Contributions: M.B.S. and L.K.R. conceived and designed this experiment; M.B.S. performed the experiments and analyzed the data as partial requirements for a master degree; M.B.S. and L.K.R. wrote the paper.

Conflicts of Interest: The authors declare no conflict of interest.

Appendix A

Table A1. Coleopteran family abundance at five forested sites in north central Kentucky with trophic guild designations including; herbivores (H), fungivores (F), predators (P), saprophages (S), xylophages (X), and parasitoids (Pa).

Coleopteran Families	Trophic Guilds	Abundance					
		Fayette	Spencer	Shelby	Anderson	Henry	Total
Elateridae	H	205	345	678	353	240	1821
Chrysomelidae	H	127	176	272	56	55	686
Curculionidae	H	81	127	154	99	113	574
Tenebrionidae	H	94	83	142	54	154	527
Mordellidae	H	48	103	148	75	59	433
Scarabaeidae	H	20	24	20	23	23	110
Phalacridae	H	2	5	26	6	12	51
Attelabidae	H	0	0	2	0	3	5
Latridiidae	F	141	33	274	230	340	1018
Corylophidae	F	21	2	73	41	94	231
Ptinidae	F	16	47	38	36	13	150
Eucnemidae	F	16	50	33	39	8	146
Erotylidae	F	14	14	8	16	17	69
Mycetophagidae	F	22	5	16	6	10	59
Tetratomidae	F	3	12	7	21	15	58
Nitidulidae	F	7	13	11	11	13	55
Cerylonidae	F	15	4	8	13	7	47
Zopheridae	F	6	10	5	13	5	39
Silvanidae	F	1	22	1	6	7	37
Melandryidae	F	5	11	10	6	4	36
Synchroidae	F	4	4	6	9	9	32
Endomychidae	F	7	1	6	4	4	22
Leiodidae	F	5	2	3	8	4	22
Cryptophagidae	F	0	1	4	3	0	8
Laemophloeidae	F	3	1	1	2	1	8
Anthribidae	F	4	1	0	1	0	6
Cucujidae	F	0	0	1	1	0	2
Pyrochoidae	F	0	0	1	0	1	2
Sphindidae	F	0	0	1	1	0	2
Throscidae	F	0	1	0	0	1	2
Salpingidae	F	1	0	0	0	0	1
Staphylinidae	P	225	224	356	443	192	1440
Histeridae	P	188	316	141	204	136	985
Trogossitidae	P	27	42	46	43	70	228
Carabidae	P	16	20	63	43	34	176
Cleridae	P	7	27	31	31	48	144
Lampyridae	P	3	14	14	13	4	48
Coccinellidae	P	5	7	4	0	2	18
Melyridae	P	1	0	1	0	4	6
Cantharidae	P	0	2	1	0	0	3
Hydrophilidae	P	0	1	0	0	0	1
Lycidae	P	0	0	0	1	0	1
Ptilodactylidae	S	139	131	480	266	40	1056
Dermestidae	S	7	26	1	6	7	47
Monotomidae	S	2	6	0	0	0	8
Scirtidae	S	0	1	0	3	0	4
Hybosoridae	S	2	0	0	1	0	3
Silphidae	S						0

Table A1. *Cont.*

Coleopteran Families	Trophic Guilds	Abundance					
		Fayette	Spencer	Shelby	Anderson	Henry	Total
Scolytinae	X	168	465	107	117	134	991
Scraptiidae	X	1	1	8	24	53	87
Cerambycidae	X	6	13	12	19	24	74
Bostrichidae	X	1	5	2	0	3	11
Buprestidae	X	0	0	2	2	5	9
Lucanidae	X	0	0	0	1	0	1
Lymexylidae	X	0	0	0	1	0	1
Passandridae	Pa	21	13	14	16	14	78
Rhipiceridae	Pa	1	0	0	1	2	4
Bothrideridae	Pa	0	3	0	0	0	3
Unidentified	–	33	22	9	21	15	100
Total		1721	2436	3241	2389	1999	11,786

References

1. Ehrenfeld, J.G. Ecosystem consequences of biological invasions. *Annu. Rev. Ecol. Syst.* **2010**, *41*, 59–80. [CrossRef]
2. Wilcove, D.S.; Rothstein, D.; Dubow, J.; Phillips, A.; Losos, E. Quantifying threats to imperiled species in the United States: Assessing the relative importance of habitat destruction, alien species, pollution, overexploitation, and disease. *Bioscience* **1998**, *48*, 607–615. [CrossRef]
3. Byers, J.E.; Reichard, S.; Randall, J.M.; Parker, I.M.; Smith, C.S.; Lonsdale, W.M.; Atkinson, I.A.; Seastedt, T.R.; Williamson, M.; Chornesky, E. Directing research to reduce the impacts of nonindigenous species. *Conserv. Biol.* **2002**, *16*, 630–640. [CrossRef]
4. Pimentel, D.; Zuniga, R.; Morrison, D. Update on the environmental and economic costs associated with alien-invasive species in the United States. *Ecol. Econ.* **2005**, *52*, 273–288. [CrossRef]
5. Aukema, J.E.; Leung, B.; Kovacs, K.; Chivers, C.; Britton, K.O.; Englin, J.; Frankel, S.J.; Haight, R.G.; Holmes, T.P.; Liebhold, A.M. Economic impacts of non-native forest insects in the continental United States. *PLoS ONE* **2011**, *6*, e24587. [CrossRef] [PubMed]
6. Kennedy, H.E., Jr. *Fraxinus pennsylvanica* Marsh. Green ash. In *Silvics of North America: Volume 2, Hardwoods*; Burns, R.M., Honkala, B.H., Eds.; Agricultural Handbook 654; USDA Forest Service: Washington, DC, USA, 1990; pp. 348–354.
7. Schlesinger, R.C. *Fraxinus americana* L. white ash. In *Silvics of North America: Volume 2, Hardwoods*; Burns, R.M., Honkala, B.H., Eds.; Agricultural Handbook 654; USDA Forest Service: Washington, DC, USA, 1990; pp. 654–665.
8. Haack, R.A.; Jendak, E.; Houping, L.; Marchant, K.R.; Petrice, T.R.; Poland, T.M.; Ye, H. The emerald ash borer: A new exotic pest in North America. *Newsl. Mich. Entomol. Soc.* **2002**, *47*, 1–5.
9. USDA APHIS. USDA Animal and Plant Health Inspection Service Cooperative Emerald Ash Borer Project: Initial County EAB Detections in North America. 2017. Available online: http://www.emeraldashborer.info/files/MultiState_EABpos.pdf (accessed on 12 December 2017).
10. Cappaert, D.D.; McCullough, D.G.; Poland, T.M.; Siegert, N.W. Emerald ash borer in North America: A research and regulatory challenge. *Am. Entomol.* **2005**, *51*, 152–165. [CrossRef]
11. Flower, C.E.; Knight, K.S.; Rebbeck, J.; Gonzalez-Meler, M.A. The relationship between the emerald ash borer (*Agrilus planipennis*) and ash (*Fraxinus* spp.) tree decline: Using visual canopy condition assessments and leaf isotope measurements to assess pest damage. *For. Ecol. Manag.* **2013**, *303*, 143–147. [CrossRef]
12. Poland, T.M.; McCullough, D.G. Emerald ash borer: Invasion of the urban forest and the threat to North America's ash resource. *J. For.* **2006**, *104*, 118–124.
13. Kashian, D.M.; Witter, J.A. Assessing the potential for ash canopy tree replacement via current regeneration following emerald ash borer-caused mortality on southeastern Michigan landscapes. *For. Ecol. Manag.* **2011**, *261*, 480–488. [CrossRef]
14. Gandhi, K.J.K.; Herms, D.A. Direct and indirect effects of alien insect herbivores on ecological processes and interactions in forests of eastern North America. *Biol. Invasions* **2010**, *12*, 389–405. [CrossRef]

15. Gandhi, K.J.K.; Herms, D.A. North American arthropods at risk due to widespread Fraxinus mortality caused by the alien emerald ash borer. *Biol. Invasions* **2010**, *12*, 1839–1846. [CrossRef]

16. Flower, C.E.; Knight, K.S.; Gonzalez-Meler, M.A. Impacts of the emerald ash borer (*Agrilus planipennis* Fairmaire) induced ash (*Fraxinus* spp.) mortality on forest carbon cycling and successional dynamics in the eastern United States. *Biol. Invasions* **2013**, *5*, 931–944. [CrossRef]

17. Levin-Nielsen, A.; Rieske, L.K. Evaluating Short Term Simulations of a Forest Stand Invaded by Emerald Ash Borer. *iForest* **2014**, e1–e6. [CrossRef]

18. Klooster, W.S.; Herms, D.A.; Knight, K.S.; Herms, C.P.; McCullough, D.G.; Smith, A.; Gandhi, K.J.K.; Cardina, J. Ash (*Fraxinus* spp.) mortality, regeneration, and seed bank dynamics in mixed hardwood forests following invasion by emerald ash borer (*Agrilus planipennis*). *Biol. Invasions* **2014**, *16*, 859–873. [CrossRef]

19. Perkins, T.D.; Vogelmann, H.W.; Klein, R.M. Changes in light intensity and soil temperature as a result of forest decline on Camels Hump, Vermont. *Can. J. For. Res.* **1987**, *17*, 565–568. [CrossRef]

20. Zhang, Q.; Liang, Y. Effects of gap size on nutrient release from plant litter decomposition in a natural forest ecosystem. *Can. J. For. Res.* **1995**, *25*, 1627–1638. [CrossRef]

21. Ulyshen, M.D.; Klooster, W.S.; Barrington, W.T.; Herms, D.A. Impacts of emerald ash borer-induced tree mortality on leaf litter arthropods and exotic earthworms. *Pedobiologia* **2011**, *54*, 261–265. [CrossRef]

22. Perry, K.I.; Herms, D.A. Effects of late stages of emerald ash borer (Coleoptera: Buprestidae)-induced ash mortality on forest floor invertebrate communities. *J. Insect Sci.* **2017**, *119*, 1–10. [CrossRef]

23. Rainio, J.; Niemelä, J. Ground beetles (Coleoptera: Carabidae) as bioindicators. *Biodivers. Conserv.* **2003**, *12*, 487–506. [CrossRef]

24. Pearce, J.L.; Venier, L.A. The use of ground beetles (Coleoptera: Carabida) and spiders (Araneae) as biodindicators of sustainable forest management: A review. *Ecol. Indic.* **2006**, *6*, 780–793. [CrossRef]

25. Gandhi, K.J.K.; Smith, A.; Hartzler, D.M.; Herms, D.A. Indirect effects of emerald ash borer-induced ash mortality and canopy gap formation on epigaeic beetles. *Environ. Entomol.* **2014**, *43*, 546–555. [CrossRef] [PubMed]

26. Perry, K.I.; Herms, D.A. Short-term responses of ground beetles to forest changes caused by early stages of emerald ash borer (Coleoptera: Buprestidae)-induced ash mortality. *Environ. Entomol.* **2016**, *45*, 616–626. [CrossRef] [PubMed]

27. Davidson, W.; Rieske, L.K. Native parasitoid response to emerald ash borer (Coleoptera: Buprestidae) and ash decline in recently invaded forests of the central United States. *Ann. Entomol. Soc. Am.* **2015**, *108*, 777–784. [CrossRef]

28. Wharton, M.E.; Barbour, R.W. *Trees and Shrubs of Kentucky*; University Press of Kentucky: Lexington, KY, USA, 1973.

29. Campbell, J.J. Historical evidence of forest composition in the Bluegrass Region of Kentucky. In Proceedings of the Seventh Central Hardwood Forest Conference, Carbondale, IL, USA, 5–8 March 1989; Rink, G., Budelsky, C., Eds.; General Technical Report, NC-135. Department of Agriculture, Forest Service, North Central Forest Experiment Station: St. Paul, MN, USA, 1989; pp. 231–246.

30. Montgomery, M.E.; Wargo, P.M. Ethanol and other host-derived volatiles as attractants to beetles that bore into hardwoods. *J. Chem. Ecol.* **1983**, *9*, 181–190. [CrossRef] [PubMed]

31. Lindelöw, Å.; Risberg, B.; Sjödin, K. Attraction during flight of scolytids and other bark- and wood-dwelling beetles to volatiles from fresh and stored spruce wood. *Can. J. For. Res.* **1992**, *22*, 224–228. [CrossRef]

32. Bouget, C.; Brustel, H.; Brin, A.; Valladares, L. Evaluation of window flight traps for effectiveness at monitoring dead wood-associated beetles: The effect of ethanol lure under contrasting environmental conditions. *Agric. For. Entomol.* **2009**, *11*, 143–152. [CrossRef]

33. Triplehorn, C.A.; Johnson, N.F. *Borror and DeLong's Introduction to the Study of Insects*; Brooks/Cole: Belmont, CA, USA, 2005; p. 888.

34. Marshall, S.A. *Insects: Their Natural History and Diversity—With a Photographic Guide to Insects of Eastern North America*; Firefly Books: Buffalo, NY, USA, 2006; p. 736.

35. Evans, A.V. *Beetles of Eastern North America*; Princeton University Press: Princeton, NJ, USA, 2014; p. 560.

36. BugGuide. Order Coleoptera: Beetles. Available online: http://bugguide.net/node/view/60/tree (accessed on 1 March 2017).

37. Hammond, P.M. Insect abundance and diversity in the Dumoga-Bone National Park, N. Sulawesi, with special reference to the beetle fauna of lowland rain forest in the Toraut region. In *Insects and the Rain Forests of South East Asia (Wallacea)*; Knight, W.J., Holloway, J.D., Eds.; Royal Entomological Society of London: London, UK, 1990; pp. 197–254.

38. Ellis, D. Taxonomic sufficiency in pollution assessment. *Marine Poll. Bull.* **1985**, *16*, 459. [CrossRef]

39. Birkhofer, K.; Bezemer, T.; Hedlund, K.; Setälä, H. Community composition of soil organisms under different wheat-farming systems. In *Microbial Ecology in Sustainable Agroecosystems*; Cheeke, T., Coleman, D., Wall, D., Eds.; CRC Press: Boca Raton, FL, USA, 2012; pp. 89–111.

40. Hoback, W.W.; Svatos, T.M.; Spomer, S.M.; Higley, L.G. Trap color and placement effects estimates of insect family-level abundance and diversity in a Nebraska salt marsh. *Entomol. Exp. Appl.* **1999**, *91*, 393–402. [CrossRef]

41. Riggins, J.J.; Davis, C.A.; Hoback, W.W. Biodiversity of belowground invertebrates as an indicator of wet meadow restoration success (Platte River, Nebraska). *Restor. Ecol.* **2009**, *17*, 495–505. [CrossRef]

42. Rohr, J.R.; Mahan, C.G.; Kim, K.C. Response of arthropod biodiversity to foundation species declines: The case of the eastern hemlock. *For. Ecol. Manag.* **2009**, *258*, 1503–1510. [CrossRef]

43. Adkins, J.K.; Rieske, L.K. Loss of a foundation forest species due to an exotic invader impacts terrestrial arthropod communities. *For Ecol. Manag.* **2013**, *295*, 126–135. [CrossRef]

44. Hammond, P.M. Species Inventory. In *Global Biodiversity: Status of the Earth's Living Resources*; Groombridge, B., Ed.; Chapman and Hall: London, UK, 1992; pp. 17–39.

45. Root, R.B. The niche exploitation pattern of the blue-gray gnatcatcher. *Ecol. Monogr.* **1967**, *37*, 317–350. [CrossRef]

46. Smith, B.; Wilson, J.B. A consumer's guide to evenness indices. *Oikos* **1996**, *76*, 70–82. [CrossRef]

47. SAS Institute. *SAS/IML 9.3 User's Guide*; SAS Institute: Cary, NC, USA, 2011.

48. Kimmerer, T.W.; Kozlowski, T.T. Ethylene, ethane, acetaldehyde, and ethanol production by plants under stress. *Plant Physiol.* **1982**, *69*, 840–847. [CrossRef] [PubMed]

49. Harmon, M.E.; Franklin, J.F.; Swanson, F.J.; Sollins, P.; Gregory, S.V.; Lattin, J.D.; Anderson, N.H.; Cline, S.P.; Aumen, N.G.; Sedell, J.R. Ecology of coarse woody debris in temperate ecosystems. *Adv. Ecol. Res.* **1986**, *15*, 302.

50. Bauer, L.S.; Liu, H.; Haack, R.A.; Petrice, T.R.; Miller, D.L. Natural enemies of emerald ash borer in southeastern Michigan. In Proceedings of the Emerald Ash Borer Research and Technology Development Meeting, Port Huron, MI, USA, 30 September–1 October 2003; Mastro, V., Reardon, R., Eds.; Comps. USDA Forest Service, Forest Health Technology Enterprise Team: Morgantown, WV, USA, 2004; pp. 33–34.

51. Williams, P.H.; Gaston, K.J. Measuring more of biodiversity: Can higher-taxon richness predict wholesale species richness? *Biol. Conserv.* **1994**, *67*, 211–217. [CrossRef]

52. Balmford, A.; Jayasuriya, A.H.M.; Green, M.J.B. Using higher-taxon richness as a surrogate for species richness: II. Local applications. *Proc. Biol. Sci.* **1996**, *263*, 1571–1575. [CrossRef]

53. Hilsenhoff, W.L. Rapid field assessment of organic pollution with a family level biotic index. *J. N. Am. Benthol. Soc.* **1988**, *7*, 65–68. [CrossRef]

54. Barbour, M.T.; Gerritsen, J.; Snyder, B.D.; Stribling, J.B. *Rapid Bioassessment Protocols for Use in Streams and Wadeable Rivers: Periphyton, Benthic Macroinvertebrates and Fish*, 2nd ed.; U.S. Environmental Protection Agency, Office of Water: Washington, DC, USA, 1999.

55. Bailey, R.C.; Norris, R.H.; Reynoldson, T.B. Taxonomic resolution of benthic macroinvertebrate communities. *J. N. Am. Benthol. Soc.* **2001**, *20*, 280–286. [CrossRef]

56. Reynoldson, T.B.; Norris, R.H.; Resh, V.H.; Day, K.E.; Rosenberg, D.M. The reference-condition: A comparison of multimetric and multivariate approaches to assess water-quality impairment using benthic macroinvertebrates. *J. N. Am. Benthol. Soc.* **1997**, *16*, 833–852. [CrossRef]

57. Ferraro, S.P.; Cole, F.A. Taxonomic level sufficient for assessing a moderate impact on macrobenthic communities in Puget Sound, Washington, USA. *Can. J. Fish. Aquat. Sci.* **1992**, *49*, 1184–1188. [CrossRef]

58. Resh, V.H.; Unzicker, J.D. Water quality monitoring and aquatic organisms: The importance of species identification. *J. Water Pollut. Control Fed.* **1975**, *47*, 9–19. [PubMed]

59. Hölldobler, B.; Wilson, E.O. *The Ants*; Harvard University Press: Cambridge, MA, USA, 1990.

60. Longcore, T. Terrestrial arthropods as indicators of ecological restoration success in coastal sage scrub (California, USA). *Restor. Ecol.* **2003**, *11*, 397–409. [CrossRef]

61. Herms, D.A.; McCullough, D.G. Emerald ash borer invasion of North America: History, biology, ecology, impacts, and management. *Annu. Rev. Entomol.* **2014**, *59*, 13–30. [CrossRef] [PubMed]

62. Duan, J.J.; Bauer, L.S.; Abell, K.J.; Ulyshen, M.D.; Van Driesche, R.G. Population dynamics of an invasive forest insect and associated natural enemies in the aftermath of invasion: Implications for biological control. *J. Appl. Ecol.* **2015**, *52*, 1246–1254. [CrossRef]

forests

MDPI

Article

Reducing Firewood Movement by the Public: Use of Survey Data to Assess and Improve Efficacy of a Regulatory and Educational Program, 2006–2015

Andrea Diss-Torrance [1,*], Kim Peterson [2] and Colleen Robinson [3]

[1]   Wisconsin Department of Natural Resources (WI DNR), P.O. Box 7291, Madison, WI 53707, USA
[2]   Alpha Technology Group LLC, 23 South Owen Drive, Madison, WI 53705-5032, USA; ptrson@gmail.com
[3]   Forest Stewards Guild, 612 W. Main St., Suite 200, Madison, WI 53703, USA; colleen@forestguild.org
*   Correspondence: andrea.disstorrance@wisconsin.gov; Tel.: +1-608-264-9247

Received: 20 December 2017; Accepted: 11 February 2018; Published: 14 February 2018

Abstract: This paper describes a program of policy management and research from 2006 through 2015. It focuses on regulator efforts to understand and address challenges presented by dispersal of forest diseases and invasive pests in firewood by the camping public. Five surveys conducted at two-year intervals informed these efforts. The first survey in 2006 benchmarked campers' awareness of forest threats by invasive species, their evaluations of firewood supplied at and near Wisconsin state parks, and their compliance with firewood movement rules which had been implemented that year. The 2008 survey tested for improvements in awareness and compliance and investigated campers' motivations. The motivation research showed that calculated, normative, and social motivations are all important to rule compliance in the camping context. Surveys in 2010, 2012, and 2014 confirmed these results and guided education and outreach efforts, adjustments to firewood movement rules for Wisconsin state parks and forests, and improvements to firewood supplies at state campgrounds. The survey sequence as a whole revealed that: (1) compliance improves dramatically in early program years and then levels off, suggesting that it may be unrealistic and cost ineffective to strive for 100% compliance in similar regulatory contexts; (2) persistence in messaging is important in building awareness and motivation; and (3) regulation and persuasion based on motivational principles can extend beyond specific situations where informing and regulating take place, suggesting that public properties can be useful venues for encouraging other types of environmentally responsible behavior.

Keywords: firewood; emerald ash borer; regulation; compliance; motivation; forest health

1. Introduction

Across North America, many invasive pests and diseases of trees are moved in firewood, including the emerald ash borer (*Agrilus planipennis* Fairmaire), Asian long horned beetle (*Anoplophora glabripennis* Motschulsky), gold spotted oak borer (*Agrilus auroguttatus* Schaeffer), and oak wilt (*Ceratocystis fagacearum* (Bretz) Hunt), as well as many others. These pests and diseases have caused millions of dollars of damage to communities and forests. Movement of firewood is an effective pathway for the accidental spread of invasive species for several reasons, the most basic of which are economic. Wood too riddled with tunnels or stained by fungi for use as lumber has some economic value as firewood, and homeowners may cut up trees that die on their properties for firewood instead of paying for disposal. Another feature of firewood that makes it an effective pathway is that it is often moved when it is moist and the organisms within it are alive. Firewood may not be used immediately at its destination, allowing the emergence of infesting organisms into new environments. Finally, quarantines on movement of firewood may not be as effective as those placed on well-organized industries such as nurseries or wood mills because so much firewood is moved by the general public, and informing the public of the law and enforcement is difficult.

In 2006, in response to the rapid spread of the emerald ash borer (EAB) and other invasive forest pests, the Wisconsin Department of Natural Resources (Wisconsin DNR) began regulating firewood allowed by state campgrounds. This regulation was linked with an education campaign designed to dissuade the public from moving firewood for any purpose—to "Buy it where you burn it". We recognized that reducing the rate of spread of wood borne invasive pests required a general change in public behavior, not just a temporary accommodation to rules encountered at state parks and forests. To determine if public awareness and behavior were improving, we surveyed campers at state parks and forests in that year, and then repeated the survey at two-year intervals through the 2014 camping season (five surveys in all). This ten-year study allowed us to observe the full development of public awareness and track motivations and behaviors of our target audience. We were able to observe responses to changes in our messaging and other external factors that influenced the decision to move firewood. We were also able to identify the point at which our messaging program appeared to reach saturation, which allowed us to economize by shifting from program development to maintenance.

2. Conceptual Basis

An important part of our research and management agenda was to address campers' motivations to bring firewood with them on their trips. Understanding motivations for environmental behavior is essential to the design of effective regulations and in persuading the public to take desired voluntary actions. A comprehensive discussion of motivations is beyond the scope of this report and may be accessed in our previous work [1,2]. Briefly, our awareness-building and persuasion efforts sought to leverage three types of motivation, the first being normative motivation. People comply with regulations based on two related sets of normative considerations. One set comprises general moral principles, which include an individual's sense of civic duty to obey laws and conform to general ideological values. A second set is more specific, and includes evaluation of rule appropriateness or value of a regulation. We measured these motivations using survey questions that captured campers' feelings about the size of threat posed by invasive forest diseases and pests, the importance of stopping firewood movement, and whether limiting firewood movement makes sense and is the right thing to do. Knowing how campers perceived these issues could allow us to direct supplemental messaging to raise awareness of the threat and recruitment to address it. Messages such as "It's not just EAB that moves in wood", "Parks take this risk seriously, so should you when getting firewood for your property", and "The wood industry takes quarantine precautions, do your part and not move firewood" were released to the public to bolster public perception of the value of not moving firewood. We also tailored education and outreach messaging to access social motivations for campers to follow firewood movement rules because: (1) people who are important to them think they should do so, (2) people who are important to them observe these rules themselves, and (3) they feel subjective pressure from people who are important to them to observe these rules. We addressed the third general motivation type, calculated motivation, by improving firewood supplies at state parks and forests in price, quality, convenience, and reliability (features that campers repeatedly told us were important to them) and by encouraging the provision of "certified-as-safe" firewood that campers could move freely throughout the state. A selective list of steps taken and their timing is provided in the appendix.

3. Materials and Methods

All five surveys conducted during the 2006 to 2015 timeframe utilized mail questionnaires administered in three waves over eight to ten-week periods. Although email questionnaires were considered, getting stable email addresses proved a challenge, and the length and format of our questionnaires further complicated effective use of this approach. ReserveAmerica, Wisconsin's state campground reservation agent, provided comprehensive lists of campers who had reserved one or more sites at a Wisconsin state park or forest during the just-completed camping year. Random samples of 800 names were selected from each of these lists; the surveys were conducted and data were collected, entered in a database, checked for accuracy and extreme values, and analyzed. Response rates are based on returns divided by questionnaires successfully delivered. The number of respondents and % response rate follow in parentheses for each survey year: 2006 (479, 62%), 2008 (495, 64%), 2010 (524, 69%), 2012 (450, 62%), and 2014 (468, 64%).

The surveys included a variety of yes/no response questions, semantic differentials, and constant sum scales to measure camper awareness, motivations, satisfaction with firewood supplies at and nearby campgrounds visited, and compliance with firewood movement rules. The questionnaire format was held mostly constant although some questions were added, removed, or modified, depending on information needs of the specific year. As a basis for the present analysis, we chose a subset of questions that were included in each survey (most appeared every year, although a few were absent in 2012), harmonized the data (e.g., adjusted for any differences in scale size or direction), merged the five data sets, and added an indicator variable to identify survey year. The merged file included 2416 records.

4. Results

4.1. Awareness

The first program goal undertaken in 2006 was to measure camper awareness of EAB and other forest diseases and pests. The Wisconsin DNR emphasized EAB because of its dramatic and highly publicized impact on both urban and rural forests in neighboring states. In subsequent surveys, we reduced the emphasis on EAB and instead included it in a list with six other invasive forest diseases and pests moved in firewood.

Results of policy and education program management during the time frame were encouraging. Data in Table 1 (row 1) show that the percentage of campers who were aware of EAB rose rapidly from 2006 to 2008 then leveled off as it approached saturation. The differences between the percentages are statistically significant.

Table 1. Camper awareness of firewood related issues [†,◊].

	2006	2008	2010	2012	2014	Test Results
Emerald ash borer	77.1%$_a$	91.9%$_b$	92.6%$_b$	94.9%$_b$	95.7%$_b$	X2 (4, 2412) = 127.123, ρ = 0.000, V = 0.230
Movement risks	67.9%$_a$	91.7%$_b$	97.7%$_c$	95.3%$_{b,c}$	96.6%$_c$	X^2 (4, 2387) = 324.342, ρ = 0.000, V = 0.369
Prohibitions	66.5%$_a$	94.1%$_{b,c}$	93.5%$_c$	89.2%$_c$	97.6%$_b$	X^2 (4, 2399) = 286.952, ρ = 0.000, V = 0.346

[†] All figures represent percentages of campers who completed the survey. Awareness was measured for multiple invasive species, although row 1 data reflect only the emerald ash borer. All three variables are based on Yes/No responses to the question regarding awareness. [◊] Each subscript letter denotes a subset of categories whose column percentages do not differ significantly from each other at the 0.05 level. Bonferroni corrections were applied to account for the fact that multiple comparisons were being made.

In addition to increasing camper awareness of threats posed by forest diseases and invasive pests, Wisconsin DNR managers worked to educate campers about risks associated with moving firewood and to ensure that campers believed these risks were legitimate. We realized that this type of awareness and recognition of risk would be somewhat slower to increase over time, as camper calculated motivations had to be overcome; for example, there is a widely held belief that firewood brought from home is cheaper and of higher quality than firewood obtained at or near campgrounds. Survey data over this period demonstrates that awareness of the link between invasive forest pests and firewood movement (risk) begins at a lower level than awareness of EAB, as expected, but then rises to levels that slightly surpass it (Table 1, row 2).

The third critical element of Wisconsin DNR's awareness building effort was for awareness of regulations related to firewood movement. Survey data show a sharp increase in the percentage of campers who were aware of the regulations from 2006 to 2008, then modest fluctuations in awareness through 2014 (Table 1, row 3). Frequent refreshment of the message that firewood is regulated on state lands likely played a role in the rapid rise and maintained a high level of awareness of said regulation. Each time the distance was reduced for compliant firewood (2006, 2007, 2010, and 2014), extra effort was made by the agency to publicize the change in regulation; media coverage amplified this effort. In 2008, EAB was found for the first time in Wisconsin. Media coverage of this event and the efforts to prevent its spread, including firewood regulation, was intensive. In the period from 2011 to 2013, there were no strong new messages related to firewood regulation put forth by the Wisconsin DNR, which may have led to the small decline in awareness of regulation during 2012. Yearly refreshment and reminder of our message was beneficial in maintaining a high level of awareness, the first requirement for compliance. The frequency at which messaging must be renewed most likely differs in other situations, but program managers should plan to refresh messaging regularly to maintain public awareness.

4.2. Motivation

4.2.1. Normative Motivation

Although increasing camper awareness is important, it is not always sufficient by itself to achieve rule compliance. Campers must also be motivated specifically for compliance as well. Given 2006 benchmarks for camper awareness, education and outreach initiatives were refined. These initiatives attempted to leverage campers' normative motivations, and the perceived threat size of forest diseases and pests was chosen as an appropriate measure. Although perceived threat size does not reveal normative motivation directly (i.e., to fulfill a moral obligation or do the right thing), we are confident that camper recognition of threats provides prima facie evidence of it. To measure this construct, we asked campers to rate the size of the perceived threat on a seven-point scale where 1 = no threat; 4 = a moderate threat; and 7 = a huge threat. Average camper ratings of threat size increased from 2006 through 2010, then declined in 2012 before rebounding slightly in 2014 (Table 2 row 1). Differences between the means were statistically significant overall, although post hoc comparisons showed that some of these differences were not significantly different at the 0.05 level (Bonferroni correction was applied). This suggests that public education and outreach efforts effectively utilized this normative motive, but that possible reductions in publicity between 2011 and 2013 (as discussed above) may have contributed to the 2012 dip in perceived threat.

Table 2. Normative motivation [†,◊].

	2006	2008	2010	2012	2014	Test Results ∨
Threat Size—Invasive Species	5.00$_a$	5.57$_b$	5.81$_c$	5.63$_{b, c}$	5.73$_{b, c}$	$F_{(4, 2313)} = 29.224$, $p = 0.000$, $\eta^2 = 0.048$
Stop Movement Importance	5.53$_a$	6.08$_b$	6.16$_b$	6.21$_b$	6.29$_b$	$F_{(4, 2324)} = 24.878$, $p = 0.000$, $\eta^2 = 0.041$

[†] Measures are based on seven-point scales where values ranged from *1 = not a threat* to *7 = A huge threat*, and from *1 = Not at all important* to *7 = Extremely important*. ◊ Each subscript letter denotes a subset of categories whose column values do not differ significantly from each other at the 0.05 level. Bonferroni corrections were applied to account for the fact that multiple post hoc tests were made. ∨ Standard deviations ranged between 1.173 and 1.439 for Threat Size and from 1.134 to 1.618 for Stop Movement Importance. Levene statistics suggested unequal variances for both measures, but sample sizes for all groups were large and approximately equal, so we feel comfortable with these results.

A more direct measure of normative motivation is the perceived importance of stopping firewood movement. Campers were asked to rate importance on a seven-point scale where 1 = *not at all important*, 4 = *neither unimportant nor important*, and 7 = *extremely important*. Overall results were statistically significant, but following a sharp increase from 2006 to 2008, ratings increased only gradually through 2014; differences from 2008 through 2014 were not statistically significant (Table 2, row 2).

4.2.2. Social Motivation

Social motivation was also investigated for its potential to influence compliance. We did not measure social motivation in 2006, and the approach used in 2008 utilized an indirect measure that produced ambiguous results. We improved the measure significantly for the surveys of 2010 and 2012, and revamped it completely for 2014. Because of the differences in measurement, we discuss social motivation for only 2014. Our measure for this survey year utilized seven-point scales where values ranged from *1 = strongly disagree* to *7 = strongly agree*. The Family Influence statement was worded "My family wants me to limit my movement of firewood". The Friends Influence statement substituted "friends" for "family" ("My friends want me to limit my movement of firewood"). Data showed that campers were motivated by these two important social groups, although to differing degrees. Nearly 61% of respondents agreed that their family wanted them to limit movement of firewood, but only 45.4% reported friend-based influence. This difference is not unexpected, and may follow from ambiguities in camper mindsets regarding family members and friends, as well as from differences in knowledge of their wishes regarding firewood movement. We hope to investigate this construct with more precision in future research.

4.2.3. Calculated Motivation

Firewood movement decisions are made only partly on the basis of perceived risk of spreading invasive pests and under varying degrees of social pressure. Previous research has shown that calculated motivations related to firewood availability at home, and to price, quality, convenience, and reliability differentials between at-home and at-campground firewood are also important, and have often worked counter to education and outreach efforts [1,3]. Specifically, some campers believe that firewood supplied at or near state parks and forests are lower in quality, higher priced, inconvenient, and/or unreliable when compared with wood from home, which is often free or already paid for, selected by the camper (so its quality is known), and readily at hand; therefore, they tend to bring firewood from home on camping trips. To investigate this issue, we asked campers questions such as: "People have told us that they bring firewood on camping trips for different firewood related reasons. Which of the following reasons apply to you?" In each study year, seven or eight reasons were listed plus an open-ended option, and respondents could select more than one reason. The most frequently cited reasons were that campers had firewood at home they wanted to use, and/or they had concerns with price, quality, or other characteristics of firewood obtained at or near campgrounds. These factors were highly correlated with compliance in all five surveys.

We hoped to see steadily declining percentages across survey years for all reasons campers brought firewood from home; Wisconsin state parks and forests worked diligently throughout the study period to persuade campers not to move firewood and to improve firewood supplies at campgrounds. However, only the reason "had wood at home" consistently declined (Table 3). In 2006, 33.2% of all respondents indicated that they brought wood because they had it at home and wanted to use it. This percentage declined in 2008, 2010, and 2012, then increased slightly to 10.5% in 2014. These changes were statistically significant overall (X^2 (4, 2384) = 120.496, ρ = 0.000, V = 0.225), but proportions for 2010 through 2014 did not differ at the 0.05 level (see table cell subscripts). These results suggest that Wisconsin DNR has been successful in persuading campers not to bring firewood from home, not to stock firewood at home for use in camping, or both.

Table 3. Calculated motivation—brought wood because they had wood at home they wanted to use $^{\diamond}$.

		2006	2008	2010	2012	2014	Total count and % across years
Yes	Count	159$_a$	94$_b$	77$_{b, c}$	39$_c$	49$_c$	418
	% answering 'Yes'	33.2%	19.0%	14.7%	9.3%	10.5%	17.5%
No	Count	320$_a$	401$_b$	447$_{b, c}$	379$_c$	419$_c$	1966
	% answering 'No'	66.8%	81.0%	85.3%	90.7%	89.5%	82.5%
Total	Count	479	495	524	418	468	2384

$^{\diamond}$ Each subscript letter denotes a subset of categories whose column values do not differ significantly from each other at the 0.05 level. Bonferroni corrections were applied to account for the fact that multiple comparisons were being made.

4.3. Compliance

Improvements in camper awareness and appeal to camper normative, social, and calculated motivations led to increases in compliance with firewood movement rules and to favorable shifts in other firewood-related behaviors. The analysis that follows distinguishes between compliance based on firewood source type, movement in bulk, and movement on camping trips.

4.3.1. Source Compliance

Our first measure of source compliance was based on primary source of firewood used by campers, which was derived from the following survey question: *"Where do you TYPICALLY get the firewood you use for camping? (Circle numbers of all that apply)"*. Campers then chose among the following options: *(1) Nowhere—we seldom if ever build a campfire; (2) Buy inside park or forest; (3) Buy within (the allowable distance) miles of a park or forest (for example, at a roadside stand or convenience store); (4) Buy in hometown at a convenience store, grocery store, or other retail place; (5) Buy in hometown from a dealer (for example, a person with access to a woodlot); (6) Cut it ourselves (on our land or family/friend's land, etc.); (7) Gather inside park or forest (from grounds or vacant campsites); (8) Gather in hometown from municipal sources (such as your city or town brush disposal site); (9) A friend or relative supplies it to us; (10) Get wood from pallets; (11) Get scrap lumber or wood left over from home projects, work, or business etc.; and (12) Other (Please describe).*

The next survey question asked which was the respondent's primary source of firewood; campers were considered compliant if they obtained firewood exclusively inside or near the places they camped (options 2, 3, and 7 above) or brought only scrap dimensional lumber left over from home or work projects, an allowable alternative (option 11). Data in Table 4 (row 1) demonstrate that camper compliance with firewood movement rules increased steadily over the study period, and proportions are statistically different for 2006, 2008, and 2010 (Bonferroni correction was applied).

A more rigorous definition of compliance included all firewood sources used, not just a primary source: if any source utilized was considered non-compliant then the camper was classified as non-compliant (row 2). As expected, percentages of compliant campers were lower across all camping seasons vis-à-vis the primary source measure but did increase similarly over time. The lower percentage of complying campers when the more rigorous definition was used shows that while most campers (77.8% in 2014) comply all of the time, some (92.2% − 77.8% = 14.4%) comply only most of the time. Comments supplied by some survey participants in the second group suggest that they may have used non-compliant sources of firewood to ensure that there was a minimal supply of dependable quality firewood on hand, to reduce the expense of camping, and/or because they wanted to use firewood acquired previously in bulk.

Table 4. Firewood movement rule compliance by campers ◊.

	2006	2008	2010	2012	2014	Test Results
% of campers whose primary source of wood was compliant	57.9%a	72.7%b	88.5%c	91.8%c	92.2%c	X^2 (4, 2296) = 259.698, p = 0.000, V= 0.336
% of campers, all of whose sources of wood were compliant	44.5%a	59.8%b	77.6%c	83.1%c	77.8%c	X^2 (4, 2324) = 219.151, p = 0.000, V = 0.307
% of campers who did not move wood in bulk for use in camping	87.7%a	91.9%a	97.5%b	97.1%b	98.7%b	X^2 (4, 2416) = 81.344, p = 0.000, V = 0.183
% of campers who did not move wood in bulk for use in a home	82.5%a	86.9%a	87.8%a	88.7%a	85.7%a	X^2 (4, 2416) = 9.360, p = 0.053, V = 0.062
Average Distance Bulk Wood Was Moved for any purpose (Miles)	55.0	33.2	24.8	25.2	20.3	F (4, 425) = 8.043 p = 0.000, η^2 = 0.071
% of campers who complied on their last camping trip [†]	55.8%a	73.2%b	93.7%c	-	93.6%c	X^2 (3, 1940) = 291.578, p = 0.000, V = 0.388

◊ Each table subscript letter denotes a subset of categories whose column values do not differ significantly from each other at the 0.05 level. Bonferroni corrections were applied to account for the fact that multiple comparisons were being made. [†] In 2014, this question asked if campers complied on *all* trips for that year.

In addition to asking about firewood sources, we asked "Have you moved any large quantities of firewood (for example, a trailer or truckload) to or from your home in the last two years?" Response options included: for camping, for use at home, for use at a family member's home, and for use at a cottage or second home. Data related to bulk movement of wood for camping and home use appear in rows 3 through 5 of Table 4. We asked about movement of large quantities because this may encourage trip non-compliance even after campers become aware of movement rules. This category of firewood is typically less expensive or free, of known quality and condition, and possibly more attractive than firewood acquired at or near campgrounds. Moreover, bulk supplies represent sunk costs that campers may be unwilling to write off after learning about firewood movement rules. Data in row 3 strongly suggest that bulk movement of wood for camping was not common at the start of regulation, and that its incidence decreased to nearly zero by 2014 (data show percentages of respondents who did *not* move firewood). Firewood regulations at campgrounds may have especially discouraged the movement of trailer loads of firewood as it is difficult to conceal such quantities, and surrender would be a significant, and thus deterring, loss. In addition, firewood regulation is not limited to state lands in Wisconsin. Federal and many county campgrounds also regulate firewood use. In a survey of private campground owners in 2011 [4], we found that 66% of respondents imposed limitations on firewood that was allowed onto their properties. In addition to camping, firewood is also moved in bulk for use in homes. Combining bulk movement for home use with cabin and second home use produced the fourth compliance measure shown in Table 4 (row 4). Our goal was to see if the reduced movement and stocking of wood for camping was generalized to firewood movement for use at residential properties. In 2006, when firewood regulation on state lands was first implemented, the proportion of campers who moved large amounts of wood for use at home was similar to the proportion that moved firewood for camping, but reductions in these percentages through 2012 were smaller than for camping-related movement, and were only borderline statistically significant. Movement of firewood in bulk for home use increased in 2014, but the change was not statistically significant and coincided with a shortage of propane fuel for heating in Wisconsin during the winter of 2013–2014, a factor that may have promoted bulk wood movement as a precautionary measure for the following winter.

The differences in reduction of bulk firewood movement for camping and home use suggest that additional approaches of reducing risk of spreading invasive pests in firewood are needed. One such approach is to educate homeowners to select wood sources that are less likely to carry pests. Wood from sources close to the home or cabin are less likely to harbor pests not already present in the area, and aging wood for two years near the place where it was cut also reduces risk. This message, which is at the heart of the distance limitation on wood entering state lands, appears to have been received and acted upon by those who moved firewood in bulk. The average distance large amounts of wood were moved decreased from 55 miles in 2006 to approximately 20 miles in 2014 (Table 4, row 5).

4.3.2. Trip Compliance

During all survey years except 2012, when we did not investigate the respondent's most recent camping trip, we asked campers the following question (or a close variant) based on allowable mileage: "For your most recent camping trip to a Wisconsin state park or forest, did you bring any firewood from more than (the allowable distance) miles or from out of state?" Movement limits were set at 50 miles for the 2008 season, 25 miles for 2010, and 10 miles for 2014. Campers could then choose either No; Yes—we brought only logs; Yes—we brought only scrap lumber (2 × 4 s etc.); or Yes—we brought both logs and scrap lumber. The two options that excluded logs defined compliance. For the 2014 survey, we asked if campers had brought firewood on any camping trip to a state park or forest. Results show that compliance rose dramatically from 2006 to 2008, and then again in 2010; differences among these percentages were all statistically significant (Table 5, Row 6).

Table 5. Compliance with firewood regulations on trips in 2014 by three groups of campers separated based on their sources of wood: Highly Compliant, all their sources of wood are compliant with regulations; Moderately Compliant, their primary source of wood is compliant but their secondary sources are not; and Minimally Compliant, whose primary source of wood is not compliant ◊.

			Camper Group			Total [†]
			Highly Compliant	Moderately Compliant	Minimally Compliant	
Compliance on any trip in 2014	Yes	Count	325_a	56_b	19_c	400
		% of the Group	98.2%	84.8%	59.4%	93.2%
	No	Count	6_a	10_b	13_c	29
		% of the Group	1.8%	15.2%	40.6%	6.8%
Total		Count	331	66	32	429

◊ Each table subscript letter denotes a subset of categories whose column values do not differ significantly from each other at the 0.05 level. Bonferroni corrections were applied to account for the fact that multiple comparisons were being made. [†] The compliance percentage shown in this column differs slightly from that reported in Table 4 due to missing data based on source compliance.

5. Discussion

Our results strongly suggest that efforts to increase public awareness of invasive forest diseases and pests and to reduce firewood movement have been effective, but several features of our approach warrant further comment. First, persistence in messaging may be instrumental in building awareness and motivation. We saw a slight dip in awareness of firewood prohibitions in 2012, during the period 2011–2013 when there was no new firewood movement information sent to the public. Managers and educators should refresh the core message periodically to maintain sensitivity and elicit appropriate responses. Second, compliance data for the study period show dramatic improvements from 2006 through 2010, and then a leveling off in 2012 and 2014. It is therefore appropriate to ask if we can do better or if there a small segment of the camping public that will not be persuaded and will never fully comply? Comparison of campers based on their typical sources of firewood provides useful insights in this matter, and three camper groups can be defined in terms of their presumed likelihood to comply as of 2014 (Table 5). The first group is considered potentially Highly Compliant because all of their sources facilitate compliance. Group two is considered only Moderately Compliant, as their primary source suggests compliance but their secondary sources indicate non-compliance. Group three is considered Minimally Compliant, as their primary source indicates non-compliance. All three groups are contrasted in terms of their association with trip compliance, which is the best indicator we have of actual firewood movement that conforms with stipulated limits.

These data are encouraging and show that only 1.8% of the Highly Compliant group moved firewood more than 10 miles on any trip taken to a state park or forest in 2014. This figure increases to 15.2% for Moderately Compliant campers and to 40.6% for Minimally Compliant campers. The differences are statistically significant, and the association between the variables is strong (X^2 (2, 429) = 78.452, ρ = 0.000, V = 0.428). Although the proportion of Minimally Compliant campers who brought wood in violation of regulations is high, the overall percentage of such violators is relatively low (i.e., only 13 of 429 campers, or 3%), and including the ten Moderately Compliant campers did not increase this percentage much (i.e., to 23 of 429, or 5.4%). Accordingly, it may be unrealistic and cost ineffective to try to stop all non-compliant firewood movement. Instead, reduction of the risk of successful introduction is a reasonable goal.

6. Conclusions

Results of five camper surveys implemented from 2006 through 2015 demonstrate that all forms of awareness of firewood movement regulations and risks in Wisconsin campers have improved. Results also confirm that calculated, normative, and social motivations can be measured and managed to increase camper compliance with firewood movement rules (selected management actions taken follow as an appendix). We also found that compliance behavior transferred from camping activity to home firewood use. Therefore, it follows that regulation and persuasion based on motivational principles can lead to changes in behavior that extend beyond specific situations where regulation takes place, and that public properties can provide a venue for encouraging other types of environmentally responsible behavior.

Acknowledgments: The authors thank the Wisconsin Department of Natural Resources Division of Forestry, Public Lands and Conservation Services Section for funding this research.

Author Contributions: Andrea Diss-Torrance and Kim Peterson conceived and designed the study; Kim Peterson developed and conducted the surveys and did the statistical analysis of the data. Colleen Robinson crafted and refined messaging based on survey results throughout the study and contributed input on each iteration of the questionnaire. Andrea Diss-Torrance and Kim Peterson wrote the paper.

Conflicts of Interest: The authors declare no conflict of interest.

Appendix A. Actions Taken to Combat Invasive Forest Diseases and Pests

Throughout the course of a ten-year study period (2006–2015), Wisconsin DNR took a variety of actions to combat invasive forest diseases and pests. The overarching goal of education and outreach efforts was to present a consistent message that limiting firewood movement was important, and that it was justified on moral grounds—it was the right thing to do. A selected listing of actions follows:

1. We improved firewood availability, quality, and reliability at state parks and forests in response to campers' complaints in our early surveys. We strove to keep costs competitive. All staffed campgrounds now provide firewood for sale, and firewood availability has been added to information provided for each campground on the Wisconsin DNR website. In addition, firewood for sale is now stored in shelters to keep it dry. Most state campgrounds provide firewood from state certified sources, many of which heat treat or age the wood thus ensuring higher quality.
2. We worked with the Wisconsin Department of Agriculture, Trade, and Consumer Protection (DATCP) to encourage/sponsor development of a firewood certification program. Vendors are certified if they process wood by heating or aging to specification. Wisconsin DATCP inspects and certifies firewood processors at their request for an annual cost of $50.
3. Wisconsin DNR supports development of the firewood industry by selling certified wood at state campgrounds, thus providing a predictable market and an opportunity for the public to try this value-added product. Since 2012, state campgrounds have limited firewood that is sold on premises to that which is either from a certified vendor or harvested from the property itself.

4. Wisconsin DNR law enforcement staff assist Wisconsin DATCP to enforce regulation of commercial and private transport of logs and firewood. Two federal quarantines of forest pests were in force in Wisconsin during the times of our surveys, for emerald ash borer and gypsy moth. The quarantines require mills, loggers, and those moving firewood to take specified precautions to prevent spread of these pests. Movement of regulated items across quarantine borders within Wisconsin is enforced by two state agencies. Wisconsin DATCP has lead authority for enforcement of these regulations and has extended its authority under agreement with Wisconsin DNR to allow wardens to enforce quarantine violations by private individuals, thereby filling a gap that had existed in enforcement.

5. Wisconsin DNR staff and researchers with the U.S. Forest Service developed a model of risk of introduction to state properties with increasing establishment of an invasive species [5]. Using this model, Wisconsin DNR adjusted firewood movement limits to state parks and forests to keep the level of risk stable as the EAB and other invasive forest pests and diseases established themselves in the state. Reductions in movement distance limits may have helped slow the spread of these species, and communication of these limits sent a strong message to campers that this threat was significant and imminent.

6. Created public service messages that utilized injunctive norms (e.g., "Don't move firewood! Buy it where you burn it."). Social science research has demonstrated that this improves regulation compliance.

7. Surveyed private campground operators (2011) to: (1) Determine their awareness of invasive pests and diseases of trees that could impact their properties and identify the steps they were taking or wanted taken by the state to protect their business (e.g., did they impose their own restrictions, how did they feel about a state-wide ban on transport, etc.). (2) Make them aware of free educational material available from Wisconsin DNR.

8. We distributed posters and 39,000+ brochures to campground operators for distribution to their camping customers. This established a consistent message for campers at both public and private campgrounds and greatly increased the dissemination of this information.

9. Suggested improvements in the listing of certified firewood vendors provided at DATCP's website. State campgrounds provided opportunities for campers to try certified firewood.

10. We surveyed representative samples of campers to discover where they look for information about environmental matters, and which of these sources they consider most authoritative and valuable. This information helps target public service messaging efficiently.

11. We sponsored and/or participated in research to identify effective treatment options for firewood with the U.S. Forest Service's Forest Products Laboratory.

12. We demonstrated the use of mechanized harvesting equipment in urban settings to increase efficiency in removing infested ash trees and make possible the utilization of these logs for the highest value products.

13. We developed precautions that people using or producing firewood can implement to prevent transmission of EAB and other invasive pests in firewood. We shared our survey results and managerial approaches with state and federal staff, working to reduce the spread of wood borne invasive pests.

References

1. Peterson, K.; Diss-Torrance, A. Motivations for rule compliance in support of forest health: Replication and extension. *J. Environ. Manag.* **2014**, *139*, 135–145. [CrossRef] [PubMed]
2. Peterson, K.; Diss-Torrance, A. Motivation for compliance with environmental regulations related to forest health. *J. Environ. Manag.* **2012**, *112*, 104–119. [CrossRef] [PubMed]
3. Peterson, K.; Diss-Torrance, A. *Wisconsin Certified Firewood Survey: Results and Analysis*; Wisconsin Department of Natural Resources: Madison, WI, USA, 2013.

Forests **2018**, *9*, 90

4. Peterson, K.; Diss-Torrance, A. *A Survey of Private Campground Operators: Benchmark Results*; Wisconsin Department of Natural Resources: Madison, WI, USA, 2013.

5. Tobin, P.C.; Diss-Torrance, A.; Blackburn, L.M.; Brown, B.D. What Does "Local" Firewood Buy You? Managing the Risk of Invasive Species Introduction. *J. Econ. Entomol.* **2010**, *103*, 1569–1576. [CrossRef] [PubMed]

forests

MDPI

Review

Progress and Challenges of Protecting North American Ash Trees from the Emerald Ash Borer Using Biological Control

Jian J. Duan [1,*], Leah S. Bauer [2], Roy G. van Driesche [3] and Juli R. Gould [4]

[1] United States Department of Agriculture, Agricultural Research Service, Beneficial Insects Introduction Research Unit, Newark, DE 19713, USA

[2] United States Department of Agriculture, Forest Service, Northern Research Station, Lansing, MI 48910, USA; lbauer@fs.fed.us

[3] Department of Environment and Conservation, University of Massachusetts, Amherst, MA 01003, USA; vandries@cns.umass.edu

[4] United States of Department of Agriculture, Animal and Plant Health Inspection Service, Plant Protection and Quarantine, Science and Technology, Buzzards Bay, MA 02542, USA; juli.r.gould@aphis.usda.gov

* Correspondence: jian.duan@ars.usda.gov; Tel.: +1-302-731-7330 (ext. 249)

Received: 11 February 2018; Accepted: 11 March 2018; Published: 15 March 2018

Abstract: After emerald ash borer (EAB), *Agrilus planipennis* Fairmaire, was discovered in the United States, a classical biological control program was initiated against this destructive pest of ash trees (*Fraxinus* spp.). This biocontrol program began in 2007 after federal regulatory agencies and the state of Michigan approved release of three EAB parasitoid species from China: *Tetrastichus planipennisi* Yang (Eulophidae), *Spathius agrili* Yang (Braconidae), and *Oobius agrili* Zhang and Huang (Encyrtidae). A fourth EAB parasitoid, *Spathius galinae* Belokobylskij (Braconidae) from Russia, was approved for release in 2015. We review the rationale and ecological premises of the EAB biocontrol program, and then report on progress in North American ash recovery in southern Michigan, where the parasitoids were first released. We also identify challenges to conserving native *Fraxinus* using biocontrol in the aftermath of the EAB invasion, and provide suggestions for program improvements as EAB spreads throughout North America. We conclude that more work is needed to: (1) evaluate the establishment and impact of biocontrol agents in different climate zones; (2) determine the combined effect of EAB biocontrol and host plant resistance or tolerance on the regeneration of North American ash species; and (3) expand foreign exploration for EAB natural enemies throughout Asia.

Keywords: *Fraxinus*; ash regeneration; *Agrilus planipennis*; biocontrol; natural enemy introductions; parasitoids; invasive pests

1. Introduction

The movement of forest insects and plant pathogens, caused by the rapidly expanding global economy, poses one of the greatest threats to the ecological sustainability of forested ecosystems throughout the world [1–3]. Despite efforts to combat this problem through improved regulatory controls in international trade, the accidental introduction of non-native forest pests in wood packaging materials such as pallets and dunnage, as well as commodities such as nursery stock, lumber, and manufactured goods, continues [4–7]. Although a relatively small proportion of these introduced species become serious invasive pests in their invaded regions, increasing numbers of forest insects and diseases are devastating natural and urban forests worldwide [8,9].

The most recent and notable example of a destructive invasive insect damaging forests in North America is the emerald ash borer (EAB), *Agrilus planipennis* Fairmaire (Coleoptera: Buprestidae), introduced from Asia during the 1990s [10–14]. This phloem-feeding beetle attacks ash trees (*Fraxinus*; Oleaceae) and was discovered as the cause of ash tree mortality in southeast Michigan, USA and nearby Ontario, Canada in 2002. Over the next several years, EAB was discovered throughout the region and well beyond, being spread primarily by human-mediated transport of infested ash materials such as firewood, nursery stock, and lumber [15–17]. Consequently, early attempts to eradicate EAB in North America were abandoned, and research, development, and implementation of EAB-management strategies were expanded.

Biological control is now the primary management tool developed to suppress EAB densities in forested ecosystems, thereby conserving or protecting the surviving and regenerating ash trees [18–20], whereas systemic insecticides are available to protect high-value ash, mainly landscape trees in urban forests [21,22]. Long-term sustainability of native ash species may also require the development of EAB-resistant or tolerant ash genotypes [23,24]. An earlier review article described progress in developing EAB biocontrol program in the U.S such as foreign exploration for natural enemies in EAB's native range Asia, host specificity testing and risk assessment for the introduced biocontrol agents, the basic biology of both introduced biocontrol agents and native North American natural enemies, and an overview of EAB biocontrol releases and research through 2014 [20]. The focus of the present paper is an overview of progress and challenges in developing, implementing, and evaluating efforts to manage EAB in forested areas using biological control. Specifically, this paper first discusses the rationale for selecting EAB as a target and the ecological premises for biocontrol, and then highlights not only recent progress made in EAB biocontrol, but also the challenges in implementing biocontrol as an EAB management tool for the conservation of North American ash species. Finally, it proposes potential solutions to overcome these challenges, including the need for expanded long-term research on EAB biocontrol as this pest continues spreading throughout North America.

2. Rationale for Selection of Emerald Ash Borer as Target for Biological Control

Emerald ash borer is a specialist herbivore attacking primarily species of *Fraxinus* in Asia including China, the Korean Peninsula, and the Russian Far East [25]. Asian ash species are more resistant to EAB than are North America species [24]. The high densities of EAB feeding in the phloem of North American ash cause tree mortality within five to seven years of EAB's invasion of new locations [26–32]. As EAB spreads further south in the United States, it was also found attacking another native tree species in the family Oleaceae, the white fringetree (*Chionanthus virginicus* L.), which is commonly planted as an ornamental in eastern states [33].

Ash trees are widely distributed and highly valued in the deciduous forests of North America [34]; however, the arrival of EAB from Asia has greatly reduced the abundance of many species of ash trees in the invaded regions in the U.S. [35]. There are 16 species of *Fraxinus* native to North America, each species adapted to different ecological habitats across a range of climates zones, soil types, and moisture gradients, with many species in western states having limited geographical distributions [36–38]. Ash trees serve as food, cover, nesting sites, and habitat for mammals, birds, arthropods, and other organisms [39,40]. The earliest infestation of EAB in southeast Michigan resulted in mortality of 99% of healthy overstory ash trees in some infested forests, demonstrating the potential of EAB to functionally extirpate ash trees from the continent [29]. Emerald ash borer has since spread to 32 states and three Canadian provinces and killed hundreds of millions of ash trees in both urban and forested areas [41,42]. As a consequence, the six species of *Fraxinus* endemic to eastern North America are listed as critically endangered by the International Union for Conservation of Nature: white ash *F. americana* L., Carolina ash *F. caroliniana* Mill., black ash *F. nigra* Marshall, green ash *F. pennsylvanica* Marshall, pumpkin ash *F. profunda* (Bush) Bush, and blue ash *F. quadrangulata* Michx [43]. The loss of ash diversity and abundance in natural forests in the earliest-invaded regions (e.g., Midwestern and Mid-Atlantic States, USA) has already harmed native plants and ash-dependent invertebrates, and altered nutrient cycling and other ecological processes [40,44–48].

Although the environmental and ecological impacts of EAB on the diverse forested ecosystems of North America are not fully understood, several estimates of its economic impacts have been made. In natural forests and timberlands of the United States, more than 7.55 billion timber-sized ash trees were valued at more than \$282 billion [35]. Moreover, ash trees were widely planted as landscape trees in urban forests, and an estimate for the undiscounted value of these trees in the United States ranged from \$20–60 billion soon after EAB's discovery [11]. A cost projection of EAB in just 25 northeastern communities of the United States for only one decade (2009–2019) to treat, remove, and replace landscape ash was \$25 billion [49], making EAB the most destructive and costly wood-boring insect to invade the United States [50].

All evidence associated with the invasion of the United States and Canada by EAB demonstrates that this invasive insect is driving ecological degradation in the forests of North America, and taking no action against EAB is not a sensible or responsible option. Initial efforts by regulatory agencies focused on the eradication of incipient EAB populations by the creation of an ash-free zones in and around newly detected infestations [51], while imposing quarantine regulations to restrict the movement of firewood of all hardwood species and materials of the genus *Fraxinus* [52,53]. Although EAB- and ash-quarantine regulations remain in place, efforts to eradicate EAB were abandoned as regulatory agencies in the United States and Canada determined that eradication of EAB was not possible [13,54]. Subsequently, efforts have shifted to developing biological control-based pest management tools and strategies to slow the spread and reduce densities of EAB using conventional and biological controls, and to develop varieties of *Fraxinus* tolerant or resistant to EAB [24,55–57].

3. The Role of Natural Enemies in Suppressing EAB in Its Native Range

Little was known about the biology and natural enemy complexes of EAB in Asia before the foreign exploration work for development of a classical biological control program against EAB in the United States [18–20]. In 2003, researchers began foreign exploration for EAB natural enemies in northeastern China, resulting in the discovery of four hymenopteran parasitoid species: (1) *Sclerodermus pupariae* Yang and Yao (Bethylidae), an ectoparasitoid of larvae, prepupae, and pupae [58,59]; (2) *Spathius agrili* Yang (Braconidae), an ectoparasitoid of late-instar larvae [26,60]; (3) *Tetrastichus planipennisi* Yang (Eulophidae), an endoparasitoid of late-instar larvae [26,61]; and (4) *Oobius agrili* Zhang and Huang (Encyrtidae), an egg parasitoid [62]. Subsequent EAB natural enemy surveys in the Russian Far East from 2008 to 2012 led to the discovery of three additional species of hymenopteran parasitoid: (5) *Spathius galinae* Belokobylskij and Strazanac (Braconidae), an ectoparasitoid of late-instar larvae [63,64]; (6) *Atanycolus nigriventris* Vojnovskaja-Krieger (Braconidae), an ectoparasitoid of late-instar larvae [63,64]; and (7) *Oobius primorskyensis* Yao and Duan (Encyrtidae), an egg parasitoid [65]. In a more recent EAB natural enemy survey in northeastern China, two species of predacious Coleoptera were found attacking late-instar larvae and pupae of EAB: (8) *Tenerus* sp. (Cleridae); and (9) *Xenoglena quadrisignata* Mannerheim (Trogossitidae) [66].

Ecological studies at field sites in northeast China and the Russian Far East revealed these insect natural enemies cause high mortality of EAB eggs and larvae and play a critical role in suppressing EAB densities in forested areas of Asia [26,64,66,67]. The abundance and contribution of individual species to EAB control varied by geographic region (Table 1). For example, *S. galinae* is the dominant EAB larval parasitoid in the Russian Far East, causing up to 63% larval parasitism in some stands, but it has not been observed in China [64]. In contrast, *T. planipennisi*, is the dominant EAB larval parasitoid in northeast China, causing an average of 40% larval parasitism, but it is observed less frequently in the Russian Far East and further south in Beijing, but never in our most southern survey site in Tianjin, China where *S. agrili* is more abundant [26,64,66,67]. To date, *O. agrili* is the only EAB egg parasitoid collected consistently in China where it was found parasitizing 12–62% of EAB eggs [67]. *Oobius primorskyensis*, the only parasitoid found attacking EAB eggs in the Russian Far East, caused about 23–44% egg parasitism (JJD, unpublished data). Currently, studies are lacking on the ecological factors that determine the structure of EAB parasitoid assemblages and dominance of different species in different regions of the beetles' native range. We suspect that climatic factors such as temperature and photoperiod, as well as synchronization of EAB and parasitoid phenology in different geographic regions, may have led to differences in the parasitoid assemblages and dominance. Nevertheless, our current knowledge of EAB parasitoid complexes in Asia strongly indicates that the success of EAB biological control in North America will require the introduction and successful establishment of a variety of species and genotypes of EAB parasitoids collected from different climatic regions of Asia.

Table 1. Natural enemy complexes and their observed attack rates on the emerald ash borer (EAB) in different regions of its native range in northeastern Asia.

Geographic Region	Natural Enemies	EAB Stage(s) Attacked	Rate of Attack (Parasitism or Predation)	References
Northeast China: Heilongjiang, Jilin, and Liaoning provinces	*Oobius agrili*	eggs	12–62%	[26,62,66]
	Oencyrtus sp.	eggs	1–2%	[19]
	Tetrastichus planipennisi	3rd and 4th instars	3–44%	[26,61,66,67]
	Spathius agrili	3rd and 4th instars	0–13%	[26,60,66]
	Atanycolus spp. Foerster (Hymenoptera: Braconidae)	3rd and 4th instars	0–23%	[66]
	Xorides sp. Latreille (Hymenoptera: Ichneumonidae)	3rd and 4th instars	0–11%	[66]
	Tenerus sp. Laporte (Coleoptera: Cleridae)	JLand pupae	0–21%	[66]
	Xenoglena quadrisignata Mannerheim (Coleoptera: Trogossitidae)	JL and pupae	0–1.2%	[66]
Northcentral China: Beijing and Tianjin cities	*Oobius agrili*	eggs	0–4.0%	[62,66]
	Tetrastichus planipennisi	3rd and 4th instars	0–7%	[26,61,66,67]
	Spathius agrili	3rd and 4th instars	44–67%	[61,66]
	Atanycolus sp.	3rd and 4th instars	0–5%	[66]
	Metapelma sp. Westwood (Hymenoptera: Eupelmidae)	3rd and 4th instars	0–4%	[66]
	Sclerodermus pupariae Yang and Yao (Hymenoptera: Bethylidae)	3rd and pupae	1–1.3%	[66]
Russia: Primorsky Kray	*Oobius primorskyensis*	egg	23–44%	JJD (unpublished data)
	Tetrastichus planipennisi	3rd and 4th instars	0–7%	[64]
	Spathius galinae	3rd and 4th instars	0–78%	[64]
	Atanycolus nigriventris Vojnovskaja-Krieger (Braconidae: Braconinae)	3rd and 4th instars	0–55%	[64]
	Atanycolus sp.	3rd and 4th instars	0–1%	[64]

4. Development of an EAB Biological Control Program in North America

To facilitate implementation of environmentally sound biological control programs, the North American Plant Protection Organization (NAPPO), with members from Canada, Mexico, and the United States, provides guidelines to petition for the release of non-indigenous entomophagous biocontrol agents in member countries [68]. The NAPPO regional standards, developed to analyze the risks and benefits of implementing a biocontrol program, are based on those developed by the International Plant Protection Convention of the Food and Agriculture Organization of the United Nations [69]. Petitions for the release of biocontrol agents are currently reviewed by Agriculture & Agri-Food Canada's (AAFC) Biological Control Review Committee (BCRC). With experts from each member country, BCRC evaluates biocontrol petitions and makes release recommendations to United States Department of Agriculture, Animal and Plant Health Inspection Service (USDA APHIS) in the United States and to Canadian Food Inspection Agency (CFIA) in Canada. Authority to release biocontrol agents in the United States may be granted by USDA APHIS after posting on the federal register, consideration of public comments, a risk-benefit analysis, and state concurrence; the release of entomophagous biocontrol agents in Canada is coordinated by AAFC and CFIA [68–70].

Of the Asiatic natural enemies discovered during foreign exploration for EAB natural enemies, three EAB parasitoid species from China (*T. planipennisi*, *S. agrili* and *O. agrili*) were proposed for introduction, and after extensive host range testing and safety evaluation, a petition for their release was approved in the United States in 2007 [71]. In Canada, releases of *T. planipennisi* and *S. agrili* were approved by CFIA in 2013 and of *O. agrili* in 2015. However, *S. agrili* has not been released in Canada because its sustained establishment is not confirmed in the United States. Two additional species of insect natural enemy collected from EAB in the Russian Far East were proposed for release in the United States in 2014: the larval ectoparasitoid *S. galinae* [72] and the egg parasitoid *O. primorskyensis* (JJD, unpublished data). While the petition for release of *S. galinae* as an EAB biocontrol agent was approved by USDA APHIS in the United States in 2015 and by CFIA in Canada in 2017 [53], BCRC recommended *O. primorskyensis* not be approved for release in the United States until reconsideration of the petition after additional research, risk-benefit analysis, and resubmission. The introduction of *A. nigriventris* from the Russian Far East was also considered, but difficulties in maintaining a viable colony of this species under quarantine laboratory conditions prevented researchers from conducting host range studies for further evaluation (JJD, unpublished data). The host ranges of the hymenopteran parasitoid *S. pupariae* and the two species of predacious beetle (*Tenerus* sp. and *X. quadrisignata*), were deemed by researchers as too broad, and they were not considered further as potential biocontrol agents.

The safety of Asiatic parasitoids petitioned for environmental releases in North America was assessed with data collected from both field surveys of other wood-boring insects in the parasitoids' native ranges (China and Russian Far East) and laboratory testing with Asian and North American wood-boring and other insects. Data from these studies show that host specificity of the released Asiatic parasitoids is highly constrained by the close phylogenetic proximity of potential nontarget hosts to EAB [19,71–74]. Field data from China and the Russian Far East show that these parasitoids do not attack other wood-boring insects in ash, such as bark beetles (Scolytidae) or longhorned beetles (Cerambycidae) [73,74]. However, host specificity studies in the laboratory further show that three of these introduced parasitoids—*O. agrili*, *S. agrili*, and *S. galinae*—do attack some Asian and North American species of *Agrilus* (Table 2). In contrast to the attack on EAB, however, their attack rate is lower on these potentially susceptible non-target *Agrilus* spp., even under laboratory conditions, which promote maximum parasitism [19,64,71–74]. Based on both laboratory and field host- range studies, the predicted non-target impact from introduction of these Asiatic parasitoids for EAB biocontrol, if any, would be a low level of parasitism of some non-target *Agrilus* species in North America.

Table 2. Non-target insect taxa tested with the Asiatic parasitoids petitioned for environmental release in North America as EAB biocontrol agents.

EAB Parasitoids from Asia	Insect Orders Tested	Insect Families Tested	Insect Species Tested	*Agrilus* Species Tested	The Only Non-Targets Attacked Were *Agrilus* Species
Oobius agrili [1]	2	6	18	6	3
Tetrastichus planipennisi [1]	3	6	14	5	0
Spathius agrili [1]	2	6	18	9	5
Spathius galinae [2]	3	6	15	6	1

[1] Data compiled from Federal Register 2007 [71]; Yang et al., 2008 [74]; Bauer et al., 2014 [19]. [2] Data compiled from Federal Register 2015 [72]; Duan et al., 2015a [73].

There are approximately 800 *Agrilus* species in North America, with about 175 species in the United States [75,76]. Based on the results of these host-range studies, it is possible that the introduced EAB parasitoids may occasionally attack some of the non-target *Agrilus* species in North America; however, recent host-finding studies show that some EAB parasitoids are attracted to volatiles from ash trees [74,77], indicating a strong affinity to *Fraxinus*, their host's food plants. Thus, it can be reasonably predicted that the level of attack on non-ash feeding *Agrilus* species, if any, would be limited. Field surveys of non-target insects associated with ash trees following field releases of introduced EAB parasitoids in Michigan and Maryland found no evidence of non-target attack from these introduced parasitoids [45]. In contrast, arthropod diversity associated with ash trees is significantly reduced because of the EAB invasion in Maryland [78]. The parasitoids introduced from Asia were selected for high EAB-host specificity, and in the aftermath of the EAB invasion of North America, the resulting conservation of *Fraxinus* and recovery of forests will produce many desirable ecological benefits [79,80].

5. Introduction and Establishment of EAB Biocontrol Agents

In 2007, after USDA APHIS issued permits for the environmental release of *O. agrili*, *S. agrili*, and *T. planipennisi* in Michigan, small numbers (a few hundred per species) were laboratory-reared and released at a few sites [18,81,82]. In subsequent years, larger numbers of parasitoids were released (tens of thousands) in additional states after USDA APHIS' EAB biocontrol mass-rearing facility in Brighton, Michigan became operational in 2010 [19,20,83]. To date, more parasitoid releases and data on recovery is ongoing in regions with more ash trees, a longer history of EAB, and where researchers, regulatory agencies, or land managers are actively involved in research or management of EAB using biocontrol. By the end of the 2017 field season, parasitoids had been released in 27 of 32 United States and two of three Canadian provinces invaded by EAB (Figures 1 and 2).

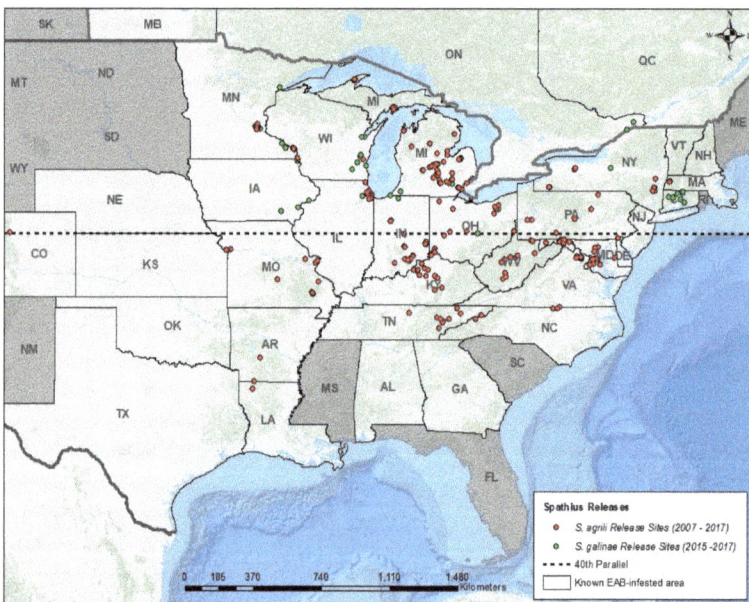

Figure 1. Map showing known regions of North America invaded by EAB, and locations where *Spathius agrili* and *S. galinae* were released by fall 2017 [84]. In the United States, releases of *S. agrili* and *S. galinae* began in 2007 and 2015, respectively. In Canada, release of *S. galinae* began in 2017. In 2013, release of *S. agrili* was limited to EAB infestations south of the 40th parallel due to lack of establishment further north. No recovery sites for *Spathius* are shown because establishment of *S. agrili* was found at only one release site, and it is too soon to confirm establishment of *S. galinae*.

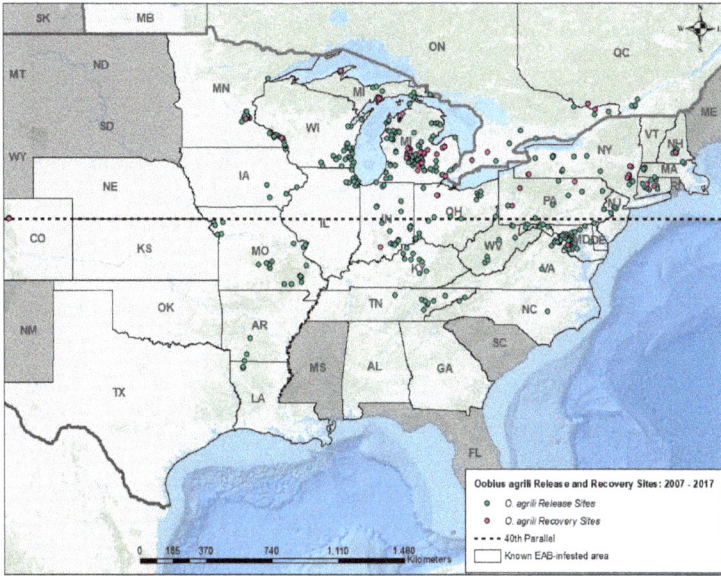

(a)

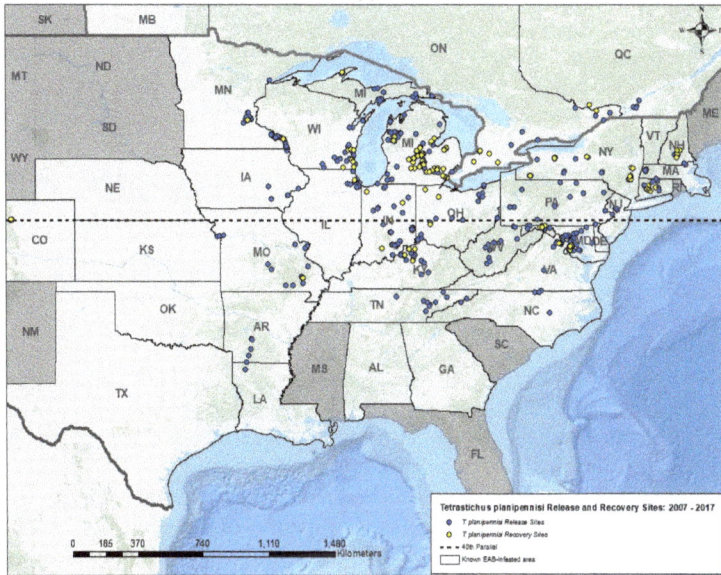

(b)

Figure 2. Maps showing known regions of North America invaded by EAB, and the release and recovery sites for (**a**) *O. agrili* and (**b**) *T. planipennisi*, EAB biocontrol agents introduced from China in North America from 2007 to 2017, using a variety of methods [83] and documented on the EAB biocontrol geospatial database [84]. In the United States, releases of *T. planipennisi* and *O. agrili* began in 2007, and in Canada, releases of these species began in 2013 and 2015, respectively. Establishment of these two biocontrol agents in EAB populations are confirmed at many early release sites.

In many regions of the United States, the three Chinese biocontrol agents have been recovered from EAB larvae and eggs one year after their release, indicating successful reproduction and overwintering in the target host. However, only *O. agrili* (Figure 2a) and *T. planipennisi* (Figure 2b) are consistently recovered two or more years after their last release, and these two species are now considered established and spreading naturally beyond their initial release sites [84]. By 2013, however, researchers found that *S. agrili* was not establishing in northern regions of the United States, and release of this species was subsequently restricted to EAB infestations south of the 40th parallel [83] (Figure 1). To date, the establishment of *S. agrili* has been confirmed at only one site in Maryland (~38th parallel), 7 years after release and at a low rate of parasitism (JJD, unpublished data). Releases of *S. galinae* began in 2015 north of the 40th parallel, and although it is too soon to confirm establishment, the results of ongoing research on EAB natural enemies in Michigan, Connecticut, Massachusetts, and New York suggest this parasitoid is establishing and spreading [85].

6. Impact of EAB Biocontrol Agents on Target Pest Populations

The question then arises whether populations of the established parasitoids can effectively reduce the invasive EAB populations to a sufficiently low level to allow for ash regeneration and recovery in the aftermath forests. The answer to this question requires long-term research and also depends on impacts of other biotic and abiotic mortality factors of EAB in the targeted forest ecosystem. For example, a population dynamics model parameterized with observed larval and egg parasitism rates (~60%) in Asia, showed that natural enemies in Asia can quickly reduce EAB populations (i.e., with a net population growth rate < 1) when accompanied by moderate to high levels of host plant resistance with no predation from avian predators [86]. When accompanied by heavy woodpecker predation (~60%) in North America (e.g., [87–89]), an addition of ~35% of larval parasitism rate is sufficient to reduce the EAB population growth rate to <1, even with limited levels of host tree resistance or tolerance [86,90]. Moreover, other factors are periodically important, such as mortality caused by fungal entomopathogens [91,92] or cold winter temperatures [93].

In the same line of analysis, key abiotic factors such as temperature can also affect the efficacy of EAB biocontrol. For example, there are regional differences in the EAB life cycle because of variation of the heat accumulation in different geographic regions. In warmer climates of the southern United States, EAB eggs and larvae develop faster, thereby reducing exposure times to egg and larval parasitoids and causing EAB population growth rates to increase. In addition, this shortened EAB life cycle may also result in asynchrony of EAB egg and larval stages with adult parasitoid phenology, causing failure of establishment, or reduced impacts of natural enemies on EAB population densities. Consequently, the population-level impact of the introduced EAB biocontrol agents, in the southern United States, may be reduced by the climatic condition that favor a shortened (one year) EAB life cycle. Thus, foreign exploration for EAB natural enemies is needed in southern Asia for biocontrol of EAB in southern regions of the United States.

Data collected recently from a long-term study conducted in Michigan may provide us with some insights into the population-level impact of these introduced biocontrol agents. The long-term study consisted of six forested sites in southern Michigan, each comprised of a release and non-release control plot, which were established between 2007 and 2010. At each release plot, small numbers of adult *O. agrili*, *S. agrili*, and *T. planipennisi* were released, and in subsequent years, infested ash trees are being sampled to estimate EAB egg and larval parasitism, and other causes of larval mortality [81,82].

During the first five years after release of the EAB parasitoids at these study sites, EAB egg parasitism by *O. agrili* averaged ~1 to 4% from 2008–2011 and then increased to ~28% by 2014 in release plots. The natural spread of *O. agrili* from the release plots to the control plots was slow and somewhat variable between sites [82]. Overall, the impact of *O. agrili* in suppressing EAB population growth, as well as the natural spread rate of this biocontrol agent, has yet to be determined, because sampling EAB eggs (1 mm in diameter and cryptically colored) from ash bark layers and crevices is labor intensive and difficult to standardize [82,94]. Moreover, parasitism of EAB eggs by *O. agrili* is patchy, thus more intensive sampling is needed to find this tiny parasitoid and quantify its impact on EAB population dynamics. Despite these challenges, researchers are confirming the establishment and relatively slow spread of *O. agrili* in this and other regions of North America (Figure 2a).

In contrast to the sampling of EAB eggs, sampling EAB for larval parasitism is done by debarking live, infested ash trees, a relatively simple and reproducible method. Using this approach, average larval parasitism by *T. planipennisi* was ~1 to 6% from 2008–2011 and increased to ~30% by 2014 in both the release and control plots [81,95,96]. As more recovery work is done in this and other regions, researchers are finding a rapid spread of *T. planipennisi* across EAB-infested sites (Figure 2b). More recent life table analyses after seven years of data collection from these six study sites revealed that *T. planipennisi* contributed significantly to the reduction of net EAB population growth rates approximately four years after its initial release [96]. Moreover, with additional larval mortality from local natural enemies of wood boring insects, such as woodpeckers and native parasitoids (primarily braconids in the genus *Atanycolus*) [97], the resource-adjusted EAB larval density (per m^2 of live phloem tissues) declined ~90% in infested ash trees at both the release and control plots between 2009 and 2014 [96,98]. The decline in the resource-adjusted EAB density may also be attributed in part to the general collapse of EAB populations following widespread mortality of the overstory ash trees. Depletion of host tree resources in a local area would cause EAB adults at some point to disperse in search of more abundant hosts [99,100]. However, many small ash trees and saplings, ranging in size from 2.5- to 15-cm Diameter at Breast Height (DBH)) are still abundant and susceptible to EAB infestation in the study sites [79,86]. For the surviving ash trees, the pest pressure they now experience is reduced, increasing prospects for their survival and reproduction [79,86].

The results from EAB field studies in the United States and Asia reveal that larval parasitism rates by *T. planipennisi* are inversely correlated ash tree diameter with 95% of larval parasitism in ash trees < 16-cm DBH [26,45,67,101]. This can be attributed to the relatively short ovipositor of *T. planipennisi* (average 2- to 2.5-mm long), limiting its ability to reach EAB larvae under the thick bark on lower boles of large-diameter ash trees [101]. In the same study, a larger parasitoid species (*Atanycolus*) with a longer ovipositor (average 4- to 6-mm long), parasitized EAB larvae in ash trees up to 57.4-cm DBH. It has been shown that *T. planipennisi* is important in protecting ash saplings and basal sprouts (2- to 6-cm DBH) from EAB in post-invasion recovering forests [79,80]. However, the protection of ash trees as they mature, will require establishment of the larger EAB biocontrol agent, *S. galinae*, which has a longer ovipositor (average 4- to 6-mm long) and capable of parasitizing EAB larvae in large-diameter ash trees [64,102]. Since releases of *S. galinae* began in several northern states in 2015, it appears to be establishing and spreading, and researchers will continue monitoring its impacts on EAB population dynamics and the health of large-diameter ash trees at study sites.

In theory, highly effective egg parasitoids from EAB's native range may protect all size-class ash trees against EAB, as they can kill the pest before its larvae bore into the ash phloem to feed. However, the current level of egg parasitism by *O. agrili* (<29%) by itself is not sufficient to protect ash trees. Introduction of a second species of EAB egg parasitoid, *O. primorskyensis*, may enhance egg parasitism in some regions of North America and improve ash tree survival [103].

7. Ash Recovery and Regeneration after EAB invasion with Biological Control

Evidence gathered in the native range of EAB has shown that EAB outbreaks in northeastern Asia are rare events in natural forests, and outbreaks occur primarily in isolated plantations and urban plantings of mostly North American ash species (*F. pennsylvanica, F. americana, F. velutina*) [26,64,86,104]. Even if EAB can occasionally cause significant ash mortality in urban plantings or plantations of North American ash in Asia, no widespread outbreaks comparable to those observed in North American forests have been recorded in forested regions of Asia [26,64]. In addition, large, relatively healthy North American ash trees, mainly *F. pennyslvanica* and *F. americana*, have been observed in forested parks in China and urban areas in the Russian Far East [26,64,67]. It is plausible that EAB parasitoids in these regions may be protecting the more susceptible North American ash species. This protection may occur at two different phases. First, saplings or trees of susceptible ash species planted in Asia maybe colonized at low levels of EAB because there are fewer EAB founders coming from the resistant native ash species in Asia, thereby delaying EAB population increase in these susceptible ash trees or saplings. Second, the abundance and diversity of EAB parasitoids in their native range may facilitate a more rapid numerical response to incipient infestations of EAB in the susceptible ash saplings or trees, resulting in direct protection at relatively low EAB densities. In post-EAB invaded forests of North America, ash trees are much scarcer than in forests prior to invasion, and established populations of the introduced EAB parasitoids may conserve surviving native ash by moderating the frequency and amplitude of future EAB outbreaks.

The EAB invasion of forests in southeast Michigan, during the 1990s, resulted in nearly 100% mortality of overstory ash trees by 2010 [29,105]. The potential for recovery of the ash canopy was assessed in this region from 2007 to 2009, and abundant regeneration of smaller height class ash trees, mainly *F. americana* and *F. pennsylvanica*, were found. However, young 1–2 year ash seedlings were much less common, and the lack of new seedlings was traced to a depleted seed bank, as few or no nearby mature ash trees existed to provide seed [29,105]. However, the results of a more recent study of regenerating *F. pennsylvanica* in this region reported abundant seed production on surviving mature ash trees during mast years, as well as on sexually mature small ash trees and basal sprouts regenerating from top-killed trees, suggesting a significant, though greatly reduced, pool of ash trees in this region [106]. In a separate study in southeast Michigan where the establishment and spread of *T. planipennisi* is now confirmed, densities of ash and other native saplings were higher and densities of weedy species lower in closer proximity to study sites where more parasitoids were released [80]. These results suggest that protection of ash saplings by *T. planipennisi* favors the recruitment of native woody species over weedy species in gaps as these forests recover from loss of the overstory ash canopy in the aftermath of the EAB invasion.

Researchers also estimated the abundance and condition of ash saplings and trees at the six long-term EAB biocontrol study sites in southern Michigan, where the sustained establishment *T. planipennisi* and *O. agrili* have been documented for nearly a decade [79]. Results of this study showed that healthy ash saplings (400–1600 per hectare) and young trees (200–900 per hectare) remained in these study sites, despite formerly high EAB densities that resulted in loss of most overstory ash trees by 2010. In addition, life table analysis of EAB population dynamics at these sites indicates that the net population growth rate of EAB is near or below replacement levels, and that the introduced biocontrol agent *T. planipennisi* reduced the pest's net population growth rates at these sites by over 50%. These findings strongly indicate that the introduced EAB parasitoids can provide significant biocontrol services, enhancing ash survival, and promoting forest recovery in North America [79,80,85].

8. Conclusions

Following its accidental introduction into the United States in the 1990s, EAB continues to spread and degrade ash communities and forested ecosystems in North America. The EAB Biocontrol Program, which started over a decade ago via the introduction and establishment of co-evolved natural enemies from the pest's native range, appears to hold promise for forests of northern regions of North America. This program has documented establishment of the egg parasitoid *O. agrili* and the larval parasitoid *T. planipennisi*, both introduced from China, in EAB populations at most release sites in northern United States and southern Canada, where surveys to document parasitoid establishment are ongoing. While the role of *O. agrili* in reducing EAB population growth requires continued evaluation, the larval parasitoid *T. planipennisi* has been shown to play a significant role in protecting ash saplings and smaller trees (DBH < 12 cm) in aftermath forests in Michigan [79,80,96]. The suppression of EAB densities is likely to spread geographically as populations of *O. agrili* and *T. planipennisi* increase and spread to new areas, protecting the regenerating ash saplings and young trees. To protect growing and surviving ash trees, however, more widespread releases and successful establishment of *S. galinae*, the largest of the EAB biocontrol agents, are needed. As EAB continues spreading through the southern and western United States, we recommend expanding EAB biocontrol research to: (1) quantify the impacts of EAB biocontrol on ash and other native tree species as forests recover in the aftermath of EAB in northern regions; (2) develop parasitoid release methods for more widespread, remote, or larger ash stands; (3) expand research on synergistic effects of EAB biocontrol and ash resistance or tolerance to EAB in native North American ash species; (4) determine parasitoid establishment in EAB populations in warmer climates; and (5) explore different regions of Asia for EAB natural enemies adapted to climate zones similar to those in the southern and western United States where EAB is invading. Over many decades, it is reasonable to assume that a diverse complex of mortality factors and lower ash density will reduce both the frequency and intensity of EAB outbreaks, permitting the growth, survival, reproduction, and conservation of *Fraxinus* species.

Acknowledgments: The authors thank Deborah Miller (USDA FS, Lansing, MI), Richard Reardon (USDA FS, Morgantown, WV, USA), Jonathan Lelito and Benjamin Slager (USDA APHIS, Brighton, MI, USA), David Jennings and Paula Shrewsbury (University of Maryland), Houping Liu (Pennsylvania Department of Conservation and Natural Resources), Deborah McCullough and Amos Ziegler (Michigan State University), Scott Salom (Virginia Tech University), Clifford Sadof (Purdue University), Lynne Rieske-Kinney (University of Kentucky), Melissa Fierke (State University of New York, College of Environmental Science and Forestry), Daniel Kashian (Wayne State University), Inés Ibáñez (University of Michigan), and Joseph Elkinton and Kristopher Abell (University of Massachusetts) for their collaborative work on EAB, which helped to facilitate and inspire the write-up of this manuscript. The authors also thank Travis Perkins (Michigan State University) for preparing the maps, the many entomologists, technicians, and student employees from USDA Agricultural Research Service (ARS), Forest Service (FS), and, Michigan State University, University of Maryland, University of Massachusetts, and the many state and provincial regulators and land managers carrying out EAB biocontrol research and implementing EAB biocontrol programs. The authors also greatly appreciate early reviews of this manuscript by Kim Hoelmer (USDA ARS, Newark, DE, USA) and Noah Koller (USDA ARS, E. Lansing, MI, USA) and funding provided by USDA FS, APHIS, and ARS.

Conflicts of Interest: The authors declare no conflict of interest.

References

1. Pimentel, D. Biological invasions of plants and animals in agriculture and forestry. In *Ecology of Biological Invasions of North America and Hawaii*; Mooney, H.A., Drake, J.A., Eds.; Ecological Studies 58; Springer: New York, NY, USA, 1986; pp. 149–162.

2. Aukema, J.E.; McCullough, D.G.; Von Holle, B.; Liebhold, A.M.; Britton, K.; Frankel, S.J. Historical accumulation of non-indigenous forest pests in the continental United States. *BioScience* **2010**, *60*, 886–897. [CrossRef]

3. Boyd, I.L.; Freer-Smith, P.H.; Gilligan, C.A.; Godfray, H.C.J. The consequence of tree pests and diseases for ecosystem services. *Science* **2013**, *342*, 1235773. [CrossRef] [PubMed]

4. Liebhold, A.M.; MacDonald, W.L.; Bergdahl, D.; Mastro, V.C. Invasion by exotic forest pests: A threat to forest ecosystems. *For. Sci. Monogr.* **1995**, *41*, 1–49.

5. Haack, R.A. Exotic bark and wood-boring Coleoptera in the United States: Recent establishments and interceptions. *Can. J. For. Res.* **2006**, *36*, 269–288. [CrossRef]

6. Haack, R.A.; Britton, K.O.; Brockerhoff, E.G.; Cavey, J.F.; Garrett, L.J.; Kimberley, M.; Lowenstein, F.; Nuding, A.; Olson, L.J.; Turner, J.; et al. Effectiveness of the international phytosanitary standard ISPM15 on reducing wood borer infestation rates in wood packaging material entering the US. *PLoS ONE* **2014**, *9*, e96611. [CrossRef] [PubMed]

7. Lovett, G.M.; Weiss, M.; Liebhold, A.M.; Holmes, T.; Leung, B.; Lambert, K.F.; Orwig, D.A.; Campbell, F.T.; Rosenthal, J.; McCullough, D.G.; et al. Nonnative forest insects and pathogens in the United States: Impacts and policy options. *Ecol. Soc. Am.* **2016**, *26*, 1437–1455. [CrossRef] [PubMed]

8. Williamson, M.; Fitter, A. The varying success of invaders. *Ecology* **1996**, *77*, 1661–1666. [CrossRef]

9. Brockerhoff, E.G.; Liebhold, A.M. Ecology of forest insect invasions. *Biol. Invasions* **2017**, *19*, 3141–3159. [CrossRef]

10. Haack, R.A.; Jendek, E.; Liu, H.; Marchant, K.R.; Petrice, T.R.; Poland, T.M.; Ye, H. The emerald ash borer: A new exotic pest in North America. *Newsl. Mich. Entomol. Soc.* **2002**, *47*, 1–5.

11. Federal Register. Emerald Ash Borer; Quarantine and Regulations. 2003. 7 CFR Part 301 [Docket Number 02-125-1]. Available online: https://www.federalregister.gov/documents/2003/10/14/03-25881/emerald-ash-borer-quarantine-and-regulations (accessed on 18 December 2017).

12. Cappaert, D.L.; McCullough, D.G.; Poland, T.M.; Siegert, N.W. Emerald ash borer in North America: A research and regulatory challenge. *Am. Entomol.* **2005**, *51*, 152–165. [CrossRef]

13. GAO. Invasive Forest Pests: Lessons Learned from Three Recent Infestations May Aid in Managing Future Efforts. Report of the United States Government Accounting Office. 2006. GAO-06-353. Available online: https://www.gao.gov/assets/250/249776.pdf (accessed on 5 January 2018).

14. Siegert, N.W.; McCullough, D.G.; Liebhold, A.M.; Telewski, F.W. Dendrochronological reconstruction of the epicentre and early spread of emerald ash borer in North America. *Divers. Distrib.* **2014**, *20*, 847–858. [CrossRef]

15. Poland, T.M.; McCullough, D.G. Emerald ash borer: Invasion of the urban forest and the threat to North America's ash resource. *J. For.* **2006**, *104*, 118–124.

16. Herms, D.A.; McCullough, D.G. Emerald ash borer invasion of North America: History, biology, ecology, impact and management. *Annu. Rev. Entomol.* **2014**, *59*, 13–30. [CrossRef] [PubMed]

17. Morin, R.S.; Liebhold, A.M.; Pugh, S.A.; Crocker, S.J. Regional assessment of emerald ash borer, *Agrilus planipennis*, impacts in forests of the Eastern United States. *Biol. Invasions* **2017**, *19*, 703–711. [CrossRef]

18. Bauer, L.S.; Liu, H.P.; Miller, D.L.; Gould, J. Developing a classical biological control program for *Agrilus planipennis* (Coleoptera: Buprestidae), an invasive ash pest in North America. *Newsl. Mich. Entomol. Soc.* **2008**, *53*, 38–39.

19. Bauer, L.S.; Duan, J.J.; Gould, J.R. Emerald ash borer *Agrilus planipennis* Fairmaire Coleoptera: Buprestidae. In *The Use of Classical Biological Control to Preserve Forests in North America*; FHTET-2013-2; Van Driesche, R., Reardon, R., Eds.; United States Department of Agriculture, Forest Service, Forest Health and Technology Enterprise Team: Morgantown, WV, USA, 2014; pp. 189–209. Available online: https://www.nrs.fs.fed.us/pubs/48051 (accessed on 30 January 2018).

20. Bauer, L.S.; Duan, J.J.; Gould, J.R.; Van Driesche, R.G. Progress in the classical biological control of *Agrilus planipennis* Fairmaire (Coleoptera: Buprestidae) in North America. *Can. Entomol.* **2015**, *147*, 300–317. [CrossRef]

21. Sadof, C.S.; Hughes, G.P.; Witte, A.R.; Peterson, D.J.; Ginzel, M.D. Tools for staging and managing emerald ash borer in the urban forest. *Arboric. Urban For.* **2017**, *43*, 15–26.

22. Mercader, R.J.; McCullough, D.G.; Storer, A.J.; Bedford, J.; Poland, T.M.; Katovich, S. Evaluation of the potential use of a systemic insecticide and girdled trees in area wide management of the emerald ash borer. *For. Ecol. Manag.* **2015**, *350*, 70–80. [CrossRef]

23. Rigsby, C.M.; Showalter, D.N.; Herms, D.A.; Koch, J.L.; Bonello, P.; Cipollini, D. Physiological responses of emerald ash borer larvae to feeding on different ash species reveal putative resistance mechanisms and insect counter-adaptations. *J. Insect Physiol.* **2015**, *78*, 47–54. [CrossRef] [PubMed]

24. Villari, C.; Herms, D.A.; Whitehill, J.G.A.; Cipollini, D.; Bonello, P. Progress and gaps in understanding mechanisms of ash tree resistance to emerald ash borer; a model for wood-boring insects that kill angiosperms. *New Phytol.* **2016**, *209*, 63–79. [CrossRef] [PubMed]

25. Bray, M.; Bauer, L.S.; Poland, T.M.; Haack, R.A.; Cognato, A.I.; Smith, J.J. Genetic analysis of emerald ash borer (*Agrilus planipennis* Fairmaire) populations in Asia and North America. *Biol. Invasions* **2011**, *13*, 2869–2887. [CrossRef]

26. Liu, H.; Bauer, L.S.; Gao, R.; Zhao, T.; Petrice, T.R.; Haack, R.A. Exploratory survey for the emerald ash borer, *Agrilus planipennis* (Coleoptera: Buprestidae), and its natural enemies in China. *Great Lakes Entomol.* **2003**, *36*, 191–204.

27. Eyles, A.; Jones, W.; Riedl, K.; Cipollini, D.; Schwartz, S.; Chan, K.; Herms, D.A.; Bonello, P. Comparative phloem chemistry of Manchurian *Fraxinus mandshurica* and two North American ash species *Fraxinus americana* and *Fraxinus pennsylvanica*. *J. Chem. Ecol.* **2007**, *33*, 1430–1448. [CrossRef] [PubMed]

28. Rebek, E.J.; Herms, D.A.; Smitley, D.R. Interspecific variation in resistance to emerald ash borer (Coleoptera: Buprestidae) among North American and Asian ash (*Fraxinus* spp.). *Environ. Entomol.* **2008**, *37*, 242–246. [CrossRef]

29. Klooster, W.S.; Herms, D.A.; Knight, K.S.; Herms, C.P.; McCullough, D.G.; Smith, A.S.; Gandhi, K.J.K.; Cardina, J. Ash (*Fraxinus* spp.) mortality, regeneration, and seed bank dynamics in mixed hardwood forests following invasion by emerald ash borer (*Agrilus planipennis*). *Biol. Invasions* **2014**, *16*, 859–873. [CrossRef]

30. Tanis, S.R.; McCullough, D.G. Differential persistence of blue ash (*Fraxinus quadrangulata*) and white ash (*Fraxinus americana*) following emerald ash borer (*Agrilus planipennis*) invasion. *Can. J. For. Res.* **2012**, *42*, 1542–1550. [CrossRef]

31. Tanis, S.R.; McCullough, D.G. Host resistance of five *Fraxinus* species to *Agrilus planipennis* (Coleoptera: Buprestidae) and effects of paclobutrazol and fertilization. *Environ. Entomol.* **2015**, *41*, 287–299. [CrossRef] [PubMed]

32. Knight, K.S.; Brown, J.P.; Long, R.P. Factors affecting the survival of ash (*Fraxinus* spp.) trees infested by emerald ash borer (*Agrilus planipennis*). *Biol. Invasions* **2013**, *15*, 371–383. [CrossRef]

33. Cipollini, D. White fringetree as a novel larval host for emerald ash borer. *J. Econ. Entomol.* **2015**, *108*, 370–375. [CrossRef] [PubMed]

34. USDA–FS (United States Department of Agriculture, Forest Service). *Silvics of Forest Trees of the United States*; Compiler; Agriculture Handbook 271; Folwells, H.A., Ed.; United States Department of Agriculture, Forest Service: Washington, DC, USA, 1965; pp. 181–196. Available online: https://catalog.hathitrust.org/Record/001507718 (accessed on 20 December 2017).

35. Nowak, D.; Crane, D.; Stevens, J.; Walton, J. *Potential Damage from Emerald Ash Borer*; United States Department of Agriculture, Forest Service, Northern Research Station: Syracuse, NY, USA, 2003; pp. 1–5. Available online: https://www.nrs.fs.fed.us/disturbance/invasive_species/eab/local-resources/downloads/EAB_potential.pdf (accessed on 18 December 2017).

36. Eyre, F.H. (Ed.) *Forest Cover Types of the United States and Canada*; Society of American Foresters: Washington, DC, USA, 1980; 148p, ISBN 13: 978-0686306979.

37. Harlow, W.M.; Harrar, E.S.; Hardin, J.W.; White, F.M. *Textbook of Dendrology*, 8th ed.; McGraw Hill Book Company: New York, NY, USA, 1996; ISBN 13: 978-0070265721.

38. USDA–NRCS. Plants Profile for *Fraxinus* in North America. 2017. Available online: https://plants.usda.gov/core/profile?symbol=fraxi (accessed on 18 December 2017).

39. Koenig, W.D.; Liebhold, A.M.; Bonter, D.N.; Hachachka, W.M.; Dicknson, J.L. Effects of the emerald ash borer on four species of birds. *Biol. Invasions* **2013**, *15*, 2095–2103. [CrossRef]

40. Wagner, D.L.; Todd, K.J. New ecological assessment for the emerald ash borer: A cautionary tale about unvetted host-plant literature. *Am. Entomol.* **2016**, *62*, 26–35. [CrossRef]

41. USDA–APHIS. Initial County EAB Detection Map. 2018. Available online: https://www.aphis.usda.gov/plant_health/plant_pest_info/emerald_ash_b/downloads/MultiState.pdf (accessed on 7 February 2018).

42. Emerald Ash Borer Information. Emerald Ash Borer Information Network. 2016. Available online: http://www.emeraldashborer.info/ (accessed on 18 December 2017).

43. IUCN (International Union for Conservation of Nature). IUCN Red List of Threatened Species. Available online: http://www.iucnredlist.org/ (accessed on 18 December 2017).

44. Flower, C.E.; Knight, K.S.; Gonzalez-Meler, M.A. Impacts of the emerald ash borer (*Agrilus planipennis*) induced ash (*Fraxinus* spp.) mortality on forest carbon cycling and successional dynamics in the eastern United States. *Biol. Invasions* **2013**, *15*, 931–944. [CrossRef]

45. Jennings, D.E.; Duan, J.J.; Bean, D.; Gould, J.R.; Kimberly, A.R.; Shrewsbury, P.M. Monitoring the establishment and abundance of introduced parasitoids of emerald ash borer larvae in Maryland, U.S.A. *Biol. Control* **2016**, *101*, 138–144. [CrossRef]

46. Stephens, J.P.; Berven, K.A.; Tiegs, S.D. Anthropogenic changes to leaf litter input affect the fitness of a larval amphibian. *Freshw. Biol.* **2013**, *58*, 1631–1646. [CrossRef]

47. Ulyshen, M.D.; Klooster, W.S.; Barrington, W.T.; Herms, D.A. Impacts of emerald ash borer-induced tree mortality on leaf litter arthropods and exotic earthworms. *Pedobiologia* **2011**, *54*, 261–265. [CrossRef]

48. Ulyshen, M.D.; Barrington, W.T.; Hoebeke, E.R.; Herms, D.A. Vertically stratified ash-limb beetle fauna in northern Ohio. *Psyche* **2012**, *2012*, 215891. [CrossRef]

49. Kovacs, F.K.; Haight, R.G.; McCullough, D.G.; Mercader, R.J.; Siegert, N.W.; Leibhold, A.M. Cost of potential emerald ash borer damage in U.S. communities, 2009–2019. *Ecol. Econ.* **2010**, *69*, 569–578. [CrossRef]

50. Aukema, J.; Leung, B.; Kovacs, K.; Chivers, C.; Britton, K.O.; Englin, J.; Frankel, S.J.; Haight, R.G.; Holmes, T.P.; Liebhold, A.M.; et al. Economic impacts of non-native forest insects in the United States. *PLoS ONE* **2011**, *6*, e24587. [CrossRef] [PubMed]

51. Taylor, R.A.J.; Bauer, L.S.; Poland, T.M.; Windell, K. Flight performance of *Agrilus planipennis* (Coleoptera: Buprestidae) on a flight mill and in free flight. *J. Insect Behav.* **2010**, *23*, 128–148. [CrossRef]

52. USDA–APHIS. Emerald Ash Borer Federal Regulations and Quarantine Notices. 2017. Available online: https://www.aphis.usda.gov/aphis/ourfocus/planthealth/plant-pest-and-disease-programs/pests-and-diseases/emerald-ash-borer/ct_quarantine (accessed on 30 January 2018).

53. CFIA. Canadian Food Inspection Agency. Emerald Ash Borer. 2018. Available online: http://www.inspection.gc.ca/plants/plant-pests-invasive-species/insects/emerald-ash-borer/eng/1337273882117/1337273975030 (accessed on 29 January 2018).

54. USDA–APHIS. Emerald Ash Borer Program Manual, *Agrilus planipennis* (Fairmaire), ver. 1.6. 2015. Available online: https://www.aphis.usda.gov/import_export/plants/manuals/domestic/downloads/emerald_ash_borer_manual.pdf (accessed on 7 February 2018).

55. Flower, C.E.; Dalton, J.E.; Knight, K.S.; Brikha, M.; Gonzalez-Meler, M.A. To treat or not to treat: Diminishing effectiveness of emamectin benzoate tree injections in ash trees heavily infested by emerald ash borer. *Urban For. Urban Green.* **2015**, *14*, 790–795. [CrossRef]

56. O'Brien, E.M. Conserving Ash (*Fraxinus*) Populations and Genetic Variation in Forests Invaded by Emerald Ash Borer Using Large-Scale Insecticide Applications. Ph.D. Thesis, The Ohio State University, Columbus, OH, USA, 2017.

57. Davidson, W.; Rieske, L.K. Establishment of classical biological control targeting emerald ash borer is facilitated by use of insecticides, with little effect on native arthropod communities. *Biol. Control* **2016**, *101*, 78–86. [CrossRef]

58. Wu, H.; Wang, X.Y.; Li, M.L.; Yang, Z.Q.; Zeng, F.Z.; Wang, H.Y.; Bai, L.; Liu, S.J.; Sun, J. Biology and mass rearing of *Sclerodermus pupariae* Yang et Yao (Hymenoptera: Bethylidae), an important ectoparasitoid of the emerald ash borer, *Agrilus planipennis* (Coleoptera: Buprestidae) in China. *Acta Entomol. Sin.* **2008**, *51*, 46–54.

59. Yang, Z.Q.; Wang, X.Y.; Yao, Y.X.; Gould, J.R.; Cao, L.M. A new species of *Sclerodermus* (Hymenoptera: Bethylidae) parasitizing *Agrilus planipennis* (Coleoptera: Buprestidae) from China, with a key to Chinese species in the genus. *Ann. Entomol. Soc. Am.* **2012**, *105*, 619–627. [CrossRef]

60. Yang, Z.Q.; Achterberg, C.V.; Choi, W.Y.; Marsh, P.M. First recorded parasitoid from China of *Agrilus planipennis*: A new species of *Spathius* (Hymenoptera: Braconidae: Doryctinae). *Ann. Entomol. Soc. Am.* **2005**, *98*, 636–642. [CrossRef]

61. Yang, Z.Q.; Yao, Y.X.; Wang, X.Y. A new species of emerald ash borer parasitoid from China belonging to the genus *Tetrastichus* (Hymneoptera: Eulophidae). *Proc. Entomol. Soc. Wash.* **2006**, *108*, 550–558.

62. Zhang, Y.Z.; Huang, D.W.; Zhao, T.H.; Liu, H.P.; Bauer, L.S. Two new species of egg parasitoids (Hymenoptera: Encyrtidae) of wood-boring beetle pests from China. *Phytoparasitica* **2005**, *53*, 253–260. [CrossRef]

63. Belokobylskij, S.A.; Yurchenko, G.I.; Strazanac, J.S.; Zaldi'var-Riveron, A.L.; Mastro, V. A new emerald ash borer (Coleoptera: Buprestidae) parasitoid species of *Spathius* Nees (Hymenoptera: Braconidae: Doryctinae) from the Russian Far East and South Korea. *Ann. Entomol. Soc. Am.* **2012**, *105*, 165–178. [CrossRef]

64. Duan, J.J.; Yurchenko, G.; Fuester, R.W. Occurrence of emerald ash borer (Coleoptera: Buprestidae) and biotic factors affecting its immature stages in the Russian Far East. *Environ. Entomol.* **2012**, *41*, 245–254. [CrossRef] [PubMed]

65. Yao, Y.X.; Duan, J.J.; Hopper, K.R.; Mottern, J.L.; Gates, M.W. A new species of *Oobius* Trjapitzin (Hymenoptera: Encyrtidae) from the Russian Far East that parasitizes eggs of emerald ash borer (Coleoptera: Buprestidae). *Ann. Entomol. Soc. Am.* **2016**, *106*, 629–638. [CrossRef]

66. Wang, X.Y.; Cao, L.M.; Yang, Z.Q.; Duan, J.J.; Gould, J.R.; Bauer, L.S. Natural enemies of emerald ash borer (Coleoptera: Buprestidae) in northeast China, with notes on two species of parasitic Coleoptera. *Can. Entomol.* **2016**, *148*, 329–342. [CrossRef]

67. Liu, H.; Bauer, L.S.; Miller, D.L.; Zhao, T.; Gao, R.; Song, L.; Luan, Q.; Jin, R.; Gao, C. Seasonal abundance of *Agrilus planipennis* (Coleoptera: Buprestidae) and its natural enemies *Oobius agrili* (Hymenoptera: Encyrtidae) and *Tetrastichus planipennisi* (Hymenoptera: Eulophidae) in China. *Biol. Control* **2007**, *42*, 61–71. [CrossRef]

68. NAPPO. (North American Plant Protection Organization) NAPPO Regional Standards for Phytosanitary Measures (RSPM). RSPM 12: Guidelines for Petition for First Release of Non-Indigenous Entomophagous Biological Control Agents. 2015. Available online: https://www.nappo.org/files/1814/4065/2949/RSPM12_30-07-2015-e.pdf (accessed on 19 January 2018).

69. Mason, P.G.; Kabaluk, J.T.; Spence, B.; Gillespie, D.R. Regulation of Biological Control in Canada. In *Biological Control Programmes in Canada 2001–2012*; Mason, P.G., Gillespie, D.R., Eds.; CABI: Wallingford, UK, 2013; pp. 1–5.

70. Montgomery, M. Understanding federal regulations as guidelines for classical biological control programs. In *Implementation and Status of Biological Control of the Hemlock Woolly Adelgid*; Onken, B., Reardon, R., Eds.; FHTET-2011-04; United States Department of Agriculture, Forest Service: Morgantown, WV, USA, 2011; pp. 25–40. Available online: http://www.fs.fed.us/nrs/pubs/jrnl/2011/nrs_2011_montgomery_001.pdf (accessed on 19 January 2018).

71. Federal Register. Availability of an environmental assessment for the proposed release of three parasitoids for the biological control of the emerald ash borer *Agrilus planipennis* in the Continental United States. *Fed. Regist.* **2007**, *72*, 28947–28948. Available online: http://www.regulations.gov/#!documentDetail;D= APHIS-2007-0060-0043 (accessed on 18 December 2017).

72. Federal Register. Availability of an environmental assessment for field release of the parasitoid *Spathius galinae* for the biological control of the emerald ash borer (*Agrilus planipennis*) in the contiguous United States. *Fed. Regist.* **2015**, *80*, 7827–7828. Available online: https://www.regulations.gov/docket?D=APHIS-2014-0094 (accessed on 18 December 2017).

73. Duan, J.J.; Gould, J.R.; Fuester, R.W. Evaluation of the host specificity of *Spathius galinae* (Hymenoptera: Braconidae), a larval parasitoid of the emerald ash borer (Coleoptera: Buprestidae) in Northeast Asia. *Biol. Control* **2015**, *89*, 91–97. [CrossRef]

74. Yang, Z.Q.; Wang, X.Y.; Gould, J.R.; Wu, H. Host specificity of *Spathius agrili* Yang (Hymenoptera: Braconidae), an important parasitoid of the emerald ash borer. *Biol. Control* **2008**, *47*, 216–221. [CrossRef]

75. Bellamy, C.L. World catalogue and bibliography of the jewel beetles (Coleoptera: Buprestidae). In *Agrilinae: Agrilina through Trachyini*; Pensoft Series Faunistica #79; Pensoft Publishers: Sofia, Bulgaria; Moscow, Russia, 2008; Volume 4, pp. 1932–2684, ISBN 9789546423214.

76. Nelson, G.H.; Walters, G.C., Jr.; Haines, R.D.; Bellamy, C.L. *A Catalog and Bibliography of the Buprestoidea of America North of Mexico*; The Coleopterists Society, Special Publication: North Potomac, MD, USA, 2008; pp. 1–274.

77. Johnson, T.D.; Lelito, J.P.; Raffa, K.F. Responses of two parasitoids, the exotic *Spathius agrili* Yang and the native *Spathius floridanus* Ashmead, to volatile cues associated with the emerald ash borer, *Agrilus planipennis* Fairmaire. *Biol. Control* **2014**, *79*, 110–117. [CrossRef]

78. Jennings, D.E.; Duan, J.J.; Bean, D.; Kimberly, A.R.; Williams, G.L.; Bells, S.K.; Shurtleff, A.S.; Shrewsbury, P.M. Effects of the emerald ash borer invasion on the community composition of arthropods associated with ash tree boles in Maryland, U.S.A. *Agric. For. Entomol.* **2017**, *19*, 122–129. [CrossRef]

79. Duan, J.J.; Bauer, L.S.; Van Driesche, R.G. Emerald ash borer biocontrol in ash saplings: The potential for early stage recovery of North American ash trees. *For. Ecol. Manag.* **2017**, *394*, 64–72. [CrossRef]

80. Margulies, E.; Bauer, L.; Ibanez, I. Buying time: Preliminary assessment of biocontrol in the recovery of native forest vegetation in the aftermath of the invasive emerald ash borer. *Forests* **2017**, *8*, 369. [CrossRef]

81. Duan, J.J.; Bauer, L.S.; Abell, K.J.; Lelito, J.P.; Van Driesche, R.G. Establishment and abundance of *Tetrastichus planipennisi* (Hymenoptera: Eulophidae) in Michigan: Potential for success in classical biocontrol of the invasive emerald ash borer (Coleoptera: Buprestidae). *J. Econ. Entomol.* **2013**, *106*, 1145–1154. [CrossRef] [PubMed]

82. Abell, K.J.; Bauer, L.S.; Duan, J.J.; Van Driesche, R.G. Long-term monitoring of the introduced emerald ash borer (Coleoptera: Buprestidae) egg parasitoid, *Oobius agrili* (Hymenoptera: Encyrtidae), in Michigan, USA and evaluation of a newly developed monitoring technique. *Biol. Control* **2014**, *79*, 36–42. [CrossRef]

83. USDA–APHIS/ARS/FS. USDA Animal Plant Health Inspection Service/Agricultural Research Service/Forest Service. Emerald Ash Borer Biological Control Release and Recovery Guidelines. 2016. Available online: https://www.aphis.usda.gov/plant_health/plant_pest_info/emerald_ash_b/downloads/EAB-FieldRelease-Guidelines.pdf (accessed on 18 January 2018).

84. MapBioControl.org. Agent Release Tracking and Data Management for Federal, State, and Researchers Releasing Biocontrol Agents for Management of the Emerald Ash Borer. 2018. Available online: http://www.mapbiocontrol.org/ (accessed on 16 January 2018).

85. Duan, J.J.; Van Driesche, R.G.; Bauer, L.S.; Reardon, R.; Gould, J.; Elkinton, J.S. *The Role of Biocontrol of Emerald Ash Borer in Protecting Ash Regeneration after Invasion*; FHAAST-2017-02; United States Department of Agriculture, Forest Service, Forest Health Assessment and Applied Sciences Team: Morgantown, WV, USA, 2017; pp. 1–10.

86. Duan, J.J.; Van Driesche, R.G.; Bauer, L.S.; Kashian, D.M.; Herms, D.A. Risk to ash from emerald ash borer: Can biological control prevent the loss of ash stands? In *Biology and Control of Emerald Ash Borer*; van Driesche, R.G., Reardon, R.C., Eds.; FHTET-2014-9; United States Department of Agriculture Forest Service, Forest Health Technology Enterprise Team: Morganton, WV, USA, 2015; pp. 153–163. Available online: https://www.nrs.fs.fed.us/pubs/49310 (accessed on 30 January 2018).

87. Lindell, C.A.; McCullough, D.G.; Cappaert, D.; Apostolou, N.M.; Roth, M.B. Factors influencing woodpecker predation on emerald ash borer. *Am. Midl. Nat.* **2008**, *159*, 434–444. [CrossRef]

88. Flower, C.E.; Long, L.C.; Knight, K.S.; Rebbeck, J.; Brown, J.S.; Gonzalez-Meler, M.A.; Whelan, C.J. Native bark-foraging birds preferentially forage in infected ash (*Fraxinus* spp.) and prove effective predators of the invasive emerald ash borer (*Agrilus planipennis* Fairmaire). *For. Ecol. Manag.* **2014**, *313*, 300–306. [CrossRef]

89. Jennings, D.E.; Gould, J.R.; Vandenberg, J.D.; Duan, J.J.; Shrewsbury, P.M. Quantifying the impact of woodpecker predation on population dynamics of the emerald ash borer (*Agrilus planipennis*). *PLoS ONE* **2013**, *8*, e83491. [CrossRef] [PubMed]

90. Jennings, D.E.; Duan, J.J.; Abell, K.J.; Bauer, L.S. Life table evaluation of change in emerald ash borer populations due to biological control. In *Biology and Control of Emerald Ash Borer*; Van Driesche, R.G., Reardon, R.C., Eds.; FHTET-2014-9; United States Department of Agriculture Forest Service, Forest Health Technology Enterprise Team: Morganton, WV, USA, 2015; pp. 139–151. Available online: https://www.nrs.fs.fed.us/pubs/49312 (accessed on 6 March 2018).

91. Liu, H.P.; Bauer, L.S. Susceptibility of *Agrilus planipennis* (Coleoptera: Buprestidae) to *Beauveria bassiana* and *Metarhizium anisopliae*. *J. Econ. Entomol.* **2006**, *99*, 1096–1103. [CrossRef] [PubMed]

92. Castrillo, L.A.; Bauer, L.S.; Houping, L.P.; Griggs, M.H.; Vandenberg, J.D. Characterization of *Beauveria bassiana* (Ascomycota: Hypocreales) isolates associated with *Agrilus planipennis* (Coleoptera: Buprestidae) populations in Michigan. *Biol. Control* **2010**, *54*, 135–140. [CrossRef]

93. Wu, H.; Li, M.L.; Yang, Z.Q.; Wang, X.Y. Research on cold hardiness of emerald ash borer and its two parasitoids, *Spathius agrili* Yang (Hym., Braconidae) and *Tetrastichus planipennisi* Yang (Hym., Eulophidae). *Chin. J. Biol. Control* **2007**, *23*, 119–122. (In Chinese)

94. Abell, K.J.; Bauer, L.S.; Miller, D.L.; Duan, J.J.; Van Driesche, R.G. Monitoring the establishment and flight phenology of parasitoids of emerald ash borer (Coleoptera: Buprestidae) in Michigan by using sentinel eggs and larvae. *Fla. Entomol.* **2016**, *99*, 667–672. [CrossRef]

95. Duan, J.J.; Bauer, L.S.; Abell, K.J.; Van Driesche, R.G. Population responses of hymenopteran parasitoids to the emerald ash borer (Coleoptera: Buprestidae) in recently invaded areas in north central United States. *BioControl* **2012**, *57*, 199–209. [CrossRef]

96. Duan, J.J.; Bauer, L.S.; Abell, K.J.; Ulyshen, M.D.; Van Driesche, R.G. Population dynamics of an invasive forest insect and associated natural enemies in the aftermath of invasion: Implications for biological control. *J. Appl. Ecol.* **2015**, *52*, 1246–1254. [CrossRef]

97. Bauer, L.S.; Duan, J.J.; Lelito, J.P.; Liu, H.P.; Gould, J.R. Biology of emerald ash borer parasitoids. In *Biology and Control of Emerald Ash Borer*; van Driesche, R.G., Reardon, R.C., Eds.; FHTET-2014-9; United States Department of Agriculture Forest Service, Forest Health Technology Enterprise Team: Morganton, WV, USA, 2015; Chapter 6, pp. 97–112. Available online: https://www.nrs.fs.fed.us/pubs/49294 (accessed on 30 January 2018).

98. Duan, J.J.; Bauer, L.S.; Abell, K.J.; Van Driesche, R.G. Natural enemies implicated in the regulations of an invasive pest: A life table analysis of the population dynamics of the invasive emerald ash borer. *Agric. For. Entomol.* **2014**, *16*, 406–416. [CrossRef]

99. Mercader, R.; Siegert, N.W.; Liebhold, A.M.; McCullough, D.G. Dispersal of the emerald ash borer, *Agrilus planipennis*, in newly colonized sites. *Agric. For. Entomol.* **2009**, *11*, 421–424. [CrossRef]

100. Siegert, N.W.; McCullough, D.G.; Williams, D.W.; Fraser, I.; Poland, T.M. Dispersal of *Agrilus planipennis* (Coleoptera: Buprestidae) from discrete epicenters in two outlier sites. *Environ. Entomol.* **2010**, *39*, 253–265. [CrossRef] [PubMed]

101. Abell, K.J.; Duan, J.J.; Bauer, L.S.; Lelito, J.P.; Van Driesche, R.G. The effect of bark thickness on the effectiveness of *Tetrastichus planipennisi* (Hymen: Eulophidae) and *Atanycolus* spp. (Hymen: Braconidae) two parasitoids of emerald ash borer (Coleop: Buprestidae). *Biol. Control* **2012**, *63*, 320–325. [CrossRef]

102. Murphy, T.C.; Van Driesche, R.G.; Gould, J.R.; Elkinton, J. Can *Spathius galinae* attack emerald ash borer larvae feeding in large ash trees? *Biol. Control* **2017**, *114*, 8–14. [CrossRef]

103. Larson, K.M.; Duan, J.J. Differences in the reproductive biology and diapause of two congeneric species of egg parasitoids (Hymenoptera: Encyrtidae) from northeast Asia: Implications for biological control of the invasive emerald ash borer (Coleoptera: Buprestidae). *Biol. Control* **2016**, *101*, 39–45. [CrossRef]

104. Wei, X.; Reardon, D.; Wu, Y.; Sun, J.H. Emerald ash borer, *Agrilus planipennis*, in China: A review and distribution survey. *Acta Entomol. Sin.* **2004**, *47*, 679–685.

105. Kashian, D.M.; Witter, J.A. Assessing the potential for ash canopy tree replacement via current regeneration following emerald ash borer-caused mortality on southeastern Michigan landscapes. *For. Ecol. Manag.* **2011**, *261*, 480–488. [CrossRef]

106. Kashian, D.M. Sprouting and seed production may promote persistence of green ash in the presence of the emerald ash borer. *Ecosphere* **2016**, *7*, e01332. [CrossRef]

![forests logo] *forests*

MDPI

Article

Methods to Improve Survival and Growth of Planted Alternative Species Seedlings in Black Ash Ecosystems Threatened by Emerald Ash Borer

Nicholas Bolton [1,2,*], Joseph Shannon [1], Joshua Davis [1], Matthew Van Grinsven [1,3], Nam Jin Noh [1,4], Shon Schooler [5], Randall Kolka [6], Thomas Pypker [7] and Joseph Wagenbrenner [1,8]

1   School of Forest Resources & Environmental Science, Michigan Technological University, Houghton, MI 49931, USA; jpshanno@mtu.edu (J.S.) joshuad@mtu.edu (J.D.); mvangrin@nmu.edu (M.V.G.); n.noh@westernsydney.edu.au (N.J.N.); jwagenbrenner@fs.fed.us (J.W.)
2   Daniel B. Warnell School of Forestry and Natural Resources, University of Georgia, Athens, GA 30602, USA
3   Department of Earth, Environment, & Geosciences, Northern Michigan University, Marquette, MI 49855, USA
4   Hawkesbury Institute for the Environment, Western Sydney University, Richmond, NSW 2753, Australia
5   Lake Superior National Estuarine Research Reserve, University of Wisconsin-Superior, Superior, WI 54880, USA; sschoole@uwsuper.edu
6   USDA (United States Department of Agriculture) Forest Service, Northern Research Station, Grand Rapids, MN, 55744, USA; rkolka@fs.fed.us
7   Department of Natural Resource Sciences, Thompson Rivers University, Kamloops, BC V2C 0C8, Canada; TPypker@tru.ca
8   USDA (United States Department of Agriculture) Forest Service, Pacific Southwest Research Station, Arcata, CA 95521, USA
*   Correspondence: Nicholas.Bolton@uga.edu

Received: 22 February 2018; Accepted: 14 March 2018; Published: 16 March 2018

Abstract: Emerald ash borer (EAB) continues to spread across North America, infesting native ash trees and changing the forested landscape. Black ash wetland forests are severely affected by EAB. As black ash wetland forests provide integral ecosystem services, alternative approaches to maintain forest cover on the landscape are needed. We implemented simulated EAB infestations in depressional black ash wetlands in the Ottawa National Forest in Michigan to mimic the short-term and long-term effects of EAB. These wetlands were planted with 10 alternative tree species in 2013. Based on initial results in the Michigan sites, a riparian corridor in the Superior Municipal Forest in Wisconsin was planted with three alternative tree species in 2015. Results across both locations indicate that silver maple (*Acer saccharinum* L.), red maple (*Acer rubrum* L.), American elm (*Ulmus americana* L.), and northern white cedar (*Thuja occidentalis* L.) are viable alternative species to plant in black ash-dominated wetlands. Additionally, selectively planting on natural or created hummocks resulted in two times greater survival than in adjacent lowland sites, and this suggests that planting should be implemented with microsite selection or creation as a primary control. Regional landowners and forest managers can use these results to help mitigate the canopy and structure losses from EAB and maintain forest cover and hydrologic function in black ash-dominated wetlands after infestation.

Keywords: EAB; *Fraxinus nigra*; underplanting; mitigation; microsite

1. Introduction

Since the confirmation of emerald ash borer ((EAB) *Agrilus planipennis* Fairmaire (Coleoptera: Buprestidae)) in 2002 [1,2], quarantine zones and other management recommendations have not slowed the pace of EAB infestation and it has spread across 31 American states and two Canadian provinces (Emerald Ash Borer Information Network 2017). It is projected that the invasive exotic insect will continue to move across North America, continuing to alter forest landscapes by killing host ash (*Fraxinus* spp.) trees [3]. While some studies indicate that there are certain ash trees that may be resistant despite the infested condition of the surrounding forest [4], EAB-induced mortality in ash species in infested forests is approximately 99% [5]. The outlook for North American ash trees is bleak as the confirmed range of EAB continues to expand. One forested ecosystem that is severely impacted by EAB's continued expansion is black ash (*Fraxinus nigra* Marsh) wetlands.

Black ash grows in three ecotypes of the Upper Great Lakes region: depressional headwater catchments, wetland complexes, and riparian corridors [6,7]. All three of these ecotypes have prolonged periods of inundation or saturation throughout the growing season, the time of year when precipitation and temperature are conducive to plant growth. These wetland forest systems provide many ecosystem services. For example, black ash forested wetlands provide habitat and food sources for game birds, small animals, and deer [7], the canopy reduces heat input into streams [8,9], and the root structure maintains soil integrity during rain events, reducing erosion and sediment deposition downstream [10,11]. Current theories predict that cover type changes after EAB infestation will lead to loss of the tree canopy on the landscape and forested wetlands in the short-term will become dominated by a robust herbaceous community [12] and in the long-term possibly a shrub layer consisting of alder (*Alnus* spp.) [13,14].

Planting alternative species within black ash wetlands may be an approach to shift forest composition towards one that will be more resilient to EAB, thereby maintaining ecosystem services provided by forested wetlands. However, artificial regeneration within northern wetlands is a difficult task because of the unique conditions and climate stresses on seedlings [15,16]. For instance, a seedling planted within the region will endure a dramatic annual temperature swing and periods of time when standing water is prevalent. A recent study in northern Minnesota investigated planting in black ash wetland complexes in tandem with forest management practices [17], and their results highlighted a low survivorship among seedlings.

In this study, we used simulated EAB infestations to determine the impacts of EAB on tree seedling survival and used the initial results to subsequently test alternative planting techniques in uninfested ash forests. Our objectives were to (i) compare survival rates among deciduous and coniferous tree seedlings in black ash wetlands where manipulated overstory treatments reflected the timing of EAB infestation, and (ii) compare microsite and herbivory treatments to inform best practices for future plantings to mitigate EAB impact on forest canopy and structure.

2. Materials and Methods

This study consisted of three black ash wetlands that were part of an overstory manipulation study located on the Ottawa National Forest (ONF) and one uninfested black ash riparian corridor located on the Superior Municipal Forest (SMF) (Figure 1). There was some overlap in alternative species planted and details for each forest are presented below.

Figure 1. Map of the Great Lakes region with the three study locations in the Ottawa National Forest (●), in the western Upper Peninsula of Michigan, and the one study location in the Superior Municipal Forest (▲), in northwestern Wisconsin. The shaded region is the Great Lakes Watershed with United States and Canadian boundaries.

2.1. Ottawa National Forest Site Description

Three depressional wetland study sites were located in the Ottawa National Forest of the western Upper Peninsula of Michigan, USA (Figure 1, ●). Study site elevations ranged from 371 to 507 m, areas ranged from 0.25 to 1.2 ha, and soils were comprised of Histosols with the depth to clay lens or bedrock between 40 and 480 cm (Table 1). Mean annual precipitation was 836 mm and mean temperatures ranged from −15.7 °C in January to 18.1 °C in July. Study site canopies were dominated by black ash with lesser amounts of red maple (*Acer rubrum* L.), yellow birch (*Betula alleghaniensis* Britton), northern white cedar (*Thuja occidentalis* L.), and balsam fir (*Abies balsamea* L. (Mill)).

Table 1. Treatment and planting years, soil type [18], elevation, and canopy characteristics on the Ottawa National Forest (ONF) and Superior Municipal Forest (SMF) study wetlands.

Site	Percent Canopy Black Ash (%)	Planting Year	Soil Type	Elevation (m)	Canopy Openness (%)
ONF Control	48	2013	Woody peat Histosol	507	19.7
ONF Girdle	88	2013	Woody peat Histosol	499	16.5
ONF Ash-Cut	38	2013	Woody peat Histosol	371	6.6
SMF	90	2015	Arnheim mucky silt loam or Udifluvents	183	Closed–open

2.2. Ottawa National Forest Study Design

Treatments in the three wetlands were an untreated control ("Control"), girdling ("Girdle"), and felling of black ash ("Ash-Cut"). All black ash greater than 2.5 cm in diameter were treated in the Girdle and Ash-Cut wetlands. This is a similar design to a sister-study [12] and our intention for the Girdle treatment was to simulate the short-term impacts of an EAB infestation, while the Ash-Cut treatment simulated the long-term impacts of EAB infestation [1].

Ottawa National Forest study wetlands were planted with ten tree species suitable for saturated soils in summer 2013 (Table 2). Seedling ages ranged from two to four years and were purchased from the USDA (United States Department of Agriculture) Forest Service J.W. Toumey Nursery in Watersmeet, MI, USA. A series of ten transects were established across each wetland and seedlings were planted in pairs in high (hummock) and low (hollow) planting microsites within 1 m every 2 m along each transect, totaling 60 trees of each species in each wetland. Seedings were measured each year of the study during the last week of July.

Table 2. Ottawa National Forest species and seedling ages and planting stock type (BR—bare root, P—plug).

Common Name	Scientific Name	Age (Years)	Stock Type
American elm	*Ulmus Americana* L.	2	P
basswood (linden)	*Tilia americana* L.	3	BR
burr oak	*Quercus macrocarpa* Michx.	3	BR
red maple	*Acer rubrum* L.	2	BR
silver maple	*Acer saccharinum* L.	4	BR
yellow birch	*Betula alleghaniensis* Britton	2	P
balsam fir	*Abies balsamea* (L.) Mill	2	BR
black spruce	*Picea marina* (Mill.) Britton	2	P
northern white cedar	*Thuja occidentalis* L.	2	P
tamarack	*Larix larcinia* K. Koch	2	BR

2.3. Superior Municipal Forest Site Description

The study area was along the riparian corridor of the Pokegama River that meanders through the Superior Municipal Forest in northwestern Wisconsin, USA (Figure 1, ▲). Soils were one of two distinct types: a sandy berm adjacent to the river that was created by deposits of coarse sediment, and clay-loams in adjacent lowland "back bays" (Table 1). The riparian corridor overstory was comprised of black ash and green ash (*Fraxinus pennsylvanica* Marsh) with lesser amounts of northern white cedar, balsam fir, and trembling aspen (*Populus tremuloides* Michx.).

2.4. Superior Municipal Forest Study Design

Tree species were chosen for their suitability in saturated or inundated soils as well as their projected range within forecasted climate models [19]. Seedling species were red maple, hackberry (*Celtis occidentalis* L.), and northern white cedar (Table 3) obtained from the Wisconsin Department of Natural Resources nursery in Hayward, WI, USA. Planting groups were established in different microsite, herbivory deterrence, and elevational conditions. The three microsite conditions were natural flat areas ("Natural"), constructed hummocks ("Con. Hummock"), and cleared soil ("Scarification"). The constructed hummocks were created by placing a shovel-blade full of local soil on top of the forest floor and then fortifying it by covering it with burlap matting. The cleared planting locations were created by removing existing vegetation with a spade.

Table 3. Superior Municipal Forest species and seedling ages and planting stock type (BR—bare root, P—plug).

Common Name	Scientific Name	Age (Years)	Stock Type
hackberry	*Celtis occidentalis* L.	2	BR
red maple	*Acer rubrum* L.	2	BR
northern white cedar	*Thuja occidentalis* L.	2	BR

The three herbivore exclusion treatments were no treatment ("Control"), herbivore repellant ("Repellant") (Plantskydd®, Tree World Plant Care Products Inc., St. Joseph, MO, USA) and fencing

("Fence"). The herbivore repellant was applied in the spring and fall each year following manufacturer instructions and fenced planting locations were 1.3 m tall. Each combination of microsite (3) and tree species (3) was replicated 36 times in a low elevation and 36 times in a high elevation planting zone, each approximately parallel to the river channel. One-third, or 12 planting groups per elevation zone, were assigned an herbivore treatment. Each of the 72 planting groups had three seedlings of each of the three species, for a total of nine seedlings per group or 648 seedlings. Seedlings were planted in fall 2015. Seedlings were measured each spring and fall for each year of the study period.

2.5. Field and Laboratory Procedures

Field measurements included seedling height and root collar diameter, microsite characteristics including hummock material (mineral soil or coarse woody debris and decay class), mortality, and disease. When cause of death was clear (e.g., fungus), it was recorded. Canopy openness for the ONF study was measured during the early morning, late evening, or under cloudy conditions in early July 2015 using hemispherical photography (Nikon P5000, Nikon FC-E8 fisheye lens, Nikon, Tokyo, Japan). Nine digital photographs were processed using WinSCANOPY software (Pro Version, 2010, Regent Instruments, Inc., Quebec, QC, Canada) [20] and were averaged for each planting site. Canopy openness for the SMF study was categorized from visual observations as one of three coverages: open, partial, or closed canopy and the canopy composition was recorded.

2.6. Analysis

Differences in seedling establishment and survivorship among groups of species, microsite, and treatment were tested for significance using contingency tables via Fisher's exact test. Analysis of variance (ANOVA) was used to assess species growth metrics, and relative height and diameter (calculated by RH/RD = $(W_2 - W_1)/(W_1/(t_2 - t_1))$); where RH = relative height, RD = relative diameter, W = size, and t = time), among treatment, microsite, herbivore deterrent, zone, and canopy openness. Significance level was 0.05 for all statistical tests. All statistical analyses were performed using R: A Language and Environment for Statistical Computing (Version 3.3.1, 2016, R Foundation for Statistical Computing, Vienna, Austria) [21].

3. Results

3.1. Ottawa National Forest

The planting year experienced elevated water tables throughout the growing season because of an unusually high snow pack and delayed snowmelt [22]. Additionally, standing water was present during the initial growing season at intermittent times due to high intensity rain storms [22].

Overall seedling survival across all treatments and microsites (n = 1800) after the first winter for the ONF planting study was 36% and after three years 22% of the planted seedlings survived. The second- and third-year survivorship was significantly higher than seedling establishment. Overall seedling survivorship from years 1–2 and years 2–3 was 75% and 87%, respectively. The hardwood species with the highest survivorship across the study period were silver maple, American elm, and basswood with 74%, 53%, and 40%, respectively (Table 4). The softwood species with the highest survivorship across the study period was northern white cedar at 23% (Table 4). None of the tamarack survived the 3-year study period. We found no statistical difference in seedling survival or growth from bare root stock or plug seedlings.

Initial survival rates for seedlings planted on hummocks and hollows were 44% and 29%, respectively. Over the course of the study, seedlings planted on hummocks survived better than those planted in hollows (Table 4). On average, there was a 19% (range: 4–47%) greater rate of survival than the corresponding paired seedling in the hollow over the 3-year span. However, of the top performing species, only silver maple did not display a preference between hummock or hollow and survived well on both microsites after three years with 76% and 72% survival, respectively. The ONF results indicate that survivorship and growth were not statistically different when canopy treatment was compared.

Table 4. Three-year mean seedling survival rate, relative height growth, and relative diameter growth across all treatments by microsite hummock and hollow for each planted species in the Ottawa National Forest study. Statistical significance indicated (*) for hummock vs. hollow comparisons within species for survival. Standard deviations are indicated by ± for height and diameter.

Species	Microsite	Survival (%)	Relative Height Growth (cm)	Relative Diameter Growth (cm)
American elm	Hummock	68 *	6.4 ± 15.0	0.1 ± 0.1
	Hollow	38	3.5 ± 10.4	0.1 ± 0.1
Basswood (linden)	Hummock	64 *	2.2 ± 16.0	0.1 ± 0.3
	Hollow	17	−0.1 ± 6.3	0.0 ± 0.2
burr oak	Hummock	38 *	−1.2 ± 7.6	0.0 ± 0.3
	Hollow	11	0.4 ± 2.4	0.0 ± 0.1
red maple	Hummock	11 *	0.2 ± 5.8	0.0 ± 0.2
	Hollow	2	0.1 ± 1.0	0.0 ± 0.0
silver maple	Hummock	76	4.2 ± 14.7	0.1 ± 0.2
	Hollow	72	7.1 ± 17.1	0.1 ± 0.3
yellow birch	Hummock	8 *	−0.3 ± 4.0	0.0 ± 0.1
	Hollow	0	-	-
balsam fir	Hummock	7 *	0.2 ± 1.4	0.0 ± 0.0
	Hollow	0	-	-
black spruce	Hummock	13 *	0.9 ± 2.9	0.0 ± 0.1
	Hollow	2	0.3 ± 1.9	0.0 ± 0.0
northern white cedar	Hummock	39 *	2.8 ± 6.1	0.1 ± 0.2
	Hollow	8	0.3 ± 2.6	0.3 ± 2.6
tamarack	Hummock	0	-	-
	Hollow	0	-	-

* Statistical significance at $p = 0.05$ level.

Average 3-year relative height growth for all the species except tamarack was 1.3 cm. Three-year relative height growth for six of these species was significantly higher for seedlings planted on hummocks compared to seedlings planted in hollows. In contrast, silver maple and burr oak relative growth rates were greater for hollow microsites than hummocks (Figure 2a). Average relative diameter growth across the study period was 0.3 cm, and northern white cedar planted on hummocks had the greatest increase in diameter, but the growth was highly variable (Table 3, Figure 2b).

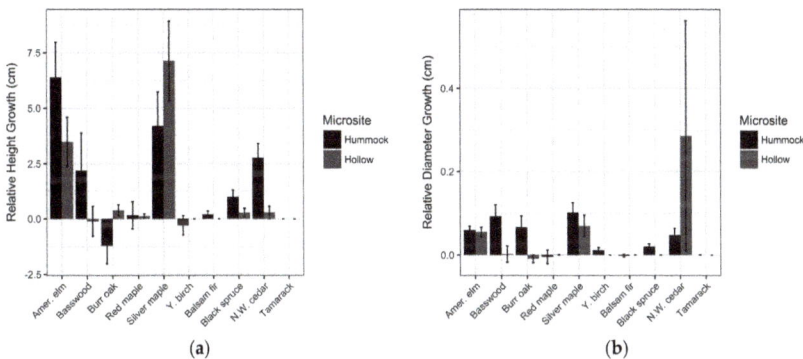

(a)   (b)

Figure 2. (**a**) Relative growth of height (cm) and (**b**) diameter (cm) of the 10 wetland-adapted tree species (American elm, basswood, burr oak, red maple, silver maple, yellow birch, balsam fir, black spruce, northern white cedar, tamarack) planted across three black ash-dominated wetlands in the Ottawa National Forest over the 3-year study period. The bars represent the mean relative growth rate for each species by microsite condition. The error bars represent ± one standard error.

3.2. Superior Municipal Forest

The growing season monthly temperature (mean 14.8 °C, range 9.4–19.4 °C) and precipitation (mean 7.3 cm, range 4.0–11.5 cm) were within the 30-year average for the Superior, Wisconsin region National Oceanic Atmospheric Administration. In contrast to the relatively low first-year survival rates on the ONF, the overall mean seedling survival across all treatments and microsites at SMF was 82% one year after planting and 54% two years after planting. Red maple had a two-year survival rate of 63%, hackberry's survival rate was 62%, and northern white cedar's survival rate was 38% (Table 5).

Table 5. Two-year mean seedling survival rate, height, and diameter across all treatments by microsite constructed hummock (CH), natural (N), and scarification (S) for each planted species in the Superior Municipal Forest study. There were no significant differences in seedling survival, relative height growth, and relative diameter growth.

Species	Microsite	Survival (%)	Relative Height Growth (cm)	Relative Diameter Growth (cm)
	CH	66	−0.1 ± 14.6	0.4 ± 5.6
hackberry	N	60	−0.9 ± 11.9	−0.6 ± 1.8
	S	58	−1.0 ± 10.4	−0.6 ± 1.7
	CH	68	12.2 ± 20.9	0.2 ± 1.8
red maple	N	57	10.9 ± 19.9	−0.5 ± 1.8
	S	63	6.4 ± 14.1	−0.6 ± 1.2
	CH	39	−1.2 ± 7.9	0 ± 1.4
northern white cedar	N	43	0.2 ± 5.6	−0.1 ± 1.3
	S	32	0.3 ± 9.7	−0.1 ± 1.9

For the SMF study, there were no statistical differences in survivorship or growth among any of our study factors: species, microsite, herbivore exclusion, and zones; therefore, we pooled the planting data and report the results here. There were no statistical differences in survivorship among browse treatments when species were pooled (mean 54%, range 39–65%). Similarly, there were no statistical differences in survivorship between the elevation zones (both 54%) despite the presence of standing water for most lower elevation (Zone 2) seedlings at the time of the 2017 measuring campaign. There were no differences among the microsite treatments when species were pooled (mean 54%, range 51–58%). Height growth for red maple was positive while hackberry showed no growth and northern white cedar decreased in height over the study period (Table 5). Average height growth for red maple was 9 cm, hackberry 0 cm, and northern white cedar −2 cm.

4. Discussion

Survival was greater for seedlings planted on hummocks when compared to seedlings planted in hollows or on cleared ground, except for silver maple at the ONF site which showed no difference between microsite conditions. Mounding has long been used in wetland forestry to establish seedlings [23] as a means to elevate seedlings out of standing water and provide a more favorable moisture regime. While the constructed hummocks in SMF were much smaller than the natural hummocks in ONF and smaller than typical mounding microsites, they still provided a marginal advantage over the hollows and cleared microsites at the two study sites.

The low survival rates on the ONF may be explained by the high amount of precipitation in the 2013 water year [24], which resulted in elevated water tables throughout the growing season and may have masked our ability to detect a difference among the treatments. The higher retention in the later years indicates that successful establishment of plantings greatly increases the probability of survival in the future. These results are similar to a study conducted on the nearby Chippewa National Forest in Minnesota [17] which showed that the successful establishment during the first growing season and winter are the major hurdles for seedling survival. Winter within the study region typically consists of high snowfall and months-long periods of below freezing temperatures.

Black ash canopy tree species loss has been determined to significantly influence water tables within black ash-dominated wetlands within northern Minnesota [25]. Black ash loss has been determined to significantly lower rates of stand transpiration in the ONF [26], significantly smaller rates of growing season drawdown within the ONF [22], and significantly higher water tables across the upper Great Lakes region [22,25] were detected in ash-dominated wetlands following a simulated EAB infestation or timber harvest. These changes subject regeneration to higher standing water levels for longer periods of time after spring inundation and after episodic summertime precipitation events. The cascading effects of forest cover loss may result in increased erosion and downstream sediment deposition. Therefore, establishing future canopy species in the understory would limit the negative environmental consequences, and provide additional time for understory vegetation to establish itself prior to exposure to the harsh environmental conditions expected following an EAB infestation.

The 4-year old silver maple seedlings had greater survival rates in both the hummocks and hollows compared to other species. The age-related height difference may explain the success of silver maple compared to the rest of the species and may have confounded the results due to the difference in planting stock. While silver maple had the highest survival rates in the ONF planting study, this species is not currently found in great numbers on this landscape, and most of the population's nearest individuals are found ~80 km to the southwest. Adaptation models suggest that future climate conditions may expand the suitable habitat for silver maple into the headwater wetlands of the upper Great Lakes region [27,28]. As global temperatures continue to rise, the cold-intolerant silver maple may shift to northerly latitudes.

American elm and basswood were also relatively successful in the ONF study. These species are commonly found along the hydric to mesic gradient near the black ash-dominated wetlands in the Great Lakes Basin. American elm is more tolerant of extended periods of inundation and saturated conditions, while basswood does not survive well when subjected to standing water [19]. If predicted future climate conditions [29] for the upper Great Lakes region come to fruition, this would put American elm at an advantage and basswood at a disadvantage because of the projected wetter and longer spring season.

Northern white cedar was the only conifer to survive at ONF in both microsite conditions, and it also had high survivorship at the SMF site. Northern white cedar is found within both black ash-dominated headwater wetlands and black ash-dominated riparian corridors. As a long-term management strategy, however, converting hardwood-dominated forests to northern white cedar may not be sustainable as northern white cedar within the region regenerates poorly and may be converted to other species [30]. Also, northern white cedar regeneration is heavily pressured by herbivores [31–33] and while our second-year results did not show a statistical difference among herbivore exclusion treatments, it may be too early to detect herbivore pressure.

Within the SMF, red maple had the highest survivorship and vigor after the first-year and based on our first year vs. third year survival rates from the ONF, we expect the survival rate for red maple to remain high. Red maple on the ONF did not fare well due to the relatively low-quality growing stock. The red maple seedlings often had missing terminal buds and were visibly less hardy when compared to the other planted seedlings. While all of the planting stock were subjected to undesirable conditions (e.g., in and out of cold storage, transport to remote study sites without temperature control) red maple's low survivorship may have been because of its small stature and frailty. Red maple is commonly found within black ash-dominated wetlands as a co-occurring species and survives in a variety of conditions [34], which indicates that red maple is a promising alternative species to plant within black ash-dominated forests. However, red maple is not very shade tolerant [35] and its success therefore will depend on release opportunities, such as those initiated by EAB infestation. As witnessed between these two study locations, if red maple were planted as an alternative species to black ash, quality growing stock and handling care will greatly enhance the success rates of planting efforts.

In a related study on the ONF, natural red maple regeneration was abundant, with density of stems ≤50 cm similar to black ash (21,944 ± 12,638 vs. 21,105 ± 13,017 stems ha^{-1}, respectively). However, the relative density of the species decreased with increasing size class. As historical data from these forests is not available, it is not clear whether this decline in density is due to legacy effects of prior growing conditions, red maple shade tolerance, poor recruitment due to current growing conditions, or some combination of these and other unidentified factors. However, this forest type is dominated by red maple elsewhere in the region [6], which suggests that a future canopy dominated by red maple is a possibility. That red maple seedlings were not negatively affected by increased herbaceous cover in our related study supports this possibility, though declines in natural regeneration may occur in the future as time since disturbance increases. The poor recruitment despite high natural regeneration indicates that the success of planting efforts may rely in part on the conditions in which the seedling establishes, and further highlights the importance of the findings in the current study.

The planting success of hackberry suggests it is a viable alternative species to ash within these systems; however, hackberry is not currently found in great numbers on this landscape, and the northernmost individuals of the defined population are found ~120 km to the southwest. As with silver maple, adaptation models suggest that future climate conditions may expand the suitable habitat for hackberry to move further north in the upper Great Lakes region [27]. In a similar study on the Chippewa National Forest, hackberry had a 52.9% survivorship over a three-year period, indicating high survival in ash-dominated wetlands [17]. While hackberry does not establish well or flourish within very wet sites [36], the hydrology of the riparian corridor may be more suitable to hackberry than the seasonal inundation in the ONF depressional wetlands.

5. Conclusions

This research includes two studies that compared plantings of wetland-adapted tree species survival and growth within black ash-dominated wetlands. In one study, seedlings were planted within black ash wetlands that underwent overstory treatments that simulated our estimated short- and long-term EAB-induced conditions. In the second study, seedlings were planted in an uninfested black and green ash-dominated riparian corridor with manipulated microsite conditions and herbivore browse exclusion treatments.

Our results indicate higher survivorship of planted seedlings when planted on hummocks in ash-dominated wetland sites in the Great Lakes region of the US. These results suggest that perching seedlings on elevated beds enhances their survivorship by providing a more stable environment. The highest surviving species we planted were silver maple, American elm, basswood, hackberry, red maple, and northern white cedar and were determined to be species well suited for alternative species plantings in ash-dominated wetlands when compared to natural regeneration within similar systems.

Acknowledgments: Funding for this work primarily came from the Great Lakes Restoration Initiative through the USDA Forest Service Northern Research Station (EPA Great Lakes Initiative Template #664: Future of Black Ash Wetlands in the Great Lakes Region) and the Wisconsin Department of Natural Resources through the Lake Superior National Estuarine Research Reserve. Additional funding came from the School of Forest Resources and Environmental Science, Ecosystem Science Center and the Center for Water and Society at Michigan Technological University. We would like to thank the Ottawa National Forest, particularly Mark Fedora, as well as the City of Superior, Wisconsin and the Superior Municipal Forest for letting us conduct this research on their lands. We would like to thank Sarah Harttung, Ashlee Lehner, and Alex Perram for assisting in data collection from the Ottawa National Forest planting sites and we would like to thank the volunteer planting crew as well as the student interns from the Lake Superior National Estuarine Research Reserve for their help at the Superior Municipal Forest planting site.

Author Contributions: N.B., J.S., S.S., J.W., R.K. and T.P. conceived and designed the experiments; N.B., J.D., J.S., M.V.G., N.J.N. and S.S. performed the experiments; N.B. and J.S. analyzed the data; and all authors contributed to writing the paper.

Conflicts of Interest: The authors declare no conflict of interest.

References

1. Haack, R.; Jendek, E.; Liu, H.; Marchant, K.; Petrice, T.; Poland, T.; Ye, H. The emerald ash borer: A new exotic Pest in North America. *Newslett. Mich. Entomol. Soc.* **2002**, *47*, 1–5.
2. Siegert, N.; McCullough, D.; Liebhold, A.; Telewski, F. Dendrochronological reconstruction of the epicentre and early spread of emerald ash borer in North America. *Divers. Distrib.* **2014**, *20*, 847–858. [CrossRef]
3. MacFarlane, D.; Meyer, S. Characteristics and distribution of potential ash tree hosts for emerald ash borer. *For. Ecol. Manag.* **2005**, *213*, 15–24. [CrossRef]
4. Marshall, J.; Smith, E.; Mech, R.; Storer, A. Estimates of *Agrilus planipennis* infestation rates and potential survival of ash. *Am. Midl. Nat.* **2013**, *169*, 179–193. [CrossRef]
5. Herms, D.; McCullough, D. Emerald ash borer invasion of North America: History, biology, ecology, impacts, and management. *Annu. Rev. Entomol.* **2014**, *59*, 13–30. [CrossRef] [PubMed]
6. Erdmann, G.; Crow, T.; Ralph, M., Jr.; Wilson, C. Managing black ash in the Lake States. In *General Technical Report NC-115*; U.S. Department of Agriculture, Forest Service, North Central Forest Experiment Station: St. Paul, MN, USA, 1987.
7. Wright, J.; Rauscher, H. *Fraxinus nigra* marsh. Black ash. *Silv. N. Am.* **1990**, *2*, 344–347.
8. Hewlett, J.; Fortson, J. Stream temperature under an inadequate buffer strip in the southeast piedmont. *J. Am. Water Resour. Assoc.* **1982**, *18*, 983–988. [CrossRef]
9. Bourque, C.A.; Pomeroy, J.H. Effects of forest harvesting on summer stream temperatures in New Brunswick, Canada: An inter-catchment, multiple-year comparison. *Hydrol. Earth Syst. Sci. Discuss.* **2001**, *5*, 599–614. [CrossRef]
10. Sheridan, J.; Lowrance, R.; Bosch, D. Management effects on runoff and sediment transport in riparian forest buffers. *Trans. Am. Soc. Agric. Eng.* **1999**, *42*, 55–64. [CrossRef]
11. Lowrance, R.; Altier, L.; Newbold, J.; Schnabel, R.; Groffman, P.; Denver, J.; Correll, D.; Gilliam, J.; Robinson, J.; Brinsfield, R. Water quality functions of riparian forest buffers in Chesapeake Bay watersheds. *Environ. Manag.* **1997**, *21*, 687–712. [CrossRef]
12. Davis, J.; Shannon, J.; Bolton, N.; Kolka, R.; Pypker, T. Vegetation responses to simulated emerald ash borer infestation in *Fraxinus nigra*-dominated wetlands of Upper Michigan, USA. *Can. J. For. Res.* **2017**, *47*, 319–330. [CrossRef]
13. Palik, B.; Ostry, M.; Venette, R.; Abdela, E. *Fraxinus nigra* (black ash) dieback in Minnesota: Regional variation and potential contributing factors. *For. Ecol. Manag.* **2011**, *261*, 128–135. [CrossRef]
14. Palik, B.; Ostry, M.; Venette, R.; Abdela, E. Tree regeneration in black ash (*Fraxinus nigra*) stands exhibiting crown dieback in Minnesota. *For. Ecol. Manag.* **2012**, *269*, 26–30. [CrossRef]
15. Ponnamperuma, F. Effects of flooding on soils. In *Flooding and Plant Growth*; Academic Press, Inc.: New York, NY, USA, 1984; pp. 9–45.
16. Roy, V.; Bernier, P.; Plamondon, A.; Ruel, J. Effect of drainage and microtopography in forested wetlands on the microenvironment and growth of planted black spruce seedlings. *Can. J. For. Res.* **1999**, *29*, 563–574. [CrossRef]
17. Looney, C.; D'Amato, A.; Palik, B.; Slesak, R. Overstory treatment and planting season affect survival of replacement tree species in emerald ash borer threatened *Fraxinus nigra* forests in Minnesota, USA. *Can. J. For. Res.* **2015**, *45*, 1728–1738. [CrossRef]
18. Staff, S.S. Natural Resources Conservation Service Web Soil Survey, United States Department of Agriculture. 2017. Available online: http://websoilsurvey.sc.egov.usda.gov/ (accessed on 26 April 2017).
19. Burns, R.; Honkala, B. *Silvics of North America: 1. Conifers; 2. Hardwoods*; United States Department of Agriculture: Washington, DC, USA, 1990.
20. *WinSCANOPY*, Pro Version ed; Regent Instruments Inc.: Quebec, QC, Canada, 2010.
21. R Development Core Team. *R: A Language and Environment for Statistical Computing*; R Foundation for Statistical Computing: Vienna, Austria, 2016.
22. Van Grinsven, M.; Shannon, J.; Davis, J.; Bolton, N.; Wagenbrenner, J.; Kolka, R.; Pypker, T. Source water contributions and hydrologic responses to simulated emerald ash borer infestations in depressional black ash wetlands. *Ecohydrology* **2017**, *10*, e1862. [CrossRef]
23. Londo, A.; Mroz, G. Bucket mounding as a mechanical site preparation technique in wetlands. *North. J. Appl. For.* **2001**, *18*, 7–13.

24. Van Grinsven, M. Implications of Emerald Ash Borer Disturbance on Black Ash Wetland Watershed Hydrology, Soil Carbon Efflux, and Dissolved Organic Matter. Ph.D. Thesis, Michigan Technological University, Houghton, MI, USA, 2015.

25. Slesak, R.A.; Lenhart, C.F.; Brooks, K.N.; D'Amato, A.W.; Palik, B.J. Water table response to harvesting and simulated emerald ash borer mortality in black ash wetlands in Minnesota, USA. *Can. J. For. Res.* **2014**, *44*, 961–968. [CrossRef]

26. Shannon, J.; Van Grinsven, M.; Davis, J.; Bolton, N.; Noh, N.; Pypker, T.; Kolka, R. Water level controls on sap flux of canopy species in black ash wetlands. *Forests* **2018**, accepted.

27. Williams, M.; Dumroese, R. Preparing for climate change: Forestry and assisted migration. *J. For.* **2013**, *111*, 287–297. [CrossRef]

28. Iverson, L.; Knight, K.S.; Prasad, A.; Herms, D.A.; Matthews, S.; Peters, M.; Smith, A.; Hartzler, D.M.; Long, R.; Almendinger, J. Potential species replacements for black ash (*Fraxinus nigra*) at the confluence of two threats: Emerald ash borer and a changing climate. *Ecosystems* **2016**, *19*, 248–270. [CrossRef]

29. Janowiak, M.; Iverson, L.; Mladenoff, D.; Peters, E.; Wythers, K.; Xi, W.; Brandt, L.; Butler, P.; Handler, S.; Shannon, P.; et al. *Forest Ecosystem Vulnerability Assessment and Synthesis for Northern Wisconsin and Western Upper Michigan: A Report from the Northwoods Climate Change Response Framework Project*; General Technical Report NRS-136; U.S. Department of Agriculture, Forest Service, Northern Research Station: Newtown Square, PA, USA, 2014; Volume 247.

30. Chimner, R.; Hart, J. Hydrology and microtopography effects on northern white-cedar regeneration in michigan's Upper Peninsula. *Can. J. For. Res.* **1996**, *26*, 389–393. [CrossRef]

31. Cornett, M.; Frelich, L.; Puettmann, K.; Reich, P. Conservation implications of browsing by *Odocoileus virginianus* in remnant upland *Thuja occidentalis* forests. *Biol. Conserv.* **2000**, *93*, 359–369. [CrossRef]

32. Rooney, T.; Waller, D. Direct and indirect effects of white-tailed deer in forest ecosystems. *For. Ecol. Manag.* **2003**, *181*, 165–176. [CrossRef]

33. Russell, F.; Zippin, D.; Fowler, N. Effects of white-tailed deer (*Odocoileus virginianus*) on plants, plant populations and communities: A review. *Am. Midl. Nat.* **2001**, *146*, 1–26. [CrossRef]

34. Abrams, M.D. The red maple paradox. *BioScience* **1998**, *48*, 355–364. [CrossRef]

35. Kobe, R.; Pacala, S.; Silander, J.; Canham, C. Juvenile tree survivorship as a component of shade tolerance. *Ecol. Appl.* **1995**, *5*, 517–532. [CrossRef]

36. Krajicek, J.; Williams, R. *Celtis occidentalis* L. Hackberry. *Silv. N. Am.* **1990**, *2*, 262.

forests

MDPI

Article

Water Level Controls on Sap Flux of Canopy Species in Black Ash Wetlands

Joseph Shannon [1,*], Matthew Van Grinsven [1,2], Joshua Davis [1], Nicholas Bolton [1,3], Nam Jin Noh [1,4], Thomas Pypker [5] and Randall Kolka [6]

[1] School of Forest Resources & Environmental Science, Michigan Technological University, Houghton, MI 49931, USA; mvangrin@nmu.edu (M.V.G.); joshuad@mtu.edu (J.D.); nwbolton@mtu.edu (N.B.); n.noh@westernsydney.edu.au (N.J.N.)
[2] Department of Earth, Environment, & Geosciences, Northern Michigan University, Marquette, MI 49985, USA
[3] D.B. Warnell School of Forestry & Natural Resources, University of Georgia, Athens, GA 30602, USA
[4] Hawkesbury Institute for the Environment, Western Sydney University, Richmond, NSW 2753, Australia
[5] Department of Natural Resource Science, Thompson Rivers University, Kamloops, BC V2C 0C8, Canada; tpypker@tru.ca
[6] Northern Research Station, USDA Forest Service, Grand Rapids, MN 55744, USA; rkolka@fs.fed.us
* Correspondence: jpshanno@mtu.edu; Tel.: +1-906-487-1831

Received: 21 February 2018; Accepted: 14 March 2018; Published: 16 March 2018

Abstract: Black ash (*Fraxinus nigra* Marsh.) exhibits canopy dominance in regularly inundated wetlands, suggesting advantageous adaptation. Black ash mortality due to emerald ash borer (*Agrilus planipennis* Fairmaire) will alter canopy composition and site hydrology. Retention of these forested wetlands requires understanding black ash's ecohydrologic role. Our study examined the response of sap flux to water level and atmospheric drivers in three codominant species: black ash, red maple (*Acer rubrum* L.), and yellow birch (*Betula alleghaniensis* Britt.), in depressional wetlands in western Michigan, USA. The influence of water level on sap flux rates and response to vapor pressure deficit (*VPD*) was tested among species. Black ash had significantly greater sap flux than non-black ash at all water levels (80–160% higher). Black ash showed a significant increase (45%) in sap flux rates as water levels decreased. Black ash and red maple showed significant increases in response to *VPD* as water levels decreased (112% and 56%, respectively). Exploration of alternative canopy species has focused on the survival and growth of seedlings, but our findings show important differences in water use and response to hydrologic drivers among species. Understanding how a replacement species will respond to the expected altered hydrologic regimes of black ash wetlands following EAB infestation will improve species selection.

Keywords: transpiration; *Fraxinus nigra*; ecohydrology; emerald ash borer; mitigation; water table; flooding; inundation

1. Introduction

Emerald ash borer (EAB, *Agrilus planipennis* Fairmaire), an exotic insect native to Asia and first detected in southeastern Michigan, USA, in 2002, has been spreading outward after introduction [1,2]. EAB has caused a regional trend of declining native ash (*Fraxinus* spp.) populations in the Great Lakes States observable since 2004 [3]. The loss of native ash throughout their ranges in North America has been shown to have important economic [4], cultural [5], and ecological [6] impacts. The northwestern periphery of the expanding infestation includes areas where black ash (*Fraxinus nigra* Marsh.) plays an important role on the landscape [7] (Figure 1). Black ash grows on wet sites with persistently high water tables or seasonal inundation [8]. These wet sites are classified as northern hardwood swamps in Michigan [9] and Wisconsin [10], and northern wet ash swamps and northern very wet ash swamps in Minnesota [11]. Black ash is the dominant canopy species in all of these classifications, which makes them of particular concern regarding EAB infestation. Current research is studying the anticipated effects of EAB infestation in these communities on vegetation composition [12,13], hydrology [14,15], suitable replacement species and planting strategies [16–18], and carbon and nitrogen cycling [18–20].

Hydrology is an important control on wetland ecotype and function [21], so that alteration to the hydrology of a wetland can lead to a change in wetland ecotype, or conversion to open water or mesic forest. Evapotranspiration is a major driver of hydrologic regimes; in the Great Lakes region, evapotranspiration accounts for 50–70% of annual precipitation [22]. In depressional wetlands of western Michigan, black ash can account for up to 70% of growing season canopy transpiration [15]. Therefore, disturbance of the forest canopy is likely to affect the hydrology of the wetland [23–25].

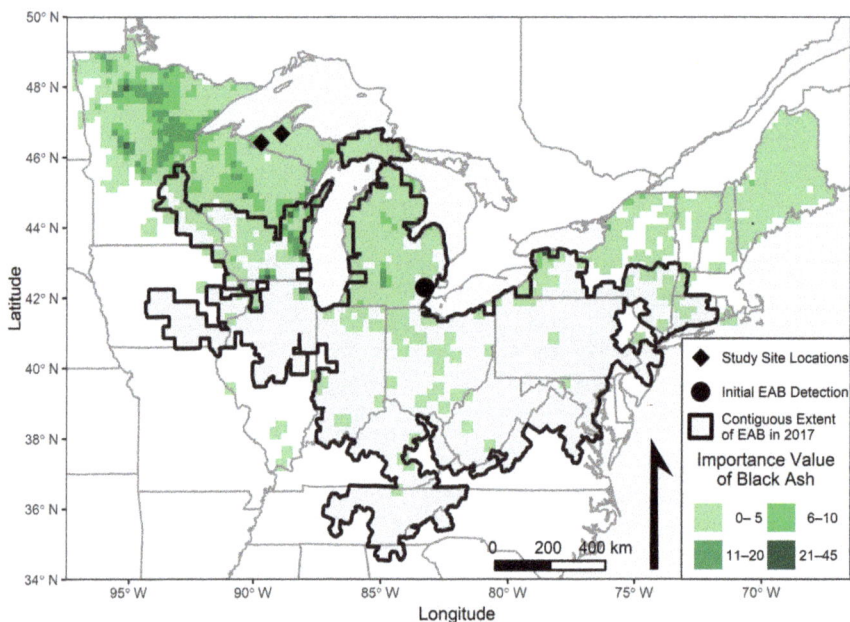

Figure 1. Extent of contiguous counties with detected emerald ash borer in 2017 [26] and the current Importance Value of black ash in the United States derived from US Forest Service Forest Inventory and Analysis data [7]. Initial Emerald ash borer (EAB) detection in Wayne County, Michigan, USA (2002) and study site locations also shown.

Recent research in black ash wetlands has shown a significant hydrologic response to the loss of black ash in the canopy following treatments aimed at simulating infestation of EAB or potential preemptive management approaches [14,15]. In both studies, water table elevation increased and/or rates of water table drawdown through the growing season were reduced following treatment when compared to unharvested control sites. These anticipated changes to the hydrology of EAB-infested black ash wetlands may result in more frequent and longer-lasting soil saturation and inundation. Slesak et al. [14] and Van Grinsven et al. [15] both present two years of post-treatment data. Longer records of the hydrologic response of these wetlands to the loss of black ash as the dominant canopy species is not available due to the relatively recent infestation of EAB in these sites.

The importance of the forest canopy in wetland function has led to studies focused on suitable replacement species for black ash with respect to the survival and growth of seedlings [16–18], but to our knowledge, none have studied the suitability of replacement species regarding the ecohydrologic function of mature individuals. The dominance of black ash as a canopy species in these wetland communities suggests adaptation to inundation and saturated soils [27]. Hypertrophied lenticels and adventitious roots have been observed on black ash (Personal Observation, [28]). Tardiff et al. [28] found that black ash regeneration could favor sexual or asexual reproduction dependent upon water levels. Green ash (*Fraxinus pennsylvanica* Marshall) has been shown to regulate stomatal openings, developing adventitious roots, hypertrophied lenticels, and aerenchyma tissues in response to inundation [29,30]. Other species, including red maple (*Acer rubrum* L.) [31], sweetgum (*Liquidambar styraciflua* L.), [32], white spruce (*Picea glauca*, (Moench) Voss) and tamarack (*Larix laricina* (Du Roi) K. Koch) [33], and American elm (*Ulmus americana* L.) [34], have shown some of these responses in seedlings in controlled settings. These adaptations as observed in seedlings may not be sufficient to sustain mature populations, as differences persist among wet-site adapted species with respect to long-term inundation and inundation during sensitive periods [35]. Mature individuals of flood-adapted species frequently continue transpiration during periods of inundation [23,32,36], but no study has examined species-level variation in transpiration under inundated conditions.

We aim to inform long-term predictions about the effects of EAB on the hydrology of black ash wetlands by examining the effect of the water table on sap flux rates and the effect of the water table on the sap flux response to atmospheric drivers. We hypothesize higher rates of sap flux in black ash than in codominants at high water levels, with the difference diminishing as water levels decrease. We also hypothesize a suppressed sap flux response to vapor pressure deficit in codominants relative to black ash as water levels increase.

2. Materials and Methods

2.1. Study Area

Six study sites were located in the Ottawa National Forest in the western Upper Peninsula of Michigan, where mean annual precipitation is 1010 mm and mean annual temperature is 4.2 °C [37]. Sites were northern hardwood swamps as described by Kost et al. [9], with the most abundant canopy species being black ash 19 m^2 ha^{-1}, yellow birch (*Betula alleghaniensis* Britt.) 3.5 m^2 ha^{-1}, and red maple 2.5 m^2 ha^{-1} [12]. The wetlands were located in landscape depressions and had an average size of 0.4 ha (range: 0.2–0.8 ha). Soils were histosols over poorly-sorted till or clay with an average depth to a confining layer of 118.8 cm. In three of the study sites, black ash stems greater than 2.54 cm diameter at breast height (DBH, 1.37 m) were girdled in the winter of 2013 to mimic an EAB infestation. Girdled black ash were not included in this study and girdling impacts on red maple and yellow birch were not addressed by this study.

2.2. Field Measures

Sap flux data were collected during the 2012, 2013, and 2014 growing seasons using 2 cm thermal heat dissipation probes [38,39] centered at DBH. To avoid potential impacts of probe age [40], new probes were installed during each growing season at a 60° offset around the circumference of the sampled stem. Probes were insulated using foam insulation and reflective wrapping to avoid the influence of external temperature gradients [41]. Voltage differentials were logged at 15-min intervals at each study site using a CR1000 datalogger and two AM 16/32B multiplexers (Campbell Scientific, Logan, UT, USA). Sample trees were selected across the range of diameter sizes observed in vegetation surveys (Table 1) [12]. Measurements across all study sites were collected on six black ash stems (at four sites), five red maple stems (at three sites), and six yellow birch stems (at four sites). Sapwood area for black ash was predicted from diameter at breast height using a linear regression fit with visually-determined sapwood area from stem cross-sections of felled ash trees. Existing empirical equations were used to calculate sapwood area for yellow birch [42] and red maple [43].

A 5 cm inner-diameter monitoring well was installed at each study site close to the wetland outlet, and a pressure transducer (Levellogger Junior M5, Solinst Canada Ltd., Georgetown, ON, Canada) recorded water levels at 15-min intervals [15]. The elevation of the root collar of each sampled stem relative to the corresponding study site monitoring well was determined by an optical survey (Leica Viva TS11 Manual Total Station, Leica Geosystems, Inc., Norcross, GA, USA).

Table 1. Number of trees instrumented for sap flux measurements along with mean diameter and diameter ranges of instrumented trees.

Species	Number of Instrumented Trees	Mean Diameter (cm)	Range of Diameters (cm)
Black Ash	6	25.58	15.00, 40.00
Red Maple	5	23.12	18.00, 34.10
Yellow Birch	6	24.62	13.00, 39.10

Daily [44] and hourly [45] meteorological data were retrieved from remote automated weather stations located in Wakefield, MI (WKFM4, UTM 16N 282871, 5146850) and Pelkie, MI (PIEM4, UTM 16N 373047, 5182028), which are within 26.1 km of our study sites. Mean daily daylight-normalized vapor pressure deficit (D_z) was calculated as the mean daily vapor pressure deficit where solar radiation was greater than or equal to 50 W m^{-2} [46,47].

2.3. Data Analysis

Sap flux densities (Q_s; m^3 m^{-2} s^{-1}) were calculated for black ash using an empirical equation derived for European ash (*Fraxinus excelsior* L.) [48], which better represents the ring-porous structure of black ash:

$$Q_s \left(\text{m}^3 \text{ m}^{-2} \text{ s}^{-1} \right) = 2.023 \left(\frac{\Delta V_m - \Delta V}{\Delta V} \right)^2 + 0.428 \left(\frac{\Delta V_m - \Delta V}{\Delta V} \right), \tag{1}$$

where Q_s is the sap flux density in m^3 m^{-2} day^{-1}, ΔV is the voltage differential across the probes, and ΔV_m is the maximum ΔV for that day.

For red maple and yellow birch, voltage differentials were converted to sap flux densities using the empirical equation from Lu ([49]), modified from Granier [38,39]:

$$Q_s \left(\text{m}^3 \text{ m}^{-2} \text{ s}^{-1} \right) = 118.9 \times 10^{-6} \times \left(\frac{\Delta V_m - \Delta V}{\Delta V} \right)^{1.231}, \tag{2}$$

where all definitions are as above.

For all species, when sapwood depth was less than the length of the probe, corrected voltage differentials were calculated according to Clearwater et al. [50]. Instantaneous sap flux density was assumed to be constant over the logging interval and daily sap flux densities (J_s; m^3 m^{-2} day^{-1}) were calculated as the sum of Q_s for each sensor. To remove the influence of phenological differences among species on measured sap flux [51], only data from June, July, and August were included in the analysis.

Daily mean water level for each well was calculated from logged 15-min data. Water level relative to the root collar of each sample tree was calculated as the monitoring well water level minus the relative elevation difference between the soil surface at the monitoring well and the root collar. For the analysis of species-level differences at high, mean, and low water levels, water levels were binned with breakpoints defined as $\mu \pm 0.5\sigma$. Breakpoints were calculated separately for each species to account for differences in relative elevation.

Response of daily sap flux to daylight-normalized vapor pressure deficit and binned mean daily water level was fit using a mixed-effects model with the lme4 package [52] in R [53,54]. Species, water table bin, and square-root transformed D_z were used as fixed effects, and each probe as the random effect:

$$J_s\left(g\ cm^{-2}\ day^{-1}\right) = Species \times WL_{bin} \times \sqrt{D_z(kPa)} + (WL_{bin} \times \sqrt{D_z(kPa)} \mid probe), \qquad (3)$$

where J_s is the cumulative daily sap flux, WL_{bin} is the binned water level, and D_z is the daylight-normalized vapor pressure deficit.

Differences in sap-flux response to D_z and water level were compared among and within species using the emmeans package [55]. For all comparisons, p-values were adjusted using the Tukey post-hoc method.

3. Results

3.1. Stem Elevations and Water Levels

In all cases, the stem root collar was at a higher elevation than the soil surface at the monitoring well. Relative root collar elevations were significantly lower in black ash (mean: 33.3 cm, standard error (se): 1.8 cm) compared with red maple (mean: 42.8 cm, se: 3.0 cm) and yellow birch (mean: 45.3 cm, se: 2.4 cm). Water levels relative to the surveyed root collar elevation ranged from −82.6 to 1.9 cm in black ash, −122.1 to −21.2 cm in red maple, and −127.0 to −18.5 cm in yellow birch (Table 2).

Table 2. Range of observed water levels (cm) for each species and water level bin. Water levels are reported as distance above (+) or below (−) the tree root collar.

Species	Minimum Observed (cm)	Low/Mean Threshold (cm)	Mean Observed (cm)	Mean/High Threshold (cm)	Maximum Observed (cm)
Black Ash	−82.6	−46.5	−37.3	−28.0	1.6
Red Maple	−122.1	−61.8	−51.7	−41.7	−21.2
Yellow Birch	−127.0	−62.0	−51.9	−41.8	−18.5

3.2. Mean Sap Flux Rates, Individual Drivers, and Interaction of Drivers

The complete model (Equation 3) had a marginal R^2 of 0.29 and a conditional R^2 of 0.80 [56], effect size and exact p-value for all pairwise comparisons can be found in Table S1. Daily sap flux rates were significantly greater in black ash than in non-black ash species at all water levels (Table 3). Within black ash, daily sap flux dropped significantly between each increasing water level bin (Table 3). There were no significant differences in daily sap flux rates among water level bins for red maple or yellow birch (Table 3). When D_z and water level are considered as separate drivers, all species showed a positive relationship between square-root transformed D_z and daily sap flux (Figure S1). Sap flux in black ash showed a negative response to increasing water levels, red maple no response to water level changes, and yellow birch a slight positive response to water level as an individual driver (Figure S1).

Table 3. Estimated marginal mean sap flux (m^3 m^{-2} day^{-1}) and standard error by species and water level bin.

Species	Low Water Level	Mean Water Level	High Water Level
Black Ash	4.00 ± 0.30 [a; 1]	3.13 ± 0.29 [a; 2]	2.76 ± 0.31 [a; 3]
Red Maple	1.63 ± 0.21 [b; 1]	1.60 ± 0.21 [b; 1]	1.49 ± 0.22 [b; 1]
Yellow Birch	1.52 ± 0.20 [b; 1]	1.56 ± 0.20 [b; 1]	1.51 ± 0.21 [b; 1]

a, b, c indicate signficance (α = 0.05) between species within a water level bin (column-wise); 1, 2, 3 indicate significance (α = 0.05) within a species across water levels (row-wise).

In black ash, the response of sap flux to D_z showed a significant reduction at high water levels compared to low and mean water levels (Figure 2, Table 4). Red maple showed a significantly stronger positive response of sap flux to D_z at lower water levels when compared to mean and high water levels. There was no significant change with water level in the response of sap flux to D_z for yellow birch (Table 4). At low and mean water levels, sap flux response to D_z was significantly greater in black ash than in yellow birch, but not greater than red maple. No difference among species was observed in response to D_z at high water levels (Table 4).

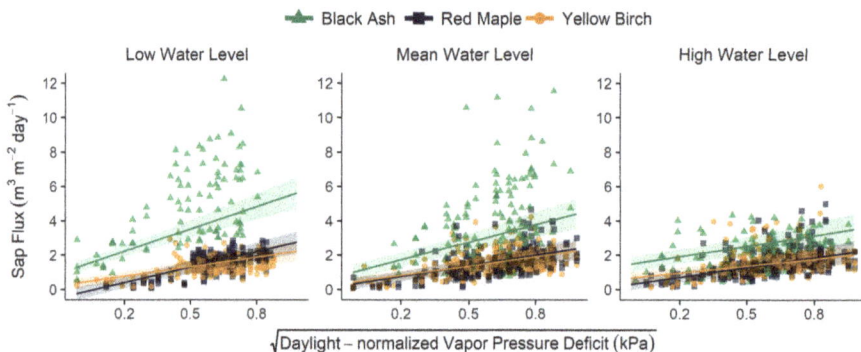

Figure 2. Observed (points) and modeled (lines) daily mean sap flux (m^3 m^{-2} day^{-1}) by square-root transformed daylight-normalized vapor-pressure deficit (kPa) within relative water level bins. Shaded regions show 95% confidence intervals. Modeled daily mean sap flux and shaded regions represent species response to D_z at the mean relative water level for each species and water level bin as fit from a single complete model.

Table 4. Marginal mean slope estimates and standard errors of sap flux (m^3 m^{-2} day^{-1}) response to square-root transformed daylight-normalized vapor pressure deficit (kPa).

Species	Low Water Level	Mean Water Level	High Water Level
Black Ash	5.23 ± 0.60 [a; 1]	4.12 ± 0.61 [a; 1]	2.46 ± 0.57 [a; 2]
Red Maple	3.60 ± 0.48 [a,b; 1]	2.47 ± 0.41 [a,b; 2]	2.30 ± 0.40 [a; 2]
Yellow Birch	2.23 ± 0.42 [b; 1]	2.17 ± 0.37 [b; 1]	1.63 ± 0.39 [a; 1]

a, b, c indicate signficance ($\alpha = 0.05$) between species within a water level bin (column-wise); 1, 2, 3 indicate significance ($\alpha = 0.05$) within a species across water levels (row-wise).

4. Discussion

4.1. Effects of Water Level on Sap Flux Rates and Response to Atmospheric Drivers

As hypothesized, we observed significantly higher sap flux rates in black ash than in codominants. However, the difference between black ash and non-black ash sap flux rates was greatest at low water levels, and lowest at high water levels, which is contrary to the second component of this hypothesis. This response was likely driven by physiological differences between black ash and non-black ash in response to low water levels or changes in water level.

This study did not examine physiological differences or physiological responses between species in these sites though adaptations have been observed in black ash and red maple (see above). Seedling studies have shown that wetland species are known to adapt in a variety of ways, including stomatal closure, hypertrophied lenticels, aerenchyma tissue formation, adventitious root growth, and reproductive resilience. In addition to species-level adaptations, the growing conditions of an individual stem may also lead to the improved tolerance of inundation. Individuals from wet-site populations, or those exposed to early inundation or continued wet soil conditions, recovered more quickly and fully from inundation treatments [57–59]. This suggests that beyond species selection for canopy replacement, stock source and early growing conditions may need to be considered to best match the current ecohydrologic function of black ash. Likely, it is a collection of these species- and individual-level adaptations that leads to black ash dominance on wet sites in the Great Lakes region and elsewhere, as Mistch and Rust [60] suggested for riparian tree growth.

Few studies have directly assessed the role of the water table on sap flux or transpiration. Using a similar study design to our own, McJannet [36] found that broad-leaved paperbark (*Melaleuca quinquenervia* (Cav.) S. T. Blake), a wetland-adapted species, showed no change in transpiration with inundation. Bald cypress (*Taxodium distichum* (L.) Rich), another species that exhibits canopy dominance in sites with regular and prolonged inundation, significantly increased sap flux during periods of inundation [61]. The results from both broad-leaved paperbark and bald cypress are in contrast to the response of black ash in this study, where a reduction in water levels caused a significant increase in sap flux rates, while inundation led to suppressed sap flux. Further research on black ash in sites less prone to inundation would help determine if the observed high sap flux rates at low water levels are the persistent state for black ash or if these rates occur only as an adaption for recovery from periods of inundation. Broad-leaved paperbark is native to eastern Australia and Oceania [8] and cannot be considered as a replacement species for black ash. The native range of bald cypress extends north to southern Illinois, USA [62], but individual cold hardiness varies, and planted individuals can survive as far north as Hayward, WI, USA [63]. No research has been conducted on bald cypress as a canopy replacement species for black ash.

We did not observe a suppressed response to atmospheric drivers in codominants relative to black ash as water levels increased, as we hypothesized. At all water levels, black ash and red maple responded similarly to D_z, and at low and mean water levels, black ash showed a significantly stronger response than yellow birch. While black ash and red maple both showed a significantly stronger positive response to D_z with decreasing water levels, the response in red maple did not lead to significantly higher sap flux rates. This suggests that red maple is adapted to respond to changing water levels, but the response is less vigorous than that of black ash.

Atmospheric conditions and energy availability are well-known drivers of sap flux [43,46,64,65]. In contrast, the results of studies on the effect of soil moisture have been more mixed, with much of the existing research focused on soil moisture deficit as a limiting factor of transpiration [43,66–68]. Specifically, within a wetland-upland transition in northern Wisconsin, Traver et al. [69] report that atmospheric drivers were more important than edaphic conditions for understanding spatial patterns of transpiration. However, significant changes in water table position, such as those observed in these systems following the removal or death of black ash [14,15], can be expected to have a greater impact than the scale of soil moisture variation often studied, as inundation can lead to the suppression of transpiration and growth, even in wetland-adapted species [27,35,70].

A study of black ash in Minnesota found that across sites with varying soil moisture regimes, mean sap flux and response to D_z were greater in sites with greater soil moisture [71]. The authors identified differences in sapwood depth and area in the black ash populations among sites, suggesting that individual adaptation or site-induced selection may play a role in the results. The design of our study, and the mixed-modeling approach in our analysis, allow us to test the effect of water level within populations of individuals rather than among populations. It is difficult to compare within-population trends to among-population trends when previous work has shown that adaptation to wet sites and inundation occurs at the species and individual levels (see above), so that systemic variations may occur at different levels.

4.2. Persistence of Hydrologic Change

The loss of black ash has been shown to lead to reduced water level drawdown during the growing season and an earlier water level rebound in the fall following senescence [15]. As a result of the inverse relationship between ash sap flux and water level, the rate of drawdown will be further reduced late in the growing season when water levels have receded from their spring peaks. The lack of an increase in sap flux in response to reduced water levels in non-ash species will result in the sustained reduction of water level drawdown, leading to earlier water level rebound in the fall. The end result will be persistently higher water levels, both intra- and inter-annually, even if a similarly stocked forest canopy of codominants becomes established. However, differences in microsite survivorship suggest that a less dense forest canopy may be expected.

Black ash consistently occupies the hollows in a hummock and hollow landscape, evidenced by black ash's significantly lower relative root collar elevations. The data presented here cannot confirm that non-black ash prefer hummocks in these sites, but observed water levels were never above non-black ash root collars. The prevalence of black ash in the hollows, paired with a significant increase in the survivorship of non-black ash seedlings planted on hummocks [18], suggest that low-lying microhabitats in these sites may not be suitable for the future growth of mature individuals of these species, precluding a significant area from potential restocking efforts. Following infestation, increased water levels will reduce the prevalence of higher elevation microsites available for seedling establishment, further increasing the potential for a less dense canopy.

Changes in the forest canopy will likely influence site hydrology in ways not examined in this study. Changes in forest canopy closure or composition can be expected to change precipitation and energy inputs to the site. As part of the ongoing research efforts to understand black ash stands and anticipated changes brought on by EAB, the impacts of simulated EAB infestation on throughfall [12] and changes in radiative energy reaching the shrub layer and vegetative ground cover have been examined. Davis et al [12] found that the loss of the ash canopy without a mature canopy replacement increased forest throughfall, though these results were confounded by high canopy heterogeneity. The effect of changes to throughfall may be masked by other inputs, as sourcewater analysis of these wetlands found groundwater contributions throughout the growing season [15]. A reduction of radiative energy reaching the forest floor following EAB-induced mortality relative to the complete removal of black ash will have the effect of suppressing the understory growth response and reducing the potential evaporative demand from the soil surface and pooled water. In this way, a non-black ash canopy with lower relative transpiration could have an even stronger negative effect on growth and evapotranspiration, further contributing to the persistence of hydrologic changes in black ash wetlands if seasonal canopy water use is less than black ash.

5. Conclusions

The loss of black ash as the dominant canopy species in wetlands in the northern Great Lakes region will have short-term and potentially longer-term hydrologic impacts. If these sites are retained as forested wetlands, either through natural regeneration or planting, the largest factor affecting the persistence of hydrologic change will be the composition and density of the replacement canopy. Some adaptation or suite of adaptations have resulted in black ash remaining historically dominant at these sites. One expression of these adaptations is an increase in sap flux with receding water levels, so that forest canopy transpiration would be reduced most at times of low water levels following the loss of ash. Reduction of forest canopy water use throughout the growing season would lead to persistently higher water levels during the growing season and a greater fall water level rebound. Increased water levels may lead to a reduction of the elevated microsites favored by current codominants, resulting in a less dense forest canopy, further reducing transpiration. A less dense forest canopy will also lead to reduced canopy interception, while still preventing additional radiative energy from reaching the forest floor and driving surface water evaporation and transpiration in the understory. These effects in combination may result in a permanent alteration of the hydrologic regime in these wetlands toward wetter conditions. Current research has focused on the survival and growth of seedlings of potential replacement species. Our research indicates that the ecohydrologic role of mature individuals may vary significantly between black ash and replacement species, regardless of seedling survival and growth. The ecohydrologic role of mature individuals should be considered when selecting a replacement to ameliorate persistent hydrologic changes. Further study of mature individuals of potential replacement species in high water table conditions is needed to inform these decisions.

Supplementary Materials: All data and code used in this analysis is available through the Knowledge Network for Biocomplexity repository [72]. The following are available online at www.mdpi.com/1999-4907/9/3/147/s1, Table S1: Effect size, reported as Cohen's d and exact Tukey-adjusted *p*-value (in parentheses) of pairwise comparisons among species within water level bins and among water level bins within species of estimated marginal means and slopes. Figure S1: Daily mean sap flux of individual probes by (A) daylight-normalized vapor-pressure deficit (D_z) and (B) mean daily water level. Linear models with 95% prediction intervals fit individually for each driver.

Acknowledgments: Funding was provided through the Great Lakes Restoration Initiative through the USDA Forest Service Northern Research Station (EPA Great Lakes Initiative Template #664: Future of Black Ash Wetlands in the Great Lakes Region). Additional funds came from the Michigan Technological University School of Forest Resources and Environmental Science, Ecosystem Science Center and the Center for Water and Society at Michigan Technological University. The USFS Ottawa National Forest provided important information regarding potential field site locations. We would like to thank Jon Bontrager, Leah Harrison, Daniel Hutchinson, Erica Jones, Jarrod Nelson, and Nicholas Schriener for their contributions to collecting these data.

Author Contributions: J.S., M.V.G., J.D., T.P., and R.K conceived and designed the experiments; J.S., M.V.G., J.D., and N.B. performed the experiments; J.S., M.V.G., and J.D. analyzed the data; J.S. wrote the paper with contributions from M.V.G., J.D., N.B., N.J.N., T.P., and R.K.

Conflicts of Interest: The authors declare no conflicts of interest.

References

1. Haack, R.; Jendek, E.; Liu, H.; Marchant, K. The emerald ash borer: A new exotic pest in North America. *Newsl. Mich. Entomol. Soc.* **2002**, *47*, 1–5. [CrossRef]

2. Herms, D.A.; McCullough, D.G. Emerald Ash Borer Invasion of North America: History, Biology, Ecology, Impacts, and Management. *Annu. Rev. Entomol.* **2014**, *59*, 13–30. [CrossRef] [PubMed]

3. Pugh, S.A.; Liebhold, A.M.; Morin, R.S. Changes in ash tree demography associated with emerald ash borer invasion, indicated by regional forest inventory data from the Great Lakes States. *Can. J. For. Res.* **2011**, *41*, 2165–2175. [CrossRef]

4. Aukema, J.E.; Leung, B.; Kovacs, K.; Chivers, C.; Britton, K.O.; Englin, J.; Frankel, S.J.; Haight, R.G.; Holmes, T.P.; Liebhold, A.M.; et al. Economic impacts of Non-Native forest insects in the continental United States. *PLOS ONE* **2011**, *6*, e24587. [CrossRef] [PubMed]

5. Willow, A.J. Indigenizing Invasive Species Management: Native North Americans and the Emerald Ash Borer (EAB) Beetle. *Cult. Agric. Food Environ.* **2011**, *33*, 70–82. [CrossRef]

6. Gandhi, K.J.K.; Herms, D.A. Direct and indirect effects of alien insect herbivores on ecological processes and interactions in forests of eastern North America. *Biol. Invasions* **2010**, *12*, 389–405. [CrossRef]

7. Prasad, A.M.; Iverson, L.R. *Little's Range and FIA Importance Value Database for 135 Eastern US Tree Species*; Northeastern Research Station, USDA Forest Service: Delaware, OH, USA, 2003; Available online: http://www.fs.fed.us/ne/delaware/4153/global/littlefia/index.html (accessed on 22 April 2016).

8. Burns, R.M.; Honkala, B.H. *Silvics of North America*; Agriculture Handbook 654; Volume 2: Hardwoods; United States Department of Agriculture (USDA), Forest Service: Washington, DC, USA, 1990.

9. Kost, M.A.; Albert, D.A.; Cohen, J.G.; Slaughter, B.S.; Schillo, R.K.; Weber, C.R.; Chapman, K.A. *Natural Communities of Michigan: Classification and Description*; Michigan Natural Features Inventory: Lansing, MI, USA, 2007.

10. Epstein, E.; Judziewicz, E.; Spencer, E. *Wisconsin Natural Heritage Inventory (NHI) Recognized Natural Communities—Working Document*; Wisconsin Natural Heritage Inventory: Madison, WI, USA, 2002.

11. Minnesota Department of Natural Resources. *Field Guide to the Native Plant Communities of Minnesota: The Prairie Parkland and Tallgrass Aspen Parklands Provinces*; Ecological Land Classification Program, Minnesota County Biological Survey, Natural Heritage, Nongame Research Program: St. Paul, MN, USA, 2005.

12. Davis, J.C.; Shannon, J.P.; Bolton, N.W.; Kolka, R.K.; Pypker, T.G. Vegetation responses to simulated emerald ash borer infestation in *Fraxinus-nigra* dominated wetlands of Upper Michigan, USA. *Can. J. For. Res.* **2017**, *47*, 319–330. [CrossRef]

13. Looney, C.E.; D'Amato, A.W.; Palik, B.J.; Slesak, R.A.; Slater, M.A. The response of Fraxinus nigra forest ground-layer vegetation to emulated emerald ash borer mortality and management strategies in northern Minnesota, USA. *For. Ecol. Manag.* **2017**, *389*, 352–363. [CrossRef]

14. Slesak, R.A.; Lenhart, C.F.; Brooks, K.N.; D'Amato, A.W.; Palik, B.J. Water table response to harvesting and simulated emerald ash borer mortality in black ash wetlands in Minnesota, USA. *Can. J. For. Res.* **2014**, *44*, 961–968. [CrossRef]

15. Van Grinsven, M.J.; Shannon, J.P.; Davis, J.C.; Bolton, N.W.; Wagenbrenner, J.W.; Kolka, R.K.; Pypker, T.G. Source water contributions and hydrologic responses to simulated emerald ash borer infestations in depressional black ash wetlands. *Ecohydrology* **2017**, *10*, e1862. [CrossRef]

16. Looney, C.E.; D'Amato, A.W.; Palik, B.J.; Slesak, R.A. Overstory treatment and planting season affect survival of replacement tree species in emerald ash borer threatened *Fraxinus nigra* forests in Minnesota, USA. *Can. J. For. Res.* **2015**, *45*, 1728–1738. [CrossRef]

17. Looney, C.E.; Amato, A.W.D.; Palik, B.J.; Slesak, R.A. Canopy treatment influences growth of replacement tree species in *Fraxinus nigra* forests threatened by the emerald ash borer in Minnesota, USA. *Can. J. For. Res.* **2017**, *192*, 183–192. [CrossRef]

18. Bolton, N.; Shannon, J.; Davis, J.; Van Grinsven, M.; Noh, N.J.; Schooler, S.; Kolka, R.; Pypker, T.; Wagenbrenner, J. Controls on alternative species seedlings survival and growth in black ash ecosystems threatened by emerald ash borer. *Forests* **2018**. in review.

19. Noh, N.J.; Shannon, J.P.; Bolton, N.W.; Davis, J.C.; Van Grinsven, M.J.; Pypker, T.G.; Kolka, R.K.; Wagenbrenner, J.W. Carbon dioxide fluxes from coarse dead wood in a black ash wetland. *Forests* **2018**, in review.

20. Van Grinsven, M.; Shannon, J.; Bolton, N.; Davis, J.; Wagenbrenner, J.; Kolka, R.; Pypker, T. Gaseous soil-carbon flux responses to simulated emerald ash borer infestations in depressional black ash wetlands. *Forests* **2018**. in review.

21. Brinson, M.M. *A Hydrogeomorphic Classification for Wetlands*; Wetlands Research Program Technical Report WRP-DE-4; East Carolina University: Greenville, NC, USA, 1993. [CrossRef]

22. Sanford, W.E.; Selnick, D.L. Estimation of Evapotranspiration Across the Conterminous United States Using a Regression With Climate and Land-Cover Data. *JAWRA J. Am. Water Resour. Assoc.* **2013**, *49*, 217–230. [CrossRef]

23. Oren, R.; Phillips, N.; Ewers, B.E.; Pataki, D.E.; Megonigal, J.P. Sap-flux-scaled transpiration responses to light, vapor pressure deficit, and leaf area reduction in a flooded *Taxodium distichum* forest. *Tree Physiol.* **1999**, *19*, 337–347. [CrossRef] [PubMed]

24. Sun, G.; McNulty, S.G.; Shepard, J.P.; Amatya, D.M.; Riekerk, H.; Comerford, N.B.; Skaggs, W.; Swift, L. Effects of timber management on the hydrology of wetland forests in the southern United States. *For. Ecol. Manag.* **2001**, *143*, 227–236. [CrossRef]

25. Sebestyen, S.D.; Verry, E.S.; Brooks, K.N. Hydrological responses to changes in forest cover on uplands and peatlands. In *Peatland biogeochemistry and watershed hydrology at the Marcell Experimental Forest*; Kolka, R.K., Sebestyen, S.D., Verry, E.S., Brooks, K.N., Eds.; CRC Press: Boca Raton, FL, USA, 2011; pp. 401–432.

26. *EDDMapS Early Detection & Distribution Mapping System*; The University of Georgia—Center for Invasive Species and Ecosystem Health: Athens, GA, USA, 2017.

27. Kozlowski, T.T.; Pallardy, S.G. Acclimation and Adaptive Responses of Woody Plants to Environmental Stresses Acclimation and Adaptive Responses of Woody Plants to Environmental Stresses. *Bot. Rev.* **2014**, *68*, 270–334. [CrossRef]

28. Tardif, J.; Dery, S.; Bergeron, Y. Sexual Regeneration of Black Ash (*Fraxinus nigra* Marsh.) in a Boreal Floodplain. *Am. Midl. Nat.* **1994**, *132*, 124–135. [CrossRef]

29. Kozlowski, T.T.; Pallardy, S.G. Stomatal Responses of *Fraxinus pennsylvanica* Seedlings during and after Flooding. *Physiol. Plant.* **1979**, *46*, 155–158. [CrossRef]

30. Gomes, A.R.S.; Kozlowski, T.T. Growth responses and adaptations of *Fraxinus pennsylvanica* seedlings to flooding. *Plant Physiol.* **1980**, *66*, 267–271. [CrossRef] [PubMed]

31. Will, R.E.; Seiler, J.R.; Feret, P.P.; Aust, W.M. Effects of Rhizosphere Inundation on the Growth and Physiology of Wet and Dry-Site *Acer-Rubrum* (Red Maple) Populations. *Am. Midl. Nat.* **1995**, *134*, 127–139. [CrossRef]

32. Pezeshki, S.R.; Chambers, J.L. Stomatal and photosynthetic response of sweet gum (*Liquidambar styraciflua*) to flooding. *Can. J. For. Res.* **1985**, *15*, 371–375. [CrossRef]

33. Reece, C.F.; Riha, S.J. Role of root systems of eastern larch and white spruce in response to flooding. *Plant Cell Environ.* **1991**, *14*, 229–234. [CrossRef]

34. Angeles, G.; Evert, R.F.; Kozlowski, T.T. Development of lenticels and adventitious roots in flooded *Ulmus americana* seedlings. *Can. J. For. Res.* **1986**, *16*, 585–590. [CrossRef]

35. Angelov, M.N.; Sung, S.-J.J.S.; Doong, R.L.; Harms, W.R.; Kormanik, P.P.; Black, C.C.; Black, C.C., Jr. Long- and short-term flooding effects on survival and sink-source relationships of swamp-adapted tree species. *Tree Physiol.* **1996**, *16*, 477–484. [CrossRef] [PubMed]

36. McJannet, D. Water table and transpiration dynamics in a seasonally inundated *Melaleuca quinquenervia* forest, north Queensland, Australia. *Hydrol. Process.* **2008**, *22*, 3079–3090. [CrossRef]

37. Arguze, A.; Durre, I.; Applequist, S.; Squires, M.; Vose, R.; Yin, X.; Bilotta, R. NOAA's 1981–2010 U.S. Climate Normals: An Overview. *Bull. Am. Meteorol. Soc.* **2012**, *93*, 1687–1697.

38. Granier, A. Une nouvelle methode pour la mesure du flux de seve brute dans le tronc des arbres. *Ann. Sci. Forestieres (Fr.)* **1985**, *42*, 193–200. [CrossRef]

39. Granier, A. Evaluation of transpiration in a Douglas-fir stand by means of sap flow measurements. *Tree Physiol.* **1987**, *3*, 309–320. [CrossRef] [PubMed]

40. Moore, G.W.; Bond, B.J.; Jones, J.A.; Meinzer, F.C. Thermal-dissipation sap flow sensors may not yield consistent sap-flux estimates over multiple years. *Trees* **2009**, *24*, 165–174. [CrossRef]

41. Lu, P.; Urban, L.; Zhao, P. Granier's thermal dissipation probe (TDP) method for measuring sap flow in trees: Theory and practice. *Acta Bot. Sin.* **2004**, *46*, 631–646.

42. Tang, J.; Bolstad, P.V.; Ewers, B.E.; Desai, A.R.; Davis, K.J.; Carey, E.V. Sap flux-upscaled canopy transpiration, stomatal conductance, and water use efficiency in an old growth forest in the Great Lakes region of the United States. *J. Geophys. Res. Biogeosci.* **2006**, *111*, 1–12. [CrossRef]

43. Bovard, B.D.; Curtis, P.S.; Vogel, C.S.; Schmid, H.P. Environmental controls on sap flow in a northern hardwood forest. *Tree Physiol.* **2005**, *25*, 31–38. [CrossRef] [PubMed]

44. Western Regional Climate Center. *RAWS USA Climate Archive*; Desert Research Institute: Reno, NV, USA, 2013.

45. Horel, J.; Splitt, M.; Dunn, L.; Pechmann, J.; White, B.; Ciliberti, C.; Lazarus, S.; Slmmer, J.; Zaff, D.; Burks, J. Mesowest: Cooperative Mesonets in the Western United States. *Bull. Am. Meteorol. Soc.* **2002**, *83*, 211–225. [CrossRef]

46. Oren, R.; Zimmermann, R.; Terbough, J. Transpiration in upper Amazonia floodplain and upland forests in response to drought-breaking rains. *Ecology* **1996**, *77*, 968–973. [CrossRef]

47. Phillips, N.; Oren, R. Intra- and Inter-Annual Variations in Transpiration of a Pine Forest. *Ecol. Appl.* **2001**, *11*, 385–396. [CrossRef]

48. Herbst, M.; Rosier, P.T.W.; Roberts, J.M.; Taylor, M.E.; Gowing, D.J. Edge effects and forest water use: A field study in a mixed deciduous woodland. *For. Ecol. Manag.* **2007**, 176–186. [CrossRef]

49. Lu, P. A direct method for estimating the average sap flux density using a modified Granier measuring system. *Aust. J. Plant Physiol.* **1997**, *24*, 701–705. [CrossRef]

50. Clearwater, M.J.; Meinzer, F.C.; Andrade, J.L.; Goldstein, G.; Holbrook, N.M. Potential errors in measurement of nonuniform sap flow using heat dissipation probes. *Tree Physiol.* **1999**, 681–688. [CrossRef]

51. Lechowicz, M.J. Why Do Temperate Deciduous Trees Leaf Out at Different Times? Adaptation and Ecology of Forest Communities. *Am. Nat.* **1984**, *124*, 821–842. [CrossRef]

52. Bates, D.; Mächler, M.; Bolker, B.; Walker, S. Fitting Linear Mixed-Effects Models Using lme4. *J. Stat. Softw.* **2015**, *67*, 1–48. [CrossRef]

53. R Core Team. *R: A Language and Environment for Statistical Computing, Version 3.4.3*; R Foundation for Statistical Computing: Vienna, Austria, 2017.

54. RStudio Team. *RStudio: Integrated Development Environment for R*; RStudio, Inc.: Boston, MA, USA, 2012.

55. Lenth, R.V. emmeans: Estimated Marginal Means, Aka Least-Squares Means. R Package Version 1.1.2. 2018. Available online: https://CRAN.R-project.org/package=emmeans (accessed on 22 April 2017).

56. Kamil, B. MuMIn: Multi-Model Inference. R Package Version 1.40.4 2018. Available online: https://CRAN.R-project.org/package=MuMIn (accessed on 22 April 2017).

57. Anella, L.B.; Whitlow, T.H. Photosynthetic Response to Flooding of *Acer rubrum* Seedlings from Wet and Dry Sites. *Am. Midl. Nat.* **2000**, *143*, 330–341. [CrossRef]

58. Yan, X.L.; Xi, B.Y.; Jia, L.M.; Li, G.D. Response of sap flow to flooding in plantations of irrigated and non-irrigated triploid poplar. *J. For. Res.* **2015**, *20*, 375–385. [CrossRef]

59. Wang, S.; Callaway, R.M.; Zhou, D.W.; Weiner, J. Experience of inundation or drought alters the responses of plants to subsequent water conditions. *J. Ecol.* **2017**, *105*, 176–187. [CrossRef]

60. Mitsch, W.J.; Rust, W.G. Tree Growth Responses to Flooding in a Bottomland in Northeastern Illinois. *For. Sci.* **1984**, *30*, 499–510.

61. Duberstein, J.A.; Krauss, K.W.; Conner, W.H.; Bridges, W.C.; Shelburne, V.B. Do hummocks provide a physiological advantage to even the most flood tolerant of tidal freshwater trees? *Wetlands* **2013**, *33*, 399–408. [CrossRef]

62. Burns, R.M.; Honkala, B.H. (Eds.) *Silvics of North America*; Agriculture Handbook 654; Volume 1: Conifers; United States Department of Agriculture (USDA), Forest Service: Washington, DC, USA, 1990.

63. Heim, M. Personal communication, 2017.

64. Ewers, B.E.; Mackay, D.S.; Gower, S.T.; Ahl, D.E.; Burrows, S.N.; Samanta, S.S. Tree species effects on stand transpiration in northern Wisconsin. *Water Resour. Res.* **2002**, *38*, 1–11. [CrossRef]

65. Adelman, J.D.; Ewers, B.E.; MacKay, D.S. Use of temporal patterns in vapor pressure deficit to explain spatial autocorrelation dynamics in tree transpiration. *Tree Physiol.* **2008**, *28*, 647–658. [CrossRef] [PubMed]

66. Oren, R.; Pataki, D.E. Transpiration in response to variation in microclimate and soil moisture in southeastern deciduous forests. *Oecologia* **2001**, *127*, 549–559. [CrossRef] [PubMed]

67. Ford, C.R.; Goranson, C.E.; Mitchell, R.J.; Will, R.E.; Teskey, R.O. Modeling canopy transpiration using time series analysis: A case study illustrating the effect of soil moisture deficit on *Pinus taeda*. *Agric. For. Meteorol.* **2005**, *130*, 163–175. [CrossRef]

68. McLaren, J.D.; Arain, M.A.; Khomik, M.; Peichl, M.; Brodeur, J. Water flux components and soil water-atmospheric controls in a temperate pine forest growing in a well-drained sandy soil. *J. Geophys. Res. Biogeosci.* **2008**, *113*. [CrossRef]

69. Traver, E.; Ewers, B.E.; Mackay, D.S.; Loranty, M.M. Tree transpiration varies spatially in response to atmospheric but not edaphic conditions. *Funct. Ecol.* **2010**, *24*, 273–282. [CrossRef]

70. Pezeshki, S.R.; Anderson, P.H. Responses of three bottomland species with different flood tolerance capabilities to various flooding regimes. *Wetl. Ecol. Manag.* **1996**, *4*, 245–256. [CrossRef]

71. Telander, A.C.; Slesak, R.A.; D'Amato, A.W.; Palik, B.J.; Brooks, K.N.; Lenhart, C.F. Sap flow of black ash in wetland forests of northern Minnesota, USA: Hydrologic implications of tree mortality due to emerald ash borer. *Agric. For. Meteorol.* **2015**, *206*, 4–11. [CrossRef]

72. Shannon, J.; Van Grinsven, M. Sap Flux and Water Levels for Black Ash Wetlands in western Michigan, USA from 2012 to 2014. *Knowl. Netw. Biocomplex.* **2018**. [CrossRef]

forests

MDPI

Article

Review of Ecosystem Level Impacts of Emerald Ash Borer on Black Ash Wetlands: What Does the Future Hold?

Randall K. Kolka [1,*], Anthony W. D'Amato [2], Joseph W. Wagenbrenner [3], Robert A. Slesak [4], Thomas G. Pypker [5], Melissa B. Youngquist [6], Alexis R. Grinde [7] and Brian J. Palik [1]

[1] USDA Forest Service Northern Research Station, Grand Rapids, MN 55744, USA; bpalik@fs.fed.us
[2] Rubenstein School of Environment and Natural Resources, University of Vermont, Burlington, VT 05405, USA; awdamato@uvm.edu
[3] USDA Forest Service Pacific Southwest Research Station, Arcata, CA 95521, USA; jwagenbrenner@fs.fed.us
[4] Minnesota Forest Resources Council, St. Paul, MN 55108, USA; raslesak@umn.edu
[5] Thompson Rivers University, Kamloops, BC V2C 0C8, Canada; tpypker@tru.ca
[6] Department of Forest Resources, University of Minnesota, St. Paul, MN 55108, USA; myoungqu@umn.edu
[7] Natural Resources Research Institute, University of Minnesota Duluth, Duluth, MN 55811, USA; agrinde@d.umn.edu
* Correspondence: rkolka@fs.fed.us; Tel.: +218-326-7115

Received: 5 March 2018; Accepted: 29 March 2018; Published: 2 April 2018

Abstract: The emerald ash borer (EAB) is rapidly spreading throughout eastern North America and devastating ecosystems where ash is a component tree. This rapid and sustained loss of ash trees has already resulted in ecological impacts on both terrestrial and aquatic ecosystems and is projected to be even more severe as EAB invades black ash-dominated wetlands of the western Great Lakes region. Using two companion studies that are simulating short- and long-term EAB infestations and what is known from the literature, we synthesize our current limited understanding and predict anticipated future impacts of EAB on black ash wetlands. A key response to the die-back of mature black ash will be higher water tables and the potential for flooding and resulting changes to both the vegetation and animal communities. Although seedling planting studies have shown some possible replacement species, little is known about how the removal of black ash from the canopy will affect non-ash species growth and regeneration. Because black ash litter is relatively high in nitrogen, it is expected that there will be important changes in nutrient and carbon cycling and subsequent rates of productivity and decomposition. Changes in hydrology and nutrient and carbon cycling will have cascading effects on the biological community which have been scarcely studied. Research to address these important gaps is currently underway and should lead to alternatives to mitigate the effects of EAB on black ash wetland forests and develop management options pre- and post-EAB invasion.

Keywords: review; hydrology; carbon; nutrients; wildlife; soil

1. Introduction

The emerald ash borer (EAB; *Agrilus planipennis* Fairmaire) is poised to decimate ash forests throughout the Great Lakes states, threatening the future of the black ash (*Fraxinus nigra* Marsh)-dominated wetland forests that occur throughout the region. Female EAB lay eggs in crevices of the black ash bark. The eggs hatch in about 2 weeks and the larvae feed on the inner phloem, cambium, and outer xylem for one to two years, girdling and subsequently killing the tree [1]. The geographic distribution of black ash includes the northeastern U.S. and the western Great Lakes (Figure 1), but the majority of black ash wetlands cover large portions of Michigan (238,000 ha), Wisconsin

(350,000 ha), and Minnesota (430,000 ha). These three states have an estimated 2.1 billion ash trees that are susceptible to EAB [2].

Black ash wetland ecosystems are an integral part of the landscape that are generally formed on poorly drained, relatively nutrient rich soils with high water tables. Soils can either be wet mineral soils or organic soils. Water tables are generally drawn down over the growing season and rebound in autumn following leaf fall. Water table elevation is an important control on black ash growth and mortality. Palik et al. [3,4] postulated that spring droughts, excessive moisture, and hydrological alterations by roads contributed to black ash crown dieback. Black ash generally dominate the canopy of these wetlands and ash density can range from approximately 40% to almost 100%. Other co-dominant trees tend to be northern white cedar (*Thuja occidentalis* L.), red maple (*Acer rubrum* L.), and American elm (*Ulmus americana* L.)—all trees that can normally survive in inundated conditions but do not thrive as well as black ash because of physiological adaptations or pathogenic constraints (e.g., Dutch elm disease).

Hydrology is the dominant factor that influences a host of ecosystem functions in black ash wetlands. Common hydrogeomorphic settings for black ash wetlands include expansive wetland complexes with perched water tables, swales associated with drainage of small catchments, central depressional areas with drainage from surrounding uplands, and linear transitional areas that form between uplands and peatlands [5]. Source inputs are predominantly from precipitation, either directly [6] or from adjacent upland drainage [7]. In most settings, water tables are typically above the surface throughout early spring, followed by drawdown below the surface during the growing season with periodic rises following rain events. Water table drawdown coincides with peak evapotranspiration following black ash leaf out [7,8], highlighting the fundamental control that this species has on animal and other plant communities.

There have been relatively few studies focused on the impacts of EAB on black ash ecosystems, largely because the spread of this invasive insect has yet to reach the large concentrations of black ash forests in the western Great Lakes region. Black ash has been shown to have similar susceptibility as green ash (*Fraxinus pennsylvanica* Marsh.) and white ash (*Fraxinus americana* L.) to EAB [9]. The discovery of EAB in St. Paul, Minnesota, in 2009, a location 100 km south of the largest concentration of black ash forests in the US, led to the establishment of a large-scale, manipulative study in north-central MN on the Chippewa National Forest focused on anticipating the impacts of EAB on black ash wetlands. This work utilized girdling treatments to simulate EAB impacts and also included an evaluation of adaptive management strategies to increase the resilience of black ash forests to EAB, such as planting non-host species as part of even-aged (clearcutting) and uneven-aged (group selection) silvicultural systems [10]. In 2010, a companion study building off this work was established in the western Upper Peninsula of Michigan on the Ottawa National Forest where the girdling treatment, a clearcut treatment, and some of the same measurements were conducted.

The form and size of black ash wetlands differ in the two locations. In Minnesota, the wetlands form in extensive areas of inundation, typically covering 10 to 150 ha, whereas in Michigan the wetlands are on the order of 1 to 2 ha that occur in depressional areas formed by glaciation. Both studies included a girdling treatment to simulate the short-term impact of EAB and a felling or harvesting treatment. The harvesting treatments in the Chippewa National Forest Minnesota study were designed to include operational harvests as part of adaptive management strategies. The Ottawa National Forest Michigan study included a harvest treatment in which all ash were felled and left in place (ash-cut) to simulate longer-term impacts of EAB infestation. In addition, the smaller size of the wetlands in the Michigan study enabled the use of a paired watershed experiment assessing the impact of the ash-cut treatment on watershed scale water flux and chemistry.

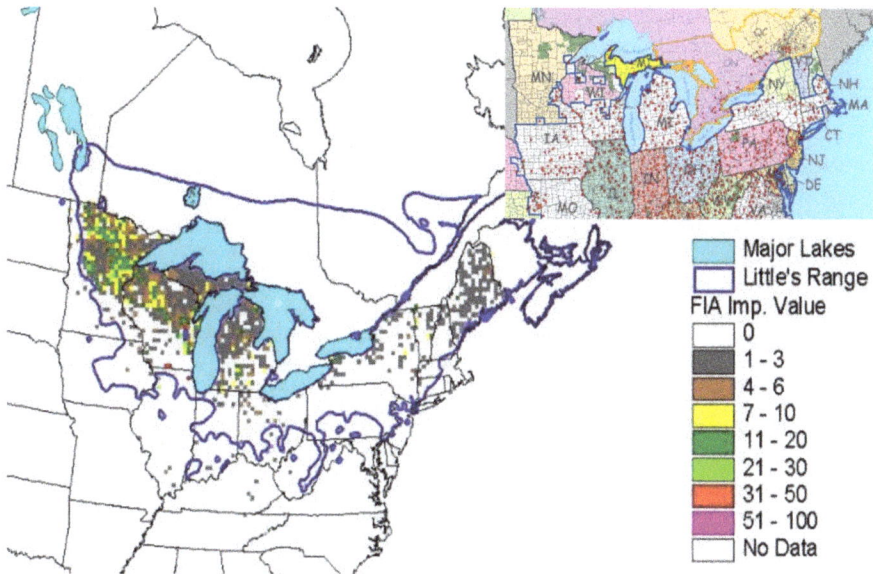

Figure 1. Geographic distribution of black ash in the United States from Forest Inventory and Analysis databases. Little's range is the geographic range of black ash based on botanical lists, forest surveys, field notes, and herbarium specimens [11]. The Forest Inventory and Analysis importance value is a measure of the abundance of a species within a stand. An importance value of 50 indicates that the stand is 50% black ash and 50% other species. Inset is the geographic distribution of emerald ash borer detected in counties as of 1 March 2018 [12].

Here, we synthesize what is known about ecosystem level impacts of EAB on black ash wetlands based, primarily, on these two study systems. Because of the adaptation of black ash to difficult growing conditions and the eradication of pole-sized and larger ash after EAB infestation, EAB impacts on black ash wetlands will likely be extreme and result in dramatic changes in hydrology and cascading effects on the plant and animal communities and nutrient cycles that influence them. Ultimately, changes in plant and animal communities will alter the carbon balance in these ecosystems. Because the structure and function of these forested wetland ecosystems rely on healthy populations of black ash, and because these systems are in imminent threat from EAB, understanding the potential ecosystem level consequences of EAB is necessary and timely. We focus our review on the western Great Lakes states because of the prevalence of black ash wetlands in this region. This review includes papers in this special issue of Forests and previously published studies that utilize either observations of post-invasion impacts or experimental simulation of potential impacts. We consider three major types of impacts on black ash wetland functions: (1) hydrological responses; (2) changes in nitrogen (N) and carbon (C) cycles; and (3) plant and animal responses.

2. Hydrological Responses of Black Ash Wetlands to EAB

Mature black ash trees can transpire up to 63 L day^{-1} during the growing season [13]. When taking into account the area of the sapwood and the basal area of the stand, mean growing season sap flux density in Minnesota ranges from 1.6 m^3 m^{-2} day^{-1} in a moderately wet site to 4.6 m^3 m^{-2} day^{-1} in a very wet site, with vapor pressure deficit and soil moisture conditions being the main controls on sap flux [13]. Sap flux density and vapor pressure deficit were positively related when soil moisture was high (full relative saturation) but negatively related when soil moisture was low (approximately 40% of relative saturation) [13]. In contrast to the results from Minnesota, mean growing season sap flux density in the Michigan study was lower in the sites with highest water tables (2.5 m^3 m^{-2} day^{-1}) as compared to the flux from the lowest water tables sites (5.2 m^3 m^{-2} day^{-1}) [14]. Although the results seem to be conflicting, the low water table (dry) conditions in upper Michigan were comparable to the high water table (very wet) site in Minnesota. For the Minnesota sites, water table depths reached nearly 100 cm below the soil surface in mid-summer during the year of the study (2012), whereas at the upper Michigan sites the minimum water table over three years of study (2012–2014) was 83 cm below the soil surface in mid-summer. There appears to be an optimum range of water table elevations where sap flux from black ash is maximized, with that range approximately 45–100 cm below the soil surface. Wetlands with mean water table levels within this range may be most susceptible to hydrologic alteration and increases in hydroperiod following ash loss.

In addition to sap flux in black ash trees, Shannon et al. [14] also measured sap flux in co-dominant species of red maple and yellow birch (*Betula alleghaniensis* Britton). Black ash had significantly higher sap flux rates (2.5–5.2 m^3 m^{-2} day^{-1}) than non-ash species (red maple and yellow birch ranged from 1.6–3.6 m^3 m^{-2} day^{-1}) in these wetland sites at all water table levels and the difference was greatest at the lowest water table levels. The high sap flux densities in black ash coupled with low sap flux among the codominant species should lead to changes in water table dynamics after EAB invasion. For the Minnesota sites, that response was seen in the first year following clearcutting, with significantly higher water tables in the clearcut treatment than the unharvested control treatment (~20–60 cm higher water table elevation) (Figure 2; [8]). During the second year, the girdling impact was evident, with a similar water table response as the clearcut treatment. Interestingly, the gap harvest treatment where about 20% of the basal area was removed led to no change in the water table response relative to the unharvested control (Figure 2). The result is consistent with past studies on hydrologic responses to forest harvest which found no change in watershed output with canopy reductions less than about 20% [15].

In contrast to the Minnesota sites, the upper Michigan sites did not see a significant increase in the water table following the harvest treatment or girdling after the first two years (Figure 3; [7]). However, the two years post-treatment were abnormally wet with greater than average snowfall and rainfall, which may have masked any treatment effects. Although mean water table levels did not vary among treatments, growing seasons rates of water table drawdown were statistically higher for the unharvested control than the ash-cut or girdle treatments [7]. Differences in landscape position may have also contributed to the different effects on water table elevation, whereby subsurface inflow from surrounding uplands would have muted the effect of reduced transpiration and local water table increases following ash loss in the Michigan sites set in localized depressions. Combined, the data from both water level studies suggest that water tables will increase following EAB invasion during normal or below normal precipitation years, with slower drawdowns during the growing season.

From a water balance perspective, some of the excess water from the lower post-EAB transpiration rates (~0.10 cm day^{-1}, [7]) will likely become part of the evaporation pool or be compensated by transpiration from replacement vegetation. However, we would still expect the overall evapotranspiration to be considerably less following EAB invasion, leading to consistently higher water tables throughout the year until a tree canopy is reestablished.

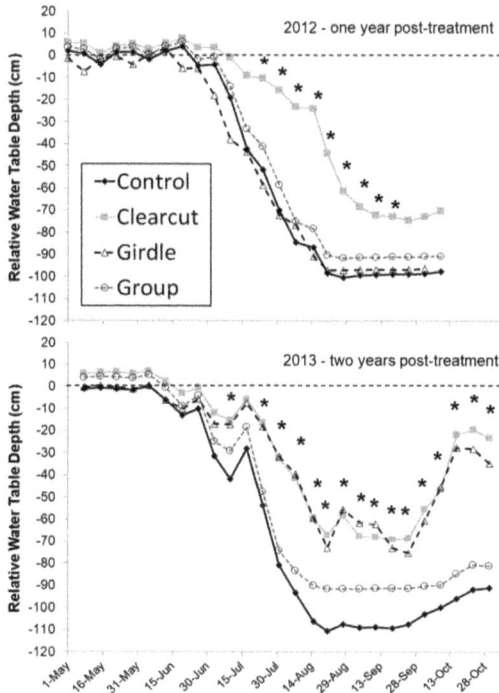

Figure 2. Weekly mean (n = 6) water table depth relative to ground surface by treatment from the Minnesota sites. The control treatment comprised unharvested reference sites, the clearcut treatment was a commercial harvest, trees were girdled in the girdled treatment, and group treatment was a group selection harvest creating gaps of approximately 0.04 ha. Each year's growing season was analyzed separately; weeks with an * indicate significant differences among treatments ($p < 0.05$). There were no differences among treatments prior to treatment application ($p > 0.2$) (reproduced from Slesak et al. [8]).

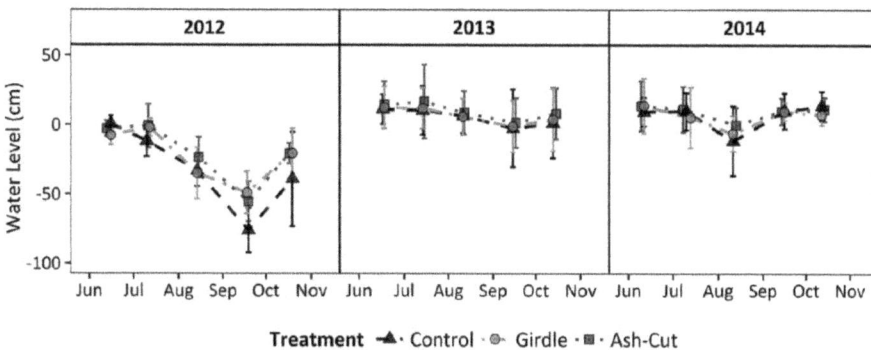

Figure 3. Mean wetland water table levels relative to the wetland soil surface, with 95% confidence intervals, during the snow-free period for the upper Michigan sites (n = 3). The year 2012 was the pre-treatment year and 2013–2014 were the first two post-treatment years (reprinted from Van Grinsven et al. [7]). The control treatment comprised unharvested reference sites, all ash trees were girdled in the girdle treatment, and all ash were cut and left on the ground in the ash-cut treatment.

3. Changes in Nitrogen (N) and Carbon (C) Cycling Resulting from EAB Invasion in Black Ash Wetlands

Historically, there has been little research on N and C cycles in black ash wetlands. Across wetlands types, black ash wetlands tend to be relatively nutrient rich, with black ash litter having some of the highest N, phosphorus (P), and cation contents of any hardwood forest species [16,17]. At the Minnesota sites with mineral soils, C to N ratios are typically between 14–15 at 0–30 cm depth, with over 75% of the total N found in the top 15 cm [18]. In the mainly organic soil sites in upper Michigan, C to N ratios were higher with a mean of 20 from 0–50 cm and the total N was relatively evenly distributed throughout the top 100 cm of the profile [19].

Research has been initiated to address the impacts on nutrient and C cycling in black ash wetlands. At the Michigan sites, Davis [20] found litter N concentrations in black ash to be greater than non-ash (i.e., red maple and yellow birch) species. Other studies have also found high concentrations of N in ash litter, leading to high turnover rates of ash litter and greater soil N availability compared to non-ash species [21,22]. Nisbet et al. [23] considered the reduction of ecosystem N following EAB invasion of ash riparian forests and speculated that (1) the terrestrial ecosystem could have dramatic changes in net primary production and species composition, and (2) lower amounts of N transport to streams could have cascading effects up the streams' trophic levels and lead to a lowering of stream productivity. Similarly, Palik et al. [24] found that black ash litter decomposed faster than upland derived species such as sugar maple (*Acer saccharum*) and trembling aspen in seasonal wetlands of northern Minnesota, suggesting loss of black ash could lead to lower quality habitat.

In upper Michigan, throughfall deposition of N was not influenced by girdling, but the ash-cut treatment had higher nitrate deposition than the unharvested control [20]. Preliminary soil data indicate that the ash-cut and girdled treatments had little influence on inorganic soil N concentrations and N availability [20]. Although litter and throughfall inputs of N are changing in the ash-cut and girdling treatments, it does not appear that enough time has passed to change soil concentrations of N. These findings are supported by those at the Minnesota sites, where no significant treatment effects on the change in soil C and N in the upper 30 cm were found three years after treatment [18].

Two years following treatment, growing season soil greenhouse gas fluxes of carbon dioxide (CO_2) were significantly higher in the girdled and ash-cut treatments than the unharvested control in upper Michigan [25]. Methane (CH_4) production was higher in the ash-cut treatment than the girdled and unharvested treatments. It appears that the addition of new organic matter to the soil surface following the manipulations and potentially higher soil temperatures because of the partial canopy removal have led to at least an initial increase in decomposition, and faster rising water tables may be leading to higher efflux rates of CH_4.

Preliminary data from the upper Michigan paired watershed experiment suggests that dissolved organic carbon (DOC) and dissolved organic nitrogen (DON) concentrations may increase in the first year following ash cutting [26]. Similar to increases in soil CO_2 fluxes, increases in the DOC and DON concentrations are likely the result of higher soil decomposition rates following the harvesting manipulation.

4. Plant Responses in Post-EAB Black Ash Wetlands

Early work evaluating plant response to EAB was associated with a post-invasion study in southeastern Michigan that included transects and stands across ash cover types with moisture regimes from xeric to mesic to hydric, with those hydric stands dominated by black ash [9,27]. The transects were established in 2004 and emanated up to 45 km from the epicenter of EAB invasion. In that study, black ash was equally susceptible as green ash (*Fraxinus pennsylvanica* Marsh.) and white ash (*Fraxinus americana* L.) to EAB invasion and subsequent mortality. However, the advancement of mortality and decline early in the invasion process was highest for black ash [9]. Smith et al. [9] postulated that the initial higher decline was a product of black ash being taxonomically most similar to Manchurian ash (*Fraxinus mandshurica* Rupr.), the species that EAB coevolved with in Asia. As a result, black ash has all the same cues for the female borer but none of the resistance, leading to higher initial decline rates. Across the same gradient of stands, all ash mortality was greater than 99% by 2009 (Figure 4; [27]).

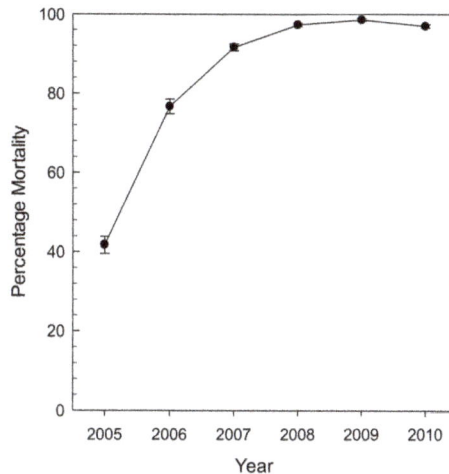

Figure 4. Percentage of ash mortality of trees >2.5 cm at 1.37 m height in subplots and >12.5 cm at 1.37 m height in main plots from 38 stands that included black, green, and white ash in the Huron River watershed in lower Michigan (reprinted with permission, [27]).

In the years shortly after EAB invasion in lower Michigan, and during the decline of black ash, there were numerous black ash seedlings and saplings in the understory, but by 2009 few to no ash seedlings germinated and the seed bank was barren of viable ash seeds [27,28]. The results from lower Michigan may be an ominous precursor of the fate of black ash across the western Great Lakes. As discussed above, the removal of black ash from the wetland canopy will lead to higher water tables and more inundation, possibly preventing the establishment of replacement canopy species. In upper Michigan, the ash-cut and girdled treatments did not lead to an increase in growth rates of the remaining canopy species that would be expected given the increase in resources (e.g., sunlight and nutrients) after canopy reduction [29]. The lack of response is presumably because of the water table dynamics discussed above. Although there were some small positive responses in non-ash sapling growth rates and a shift towards red maple and yellow birch in the seedling layer, the predominant response was a near doubling of the herbaceous species canopy area 2–3 growing seasons after the ash-cut and girdling treatments [29]. Moreover, Davis et al. [29] reported increased heights of the herbaceous layer, upwards of greater than 0.5 m post-treatment. The plant responses were most prominent in the obligate wetland plant species, again indicating that changes in water tables were affecting the plant community composition [29]. Studies in Minnesota indicated a similar increase in the height of the herbaceous cover in the clearcut, girdled, and group selection treatments, but not a difference in herbaceous cover [30]. A grass species (*Calamagrostis canadensis* (Michx.) Beauv.) was an indicator of the clearcut treatment, with a forb (*Arisaema triphyllum* (L.) Schott), fern (*Matteuccia struthiopteris* L. Tod), and sedge (*Carex radiata* (Wahlenb.)) indicators of the girdled treatment. However, unlike the upper Michigan study, Looney et al. [30] did not find an increase in obligate wetland herbaceous plants in the clearcut, girdled, or group selection treatments. In contrast to the lower Michigan studies, where an actual EAB invasion occurred and few to none of the black ash seedlings germinated about 7 years after invasion, in both upper Michigan and Minnesota black ash seedlings and saplings dominated in the woody component of the understory in the harvest treatments 2–3 years after the simulated EAB invasions. It appears time since invasion and the actual presence of EAB are important factors that will dictate the response of the vegetation community following EAB invasion.

A primary objective of the Minnesota study that was also adopted by the upper Michigan study was to investigate potential species to plant in advance of EAB to reduce vulnerability of black ash wetlands to EAB impacts (general trends summarized in Table 1: [10,31,32]). The upper Michigan investigation also included an associated study assessing alternative species establishment in green and black ash-dominated riparian areas near Superior, Wisconsin. In the Minnesota study, most of the 12 species planted had both fall and spring plantings. Across all treatments the species that had the highest rate of survival after three growing seasons were swamp white oak (*Quercus bicolor* Willd.), American elm, and the non-native Manchurian ash, all with greater than 70% survival [10]. The species that had the poorest survival were native species, currently found in these forests, including yellow birch, trembling aspen (*Populus tremuloides* Michx.), and tamarack (*Larix laricina* (Du Roi) K. Koch), all with less than 10% survival. Although within individual species there were some small differences in survival rate depending on the season of planting, across all species there were no survival differences between fall and spring planting [10]. Similarly, there were differences in survival among species across treatments but overall the clearcut treatment had the lowest survival, and the girdled and group selection treatments had no differences when compared to the unharvested controls.

Table 1. Mean under planting survivability (%) after three growing seasons from three planting studies in uninfested black ash wetlands. Minnesota results [10] are the mean values from the control stands and the Michigan and Wisconsin results [32] are from the hummock planting locations. NA is not applicable, as those species were not planted in the individual study.

Common Name	Latin Name	Minnesota	Michigan	Wisconsin
American elm	*Ulmus americana*	93	68	NA
Swamp white oak	*Quercus bicolor*	83	NA	NA
Hackberry	*Celtis occidentalis*	77	NA	66
Silver maple	*Acer saccharinum*	NA	76	NA
Manchurian ash	*Fraxinus mandshurica*	74	NA	NA
Red maple	*Acer rubrum*	33	11	68
Basswood	*Tilia americana*	NA	64	NA
N. white cedar	*Thuja occidentalis*	16	39	39
Bur oak	*Quercus macrocarpa*	NA	38	NA
Balsam poplar	*Populus balsamifera*	29	NA	NA
Black spruce	*Picea mariana*	17	13	NA
Yellow birch	*Betula alleghaniensis*	4	8	NA
Tamarack	*Larix laricina*	8	0	NA
Balsam fir	*Abies balsamea*	NA	7	NA
Eastern cottonwood	*Populus deltoides*	5	NA	NA
Quaking aspen	*Populus tremuloides*	0.3	NA	NA

Further analysis of the growth rates of 10 of the 12 planted species in the Minnesota study indicated that the seedlings in the clearcut treatment had the greatest diameter growth and the seedlings in both the clearcut and girdled treatments had the greatest height growth [31]. Even though seedlings in the clearcut treatment had the lowest survival, those seedlings that did survive were growing at a faster rate than the survivors in the other treatments, likely because of increased light availability following cutting [31]. Within species and across treatments, balsam poplar (*Populus balsamifera* L.), eastern cottonwood (*Populus deltoides* W. Bartram ex Marshall), and tamarack tended to have the greatest height and diameter growth while Manchurian ash, black spruce (*Picea mariana* Mill. britton sterns & poggenb), American elm, and common hackberry (*Celtis occidentalis* L.) tended to have the lowest growth rates [31]. Individual species growth rates do not necessarily coincide with survival rates, indicating seedling establishment is critical. Once established, species tolerance to light and hydrological conditions likely dictates growth rates. Contrary to the Minnesota study, upper Michigan seedling growth rates did somewhat track seedling survival rates, with the greatest height growth for American elm and silver maple, and least height growth for those species with the lowest survival [32].

The upper Michigan study and associated Wisconsin study focused on the effect of microsite conditions on seedling establishment. For the 10 species studied in upper Michigan, 8 species had greater survival on hummocks than in hollows and 5 of 7 species that had some survival in both locations also had greater height growth when planted on hummocks [32]. For the three species studied in Wisconsin, both hackberry and red maple tended to have greater survival on constructed hummocks, whereas northern white cedar tended to have the highest survival on unmanipulated planting sites. Overall, it is quite clear that microsites that are elevated above the water table have the greatest survival and growth rates, a result that is consistent with other studies in forested wetlands [33]. For the upper Michigan sites, silver maple (*Acer saccharinum* L.), American basswood (*Tilia americana* L.), and American elm all had greater than 40% survival rates after 3 years across microsites, while tamarack, red maple, yellow birch, balsam fir (*Abies balsamea* (L.) Mill.), and black spruce all had less than 10% survival [32]. Although the species selected across studies were anticipated to survive and grow in wet conditions, species such as American elm, hackberry, silver maple, and red maple are not obligate wetland species and yet they outperformed obligate species such as yellow birch and northern white cedar. Other factors such as the influence of microsite elevation appear to be critical when considering replacement species for black ash.

Across studies it appears that American elm and some more southerly distributed species did well, such as hackberry, swamp white oak, and eastern cottonwood in Minnesota, silver maple and bur oak (*Quercus macrocarpa* Michx.) in upper Michigan, and hackberry in Wisconsin. Modeling of black ash replacement species that considered climate change and migration of species indicated that both species that are currently present in black ash wetlands and are adapted to future climate (e.g., American elm, red maple, and American basswood) and species that do not currently co-occur with black ash but are not much further south (e.g., silver maple, eastern cottonwood, and hackberry) have great potential [34]. Although the modeling was focused on replacement species for black ash in Minnesota, the results also agree with the short-term field study results in both Wisconsin and upper Michigan. Based on early seedling survival and growth results, and modeling, the effects of climate change and associated changes in hardiness zones on canopy replacement species also should be considered when planting black ash wetlands.

5. Animal Responses in Post-EAB Black Ash Wetlands

Black ash wetlands provide important habitat for many wildlife species. The majority of black ash wetlands are uneven-aged stands [35], thus providing compositional and structural complexity that is important for wildlife. For example, young ash trees are a food source for deer and moose while mature trees provide nesting, roosting, or denning cavities for many bird and mammal species [36]. Despite the dominance of black ash wetlands on the landscape and the documented use of these forests by wildlife, minimal research has focused on the effects of EAB invasion on wildlife. Changes to hydrology and shifts in vegetation composition that occur after the die-back of black ash by EAB will impact wildlife, however, the magnitude and relative degree of impact to individual species and species diversity is unclear.

One of the only published studies on the impacts of EAB on wildlife focused on the indirect effects of EAB-induced mortality to ground beetle community assemblages (Coleoptera: Carabidae) in infected ash stands in Michigan [37]. The results of this study indicated that species diversity decreased in ash stands in response to tree mortality and increasing canopy gap size [37]. Hydric black ash wetlands had the most unique assemblage of ground beetles, with numerous beetle species only found in the hydric moisture category, but also had moderate turnover (25%) of species from 2006–2007 [37]. Overall, the results indicated that beetle communities and diversity are susceptible to changes in environmental conditions caused by EAB invasion [37].

A comprehensive study is currently underway in Minnesota's black ash wetlands to investigate the potential effects of ash loss on bird, mammal, and herptile communities [36,38]. The reported environmental changes, such as an increase in ponding and tree mortality, that are known to occur in EAB infested stands will likely impact taxa and specific species in different ways. For example, many amphibians and invertebrates preferentially select open canopy wetlands because higher water temperatures and algal productivity result in rapid larval growth. Therefore, increases in ponding and longer hydroperiods may increase species richness and species abundance of pond-breeding amphibians and invertebrates because more species will be able use the wetlands for longer periods of time during the breeding season. Longer hydroperiods may also result in higher rates of juvenile recruitment for many amphibian species because individuals will be able to reach metamorphosis before pond drying occurs. Preliminary results from surveys of experimental black ash stands in Minnesota suggest that loss of ash could increase amphibian diversity and the biomass of some invertebrate taxa—namely, Coleoptera, Diptera, and Odonata [39]. However, certain functional feeding groups of macroinvertebrates may have lower abundance following EAB mortality because black ash litter provides a higher quality food resource than other tree species (reviewed in [40]).

Black ash wetlands provide a combination of structural complexity and high functional diversity of plants and invertebrates that are beneficial for many forest bird species. Black ash wetlands have diverse understory and complex canopy structure that provide unique micro-habitats that are beneficial for breeding birds. For example, structural features common in black ash stands such as large trees and coarse woody debris provide foraging opportunities, potential nest sites, and suitable singing perches for forest birds [41]. Preliminary data from Minnesota indicates that several bird species commonly associated with mature forests such as brown creeper (*Certhia americana*), northern parula (*Setophaga americana*), and winter wren (*Troglodytes hiemalis*) were common in black ash wetlands, however, species that require canopy openings such as chestnut-sided warbler (*Setophaga pensylvanica*) were also commonly observed [36]. These factors ultimately increased abundance, species richness, and species diversity in black ash wetlands compared to neighboring upland forests. Further, data from Minnesota suggests that black ash wetlands provide important stop-over habitat for migrating species including rusty blackbirds (*Euphagus carolinus*), which are one of North America's most rapidly declining bird species [42]. The transition of black ash wetlands to shrub- and sedge-dominated wetlands will result in the loss of important micro-habitats that are often limited across the forested landscape and may also reduce availability of critical habitat for migrating birds.

Recent changes in land use in the Great Lakes region has made loss of forest habitat and forest homogenization a growing threat to wildlife biodiversity [41,43]. Loss of forested wetlands across the landscape will likely intensify the impacts of current stressors such as climate change and habitat loss on wildlife populations [44,45]. For example, recent research suggests that diverse forests that contain an assortment of tree sizes may buffer the effects of warming temperatures for some wildlife species [46]. In Minnesota, preliminary results of camera trap surveys documented 12 mammal species, all of which, with the exception of ermine (*Mustela erminea*), were significantly more likely to use forested black ash stands compared to experimental EAB simulated stands (clearcuts) [36]. The difference in habitat use was particularly pronounced for forest-dependent species such as American marten (*Martes americana*) and fisher (*Martes pennanti*) [36]. These results suggest mammals are actively avoiding open areas; therefore, the conversion of forested wetlands to open wetlands will impact forest connectivity and exacerbate impacts of habitat loss for forest dependent species. Further, because habitat connectivity is one of the most important factors in maintaining biological diversity [47], the long-term impacts of EAB on wildlife diversity will likely be substantial.

Overall, the loss of black ash over small spatial scales may result in turnover of wildlife communities from forest dependent species to open-canopy and wetland associated species (e.g., forest bird species may be replaced with wetland associated species such as waterfowl). Although net changes in wildlife biodiversity may be minimal or potentially increase for some taxa (e.g., amphibians), the long-term, large-scale impacts of EAB on forest-associated wildlife will likely be significant [40]. Adaptive management strategies that focus on planting replacement tree species that maintain long-term compositional and structural complexity within these wetland systems will help maintain wildlife diversity.

6. What Do These Ecosystem Responses Mean for the Future of Black Ash Wetlands?

The combined effects of the susceptibility of black ash to EAB, the current rate of spread of EAB, and warming conditions in the western Great Lakes, especially in the winter, will enhance the spread of EAB in the future [48]. It is likely that black ash wetlands will be highly impacted, reducing their abundance in the region. As water tables rise because of reduced black ash transpiration, we anticipate major changes in vegetation and associated animal communities. Concomitant changes in the amount and timing of precipitation could affect water table dynamics, but predictions for the region indicate higher annual precipitation in the future [49], potentially exacerbating the conversion of forests to non-forest conditions and resulting in more water available for runoff to streams or increases in water table elevations. While annual precipitation is predicted to increase in the future, the number of consecutive dry days (i.e., longer droughts) is also predicted to increase over much of the western Great Lakes. Although precipitation and drought patterns are very uncertain, possible increases in drought length and frequency may enhance the establishment of less water-tolerant non-ash tree species in former black ash wetlands.

Removing black ash from the wetland ecosystem may lower N inputs, which will likely have important consequences for plant growth and composition, microbial dynamics, decomposition, carbon sequestration, and other essential ecosystem functions. Ultimately, those changes in N will affect the lowest trophic levels of the animal community, with the potential to cascade up the food chain [24,44,45].

Black ash wetlands are long-term sinks for C based on the relatively high C present in the soil, both in mineral and organic soil wetlands. Disruption of the hydrology and plant community by EAB will likely change short- and long-term C dynamics. In mineral soil wetlands, the forest comprises a much larger fraction of the total ecosystem C than in organic soil wetlands [50]. We hypothesize that if black ash wetlands convert to shrub or grass/sedge wetlands because of higher water tables following EAB invasion, both mineral and organic soil wetlands will likely be short-term sources of C as fresh downed-wood decomposes, releasing CO_2 to the atmosphere. Higher water tables created by decreases in evapotranspiration following the loss of black ash after EAB invasion would also lead to a short-term increase in CH_4 emissions shortly after EAB invasion. Longer term, the high water tables would be expected to lead to slower litter and soil decomposition and stabilization of CH_4 emissions, lower emissions of CO_2, and an overall ecosystem sink for C, mainly through accumulation in the soil C pool.

7. Needs for Future Research

It is apparent from the paucity of research on black ash wetlands that we are only beginning to understand the dynamics of this ecosystem and specifically the ecological impacts post-EAB. Immediate research is needed to assess the impacts of EAB on native plant and animal biodiversity, ecosystem functions related to nutrient and C cycling, and hydrology, as well as to evaluate possible mitigation strategies for increasing the resilience of these imperiled ecosystems to EAB. Although we have some data on transpiration, no evaporation or groundwater flow data exist to close the water budget. A better understanding of the water budget will help us predict the possible conversion to non-forest conditions. While we have good evidence that removing all or most of the canopy will raise water tables, we have indications that canopy mortality of 20% or less can moderate water table impacts. There is a need to measure water tables and other components of the water budget across a range of canopy removal percentages to ascertain the level of mortality that causes significant changes in hydrologic regime. If that number, for example, is 40% of the canopy basal area, then those wetlands with less than 40% of the canopy in black ash would be more resilient to hydrologic changes following EAB infestation.

Other critical information gaps related to hydrologic response of black ash wetlands following EAB exist. Notably, the relative importance of water budget components to changes in hydrologic regime across black ash wetland hydrogeomorphic settings remains unknown. Different hydrogeomorphic settings likely have different mechanisms and alternative outcomes are possible. For example, the depth to confining layer is thought to be an important variable influencing response in perched systems [51], but the threshold depth where the hydrologic regime shifts is unknown. For depressional and swale settings, contributing area is likely a primary controlling factor and it may be that at certain scales (small or large) the effect of reduced black ash transpiration, because of loss of ash, on water table dynamics will be negligible. The lack of understanding of how black ash interacts with hydrology demonstrates the need for studies across the range of hydrogeomorphic settings that black ash encompasses. Ultimately, a greater understanding of hydrologic response over time is needed to assess the potential for ecosystem recovery. Data from the Minnesota sites indicates limited hydrologic recovery for at least five years after disturbance [51], but the longer-term response is unclear.

In addition to more detailed information on hydrologic regimes, we need better insight on how the replacement by herbaceous- or shrub-dominated vegetation will impact short- and long-term nutrient (N, P, and cations) and carbon cycles. If former black ash wetlands become sources of CO_2 and CH_4 to the atmosphere, or even if there is a decrease in long-term storage capacity, conversion to other wetland types could have important feedbacks to our global climate. Moreover, developing restoration strategies to recover tree cover in these areas to offset these impacts is a critical need.

Other than the ground beetle study that occurred in black ash wetlands long after EAB infestation, we have no published data on animal responses. Black ash wetlands are critical habitats for invertebrates and amphibians as well as the species at higher trophic levels that depend on those communities (e.g., reptiles, birds, and mammals) [24]. We need studies that assess the response of these animal communities to EAB invasion to both develop an understanding of the temporal animal responses and create management approaches that consider effects on both floral and faunal communities. This assessment could be done using the simulated infestation sites or a chronosequence of pre- and post-infested black ash stands.

Given the ongoing spread of EAB into regions where black ash often dominate wetland forests, the need for additional research into the effects of EAB on ecosystems is urgent. As reviewed above, there have been great developments in our understanding of black ash wetlands in the past ten years, particularly with respect to how EAB will affect these forests and some of the subsequent physical and ecological changes that will follow infestation. Assessing the impacts of the loss of black ash on biota is the strongest outstanding need. We also still lack information on how infestation will alter physical and ecological processes across the range of black ash systems. Research to address these fundamental needs and existing mitigation strategies is currently underway and should provide critical information for further refinement of mitigation and management strategies.

Acknowledgments: The authors would like to thank Wendy Klooster for providing Figure 4, as well as all the students at the Michigan Technological University, the University of Minnesota, and Virginia Polytechnic Institute and State University.

Author Contributions: R.K. led the development of this synthesis and A.D., J.W., R.S., T.P, M.Y., A.G. and B.P. contributed to the writing of the paper.

Conflicts of Interest: The authors declare no conflict of interest.

References

1. Herms, D.A.; McCullough, D.G. Emerald ash borer invasion of North America: history, biology, ecology, impacts, and management. *Ann. Rev. Entomol.* **2014**, *59*, 13–30. [CrossRef] [PubMed]
2. USDA. Forest Service Forest Inventory and Analysis Program. Available online: https://apps.fs.usda.gov/DATIM/Default.aspx (accessed on 13 November 2017).
3. Palik, B.J.; Ostry, M.E.; Venette, R.C.; Abdela, E. *Fraxinus nigra* (black ash) dieback in Minnesota: Regional variation and potential contributing factors. *For. Ecol. Manag.* **2011**, *261*, 128–135. [CrossRef]
4. Palik, B.J.; Ostry, M.E.; Venette, R.C.; Abdela, E. Tree regeneration in black ash (*Fraxinus nigra*) stands exhibiting crown dieback in Minnesota. *For. Ecol. Manag.* **2012**, *269*, 26–30. [CrossRef]

5. Minnesota Department of Natural Resources. Native Plant Communities for Northern Very Wet Ash Swamp. 2018. Available online: http://files.dnr.state.mn.us/natural_resources/npc/wet_forest/wfn64.pdf (accessed on 30 March 2018).

6. Lenhart, C.; Brooks, K.; Davidson, M.; Slesak, R.; D'Amato, A. Hydrologic source characterization of black ash wetlands: Implications for EAB response. In *Riparian Ecosystems IV: Advancing Science, Economics and Policy, Proceedings of the American Water Resources Association Summer Specialty Conference, Denver, CO, USA, 27–29 June 2012*; American Water Resources Association (AWRA): Middelburg, VA, USA, 2012.

7. Van Grinsven, M.J.; Shannon, J.P.; Davis, J.C.; Bolton, N.W.; Wagenbrenner, J.W.; Kolka, R.K.; Pypker, T.G. Source water contributions and hydrologic responses to simulated emerald ash borer infestations in depressional black ash wetlands. *Ecohydrology* **2017**. [CrossRef]

8. Slesak, R.A.; Lenhart, C.F.; Brooks, K.N.; D'Amato, A.W.; Palik, B.J. Water table response to harvesting and simulated emerald ash borer mortality in black ash wetlands in Minnesota, USA. *Can. J. For. Res.* **2014**, *44*, 961–968. [CrossRef]

9. Smith, A.; Herms, D.A.; Long, R.P.; Gandhi, K.J.K. Community composition and structure had no effect on forest susceptibility to invasion by the emerald ash borer (Coleoptera: Buprestidae). *Can. Entomol.* **2015**, *147*, 318–328. [CrossRef]

10. Looney, C.E.; D'Amato, A.W.; Palik, B.J.; Slesak, R. Overstory treatment and planting season affect survival of replacement tree species in emerald ash borer threatened *Fraxinus nigra* forests in Minnesota, USA. *Can. J. For. Res.* **2015**, *45*, 1728–1738. [CrossRef]

11. Little, L., Jr. *Checklist of United States Trees (Native and Naturalized)*; Agricultural Handbook 541; U.S. Department of Agriculture: Washington, DC, USA, 1979; 375p.

12. USDA Cooperative Emerald Ash Borer Project. Available online: http://www.emeraldashborer.info/documents/MultiState_EABpos.pdf (accessed on 26 March 2018).

13. Telander, A.C.; Slesak, R.A.; D'Amato, A.W.; Palik, B.J.; Brooks, K.N.; Lenhart, C.F. Sap flow of black ash in wetland forests of northern Minnesota, USA: Hydrologic implications of tree mortality due to emerald ash borer. *Agric. For. Meteorol.* **2015**, *206*, 4–11. [CrossRef]

14. Shannon, J.; van Grinsven, M.; Davis, J.; Bolton, N.; Noh, N.J.; Wagenbrenner, J.; Pypker, T.; Kolka, R. Water level controls on sap flux of canopy species in black ash (*Fraxinus nigra*) wetlands. *Forests* **2018**, *9*, 147. [CrossRef]

15. Stednick, J.D. Monitoring the effects of timber harvest on annual water yield. *J. Hydrol.* **1996**, *176*, 79–95. [CrossRef]

16. Ferrari, J. Fine-scale patterns of leaf litterfall and nitrogen cycling in an old-growth forest. *Can. J. For. Res.* **1999**, *29*, 291–302. [CrossRef]

17. Pastor, J.; Post, W.M. Influence of climate, soil moisture, and succession on forest carbon and nitrogen cycles. *Biogeochemistry* **1986**, *2*, 3–27. [CrossRef]

18. Slesak, R.A. (Minnesota Forest Resources Council, St. Paul, MN, USA). Unpublished data, 2015.

19. Kolka, R.K. (USDA Forest Service Northern Research Station, Grand Rapids, MN, USA). Unpublished data, 2016.

20. Davis, J. Vegetation Dynamics and Nitrogen Cycling Responses to Simulated Emerald Ash Borer Infestation in *Fraxinus Nigra*-dominated Wetlands of Upper Michigan, USA. Ph.D. Thesis, Michigan Technological University, School of Forest Resources and Environmental Science, Houghton, MI, USA, 17 November 2016.

21. Langenbruch, C.; Helfrich, M.; Flessa, H. Effects of beech (*Fagus sylvatica*), ash (*Fraxinus excelsior*) and lime (*Tilia spec.*) on soil chemical properties in a mixed deciduous forest. *Plant. Soil* **2012**, *352*, 389–403. [CrossRef]

22. Vesterdal, L.; Schmidt, I.; Callesen, I.; Nilsson, L.; Gundersen, P. Carbon and nitrogen in forest floor and mineral soil under six common European tree species. *For. Ecol. Manag.* **2008**, *255*, 35–48. [CrossRef]

23. Nisbet, D.; Kreutzweiser, D.; Sibley, P.; Scarr, T. Ecological risks posed by emerald ash borer to riparian forest habitats: A review and problem formulation with management implications. *For. Ecol. Manag.* **2015**, *358*, 165–173. [CrossRef]

24. Palik, B.J.; Batzer, D.P.; Kern, C. Upland forest linkages to seasonal wetlands: Litter flux, processing, and food quality. *Ecosystems* **2005**, *8*, 1–11. [CrossRef]

25. Van Grinsven, M.J.; Shannon, J.P.; Bolton, N.W.; Davis, J.C.; Noh, N.J.; Wagenbrenner, J.W.; Kolka, R.K.; Pypker, T.G. Response of Black Ash Wetland Gaseous Soil Carbon Fluxes to a Simulated Emerald Ash Borer Disturbance. *Forests*, **2018**, in review.

26. Van Grinsven, M.J.; Shannon, J.P.; Noh, N.J.; Kane, E.S.; Bolton, N.W.; Davis, J.C.; Wagenbrenner, J.W.; Sebestyen, S.D.; Kolka, R.K.; Pypker, T.G. Stream Water, Total Nitrogen and Carbon Load Responses to a Simulated Emerald Ash Borer Infestation in Black Ash Dominated Headwater Wetlands. In Proceedings of the American Geophysical Union Annual Conference, New Orleans, LA, USA, 11–15 December 2017.

27. Klooster, W.S.; Herms, D.A.; Knight, K.S.; Herms, C.P.; McCullough, D.G.; Smith, A.; Gandhi, K.J.K.; Cardina, J. Ash (*Fraxinus* spp.) mortality, regeneration, and seed bank dynamics in mixed hardwood forests following invasion by emerald ash borer (*Agrilus planipennis*). *Biol. Invasions* **2014**, *16*, 859–873. [CrossRef]

28. Kashian, D.M.; Witter, J.A. Assessing the potential for ash canopy tree replacement via current regeneration following emerald ash borer-caused mortality on southeastern Michigan landscapes. *For. Ecol. Manag.* **2011**, *261*, 480–488. [CrossRef]

29. Davis, J.C.; Shannon, J.P.; Bolton, N.W.; Kolka, R.K.; Pypker, T.G. Vegetation responses to simulated emerald ash borer infestation in *Fraxinus nigra* dominated wetlands of Upper Michigan, USA. *Can. J. For. Res.* **2017**, *47*, 319–330. [CrossRef]

30. Looney, C.E.; D'Amato, A.W.; Palik, B.J.; Slesak, R.A.; Slater, M.A. The response of *Fraxinus nigra* forest ground-layer vegetation to emulated emerald ash borer mortality and management strategies in northern Minnesota, USA. *For. Ecol. Manag.* **2017**, *389*, 352–363. [CrossRef]

31. Looney, C.E.; D'Amato, A.W.; Palik, B.J.; Slesak, R. Canopy treatment influences growth of replacement tree species in *Fraxinus nigra* forests threatened by emerald ash borer in Minnesota, USA. *Can. J. For. Res.* **2017**, *47*, 183–192. [CrossRef]

32. Bolton, N.; Shannon, J.; Davis, J.; van Grinsven, M.; Noh, N.J.; Schooler, S.; Kolka, R.; Pypker, T.; Wagenbrenner, J. Methods to improve alternative species seedlings survival and growth in black ash ecosystems threatened by emerald ash borer. *Forests* **2018**, *9*, 146. [CrossRef]

33. Palik, B.; Haworth, B.K.; David, A.J.; Kolka, R.K. Survival and growth of northern white-cedar and balsam fir seedlings in riparian management zones in northern Minnesota, USA. *For. Ecol. Manag.* **2015**, *337*, 20–27. [CrossRef]

34. Iverson, L.; Knight, K.S.; Prasad, A.; Herms, D.A.; Matthews, S.; Peters, M.; Smith, A.; Hartzler, D.M.; Long, R.; Almendinger, J. Potential species replacements for black ash (*Fraxinus nigra*) at the confluence of two threats: Emerald ash borer and a changing climate. *Ecosystems* **2016**, *19*, 248–270. [CrossRef]

35. Erdmann, G.G.; Crow, T.R.; Peterson, R.M.; Wilson, C.D. *Managing Black Ash in the Lake States*; General Technical Report NC-115; U.S. Department of Agriculture, Forest Service, North Central Forest Experiment Station: St. Paul, MN, USA, 1987.

36. Grinde, A.R.; (University of Minnesota Duluth, Duluth, MN, USA). Unpublished data, 2017.

37. Gandhi, K.J.K.; Smith, A.; Hartzler, D.M.; Herms, D.A. Indirect effects of emerald ash borer-induced ash mortality and canopy gap formation on epigaeic beetles. *Environmental Entomology* **2014**, *43*, 546–555. [CrossRef] [PubMed]

38. Youngquist, M.B.; (University of Minnesota, St. Paul, MN, USA). Unpublished data, 2017.

39. Youngquist, M.B.; (University of Minnesota, St. Paul, MN, USA). Unpublished data, 2016.

40. Youngquist, M.B.; Eggert, S.L.; D'Amato, A.W.; Palik, B.J.; Slesak, R.A. Potential effects of foundation species loss on wetland communities: A case study of black ash wetlands threatened by emerald ash borer. *Wetlands* **2017**, *37*, 787–799. [CrossRef]

41. Niemi, G.J.; Howe, R.W.; Sturtevant, B.R.; Parker, L.R.; Grinde, A.R.; Danz, N.P.; Nelson, M.D.; Zlonis, E.J.; Walton, N.G.; Gnass Giese, E.E.; et al. *Analysis of Long Term Forest Bird Monitoring from National Forests of the Western Great Lakes Region*; General Technical Report NRS-159; U.S. Department of Agriculture, Forest Service, Northern Research Station: Newtown Square, PA, USA, 2016.

42. Sauer, J.R.; Niven, D.K.; Hines, J.E.; Ziolkowski, D.J., Jr.; Pardieck, K.L.; Fallon, J.E.; Link, W.A. *The North American Breeding Bird Survey, Results and Analysis 1966–2015*; Version 2.07.2017; USGS Patuxent Wildlife Research Center: Laurel, MD, USA, 2017.

43. Schulte, L.A.; Mladenoff, D.J.; Crow, T.R.; Merrick, L.C.; Cleland, D.T. Homogenization of northern US Great Lakes forests due to land use. *Landsc. Ecol.* **2007**, *22*, 1089–1103. [CrossRef]

44. Grinde, A.R.; Niemi, G.J. A synthesis of species interactions, metacommunities, and the conservation of avian diversity in hemiboreal and boreal forests. *J. Avian Biol.* **2016**, *47*, 706–718. [CrossRef]

45. Grinde, A.R.; Niemi, G.J.; Sturtevant, B.R.; Panci, H.; Thogmartin, W.; Wolter, P. Importance of scale, land cover, and weather on the abundance of bird species in a managed forest. *For. Ecol. Manag.* **2017**, *405*, 295–308. [CrossRef]

46. Betts, M.G.; Phalan, B.; Frey, S.J.K.; Rousseau, J.S.; Yang, Z. Old-growth forests buffer climate-sensitive bird populations from warming. *Divers. Distrib.* **2018**, *24*, 439–447. [CrossRef]

47. Correa Ayram, C.A.; Mendoza, M.E.; Etter, A.; Salicrup, D.R.P. Habitat connectivity in biodiversity conservation: A review of recent studies and applications. *Prog. Phys. Geogr.* **2016**, *40*, 7–37. [CrossRef]

48. Climate Central. Researching and Reporting the Science and Impacts of Climate Change. Available online: http://www.climatecentral.org/gallery/maps/heres-where-winters-are-warming-the-most (accessed on 30 March 2018).

49. Pryor, S.C.; Scavia, D.; Downer, C.; Gaden, M.; Iverson, L.; Nordstrom, R.; Patz, J.; Robertson, G.P. Chapter 18: Midwest. In *Climate Change Impacts in the United States: The Third National Climate Assessment*; Melillo, J.M., Richmond, T.C., Yohe, G.W., Eds.; U.S. Global Change Research Program: Washingtion, DC, USA, 2014; pp. 418–440.

50. Bridgham, S.D.; Megonigal, J.P.; Keller, J.K.; Bliss, N.B.; Trettin, C. The carbon balance of North American wetlands. *Wetlands* **2006**, *26*, 889–916. [CrossRef]

51. Diamond, J.S.; McLaughlin, D.; Slesak, R.A.; D'Amato, T.D.; Palik, B.J. Ecohydrologic response of black ash wetlands to emerald ash borer infestation and potential mitigation strategies. *Ecol. Appl.* **2018**, in press.

forests

MDPI

Article

Neighboring Tree Effects and Soil Nutrient Associations with Surviving Green Ash (*Fraxinus pennsylvanica*) in an Emerald Ash Borer (*Agrilus planipennis*) Infested Floodplain Forest

Rachel H. Kappler [1,*], Kathleen S. Knight [2], Jennifer Koch [2] and Karen V. Root [1]

[1]   Bowling Green State University, Bowling Green, OH 43403, USA; kvroot@bgsu.edu
[2]   USDA Forest Service, Delaware, OH 43805, USA; ksknight@fs.fed.us (K.S.K.); jkoch@fs.fed.us (J.K.)
*   Correspondence: rackapp@bgsu.edu; Tel.: +1-517-614-6044

Received: 2 March 2018; Accepted: 2 April 2018; Published: 4 April 2018

Abstract: Few ash trees (*Fraxinus* spp.) have survived the initial devastation that emerald ash borer beetle (EAB) (*Agrilus planipennis*) has caused in natural populations. We studied green ash (*Fraxinus pennsylvanica*) trees in a floodplain population after >90% of ash had died from EAB infestation. We examined the relationship among the canopy health classes of surviving ash trees and their nearest neighboring trees (within 6 m) and available soil nutrients. A subset of focal ash trees was randomly selected within health classes ranging from healthy to recently deceased. Focal trees with the healthiest canopy class had significantly fewer ash neighbors compared to declining health classes. Other species of tree neighbors did not have a significant impact on surviving ash tree canopy health. Nutrients in soils immediately surrounding focal trees were compared among health classes. Samples from treeless areas were also used for comparison. There was a significantly greater amount of sulfur (ppm) and phosphorus (mg/kg) in ash tree soil compared to treeless area soil. The relationships between these soil nutrient differences may be from nutrient effects on trees, tree effects on nutrients, or microsite variation in flooded areas. Our data do not directly assess whether these ash trees with healthier canopies have increased resistance to EAB but do indicate that at neighborhood scales in EAB aftermath forests, the surviving ash trees have healthier canopies when separated at least 6 m from other ash trees. This research highlights scale-dependent neighborhood composition drivers of tree susceptibility to pests and suggests that drivers during initial infestation differ from drivers in aftermath forests.

Keywords: forest composition; insect pest; susceptibility; sulfur; phosphorus; lingering ash

1. Introduction

The impacts of the loss of tree species to forest pests and diseases highlight the need to understand the drivers of tree susceptibility to these threats. Two potential drivers of tree susceptibility to pests and diseases are tree neighborhood composition and soil nutrient availability. Tree neighborhood composition is influenced by tree interactions, such as competition, facilitation, and spread of pests and diseases. Plant composition is hypothesized to be based on a tradeoff between plants competing at high quality soil nutrient sites and tolerating poor quality soil nutrient sites [1]. Tree composition can change rapidly with the introduction or increase in tree pests and diseases. Insects that differ in their biology may exhibit different relationships with the density of their host. Some insects have greater impacts on host trees when host density is high, a phenomenon called the Resource Concentration Hypothesis [2]. Others have lower impact at high host density, supporting the Resource Dilution Hypothesis [3]. Emerald ash borer, (EAB) (*Agrilus planipennis*) has shown patterns in accordance with the resource dilution hypothesis during the initial outbreak, causing more rapid mortality in natural tree stands with low ash densities [4]. Therefore, tree composition of an area can influence ash survival, and the spatial distribution of ash may be related to the influence of soil nutrient availability and tree composition.

Soil nutrient availability may influence tree survival of pests and diseases. Some have hypothesized that trees fend off and/or survive pests and disease better in resource limited environments because in those conditions, plants put more energy into creating chemical defenses [5]. However, research results on effects of soil nutrients on defenses have varied, as the degree of defense that trees exhibit can be attributed to multiple factors, both environmental and genetic. Research results from pine weevil (*Hylobius abietis*) studies that added soil nutrients to trees for protection showed no difference in pest damage [6] or an increase in pest damage [7]. In another study, addition of soil nutrients to aspen seedlings showed varying changes in three foliar defense chemicals [8]. Variations in the degree of defense that trees exhibit can be attributed to multiple factors, including both genetic and environmental variation as well as interactions between the two. Unfortunately, tree defenses are usually not adapted to non-native diseases or pests.

The invasive EAB beetle has devastated populations of North American ash species. EAB is a specialist beetle from Asia that feeds and reproduces on all *Fraxinus* species, and was introduced near Detroit, MI, USA [9,10]. EAB spreads by flight and had been estimated to have a maximum cumulative flight distance of 9.8 km over a female beetle's life span [11]. EAB seeks out ash using both visual and olfactory cues [12]. EAB scale of dispersal from host trees has been estimated as 100 m to 200 m, dependent on ash phloem abundance, indicating that the scale of dispersal may be different in an aftermath forest which contains a different amount of ash phloem abundance [13]. There are multiple indicators of EAB damage including development of basal or epicormics sprouts, woodpecker holes, and EAB exit holes, although EAB exit holes are usually the last indicator to be seen on the tree trunk [10]. The damage EAB create usually starts in the upper canopy, and research has shown that ash canopy health is highly indicative of the amount of EAB damage done [14].

Green ash (*Fraxinus pennsylvanica*) is a deciduous tree that can grow up to 20 m tall and has small (50 mm × 6 mm) winged seeds that are wind and water dispersed [15]. The lifespan of green ash in natural stands averages 65 years [16]. In natural areas, green ash is typically found in bottomland forests, but is adapted to a variety of areas across the Eastern United States [17]. This species is tolerant of several environmental stressors including high salinity, flooding, drought, and high alkalinity [9].

While EAB typically kills most of the mature ash trees in forest settings [4] a small number of surviving ash remain. The term lingering ash refers to healthy ash trees with a diameter at breast height (DBH) >10 cm that have survived for at least two years after the initial ash mortality rate reached 95% from EAB. Although some lingering ash trees may simply be the last to be infested, others have been shown to have rare phenotypes that increase their resistance [18]. Despite the ability of lingering ash to remain healthy longer, they are still vulnerable to infestation by EAB. Lingering ash trees were first identified in the Oak Openings Preserve Metropark of Northwest Ohio and Indian Springs Metropark of Southeast Michigan in 2009 [19].

The Oak Openings Region is a mixed disturbance landscape containing natural ecosystems in a mosaic of small to large remnant habitat patches, surrounded by a matrix of agriculture and urban development. Prior to human settlement the area was composed of oak savanna, oak woodland, oak barrens, wet prairie, floodplain forests, and surrounded by black swamp forest [20]. Composition of Oak Openings Preserve Metropark floodplain includes sections that are considered silver maple-elm-cottonwood forest and maple-ash-elm forest [21]. Multiple studies have included this ash population in investigations related to EAB [4,14,18,19]. EAB is still present in the floodplain which contains a remnant cohort of lingering green ash within other smaller size classes of green ash that may not have been infested when the initial infestation occurred. Floodplain forest soil types consist of loam and sandy variations [22]. This region is a biodiversity hotspot that is undergoing large changes from the EAB invasion and other factors.

The objectives of this study were to determine if neighboring tree composition and/or soil nutrient variables differed among ash canopy health classes. Specifically, we compared the effects of ash tree neighbors, other tree species neighbors, and soil nutrients within the A horizon of ash tree root area on ash canopy health. We expected that healthier ash would be in locations with fewer tree competitors in soils with more available limiting nutrients important for growth, such as phosphorus. Investigation of this study site will help us better understand changes in a natural forest after the initial EAB decline in ash trees and may provide insights into potential restoration and land management options that could improve natural ash remnants.

2. Materials and Methods

An ash survey was conducted in the summer yearly from 2010 to 2017 at the Oak Openings Preserve Metropark Swan Creek floodplain in Ohio, which is approximately 1.23 km^2 (41.582133, −83.861483 to 41.538543, −83.824938). For each ash the following data were recorded: DBH (cm), canopy health class, crown ratio (the ratio of crown length to total tree height), presence of flowers or seeds, and signs of EAB. Signs of EAB included bark splitting, EAB exit holes and woodpecker feeding holes on the trunk between 1.25 and 1.75 cm from the ground, and the presence of basal and epicormic branching. Bark splitting and presence of basal and epicormic branching indicate tree response to stress and damage, while woodpecker holes and EAB holes indicate EAB presence. Woodpecker holes and EAB exit holes higher in the tree may appear earlier in the infestation of the tree but were not visible from the ground and thus were not counted. Therefore, the lack of exit holes or woodpecker holes on the lower trunk does not mean the tree was uninfested. Canopy health class was categorized from 1 to 5 based on thinning and dieback, with 1 having a full/healthy canopy, 2 having thinning of leaves but no dieback, 3 having a canopy with <50% dieback, 4 having a canopy with >50% dieback, and 5 having no canopy leaves, but epicormics sprouts may be present [23,24].

A subset of focal ash trees was chosen using stratified random sampling of ash trees surveyed in 2016, stratified based on their canopy health class. With trees rated 5, we only kept those most recently deceased (2016) and removed others rated 5. We checked that the selected trees were at least 50 m from each other (checked in ArcGIS with their GPS points) to reduce spatial autocorrelation. When checking for spatial autocorrelation if one tree had to be removed, we gave preference to keep trees with more years of data. We included up to 10 trees within each canopy health class, but the spatial rule constrained some canopy health class categories to fewer trees. Thus, each ash canopy health class included seven to ten trees (*n* = 44; class 1 = 9, class 2 = 10, class 3 = 7, class 4 = 10, class 5 = 8).

Distance to nearest living neighbor trees and their species identity were recorded. Nearest neighbor trees were defined as any tree or woody shrub over 1.37 m tall within a 6-m radius of the focal ash tree. The neighborhood of each ash tree was set at a radius of 6 m since effects from other trees have been shown to occur at smaller spatial scales [25,26].

To assess available nutrients, soil samples were taken once during June 2016 at each selected focal ash tree (*n* = 44). A galvanized 1-inch diameter soil corer was used to collect samples from the first 6 inches of the soil, removing surface organic material. Soil samples taken at each focal ash tree consisted of 4 sub-sampling points located 2 m from the tree trunk at cardinal directions. Soil samples from treeless floodplain sites were taken as a comparison (*n* = 8). Treeless samples had 4 sub-sampling points at cardinal directions 4 m from the center of an 8-m diameter circular area where no trees were present. The 4 sub-samples from one tree were homogenized in one plastic bag and kept cool till it could later be air dried. The soil corer was cleaned with distilled water and wiped dry after each sample was collected. Standard analyses of soils were performed (Brookside Laboratories, New Breman, OH, USA) including: pH in water, base saturation of cations (%), organic matter (%, based on the loss on ignition method), estimated nitrogen release (#'s N/acre, estimate of amount released annually through organic matter decomposition, based on the loss on ignition method), Bray II phosphorus (mg/kg), and total exchange capacity (meq/100 g). Percentages of the following nutrients pertaining to the total exchange capacity of the soil were analyzed: potassium, calcium, magnesium, sodium, hydrogen, other bases. Mehlich III extractables analyzed included: potassium (mg/kg), phosphorus (mg/kg), calcium (mg/kg), magnesium (mg/kg), sodium (mg/kg), zinc (mg/kg), copper (mg/kg), manganese (mg/kg), iron (mg/kg), aluminum (mg/kg), sulfur (ppm), and boron (mg/kg). The type of soil present at each tree site was identified from Lucas County soil survey data (2003) created by the United States Department of Agriculture, Natural Resources Conservation Service viewed in ArcMap 10.2 (ESRI, Redlands, CA, USA).

Data collected were not normally distributed; therefore, we used a nonparametric Wilcoxon rank sum test in JMP (SAS Institute, Cary, NC, USA) to examine the relationship between measured variables and focal ash canopy health classes. These variables were the total number of nearest neighbors (all species), number of nearest living ash trees, and select uncorrelated individual soil nutrients. A post-hoc paired Wilcoxon test was performed among ash canopy health classes and significant variables (Appendix A). A Spearman's correlation was used to compare the nutrient variables with each other, and those highly correlated ($\tilde{n} > 0.70$) with another nutrient were removed. The ones that were removed were chosen because they correlated with estimated nitrogen or phosphorus, identified as important nutrients in the literature. A Spearman's correlation test was also used to assess relationships between the ten select nutrient variables (pH, estimated nitrogen release, sulfur, phosphorus, Brays II phosphorus, sodium, boron, magnesium, copper, aluminum), number of ash neighbors, and total number of neighbors to check for strong correlations. To assess if tree presence alone had any effect on select soil nutrients, a Wilcoxon rank sum test was used to compare between ash tree sites and treeless sites. We used a Bonferroni correction for multiple comparisons to establish a cut-off for statistical significance.

3. Results

In the floodplain forest, the most prominent neighbor tree species found (from most to least numerous): green ash, Eastern cottonwood (*Populus deltoides*), boxelder (*Acer negundo*), willow (*Salix* spp.), American elm (*Ulmus americana*), buttonbush (*Cephalanthus occidentalis*), hawthorn (*Crataegus* sps.), silver maple (*Acer saccharinum*), and oak (*Quercus* spp.) (Table 1). For the neighbor analysis, the highest number of neighbors any one ash tree had in the 6-m radius was seven trees. Over all the 44 focal trees, 54% of the neighboring trees were green ash.

Table 1. The total abundance of each neighboring tree species around our sampled 44 ash trees (6 m radius), in order from most to least numerous.

All Neighboring Species	Total Abundance
Green Ash	39
Eastern cottonwood	10
Boxelder	10
Hawthorne	4
Spicebush laurel	4
Willow spp.	1
Maple spp.	1
American elm	1
Black Walnut	1
Oak spp.	1

The distribution of total living ash tree neighbors differed significantly among the ash canopy health classes ($p = 0.02$) (Figure 1). Ash with healthier canopies was usually found with no ash neighbors within a 6-m radius, while other ash health classes had on average one or two neighbors. No difference was found when comparing total number of neighboring trees in each ash canopy health class. Total number of neighbors was positively correlated with number of ash neighbors ($\tilde{n} = 0.78$, $p < 0.001$).

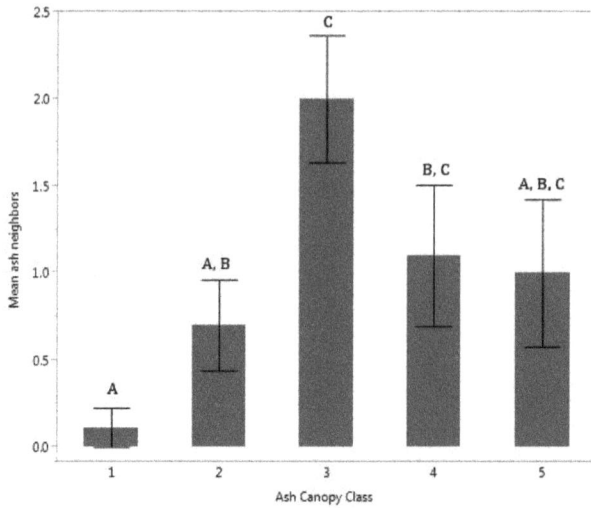

Figure 1. The mean number of ash neighbors within 6 m of focal ash trees within each ash canopy health class. Error bars represent ± one standard error Different letters (A, B, C) indicate significant differences in mean values, mean values with the same letters were similar and no statistically significant differences were observed for these samples. Ash canopy health class ratings are 1–5, where 1 is a full healthy crown to class 5, which is a dead crown.

Focal ash trees were typically 15–25 cm DBH (Table 2), much larger than the minimum susceptible size of 2 cm for EAB attack. The number of EAB holes and woodpecker holes generally increased with canopy health classes. Many of the healthy trees exhibited a very large crown ratio averaging 81% for the trees with a canopy health class rating of 1, typical of trees growing in open conditions. Presence of basal sprouts, epicormic sprouts and splitting increased with canopy health classes, although sprouts decreased from category 4 to 5 with the death of the trees.

Table 2. Measured variables for focal ash trees in five canopy health classes, range of values are given as well as average in parentheses for DBH, number of EAB and woodpecker holes, and crown ratio (%). Basal sprouts, epicromic sprouts, and splitting were presence/absence data; therefore, percent of trees with the variable present is reported.

Ash Variable	Canopy Class 1	Canopy Class 2	Canopy Class 3	Canopy Class 4	Canopy Class 5
DBH (cm)	13.5–30.2 (19.7)	13.0–24.7 (19.5)	14.2–21.3 (16.7)	17.2–26.6 (21.3)	13.5–23.8 (16)
# EAB holes	0	0–7 (0.7)	0–2 (0.67)	0–11 (3.4)	0–9 (2)
# Woodpecker holes	0	0–1 (0.3)	0–11 (4.34)	0–30 (15.8)	6–20 (17.1)
Crown Ratio %	70–90 (81)	65–80 (72.5)	50–70 (58)	10–50 (31)	0
Basal sprouts % present	22	22	83	80	75
Epicormic sprouts % present	22	10	67	90	75
Splitting % present	44	70	100	100	100

While the Oak Openings site contains sloan loam, udorthents loam, granby loam, Oakville fine sand, Ottokee fine sand, Dixboro fine sand, Tedrow fine sand, the floodplain soil was made up of saturated loam (So) for all the sampled ash trees. No significant differences were found in available soil nutrients among the ash canopy health classes. No significant correlations were found between EAB symptoms and soil nutrients or neighboring tree abundance. The soil samples had relatively neutral pH and only sulfur and phosphorus varied in abundance between treeless sites and ash samples (Table 3).

Table 3. The mean, minimum (min) and maximum (max) for nutrient amounts with significant test results.

Nutrient	Treeless			Ash		
	Min	Mean	Max	Min	Mean	Max
S *,† (ppm)	10	13	17	15	28	67
P *,† (mg/kg)	19	31	45	18	68	115
Bray II P † (mg/kg)	50	59	81	45	76	127
Cu *,† (mg/kg)	3.60	5.75	9.32	3.08	7.05	10.24
Mn *,† (mg/kg)	37	49	82	21	58	99
Al *,† (mg/kg)	394	482	650	265	470	588
Na † (mg/kg)	16	25	40	25	44	84
Estimated Nitrogen Release † (#'s N/acre)	89	97	107	86	108	126
pH †	7.0	7.4	7.8	6.8	7.6	7.9
B *,† (mg/kg)	0.90	1.09	1.27	0.77	1.19	1.67
Total Exchange Capacity (meq/100 g)	12.62	16.96	24.88	14.23	24.73	30.61
Ca ** (%)	72.04	77.19	83.1	72.98	81.95	87.66
Other Bases ** (%)	3.60	3.99	4.40	3.50	3.83	4.60
H ** (%)	0	0	0	0	0.07	3
K ** (%)	0.92	1.23	1.51	0.64	1.17	1.61
Mg ** (%)	11.69	16.96	22.05	7.14	12.21	20.12
Na ** (%)	0.47	0.62	0.80	0.55	0.77	1.30
Ca * (mg/kg)	1900	2636	4135	2239	4059	5065
K * (mg/kg)	56	82	128	38	115	170
Fe * (mg/kg)	225	284	357	199	355	484
Mg * (mg/kg)	273	335	421	152	359	585
Zn * (mg/kg)	5.28	6.51	8.89	6.21	10.51	16.00
Organic Matter (%)	3.90	4.83	6.32	3.61	6.67	11.65

The nutrient amounts are separated by those found in soil samples taken from treeless sites and ash tree samples ($n = 52$). Ash tree soil samples were taken from all health classes, including recently dead individuals. Nutrients denoted with † were used in statistical analyses. Nutrients denoted with * are Mehlich III extractable elements, and ** represents the percent of a given element found in the soils total exchange capacity and are reported as received by the Brookside Laboratory.

Sulfur ($p < 0.0001$) and phosphorus (mg/kg) ($p < 0.0001$) were found to be higher in soil sampled near ash trees compared to treeless sites (Figure 2).

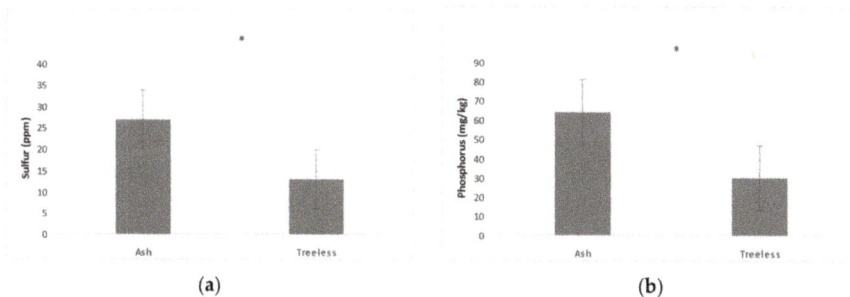

Figure 2. Differences in soil nutrient analysis between ash tree soil samples and treeless soil samples were found to be significantly different (* represent $p < 0.0001$) for: (**a**) Sulfur (ppm) and; (**b**) Phosphorus (mg/kg).

We also found two general trends in correlations between ash tree soil nutrients and tree neighbor variables. There was a positive correlation between Bray II phosphorus and total number of all neighbors ($\bar{n} = 0.38$, $p = 0.03$), and a negative correlation between copper and number of ash neighbors ($\bar{n} = -0.42$, $p = 0.01$).

4. Discussion

This study tested the relationship between neighboring tree composition, soil nutrients and canopy health class of ash after peak EAB infestation. We expected that healthier ash would be in locations with fewer tree competitors and more available nutrients important for growth, such as phosphorus. We found that the healthiest ash trees, in ash canopy health class 1, had few or no neighboring ash trees within a 6-m radius (Figure 1). Nutrient composition in the floodplain did not differ between ash canopy health classes but differed between ash tree and treeless sites in the amount of phosphorus and sulfur (Figure 2). General trends were found between copper and fewer ash neighbors, as well as phosphorus and an increase in all neighbor species. These results suggest that specific tree neighborhood composition in an EAB aftermath forest may drive ash tree susceptibility to pests.

There may be multiple reasons for the difference in phosphorus and sulfur in the soil between ash and treeless sites. Treeless sites could have anoxic conditions for a portion of the year from flooding inundation; these sites were only partially dry when samples were taken. Partial drying of previously inundated sediments will result in increased sediment affinity for phosphorus and may have resulted in the observed reduction in phosphorus at treeless sites [27]. In addition, 50% of phosphorus in areas where it is limited can be immobilized by microbes, further limiting availability to plants [28]. The amount of sulfate adsorption in the soil also increases with the amount of clay, and sites with higher clay content are more susceptible to flood due to poor drainage, like our treeless sites [29]. Organic sulfur has been shown to be highly soluble in adjacent stream slopes, leaving on average <5% unabsorbed after 24 h in a mobilization experiment [30]. Trees may not be able to establish at these soil sites as a result of nutrient adsorption to soil particles. Ash trees with fewer ash neighbors had slightly more copper in their samples, which is an important micronutrient for photosynthesis, metabolism, and potentially nitrogen fixation [31]. Ash trees with fewer neighbors were shown to have a healthier canopy, but the correlation with copper may not be the driver as variation in copper was small. The differences we found in nutrients (Table 3) may be driven by a number of factors. With no difference in soil nutrients between the ash canopy health classes, we suspect the differences found may simply be related to whether trees were present or not, and site heterogeneity.

Despite increased likelihood of EAB attacks due to ash neighbor proximity, the impact of these attacks and the degree of tolerance to them could have varied depending on various environmental factors and the specific genotype of each separate lingering ash tree, which would also contribute to variation in the number of trunk exit holes. There were exit holes in all canopy classes except class 1 (Table 2). The small number of exit holes indicated trees across most canopy classes are still being attacked but may also reflect the current low level EAB population. It is possible that the number of EAB exit holes on the lower trunk do not reflect the number in the canopy, which were not counted. For example, one study has shown locations where larvae development increased on stems/branches up to 13 cm in diameter and occurred at certain bark thickness (1.5–5 mm) [32]. Another study revealed that two mid-canopy branches (sampling two 25 cm sections each) were 18 times more likely to allow detection of low density EAB larvae than a trunk window (25 cm wide by 25% circumference of trunk bark removal above 1.3 m) [33]. The number of exit holes may not be representative of the amount of EAB feeding damage within the tree because in lingering ash trees, host defenses may have prevented larvae from becoming adults. Even if a higher proportion of larvae was killed, larval feeding prior to death still caused damage, impacting canopy development.

We found that the number of neighboring ash trees was related to ash canopy health (Figure 1). One potential explanation for this finding includes intraspecific competition: this floodplain was dominated by ash, leaving less chance for other tree species interactions. Local conspecifics can also have effects on pests or pathogens, with EAB the most likely culprit. According to the resource concentration theory, live ash near each other are more likely to encounter EAB [2]. In our case, this theory works in that ash with few or no neighbors are healthier. However, low density ash tree stands, which presumably would have had fewer ash neighbors, died from EAB faster than ash in high density stands, supporting the resource dilution theory [4]. There are two possible explanations for this discrepancy, which may be interconnected: that the relationship between ash neighbors and ash health is: (1) scale-dependent, with resource concentration theory operating at neighborhood scales and resource dilution theory operating at stand scales or (2) invasion phase dependent, with resource dilution theory operating during the initial invasion of EAB and resource concentration theory operating during the aftermath phase.

There are examples of scale-dependence in tree density and pest density. Observing the neighborhood composition at a small scale surrounding individual ash has revealed a specific distance at which ash experience conspecific interactions. Female and male Asian ash trees (*F. manshurica*) have been shown to have negative effects on same sex ash trees at distances under 10 m, suggesting intrasexual competition in ash at a small spatial scale [25]. In insect studies, a honeylocust tree (*Gleditsia tricanthos*) study showed pest susceptibility varied at different scales; density of honeylocust had effects on three pest species (honeylocust plant bug, honeylocust spider mite, mimosa webworm) abundances at the largest scale (100 m), whereas only one pest (non-native mimosa webworm) had a slightly reduced abundance from increased honeylocust density at the smallest scale (10 m) [34]. Scale of response to forest cover by native long horn beetle species had a wide range that varied by species, indicating beetle spatial response should not be assumed based on similar species [35]. EAB scale of dispersal from host trees has been shown to be mostly within 100 m, and up to 200 m, dependent on ash phloem abundance [13]. These studies show how both tree density and pest density have scale-dependent interactions. Additional studies in multiple spatial scales would be needed to fully understand the scale-dependence of EAB-ash interactions in an aftermath forest.

It is also possible that different processes operate during different phases: the initial phase is characterized by high EAB populations, highly susceptible ash populations, and rapid ash mortality. Although female EAB are efficient at locating healthy green ash, they prefer stressed, but not dying trees [36]. During the initial invasion of EAB, the majority of ash trees were stressed. The aftermath phase is characterized by low EAB populations, small ash seedlings and saplings, and lingering ash populations that may exhibit various defense mechanisms and levels of resistance to EAB infestation. In an aftermath forest, a major selection event has occurred where susceptible trees have been killed by EAB, and the remaining trees are likely to possess genetic variations that may favor their survival of EAB. Research has shown that some lingering ash genotypes (~5% of the initial ones tested) had some type of measurable defense response including killing a high proportion of larvae and reduced feeding preference by adult EAB [18]. At the time this study was initiated, the lingering ash trees had survived 3 years longer than when the Koch et al. studies were performed [18]. During these additional years, some of the ash trees continued to decline and die, so additional selection for trees with defenses against EAB occurred, making it likely that a higher percentage of the surviving trees included in our study have some level of defense against EAB. Therefore, the surviving ash trees in our study may lack cues that attract females to them, thus the mechanism by which EAB females choose to feed and lay eggs may now be more dependent on proximity to ash neighbor trees that have such cues. This also appears to be playing a role in the extended survival reported in blue ash [36]. Feeding bioassays have shown blue ash is less preferred by adult beetles than green and white ash [37]. Recent egg bioassay experiments conducted on mature blue and green ash trees growing in natural forests found that when eggs were placed directly onto the trees, larvae developed equally well in both species, leading the authors to conclude that extended survival of blue ash was due to adult beetle preference (feeding, oviposition or both) [38]. Taken together, these results provide support for the hypothesis that the discrepancy between the results in our current study (that support the resource concentration theory) and the results reported by Knight et al. [4], which support the resource dilution theory, may be due to an EAB invasion phase dependent relationship between ash health and ash neighbors. Additional studies conducted during these different phases are necessary to further evaluate and confirm this relationship.

5. Conclusions

Our findings indicated that neighboring conspecific trees have an influence on individual ash tree canopy health. Regardless of the mechanism responsible, this result provides the first suggestion that silvicultural interventions could potentially play a role in ash conservation. For example, thinning other ash trees (perhaps less healthy or smaller trees) around a healthy surviving ash tree may be beneficial. Experimental studies are needed to determine whether this management strategy would have the desired result. This research suggests that consideration of the tree neighborhood can be important for projects seeking tree resistance to forest pests. It also suggests that the remaining ash are different from those in the pre-EAB population, supporting the idea that selection and breeding of these trees to further improve EAB-resistance are important for projects seeking to increase resistance. Continued monitoring and field experiments are needed to better understand the drivers of ash tree survival in EAB aftermath landscapes, to inform management and tree resistance breeding strategies, and to ultimately ensure the future of ash species in natural areas.

Acknowledgments: We appreciate the opportunity to study at the Toledo Metroparks. This work would not be possible without the support of Bowling Green State University and the US Forest Service Forest Health Protection STDP grant NA-2014-04. We thank Robert Long for helpful comments on an earlier version of the manuscript. We would like to thank Taylor Peeps, Rachel Bienemann, and Anthony Kappler for assisting with data collection.

Author Contributions: R.H.K. and K.V.R. conceived and designed the experiments; R.H.K. and K.S.K. collected the data; R.H.K., K.S.K. and K.V.R. analyzed the data; R.H.K., K.S.K., J.K. and K.V.R. wrote the paper.

Conflicts of Interest: The authors declare no conflict of interest. The founding sponsors had no role in the design of the study; in the collection, analyses, or interpretation of data; in the writing of the manuscript, and in the decision to publish the results.

Appendix A

Nonparametric Comparisons For Each Pair Using Wilcoxon Method

q*	Alpha
1.95996	0.05

Level	- Level	Score Mean Difference	Std Err Dif	Z	p-Value	Hodges-Lehmann	Lower CL	Upper CL	
3	1	6.47500	2.219667	2.91710	0.0035*	2.00000	1.00000	3.000000	
3	2	5.25000	2.377782	2.20794	0.0272*	2.00000	0.00000	3.000000	
4	1	5.25000	2.422902	2.16682	0.0302*	1.00000	0.00000	2.000000	
2	1	3.30000	2.230176	1.47970	0.1390	0.00000	0.00000	1.000000	
5	1	3.26250	2.118198	1.54022	0.1235	0.00000	0.00000	2.000000	
4	2	2.10000	2.552405	0.82275	0.4106	0.00000	-1.00000	1.000000	
5	2	0.90000	2.342227	0.38425	0.7008	0.00000	-1.00000	2.000000	
5	4	-0.64773	2.480607	-0.26112	0.7940	0.00000	-2.00000	1.000000	
5	3	-3.37500	2.133339	-1.58203	0.1136	-1.00000	-3.00000	1.000000	
4	3	-3.92045	2.537863	-1.54479	0.1224	-1.00000	-3.00000	1.000000	

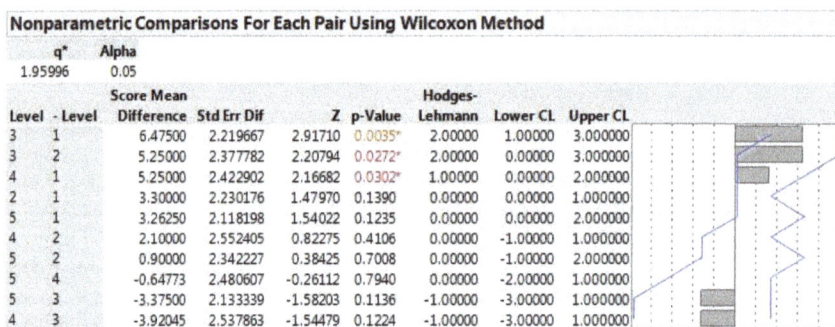

Figure A1. Results from the post-hoc paired Wilcoxon test among ash canopy health classes. Levels 1–5 represent the canopy health classes 1–5.

References

1. Grime, J.P. Evidence for the existence of three primary strategies in plant and its relevance to ecological and evolutionary theory. *Am. Nat.* **1977**, *111*, 1169–1194. [CrossRef]
2. Root, R.B. Organization of a plant-arthropod association in simple and diverse habitats: The fauna of collards (*Brassica oleracea*). *Ecol. Monogr.* **1973**, *43*, 95–124. [CrossRef]

3. Otway, S.J.; Hector, A.; Lawton, J.H. Resource dilution effects on specialist herbivores in a grassland biodiversity experiment. *J. Anim. Ecol.* **2005**, *74*, 234–240. [CrossRef]

4. Knight, K.S.; Brown, J.P.; Long, R.P. Factors affecting the survival of ash (*Fraxinus* spp.) trees infested by emerald ash borer (*Agrilus planipennis*). *Biol. Invasions* **2013**, *15*, 371–383. [CrossRef]

5. Herms, D.A.; Mattson, W.J. The dilemma of plants: To grow or defend. *Q. Rev. Biol.* **1992**, *67*, 283–335. [CrossRef]

6. Wallertz, K.; Petersson, M. Pine weevil damage to Norway spruce seedlings: Effects of nutrient loading, soil inversion and physical protection during seedling establishment. *Agric. For. Entomol.* **2011**, *13*, 413–421. [CrossRef]

7. Zas, R.; Sampredo, L.; Moreira, X.; Martíns, P. Effect of fertilization and genetic variation on susceptibility of Pinus radiata seedlings to Hylobius abietis damage. *Can. J. For. Res.* **2008**, *38*, 63–72. [CrossRef]

8. Rubert-Nason, K.F.; Couture, J.J.; Major, I.T.; Constabel, C.P.; Lindroth, R.L. Influence of Genotype, environment, and gypsy moth herbivory on local and systemic chemical defenses in trembling aspen (*Populus tremuloides*). *J. Chem. Ecol.* **2015**, *41*, 651–661. [CrossRef] [PubMed]

9. MacFarlane, D.W.; Meyer, S.P. Characteristics and distribution of potential ash tree hosts for emerald ash borer. *For. Ecol. Manag.* **2005**, *213*, 15–24. [CrossRef]

10. Herms, D.A.; McCullough, D.G. Emerald ash borer invasion of North America: History, biology, ecology, impacts, and management. *Annu. Rev. Entomol.* **2014**, *59*, 13–30. [CrossRef] [PubMed]

11. Taylor, R.A.J.; Bauer, L.S.; Poland, T.M.; Windell, K.N. Flight performance of *Agrilus planipennis* (Coleoptera: Buprestidae) on a flight mill and in free flight. *J. Insect Behav.* **2010**, *23*, 128–148. [CrossRef]

12. Poland, T.M.; Chen, Y.; Koch, J.; Pureswaran, D. Review of the emerald ash borer (Coleoptera: Buprestidae), life history, mating behaviours, host plant selection, and host resistance. *Can. Entomol.* **2015**, *147*, 252–262. [CrossRef]

13. Siegert, N.W.; Mercader, R.J.; McCullough, D.G. Spread and dispersal of emerald ash borer (Coleoptera: Buprestidae): Estimating the spatial dynamics of a difficult-to-detect invasive forest pest. *Can. Entomol.* **2015**, *147*, 338–348. [CrossRef]

14. Flower, C.E.; Knight, K.S.; Gonzalez-Meler, M.A. Impacts of the emerald ash borer (*Agrilus planipennis* Fairmaire) induced ash (*Fraxinus* spp.) mortality on forest carbon cycling and successional dynamics in the Eastern United States. *Biol. Invasions* **2013**, *15*, 931–944. [CrossRef]

15. Brakie, M. Plant Guide for green ash (*Fraxinus pennsylvanica*). In *USDA-Natural Resources Conservation Service*; East Texas Plant Materials Center: Nacogdoches, TX, USA, 2013.pennsylvanica. In *USDA-Natural Resources Conservation Service*; East Texas Plant Materials Center: Nacogdoches, TX, USA, 2013.

16. Kennedy, H.E., Jr. *Fraxinus pennsylvanica* Marsh. Available online: http://www.na.fs.fed.us/pubs/silvics_manual/volume_2/fraxinus/pennsylvanica.htm (accessed on 26 June 2016).

17. Stewart, H.A.; Krajicek, J.E. Ash, an American wood. In *American Woods Series FS216*; USDA Forest Service: Washington, DC, USA, 1973; p. 7.

18. Koch, J.L.; Carey, D.W.; Mason, M.E.; Poland, T.M.; Knight, K.S. Intraspecific variation in *Fraxinus pennsylvanica* responses to emerald ash borer (*Agrilus planipennis*). *New For.* **2015**, *45*, 995–1011. [CrossRef]

19. Knight, K.S.; Herms, D.; Plumb, R.; Sawyer, E.; Spalink, D.; Pisarczyk, E.; Wiggin, B.; Kappler, R.; Menard, K. Dynamics of surviving ash (*Fraxinus* spp.) populations in areas long infested by emerald ash borer (*Agrilus planipennis*). In Proceedings of the 4th International Workshop on the Genetics of Host-Parasite Interactions in Forestry: Disease and Insect Resistance in Forest Trees, Eugene, OR, USA, 31 July–5 August 2011; PSW-GTR-240; Tech. Cords. Sniezko, R.A., Yanchuk, A.D., Kliejunas, J.T., Palmieri, K.M., Alexander, J.M., Frankel, S.J., Eds.; Department of Agriculture, Forest Service, Pacific Southwest Research Station: Albany, CA, USA, 2012; pp. 143–152.

20. Brewer, L.G.; Vankat, J.L. Description of vegetation of the oak openings of Northwestern Ohio at the time of Euro-American settlement 1. *Ohio J. Sci.* **2004**, *104*, 76–85.

21. Faber-Langendoen, D. Plant Communities of the Midwest. In *Classification in an Ecological Context*; Association for Biodiversity Information: Arlington, VA, USA, 2001; p. 5.

22. USDA NRCS WSS. Available online: https://websoilsurvey.sc.egov.usda.gov/App/WebSoilSurvey.aspx (accessed on 20 June 2017).

23. Knight, K.S.; Flash, B.P.; Kappler, R.H.; Throckmorton, J.A.; Grafton, B.; Flower, C.E. *Monitoring Ash (Fraxinus spp.) Decline and Emerald Ash Borer (Agrilus planipennis) Symptoms in Infested Areas*; Service General Technical Report NRS-139; United States Department of Agriculture Forest Northern Research Station: Delaware, OH, USA, 2014.

24. Smith, A. Effects of Community Structure on Forest Suceptibility and Responce to the Emerald Ash Borer Invasion of the Huron River Watershed in Southeast Michigan. Master's Thesis, Ohio State University, Columbus, OH, USA, 2006; p. 122.

25. Zhang, C.; Zhao, X.; Gao, L.; Gadow, K.V. Gender, neighboring competition and habitat effects on the stem growth in dioecious *Fraxinus mandshurica* trees in a northern temperate forest. *Ann. For. Sci.* **2009**, *66*, 812. [CrossRef]

26. Canham, C.D.; LePage, P.T.; Coates, K.D. A neighborhood analysis of canopy tree competition: Effects of shading versus crowding. *Can. J. For. Res.* **2004**, *34*, 778–787. [CrossRef]

27. Baldwin, D.S.; Mitchell, A.M. The effects of drying and re-flooding on the sediment and soil nutrient dynamics of lowland river-floodplain systems: A synthesis. *Regul. Rivers Res. Manag.* **2000**, *16*, 457–467. [CrossRef]

28. Cross, A.F.; Schlesinger, W.H. A literature review and evaluation of the Hedly fractionation: Applications to the biogeochemical cycle of soil phosphorus in natural ecosystems. *Geoderma* **1995**, *64*, 197–214. [CrossRef]

29. Neller, J.R. Extractable sulfate-sulfur in soils of Florida in relation to amount of clay in the profile. *Soil Sci. Soc. Am. J.* **1959**, *23*, 346–348. [CrossRef]

30. Dail, D.B.; Fitzgerald, J.W. S Cycling in soil and stream sediment: Influence of season and in situ concentrations of carbon, nitrogen and sulfur. *Soil Biol. Biochem.* **1999**, *31*, 1395–1404. [CrossRef]

31. Brady, N.C.; Weil, R.R. Soil phosphorus, potassium, and micronutrients. In *Elements of the Nature and Properties of Soils*; Prentice-Hall: Upper Saddle River, NJ, USA, 1999; pp. 421–422.

32. Timms, L.L.; Smith, S.M.; DeGroot, P. Patterns in the within-tree distribution of the emerald ash borer *Agrilus planipennis* (Fairmaire) in young, green-ash plantations of South-Western Ontario, Canada. *Agric. For. Entomol.* **2006**, *8*, 313–321. [CrossRef]

33. Ryall, K.L.; Fidgen, J.G.; Turgeon, J.J. Detectability of the emerald ash borer (Coleoptera: Buprestidae) in asymptomatic urban trees by using branch samples. *Environ. Entomol.* **2011**, *40*, 679–688. [CrossRef] [PubMed]

34. Sperry, C.E.; Chaney, W.R.; Shago, G.; Sadof, C.S. Effects of tree density, tree species diversity and percentage of hardscape on tree insect pests of honeylocust. *JOA* **2001**, *27*, 263–271.

35. Holland, J.D.; Bert, D.G.; Fahrig, L. Determining the spatial scale of species response to habitat. *Bioscience* **2004**, *54*, 227–233. [CrossRef]

36. Tanis, S.R.; McCullough, D.G. Differential persistence of blue ash and white ash following emerald ash borer invasion. *Can. J. For. Res.* **2012**, *42*, 1542–1550. [CrossRef]

37. Puraswaran, D.S.; Poland, T.M. Host selection and feeding preference of *Agrilus planipennis* (Coleoptera: Buprestidae) on ash (*Fraxinus* spp.). *Environ. Entomol.* **2009**, *38*, 757–765. [CrossRef]

38. Peterson, D.; Duan, J.J.; Yaninek, J.S.; Ginzel, M.; Sadof, C. Growth of larval *Agrilus planipennis* (Coleoptera: Buprestidae) and fitness of *Tetrastichus planipennisi* (Hymenoptera: Eulophidae) in blue ash (*Fraxinus quadrangulate*) and green ash (*F. pennsylvanica*). *Environ. Entomol.* **2015**, *44*, 1–10. [CrossRef] [PubMed]

forests

MDPI

Article

Evidence of Ash Tree (*Fraxinus* spp.) Specific Associations with Soil Bacterial Community Structure and Functional Capacity

Michael P. Ricketts [1,*], Charles E. Flower [1,2], Kathleen S. Knight [2] and Miquel A. Gonzalez-Meler [1]

[1] Biological Sciences Department, Ecology and Evolution, University of Illinois at Chicago, 845 W. Taylor St., Chicago, IL 60607, USA; charlesflower@fs.fed.us (C.E.F.); mmeler@uic.edu (M.A.G.-M.)
[2] USDA Forest Service, Northern Research Station, 359 Main Rd, Delaware, OH 43015, USA; ksknight@fs.fed.us
* Correspondence: rickett4@uic.edu; Tel.: +1-309-229-3270

Received: 1 March 2018; Accepted: 3 April 2018; Published: 5 April 2018

Abstract: The spread of the invasive emerald ash borer (EAB) across North America has had enormous impacts on temperate forest ecosystems. The selective removal of ash trees (*Fraxinus* spp.) has resulted in abnormally large inputs of coarse woody debris and altered forest tree community composition, ultimately affecting a variety of ecosystem processes. The goal of this study was to determine if the presence of ash trees influences soil bacterial communities and/or functions to better understand the impacts of EAB on forest successional dynamics and biogeochemical cycling. Using 16S rRNA amplicon sequencing of soil DNA collected from ash and non-ash plots in central Ohio during the early stages of EAB infestation, we found that bacterial communities in plots with ash differed from those without ash. These differences were largely driven by Acidobacteria, which had a greater relative abundance in non-ash plots. Functional genes required for sulfur cycling, phosphorus cycling, and carbohydrate metabolism (specifically those which breakdown complex sugars to glucose) were estimated to be more abundant in non-ash plots, while nitrogen cycling gene abundance did not differ. This ash-soil microbiome association implies that EAB-induced ash decline may promote belowground successional shifts, altering carbon and nutrient cycling and changing soil properties beyond the effects of litter additions caused by ash mortality.

Keywords: soil bacteria; 16S rRNA; ash tree; emerald ash borer; forest disturbance; invasive species

---

1. Introduction

Anthropogenic disturbances to Earth's ecosystems have the potential to alter the abundances and distributions of organisms worldwide, [1,2] and therefore the structure and function of their environments [3–6]. Such disturbances include warming air temperatures, changing precipitation patterns, severe weather events, atmospheric nutrient deposition, or the introduction of invasive species. In temperate forest ecosystems of eastern North America, ash trees (*Fraxinus* spp.) have suffered significant declines over the past two decades due to the infestation of the invasive emerald ash borer (EAB; *Agrilus planipennis*), a wood boring beetle introduced from Asia [7,8]. The EAB selectively deposits eggs on the bark of ash trees where hatched larvae burrow into cambial tissue to feed, creating serpentine galleries and severing the distribution of water and nutrients between the roots and shoots [9]. This results in ~99% ash tree mortality within two to five years after infestation [10,11] and complete mortality within a stand in roughly five to seven years [12]. Ash trees are widely distributed throughout North America and are a major component of forest and urban tree communities, representing roughly 2.5% of the aboveground biomass stocks in the US and storing

~0.303 Pg of carbon (C) [13–16]. The widespread decline of ash has multiple cascading effects on ecosystem productivity, structure, and function, as the transformation from live standing biomass to fallen trees [17], plant litter, and soil organic matter (SOM) unfolds. Specifically, rapidly reduced water flux and plant respiration, coupled with large inputs of coarse woody debris and altered tree community composition, may significantly alter ecosystem hydrology, C and nutrient dynamics, forest tree community succession, edaphic factors, and belowground microbial community structure and function [9,18–20].

Soil microorganisms play a key role in the decomposition of SOM and regulation of nutrient availability to plants [21,22], both of which have important implications for ecosystem biogeochemical cycling and net primary productivity (NPP) [23]. Microbial functional responses to disturbances or environmental shifts, such EAB-induced ash decline, are dependent on the microbial community's resilience to change and the degree of functional redundancy within the community [24]. While functional redundancies often exist between microbial taxa, large shifts in microbial community structure may result in the altered functional capacity of the community to access and degrade SOM or perform nutrient transformations and mobilization [24–27]. Thus, identifying factors that influence microbial community structure is important to understanding potential changes in the functions of decomposers. A variety of edaphic factors are thought to influence soil microorganisms, including pH, C-availability, moisture, O_2 availability, and bulk density [28]. In particular, soil pH has been shown to be one of the governing forces driving soil microbial community structure [29–31]. Aboveground vegetation may also influence belowground microbial community structure, with specific plant species associating with (and even recruiting) unique microbial assemblages [32–34]. These above-belowground associations are most often studied at the community or ecosystem level (e.g., forest vs. grassland, deciduous vs. coniferous forests), while soil microbial associations with individual plant species or genera remain poorly understood.

This study aimed to examine soil microbial community associations with ash trees to better understand belowground consequences of EAB disturbance. Microbial functional potentials were estimated with respect to nutrient and C-cycling processes that, in turn, may affect forest recovery trajectories. If soil microbes exhibit a different community structure under stands with ash trees when compared to stands without ash trees, this would suggest a strong, genera specific relationship between the presence of, decline of, or mortality of ash trees and soil microbial communities. If instead belowground microbial communities are similar across the heterogeneous forest landscape, this would indicate a whole forest, community level influence governed by varying degrees of environmental, physical, and edaphic factors. To address these competing hypotheses, we used 16S rRNA metagenomic sequencing methods, which specifically target bacterial and archaeal organisms, to analyze archived soil DNA samples collected from paired ash and non-ash forest plots in 2011 during the early stages of EAB infestation. If differences were observed in the soil bacterial community structure between ash and non-ash plots, then we expected the functional potential to cycle C and nutrients to reflect the specific differences in the bacterial community. This work provides a unique snapshot of soil bacterial communities, their functional potentials, and their associations with dominant tree genera, during the early stages of EAB disturbance in an ash-dominated forest near the core area of infestation.

2. Materials and Methods

2.1. Site Description

In 2011, four forest sites, Bohannan Nature Preserve (BHN), Kraus Nature Preserve (KRS), Seymour Woods State Nature Preserve (SYM), and Stratford Ecological Center (STR), were selected within Delaware County, Ohio (Figure 1 and Table 1). These sites are secondary successional forests largely dominated by ash (*Fraxinus americana* L., *F. pennsylvanica* Marshall and *F. quadrangulata*). Other canopy tree genera include maple (*Acer saccharinum*, *A. saccharum*, *A. rubrum*), oak (*Quercus palustris*, *Q. rubra*,

Q. alba.), beech (*Fagus grandifolia*), shagbark hickory (*Carya ovata*), cottonwood (*Populus deltoids*), elm (*Ulmus americana, U. rubrum*), black cherry (*Prunus serotina*), black walnut (*Juglans nigra*), and willow (*Salix* spp.). In each site, we randomly established two or three "ash" plots (11.28 m radius), which contained ash trees as a major component of the canopy (48.8 ± 4.8% (mean ± S.E.) of total basal area), and two or three "non-ash" plots, which did not contain ash trees as a major component of the understory or canopy (defined as <5% of total basal area; see Table 1 and Table S3 for details). Ash and non-ash plots were located between 50–100 m away from one another and were selected to represent similar topography, soil type, and moisture regimes. Within each plot, trees >10 cm in diameter at breast height were identified and measured and the total basal area (BA) per hectare (m^2/ha), number of stems per hectare (#/ha), and relative tree dominance (%) by BA were calculated (Tables 1 and S3).

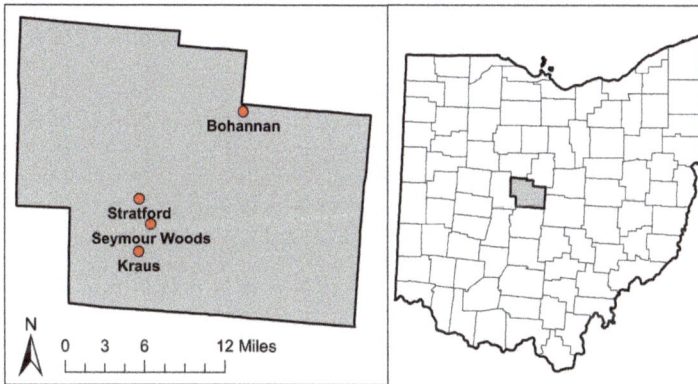

Figure 1. Map of study sites within Delaware County, Ohio.

By 2011, EAB had reached forests of central Ohio and ash trees had begun to exhibit visual symptoms of infestation at our sites. While this may not be ideal for establishing baseline associations with healthy ash trees, we were able to collect samples in the early stages of EAB infestation before complete ash mortality occurred, which is rapidly becoming more difficult to find in high-density ash tree forests. To quantify the health of trees within the plots, we used ash tree canopy condition (AC), a metric for tracking the health of ash trees exposed to EAB, which is correlated to EAB densities and tree physiology [35,36]. This assessment is a non-linear five-point categorical scale which assigns healthy trees a value of 1 and standing dead trees a value of 5. At the plot-level, ash canopy health was calculated as the mean AC of all ash trees within a plot. To account for the potential effects associated with ash trees in later states of decline, we performed a separate analysis which removed all sites that contained any plots with mean AC scores > 3, resulting in two sites consisting of six ash (AC = 2.42 ± 0.30) and four non-ash plots (Table S1).

Table 1. Summary of site characteristics. The total basal area (BA) is the mean ± standard error of all plots, while the relative BA of ash trees is from only ash plots.

Forest	Soil Type [1]	Number of Plots (Ash/Non-ash)	BA (m^2/ha)	Relative BA of Ash Trees (%)
Bohannan (BHN)	Cardington silt loam	3/2	37.7 ± 2.5	49.3 ± 5.7
Kraus (KRS)	Glynwood silt loam	3/2	34.7 ± 3.0	63.2 ± 4.4
Seymour (SYM)	Blount silt loam	2/2	26.0 ± 3.0	46.5 ± 6.0
Stratford (STR)	Glynwood silt loam	3/3	33.9 ± 5.0	35.5 ± 13.0

[1] Primary soil type ascertained from NRCS web soil survey.

2.2. Soil Collection and Characterization

To characterize potential associations between ash trees and soil bacterial communities, we randomly selected 30 locations in each plot and extracted 0–10 cm soil cores with a 1.9 cm diameter soil probe (Oakfield Model L tube sampler soil probe), which was cleaned and sterilized with 100% ethanol between plots. Soils were sampled in late July during the peak period of NPP. Roots were removed and soil samples from each plot were homogenized on site, placed in a cooler with dry ice, and stored at −80 °C until DNA extraction. Soil subsamples were analyzed for pH and a variety of solubilized soil minerals (Ca, K, Mg, P, Al, B, Cu, Fe, Mn, Na, S, and Zn) by the University of Maine Soils Lab using a modified Morgan nutrient extraction procedure and a TJA Model 975 AtomComp ICP-AES. Soil C and nitrogen (N) concentrations (%) were measured at the University of Illinois at Chicago (UIC) Stable Isotope lab using a Costech elemental analyzer (Valencia, CA, USA). Prior to analysis, samples were dried until no mass lost in a 60 °C oven, pulverized using a ball mill, and ~5mg of sample was placed into a tin capsule.

2.3. DNA Extraction, Sequencing, Quality Control and Bioinformatics

DNA was extracted from ~0.25 g of each soil sample using MoBio's PowerSoil®-htp 96 Well Soil DNA Isolation Kit as per the manufacturer's protocol. Amplification of the V4 region of the 16S SSU rRNA gene was performed using PCR primers 515F/806R following protocols outlined by the Earth Microbiome Project [37]. Final amplicon DNA concentrations were quantified using the PicoGreen® dsDNA Assay Kit and amplicons were sequenced using an Illumina MiSeq instrument (2 × 150 bp paired-end). All sequences have been deposited in the NCBI Sequence Read Archive under SRA study #SRP136455. Initial sequence data quality filtering, paired-end assembly, demultiplexing, closed reference operational taxonomic unit (OTU) picking, and phylogenetic assignments were performed using the QIIME software package version 1.9.1 (http://qiime.org/) [38]. OTU abundance data was normalized to account for estimated 16S rRNA gene copy number within each OTU assignment using the python script *normalize_by_copy_number.py* from the PICRUSt software package [39]. OTU picking identified 9387 OTU's, with an average of 2283 ± 146 OTU's per sample. In total, there were 39 phyla identified, the 10 most abundant of which encompassed 98% of all bacteria/archaea. Sequences were rarefied at 5900 sequences per sample for diversity analysis. More detailed methods can be found in Ricketts et al., 2016 [25].

The genetic functional potential of bacterial/archaeal communities was determined by estimating gene abundance using the PICRUSt software package version 1.1.0 (http://picrust.github.io/picrust/) [39]. Genetic pathways necessary for biogeochemical metabolisms were selected based on the KEGG ortholog hierarchical system, which is a knowledge database dedicated to linking genomic information to cellular and metabolic functional pathways [40]. This framework allows individual gene abundance data to be collated into broader functional groups, providing a more practical basis for functional gene analysis. We focused our analysis specifically on the energy metabolism and carbohydrate metabolism level 2 KEGG groups. Within these groups, all level 3 KEGG metabolic pathways, organized at a finer functional scale, were also analyzed.

2.4. Statistical Analyses

Bacterial community differences were explored by examining Hellinger transformed abundance data in two ways. First, the bacterial abundance differences of the 10 most abundant phyla (98.1% of total bacteria), the 20 most abundant classes (93.8% of total bacteria), and the 30 most abundant orders (90.9% of total bacteria), were analyzed between ash and non-ash plots using Mann–Whitney *U* tests and between sites using Kruskal-Wallis and posthoc Nemenyi tests, both with a significance threshold of $p < 0.05$, using the R statistical program [41]. Second, overall bacterial community structure differences between ash and non-ash plots and between sites, were analyzed by comparing Bray-Curtis dissimilarity matrices of Hellinger transformed bacterial abundances using adonis tests (similar to

PERMANOVA) in R with 99,999 permutations. Assumptions of the adonis test were verified using the *betadisper* function in the R package vegan [42], which tests the multivariate homogeneity of group dispersions (variances). A non-metric multidimensional scaling (NMDS) plot (stress = 0.080, Shepard plot non-metric R^2 = 0.994) was created using the R package phyloseq [43] and the same Bray–Curtis dissimilarity matrices to visualize differences in bacterial community structure between ash and non-ash plots and sites.

All other variables, including AC, BA, stem density, relative tree dominance, bacterial and tree alpha-diversities (Shannon diversity index), and soil factors, were analyzed for differences between ash and non-ash plots using Mann-Whitney U tests ($p < 0.05$) and for differences between sites using Kruskal-Wallis with the posthoc Nemenyi tests ($p < 0.05$). Euclidean distance matrices constructed from each variable using the *dist* function in the R package vegan [42] were compared to the soil bacterial community Bray-Curtis distance matrix (described above) using Mantel tests ($p < 0.05$) to determine how strongly each variable correlated with (or influenced) bacterial community structure. In addition, the overall soil environment was analyzed by combining all soil variables into a single Euclidian dissimilarity matrix, which was tested for ash vs. non-ash differences and site differences using adonis tests and effects on bacterial community structure using a Mantel test. To better understand the effects of EAB-induced tree stress on bacterial community structure within ash plots, linear relationships between mean AC and the ten most abundant bacterial phyla were analyzed and a Mantel test for mean AC (as described above) was performed using only ash plots.

Ash vs. non-ash differences in PICRUSt estimated functional gene abundances for the selected KEGG ortholog groups were tested in STAMP [44] using Welch's two-tailed *t*-test. To assess the significance and adjust for potential false discoveries, we utilized the Benjamini-Hochberg procedure where original *p*-values were ranked in order of significance, multiplied by the number of comparisons (Lvl 2 n = 64, Lvl 3 n = 328), and divided by their respective rank numbers to obtain a corrected *p*-value (*q*-value). The significance threshold used was $q < 0.05$. In addition, Pearson's correlations were used to determine relationships between Hellinger transformed bacterial phyla abundance and level 3 KEGG ortholog functional group gene abundance. To account for potential false discoveries here, we used the more conservative Bonferroni adjustment, where original *p*-values are simply multiplied by the number of comparisons (n = 240) and assigned a threshold of $p < 0.05$. It is important to remember that relationships between bacterial abundance and gene abundance are predetermined by algorithms used by the PICRUSt software, as all estimated gene abundance information is directly derived from bacterial abundance data in combination with genomic databases. However, it does provide information on inherent functional relationships within each bacterial phylum and reveals potential differences in function as a result of abundance differences in individual bacterial taxonomic groups.

3. Results

3.1. Environmental and Site Differences

The overall soil environment was similar between ash and non-ash plots (adonis H = 0.098, p = 0.065), but differed across sites (adonis H = 0.301, p = 0.003). Specifically, only two of the 16 soil variables, Cu (W = 12.5, p = 0.006) and Fe (W = 18, p = 0.016), differed between ash and non-ash plots (Table 2), where Cu and Fe were both greater in non-ash plots. Between sites, the %C (H = 11.51, p = 0.009), %N (H = 12.96, p = 0.005), C:N (H = 10.15, p = 0.017), P (H = 12.35, p = 0.006), Al (H = 9.71, p = 0.021), and Zn (H = 9.79, p = 0.020) were different (Table 2). Posthoc tests revealed both %C and %N to be significantly lower at SYM compared to the other sites, while C:N remained constant across sites, with the exception of being significantly lower at BHN. Similarly, soil P, Al, and Zn were lower at SYM (Table S2).

Analysis of non-soil variables revealed ash tree health (mean AC) to be variable between sites (H = 9.24, p = 0.026; Table 2). Total BA (m^2/ha) did not differ between ash and non-ash plots or between sites, although it was somewhat lower at SYM where the stem density (#/ha) was highest (H = 8.78,

$p = 0.032$) due to a large number of small trees (Tables S2 and S3). Of the five most abundant tree genera, only oak species relative dominance differed between ash and non-ash plots ($p = 0.003$) and only beech tree relative dominance differed between sites ($p = 0.007$; Table 2). Oak trees had a higher relative dominance in non-ash plots vs. ash plots and beech trees were more dominant in KRS than any of the other sites. Tree community alpha-diversity was not different between plots ($W = 60.5$, $p = 0.425$) or sites ($H = 5.67$, $p = 0.129$) and did not correlate with the soil bacterial community (mantel r-statistic $= -0.048$, $p = 0.631$; Table 2 and Table S2).

Table 2. Summary of statistical results. Adonis tests were used to analyze differences in overall bacterial community structure and overall soil chemical characteristics between categorical variables (**a**). Continuous variables were analyzed individually (**b**) for differences between ash and non-ash plots (Mann-Whitney U test), differences in forest sites (Kruskal-Wallis), and for correlations between overall bacterial community structure and individual variables (Mantel test). Alpha diversity was calculated using the Shannon diversity index (H). Text in bold and italics represents a significant result ($p < 0.05$).

(a)	Adonis Test			
	Bacterial Community		**Soil Environment**	
Categorical Variables	R^2	*p*-value	R^2	*p*-value
Ash *vs.* Non-ash	*0.334*	*0.002*	0.098	0.066
Forest site	0.140	0.502	*0.301*	*0.003*

(b)	**Mann-Whitney U Test** (Ash *vs.* Non-Ash)		**Kruskal-Wallis Test** (Forest Site; df = 3)		**Mantel Test** (Bacterial Community)	
Continuous Variables	W	*p*-value	H	*p*-value	*r*-statistic	*p*-value
Mean AC (ash only)	-	-	*9.24*	*0.026*	−0.060	0.620
Mean Stems (#/ha)	75.5	0.051	*8.78*	*0.032*	−0.127	0.870
Mean BA (m²/ha)	69	0.152	5.23	0.156	0.060	0.261
Ash (%)	-	-	1.19	0.755	*0.264*	*0.007*
Maple (%)	49	1.000	4.42	0.220	0.041	0.306
Oak (%)	*11*	*0.003*	4.13	0.247	0.030	0.338
Beech (%)	57	0.570	*12.09*	*0.007*	0.182	0.097
Hickory (%)	53.5	0.743	3.03	0.387	0.028	0.334
α-diversity (tree)	60.5	0.425	5.67	0.129	−0.048	0.631
α-diversity (bacteria)	54	0.766	4.07	0.254	0.039	0.329
Soil pH	73	0.080	3.88	0.275	*0.289*	*0.006*
%C	49.5	1.000	*11.51*	*0.009*	−0.173	0.981
%N	58	0.541	*12.96*	*0.005*	−0.175	0.986
C:N	34.5	0.270	*10.15*	*0.017*	−0.134	0.911
Ca	69	0.152	4.53	0.210	*0.304*	*0.007*
K	42	0.603	3.71	0.295	−0.030	0.594
Mg	67	0.201	4.81	0.186	*0.274*	*0.011*
P	49	1.000	*12.35*	*0.006*	−0.088	0.846
Al	29	0.131	*9.71*	*0.021*	*0.177*	*0.045*
B	40	0.494	3.64	0.303	−0.075	0.708
Cu	12.5	*0.006*	0.32	0.957	0.047	0.304
Fe	*18*	*0.016*	1.49	0.685	0.273	0.010
Mn	48	0.941	2.11	0.550	−0.143	0.921
Na	56	0.656	6.37	0.095	0.002	0.439
S	24	0.056	0.88	0.831	−0.143	0.924
Zn	40	0.503	*9.79*	*0.020*	0.083	0.241

3.2. Bacterial Community Differences

Soil bacterial community structure (i.e., beta-diversity) differed between ash and non-ash plots (adonis $R^2 = 0.334$, $p = 0.002$), but not between sites (adonis $R^2 = 0.140$, $p = 0.501$; Figure 2 and Table 2). Ash tree relative dominance was the only tree genera to show a significant correlation with bacterial community structure (mantel *r-statistic* = 0.264, *p* = 0.007). Although the overall soil environment did not show a strong relationship with bacterial community structure (mantel *r-statistic* = 0.053, *p* = 0.305), certain individual soil variables did, including soil pH (mantel *r-statistic* = 0.289, *p* = 0.006), Ca (mantel *r-statistic* = 0.304, *p* = 0.007), Mg (mantel *r-statistic* = 0.274, *p* = 0.011), and Al (mantel *r-statistic* = 0.177, *p* = 0.045; Table 2). It should be noted that Mg, Ca, and Al are all highly correlated with soil pH (>0.79, *p* < 0.001).

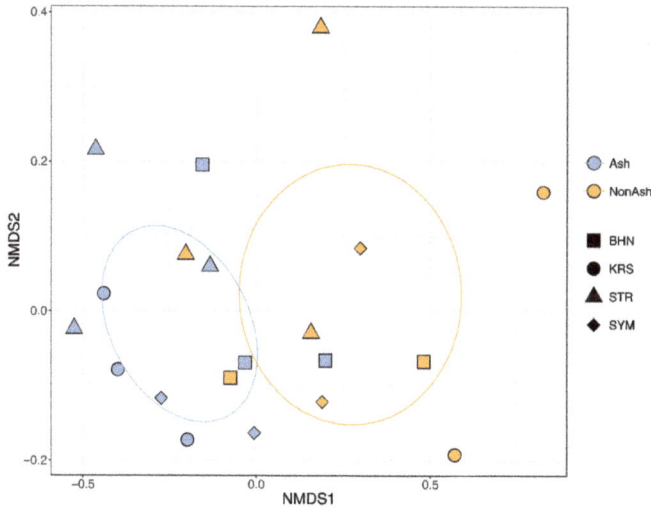

Figure 2. Non-metric multidimensional scaling (NMDS) plot where each point represents the bacterial/archaeal community structure of a sample (stress = 0.080, Shepard plot non-metric R^2 = 0.994). Color indicates ash *vs.* non-ash plots and shape indicates forest site. Ellipses represent 95% confidence intervals of centroids for ash and non-ash plots. Bacterial/archaeal community structures differed significantly between ash and non-ash plots (adonis *p* = 0.002).

We also found significant differences between ash and non-ash plots in the relative abundances of seven out of 10 of the most abundant bacterial phyla (Figure 3); however, between forest sites, there were no abundance differences in any of the phyla. Likewise, EAB-induced tree stress (i.e., mean AC) did not affect bacterial abundances (Figure 4). All phyla were less abundant in non-ash plots, except Acidobacteria and Elusimicrobia, which were more abundant in non-ash plots (*p* = 0.004 and *p* = 0.261 respectively). At finer taxonomic levels, these differences were not as noticeable, with only two out of 20 of the most abundant classes and two out of 30 of the most abundant orders showing significant differences between ash and non-ash plots (Figures S1 and S2). Interestingly, all four of these differences were in the Actinobacteria phylum, which were more abundant in the ash plots. Soil bacterial alpha-diversity did not vary between ash and non-ash plots (*W* = 54, *p* = 0.766) or between sites (*H* = 4.07, *p* = 0.254) and showed no relationship with bacterial community structure (mantel *r-statistic* = 0.264, *p* = 0.007; Tables 2 and S2).

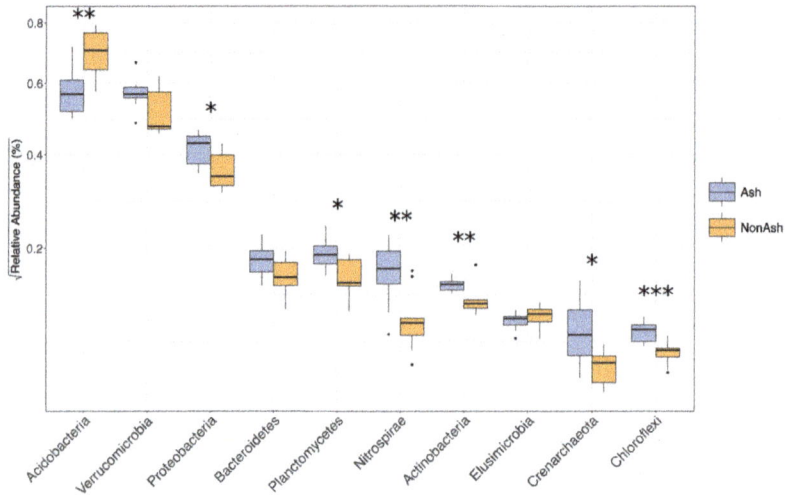

Figure 3. Boxplot comparing the average Hellinger transformed abundances of the 10 most abundant bacterial/archaeal phyla between ash (blue) and non-ash (orange) plots. Mann-Whitney U-test significance is denoted by asterisks, where * = $p < 0.05$, ** = $p < 0.01$ and *** = $p < 0.001$.

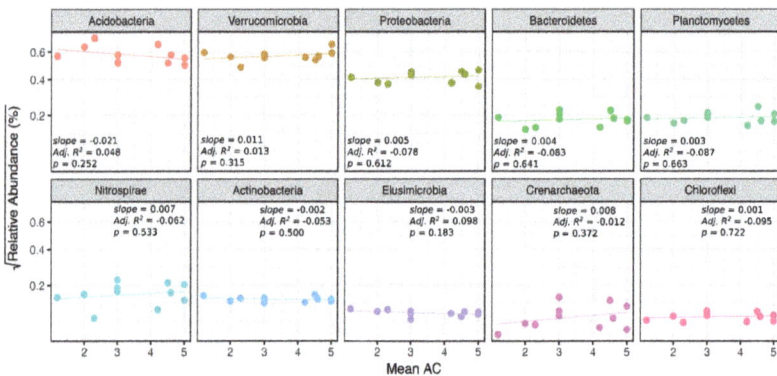

Figure 4. Linear relationships between canopy tree health (mean AC) of ash plots only ($n = 11$) and Hellinger transformed abundances of the 10 most abundant bacterial phyla.

3.3. Bacterial Functional Differences

Bacterial community differences between ash and non-ash plots resulted in estimated functional potential differences. At KEGG level 2 (see methods), differences in PICRUSt-estimated functional gene abundances were found in both energy metabolism (ash > non-ash; $d = 1.13$, $q = 0.047$) and carbohydrate metabolism (non-ash > ash; $d = -1.68$, $q = 0.015$; Figure 5). At KEGG level 3 within the energy metabolic pathways, three of the nine ortholog groups (carbon fixation pathways in prokaryotes, $d = 1.82$, $q = 0.060$; methane metabolism, $d = 1.80$, $q = 0.048$; and carbon fixation in photosynthetic organisms, $d = 1.56$, $q = 0.018$) were significantly more abundant in ash plots than non-ash. In contrast, four of the nine groups (sulfur metabolism, $d = -1.66$, $q = 0.018$; photosynthesis, $d = -1.37$, $q = 0.029$; oxidative phosphorylation, $d = -1.37$, $q = 0.029$; and photosynthesis proteins, $d = -1.27$, $q = 0.042$) were more abundant in non-ash plots (Figure 5b). Nitrogen metabolism capacity was not different in ash *vs.* non-ash plots.

Within the KEGG carbohydrate metabolic pathways, seven out of 15 ortholog groups were significantly more abundant in non-ash plots (Figure 5b). These include pentose and glucuronate interconversions ($d = -1.74$, $q = 0.037$), galactose metabolism ($d = -1.71$, $q = 0.023$), ascorbate and aldarate metabolism ($d = -1.68$, $q = 0.020$), starch and sucrose metabolism ($d = -1.70$, $q = 0.018$), inositol phosphate metabolism ($d = -1.67$, $p = 0.018$), amino sugar and nucleotide sugar metabolism ($d = -1.65$, $q = 0.018$), and the pentose phosphate pathway ($d = -1.36$, $q = 0.023$). However, four out of the 15 groups were significantly more abundant in ash plots, including the tricarboxylic acid (TCA) cycle (a.k.a. Krebs cycle; $d = 1.74$, $q = 0.027$), pyruvate metabolism ($d = 1.66$, $q = 0.018$), butanoate metabolism ($d = 1.61$, $q = 0.018$), and glycolysis/gluconeogenesis ($d = 1.38$, $q = 0.025$).

a)   KEGG Level 2

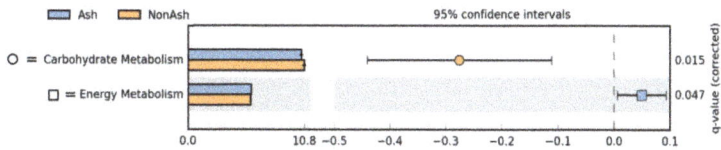

b)   KEGG Level 3

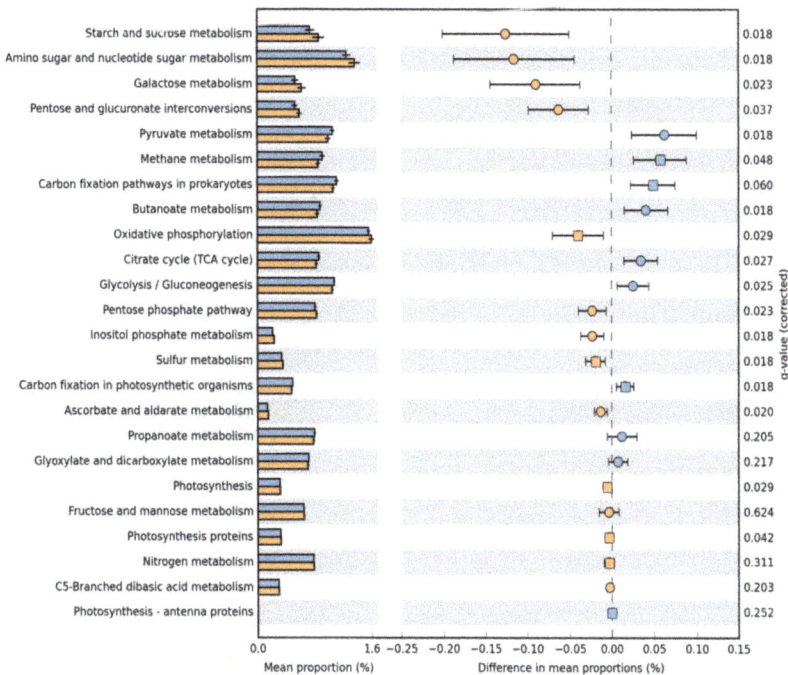

Figure 5. Functional gene abundance comparisons between ash (blue) and non-ash (orange) plots at KEGG levels 2 (**a**) and 3 (**b**). Extended bar graphs (left) show differences in the mean proportions of functional genes required for biogeochemical cycling and are ordered by decreasing effect size (right), calculated by subtracting non-ash from ash mean proportions. The color of the effect size markers indicate in which plots gene abundance was greater and the shape indicates KEGG grouping, where circles represent carbohydrate metabolism and squares represent energy metabolism. Error bars represent 95% Welch's inverted confidence intervals. Welch's two-tailed t-test was used with Benjamini-Hochberg FDR procedure to obtain corrected q-values. All statistics and graphics were produced using STAMP software.

General patterns in the correlation relationships between bacterial phyla and functional roles reveal that Acidobacteria specializes in unique functional roles compared to other phyla (Figure 6). Acidobacteria, the most abundant phylum and with large differences between ash and non-ash plots, was positively correlated with many of the KEGG level 3 functional groups, including those that were significantly higher in non-ash plots (Figure 5). Specifically, Acidobacteria relative abundance correlated with starch and sucrose metabolism ($r = 0.810$, $p = 0.004$), amino sugar and nucleotide sugar metabolism ($r = 0.821$, $p = 0.002$), galactose metabolism ($r = 0.799$, $p = 0.006$), inositol phosphate metabolism ($r = 0.817$, $p = 0.003$), and sulfur metabolism ($r = 0.755$, $p = 0.029$). Although Bacteroidetes was not one of the seven phyla which differed between ash and non-ash plots, it did have the most corollary relationships with the KEGG functional groups we analyzed (13 out of 24 with $r > 0.750$ and $p < 0.05$).

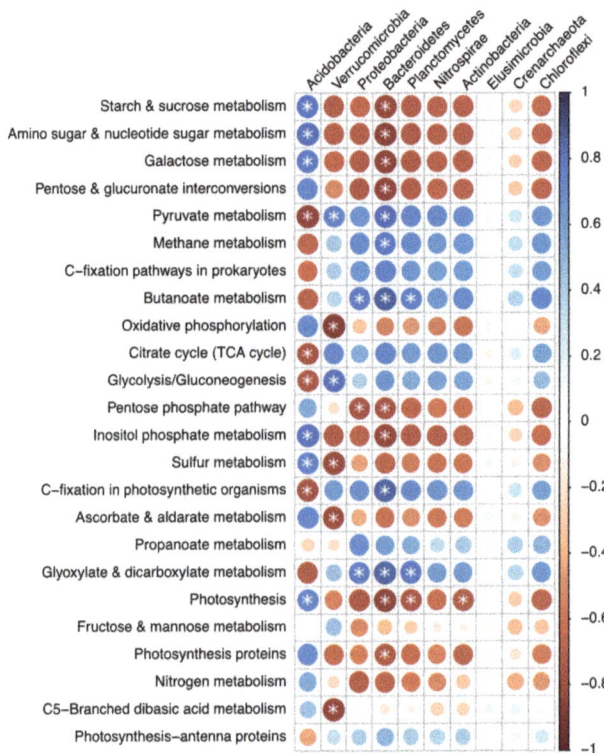

Figure 6. Pearson's correlation matrix comparing the ten most abundant bacterial phyla to level 3 KEGG functional categories, ordered as in Figure 5. Circle color indicates either a positive (**blue**) or negative (**red**) correlation and circle size and shading are proportional to correlation coefficients regardless of statistical significance. Bonferroni adjusted significance ($p < 0.05$) is indicated by white asterisks.

4. Discussion

Here, we present evidence that plots containing ash trees at varying stages of EAB-induced decline have different belowground bacterial and functional characteristics than non-ash plots, in spite of having similar soil environmental factors (Tables 1 and 2). These soil bacterial community differences between ash and non-ash plots (Figure 2), which were largely driven by Acidobacteria relative abundance (Figure 3), suggest that in temperate forest ecosystems, ash trees may exhibit a genera specific relationship with soil microorganisms and contribute to shaping soil bacterial community assemblages, which may influence specific functional capacities. The estimated functional data suggest that soil communities in ash plots may have different functional capabilities from those in non-ash plots with respect to C and P metabolism, but not with N metabolism (Figure 5). Based on these results and because of the inherent linkage between above- and belowground communities, the loss of ash trees to EAB infestation will likely drive changes in soil microbial communities that lead to altered C and nutrient cycling in this forest ecosystem beyond the expected increase in litter inputs. These fundamental biogeochemical and successional shifts may make this ecosystem susceptible to invasive plant species or pathogenic microorganisms [45].

Although the direct effects of tree decline on the belowground community were not explicitly evaluated in this study, the degree of EAB disturbance severity, as indicated by AC, did not affect the overall soil bacterial community structure (Mantel test—Table 2) or the individual abundances of major bacterial phyla within the ash stands (Figure 4). Likewise, the removal of sites with severely affected ash trees from the analysis (AC > 3) did not alter the results (Table S1). This indicates that ash associated bacterial communities may persist throughout EAB infestation and the eventual ash tree mortality. Changes in the microbial community may be expected some years after ash mortality is completed, depending on the species that occupy the newly available niche. The ash legacy ecosystem effects on soil properties deserve further investigation.

Other studies have reported that dominant tree genera may contribute to shaping soil microbial communities [46–48], but to our knowledge, few studies have investigated soil microbial community associations with ash trees specifically. The mechanisms by which trees exert influence on soil communities are generally attributed to direct and persistent inputs to the soil environment, likely from the chemical nature of litter deposition and root exudates. However, while there were obvious differences in bacterial community structure between ash and non-ash plots in our study (Figures 2 and 3), determining causation can be challenging. A variety of biotic and abiotic factors may contribute to shaping the soil microbiome at a given site. For example, the presence/absence of other non-ash tree species within the plots may confound the interpretation of results. Oak tree relative dominance was low in the plots with ash trees and was higher in plots without ash trees (Tables 2 and S3). These results may indicate that the bacterial community differences we see between ash and non-ash plots could also be due to oak tree influence. However, results from the Mantel test analysis suggest that oak tree dominance did not have an effect on bacterial community structure ($p = 0.338$), while ash tree dominance did ($p = 0.007$), providing a stronger case for soil bacterial association with ash trees specifically. Likewise, bacterial community structure has been shown to be highly influenced by soil pH [29–31], which along with other correlated soil variables (Mg, Al, and Ca), is supported by our data (Table 2). The most abundant phylum in these sites was Acidobacteria, which are known to prefer acidic environments [49]. This phylum had a 1.5-fold greater relative abundance in non-ash plots when compared to ash plots (Figure 3) and may very well be driving the overall soil bacterial community structure differences at these sites. While soil pH was only marginally statistically different between ash and non-ash plots ($W = 73$, $p = 0.080$), it was more acidic in non-ash plots where Acidobacteria were more abundant. So, while ash trees are tolerant of a wide range of soil pH values, including very acidic ones [50], it is possible that soil pH may be contributing to both bacterial and tree community structure.

Besides being the most abundant phyla in these soils and a major driver of bacterial community structure, Acidobacteria exhibit a number interesting patterns. Overall, our data reveal opposite trends in Acidobacteria relative abundance (ash *vs.* non-ash) and functional correlations when compared to eight of the nine remaining most abundant bacterial phyla (Figures 3 and 6). Acidobacteria were found to be more abundant in non-ash plots, while the other eight phyla were more abundant in ash plots (Figure 3). This pattern also holds true for correlations made with functional gene abundances, where a positive correlation with Acidobacteria often occurred alongside a negative correlation with the other phyla and vice versa (Figure 6). Our data suggests that Acidobacteria correlate positively with the breakdown of complex sugars leading to glycolysis (i.e., starch, sucrose, galactose and amino sugar metabolisms), while other phyla, such as Proteobacteria, Verrucomicrobia, and Bacteroidetes, correlate positively with enzymes tied more closely to the TCA cycle (i.e., glycolysis/gluconeogenesis and pyruvate, glycoxylate, dicarboxylate, and butanoate metabolisms). Even though the relative abundances of some major phyla (e.g., Verrucomicrobia and Bacteroidetes) did not differ greatly between ash and non-ash plots (Figure 3) and were highly correlated with the above-mentioned functions (Figure 6), the ash vs. non-ash differences in these same functional groups were still significant (Figure 5). This suggests that the combined directional relationships of non-Acidobacteria phyla with these functions may also contribute to ash vs. non-ash functional differences; however, Acidobacteria remain the most likely driver of relative abundance and functional differences. Acidobacteria are typically aerobic heterotrophs capable of utilizing a range of C sources from simple sugars to hemicellulose, cellulose, and chitin. Although this group is able to reduce nitrate and nitrite [49,51], it is incapable of N_2 fixation or nitrification and overall N metabolism was not affected by Acidobacteria abundance differences in this study, indicating some degree of functional redundancy within the bacterial community for N cycling. However, inositol phosphate and sulfur metabolic capacities, which are indicative of organic phosphorus (P) and sulfur (S) cycling capacities, respectively, are both positively correlated with Acidobacteria and are greater in non-ash forest plots when compared to ash plots (Figure 5). Phosphatases are enzymes which extract P from organic sources and their activity varies according to climate variables, soil C and N, and organic-P (as opposed to available-P measured in this study) [52]. As climate, soil C, and soil N did not vary between ash and non-ash plots, organic-P appears to be a proportionally larger source of microbial P in non-ash forest stands. Because a substantial amount of organic-P is thought to be in microbial biomass [53], this enhanced capacity to access organic-P in non-ash plots may indicate a relative difference in P availability between ash and non-ash plots via solubilisation, mobilization, and/or microbial turnover [54]. Based on our results, if future soil bacterial communities in ash forests become more similar to those in non-ash plots in the wake of EAB infestation, then these differences in P metabolism may be an indicator of future soil transformations. It also highlights the potential role of Acidobacteria in the biogeochemical cycling of nutrients in this forest system. Therefore, future abundance shifts in this phyla due to ash tree decline as a result of EAB could result in alterations of both soil C and nutrient dynamics that will go beyond the addition of dead ash woody litter, which is currently the subject of ongoing investigations.

While our results suggest that ash trees may contribute to shaping soil bacterial community structure and the loss of ash due to EAB infestation may lead to belowground alterations, this may not hold true for all tree species and/or may not affect the bacterial community over time. Ecosystem responses of soil microbes to disturbance remain poorly understood and above- belowground associations may vary across the plant kingdom. For example, Ferrenberg et al., 2014 [55] found that soil bacterial communities remained stable over a five year chronosequence following coniferous tree mortality due to bark beetle in the Rocky Mountains. Ecological resilience in the belowground environment, where the slow turnover of the plant-derived soil C may have a long legacy of the vegetation history of the site, may retain structural and functional attributes long after the removal of trees from the system. Therefore, collecting data on specific above- belowground relationships, as done here, is imperative to understanding if and how communities may respond to the loss of a given species or genera.

Research is underway to track the successional trajectory of bacterial communities over time in the wake of ash decline. If soil bacterial communities are resilient to disturbance, driven by edaphic factors that have long-term legacy effects and are not directly influenced by live ash trees, then the loss of ash trees in temperate forests may not affect bacterial community structure (Figure 7; Scenario 1). However, if instead ash trees form unique assemblages with their belowground bacterial community and the ecological memory of the soil environment is short-lived, then the loss of ash trees will likely cause major shifts in microbial community structure and, in consequence, ecosystem function. The successional trajectory of these communities could either become more similar to those in non-ash plots (Figure 7; Scenario 2), or progress into an unknown community structure potentially driven by incoming replacement plant species (Figure 7; Scenario 3). The resilience of belowground communities and the functions they perform after disturbance will ultimately govern the future states of overall ecosystem biogeochemical cycling and aboveground community structure.

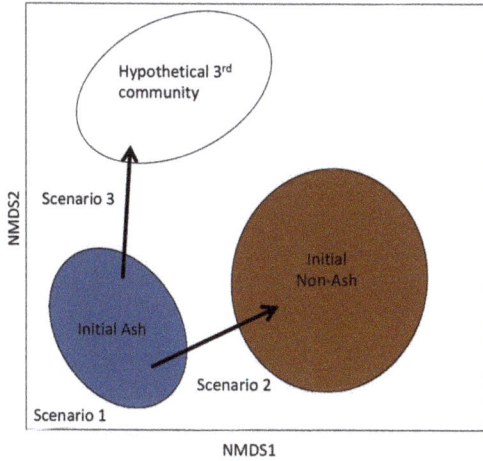

Figure 7. Theoretical diagram representing possible successional trajectories of bacterial communities over time in forests suffering from ash decline as a result of EAB infestation, where in Scenario 1 the communities stay the same, in Scenario 2 they become more similar to communities in non-ash plots, and in Scenario 3 they develop a community structure different than in either ash or non-ash plots. NMDS ordination space represents hypothetical differences in bacterial community structure based on Figure 2.

5. Conclusions

Using archived DNA samples extracted from forest soils which were collected in the early stages of EAB infestation, we compared the bacterial community structures of plots containing ash trees to those that did not contain ash trees and found that they were different. This indicates that either ash trees directly or indirectly associate with, or influence, belowground microbial organisms. However, co-occurring factors such as soil pH, correlations with other tree species, or the active decline of ash tree health cannot be fully ruled out as contributing driving forces of bacterial community structure. Estimated functional gene abundances within the soil community were also different between ash and non-ash plots as a result of phylogenetic community differences. Specifically, greater relative abundances of Acidobacteria in non-ash plots may drive increases in sugar metabolisms which lead to glycolysis, but decrease functional pathways more tightly linked to the TCA cycle, likely altering C dynamics. Although N cycling was not affected by these bacterial abundance differences, both P and S metabolic potential was elevated in non-ash plots. While we are unable to determine how the loss of ash trees due to EAB will affect belowground community structure and function over time, we provide a foundational framework to predict future successional trajectories and establish a context within which to generate new hypotheses.

Supplementary Materials: The following are available online at http://www.mdpi.com/1999-4907/9/4/187/s1. Table S1: Analysis of "healthy ash" sites, Table S2: Forest site characteristics, Table S3: Tree health, dominance and diversity, Figure S1: Twenty most abundant bacterial classes, Figure S2: Thirty most abundant bacterial orders. Additional files include "R_scripts.R" which contains R code for statistical analysis and figure production, and "Code.txt", which contains computer scripts for bioinformatics using QIIME and PICRUSt.

Acknowledgments: This research was supported by the National Science Foundation DGE-0549245, "Landscape Ecological and Anthropogenic Processes" (C.E.F.), the UIC Hadley grant (M.P.R.), the US Forest Service grant

15-JV-11242302-038 (M.A.G-M.), and the Stable Isotope Lab at UIC. Open access publishing fees were supported by the Research Open Access Publishing (ROAAP) Fund of UIC. The authors thank D. Johnston for assistance in the conception and sample collection phase of this research, as well as K. Costilow for assistance with sample collection. We would also like to thank J. Dalton, C. Whelan, E. Dias de Oliveira, and S. O'Brien for comments and advice.

Author Contributions: M.P.R. analyzed the data and wrote the paper; C.E.F. conceived and designed sample collection and collected soil samples; K.S.K. provided the plot network and contributed materials; and M.A.G.-M. contributed reagents/materials/analysis tools, provided feedback on experiments and edited the manuscript.

Conflicts of Interest: The authors declare no conflict of interest.

References

1. Parmesan, C.; Yohe, G. A globally coherent fingerprint of climate change impacts across natural systems. *Nature* **2003**, *421*, 37–42. [CrossRef] [PubMed]

2. Settele, J.; Scholes, R.J.; Betts, R.A.; Bunn, S.; Leadley, P.; Nepstad, D.; Overpeck, J.T.; Toboada, M.A. Chapter 4—Terrestrial and inland water systems. In *Climate Change 2014: Impacts, Adaptation, and Vulnerability. Part A: Global and Sectoral Aspects*; Contribution of Working Group II to the Fifth Assessment Report of the Intergovernmental Panel on Climate Change; Field, C.B., Barros, V.R., Dokken, D.J., Mach, K.J., Mastrandrea, M.D., Bilir, T.E., Chatterjee, M., Ebi, K.L., Estrada, Y.O., Genova, R.C., et al., Eds.; Cambridge University Press: Cambridge, UK; New York, NY, USA, 2014; pp. 271–359, ISBN 1102700991936.

3. Cramer, W.; Bondeau, A.; Woodward, F.I.; Prentice, I.C.; Betts, R.A.; Brovkin, V.; Cox, P.M.; Fisher, V.; Foley, J.A.; Friend, A.D.; et al. Global response of terrestrial ecosystem structure and function to CO_2 and climate change: Results from six dynamic global vegetation models. *Glob. Chang. Biol.* **2001**, *7*, 357–373. [CrossRef]

4. Drewniak, B.; Gonzalez-Meler, M.A. Earth system model needs for including the interactive representation of nitrogen deposition and drought effects on forested ecosystems. *Forests* **2017**, *8*, 267. [CrossRef]

5. Gonzalez-Meler, M.A.; Rucks, J.S.; Aubanell, G. Mechanistic insights on the responses of plant and ecosystem gas exchange to global environmental change: Lessons from Biosphere 2. *Plant Sci.* **2014**, *226*, 14–21. [CrossRef] [PubMed]

6. Mcnickle, G.G.; Gonzalez-Meler, M.A.; Lynch, D.J.; Baltzer, J.L.; Brown, J.S. The world's biomes and primary production as a triple tragedy of the commons foraging game played among plants. *Proc. R. Soc. B Biol. Sci.* **2016**, *283*, 20161993. [CrossRef] [PubMed]

7. Cappaert, D.; McCullough, D.G.; Poland, T.M.; Siegert, N.W. Emerald Ash Borer in North America: A Research and Regulatory Challenge. *Am. Entomol.* **2005**, *51*, 152–165. [CrossRef]

8. Wang, X.Y.; Yang, Z.Q.; Gould, J.R.; Zhang, Y.N.; Liu, G.J.; Liu, E. The Biology and Ecology of the Emerald Ash Borer, *Agrilus planipennis*, in China. *J. Insect Sci.* **2010**, *10*, 1–23. [CrossRef] [PubMed]

9. Flower, C.E.; Lynch, D.J.; Knight, K.S.; Gonzalez-Meler, M.A. Biotic and abiotic drivers of sap flux in mature green ash trees (*Fraxinus pennsylvanica*) experiencing varying levels of emerald ash borer (*Agrilus planipennis*) infestation. *Forests* **2018**, in press.

10. McCullough, D.G.; Katovich, S.A. Emerald ash borer. *Pest Alert* **2004**, 4–5. [CrossRef]

11. Knight, K.S.; Robert, P.; Rebbeck, J. *How Fast Will Trees Die? A Transition Matrix Model of Ash Decline in Forest Stands Infested by Emerald Ash Borer*; U.S. Department of Agriculture, Forest Service: Washington, DC, USA, 2008; pp. 28–29.

12. Costilow, K.C.; Knight, K.S.; Flower, C.E. Disturbance severity and canopy position control the radial growth response of maple trees (*Acer* spp.) in forests of northwest Ohio impacted by emerald ash borer (*Agrilus planipennis*). *Ann. For. Sci.* **2017**, *74*. [CrossRef]

13. Birdsey, R.A. *Carbon Storage and Accumulation in United States Forest Ecosystems*; General Technical Report WO-59; United States Department of Agriculture, Forest Service: Washington, DC, USA, 1992.

14. Birdsey, R.A.; Heath, L.S. Carbon Changes in U.S. Forests. In *Productivity of America's Forests and Climate Change*; General Technical Report RM-GTR-271; Joyce, L.A., Ed.; United States Department of Agriculture, Forest Service, Rocky Mountain Forest and Experiment Station: Fort Collins, CO, USA, 1995; pp. 56–70.

15. Goodale, C.L.; Apps, M.J.; Birdsey, R.A.; Field, C.B.; Heath, L.S.; Houghton, R.A.; Jenkins, J.C.; Kohlmaier, G.H.; Liu, S.; Nabuurs, G.; et al. Forest Carbon Sinks in the Northern Hemisphere. *Ecol. Appl.* **2002**, *12*, 891–899. [CrossRef]

16. Flower, C.E.; Knight, K.S.; Gonzalez-Meler, M.A. Impacts of the emerald ash borer (*Agrilus planipennis* Fairmaire) induced ash (*Fraxinus* spp.) mortality on forest carbon cycling and successional dynamics in the eastern United States. *Biol. Invasions* **2013**, *15*, 931–944. [CrossRef]

17. Higham, M.; Hoven, B.M.; Gorchov, D.L.; Knight, K.S. Patterns of Coarse Woody Debris in Hardwood Forests across a Chronosequence of Ash Mortality Due to the Emerald Ash Borer (*Agrilus planipennis*). *Nat. Areas J.* **2017**, *37*, 406–411. [CrossRef]

18. Lovett, G.M.; Canham, C.D.; Arthur, M.A.; Weathers, K.C.; Fitzhugh, R.D. Forest ecosystem responses to exotic pests and pathogens in eastern North America. *Bioscience* **2006**, *56*, 395–405. [CrossRef]

19. Telander, A.C.; Slesak, R.A.; D'Amato, A.W.; Palik, B.J.; Brooks, K.N.; Lenhart, C.F. Sap flow of black ash in wetland forests of northern Minnesota, USA: Hydrologic implications of tree mortality due to emerald ash borer. *Agric. For. Meteorol.* **2015**, *206*, 4–11. [CrossRef]

20. Flower, C.E.; Gonzalez-Meler, M.A. Responses of Temperate Forest Productivity to Insect and Pathogen Disturbances. *Annu. Rev. Plant Biol.* **2015**, *66*, 547–569. [CrossRef] [PubMed]

21. Hopkins, F.; Gonzalez-Meler, M.A.; Flower, C.E.; Lynch, D.J.; Czimczik, C.; Tang, J.; Subke, J.A. Ecosystem-level controls on root-rhizosphere respiration. *New Phytol.* **2013**, *199*, 339–351. [CrossRef] [PubMed]

22. Cheng, W.; Parton, W.J.; Gonzalez-Meler, M.A.; Phillips, R.; Asao, S.; McNickle, G.G.; Brzostek, E.; Jastrow, J.D. Synthesis and modeling perspectives of rhizosphere priming. *New Phytol.* **2014**, *201*, 31–44. [CrossRef] [PubMed]

23. Van Der Heijden, M.G.A.; Bardgett, R.D.; Van Straalen, N.M. The unseen majority: Soil microbes as drivers of plant diversity and productivity in terrestrial ecosystems. *Ecol. Lett.* **2008**, *11*, 296–310. [CrossRef] [PubMed]

24. Allison, S.S.D.; Martiny, J.B.H. Resistance, resilience, and redundancy in microbial communities. *Proc. Natl. Acad. Sci. USA* **2008**, *105*, 11512–11519. [CrossRef] [PubMed]

25. Ricketts, M.P.; Poretsky, R.S.; Welker, J.M.; Gonzalez-Meler, M. Soil bacterial community and functional shifts in response to altered snowpack in moist acidic tundra of Northern Alaska. *Soil* **2016**, *2*, 459–474. [CrossRef]

26. Bailey, V.L.; Fansler, S.J.; Stegen, J.C.; McCue, L.A. Linking microbial community structure to β-glucosidic function in soil aggregates. *ISME J.* **2013**, *7*, 2044–2053. [CrossRef] [PubMed]

27. Schimel, J.P.; Schaeffer, S.M. Microbial control over carbon cycling in soil. *Front. Microbiol.* **2012**, *3*, 1–11. [CrossRef] [PubMed]

28. Fierer, N. Embracing the unknown: Disentangling the complexities of the soil microbiome. *Nat. Rev. Microbiol.* **2017**, *15*, 579–590. [CrossRef] [PubMed]

29. Fierer, N.; Jackson, R.B. The diversity and biogeography of soil bacterial communities. *Proc. Natl. Acad. Sci. USA* **2006**, *103*, 626–631. [CrossRef] [PubMed]

30. Lauber, C.L.; Hamady, M.; Knight, R.; Fierer, N. Pyrosequencing-Based Assessment of Soil pH as a Predictor of Soil Bacterial Community Structure at the Continental Scale. *Appl. Environ. Microbiol.* **2009**. [CrossRef] [PubMed]

31. Cho, S.J.; Kim, M.H.; Lee, Y.O. Effect of pH on soil bacterial diversity. *J. Ecol. Environ.* **2016**, *40*, 10. [CrossRef]

32. Schlatter, D.C.; Bakker, M.G.; Bradeen, J.M.; Kinkel, L.L. Plant community richness and microbial interactions structure bacterial communities in soil. *Ecology* **2015**, *96*, 134–142. [CrossRef] [PubMed]

33. Bakker, M.G.; Bradeen, J.M.; Kinkel, L.L. Effects of plant host species and plant community richness on streptomycete community structure. *FEMS Microbiol. Ecol.* **2013**, *83*, 596–606. [CrossRef] [PubMed]

34. Prescott, C.E.; Grayston, S.J. Tree species influence on microbial communities in litter and soil: Current knowledge and research needs. *For. Ecol. Manag.* **2013**, *309*, 19–27. [CrossRef]

35. Flower, C.E.; Knight, K.S.; Rebbeck, J.; Gonzalez-Meler, M.A. The relationship between the emerald ash borer (*Agrilus planipennis*) and ash (*Fraxinus* spp.) tree decline: Using visual canopy condition assessments and leaf isotope measurements to assess pest damage. *For. Ecol. Manage.* **2013**, *303*, 143–147. [CrossRef]

36. Smith, A. Effects of Community Structure on Forest Susceptibility and Response to the Emerald Ash Borer Invasion of the Huron River Watershed in Southeast Michigan. Doctor's Dissertation, Ohio State University, Columbus, OH, USA, 2006.

37. Gilbert, J.A.; Jansson, J.K.; Knight, R. The Earth Microbiome project: Successes and aspirations. *BMC Biol.* **2014**, *12*, 69. [CrossRef] [PubMed]

38. Caporaso, J.; Kuczynski, J.; Stombaugh, J. QIIME allows analysis of high—Throughput community sequencing data. *Nat. Methods* **2010**, *7*, 335–336. [CrossRef] [PubMed]

39. Langille, M.G.I.; Zaneveld, J.; Caporaso, J.G.; McDonald, D.; Knights, D.; Reyes, J.A.; Clemente, J.C.; Burkepile, D.E.; Vega Thurber, R.L.; Knight, R.; et al. Predictive functional profiling of microbial communities using 16S rRNA marker gene sequences. *Nat. Biotechnol.* **2013**, *31*, 814–821. [CrossRef] [PubMed]

40. Kanehisa, M.; Goto, S. KEGG: Kyoto encyclopedia of genes and genomes. *Nucleic Acids Res.* **2000**, *28*, 27–30. [CrossRef] [PubMed]

41. R Core Team. R: A language and environment for statistical computing. R Foundation for Statistical Computing, Vienna, Austria. 2013. Available online: http://www.R-project.org/ (accessed on 3 March 2018).

42. Oksanen, J.; Blanchet, F.G.; Friendly, M.; Kindt, R.; Legendre, P.; Mcglinn, D.; Minchin, P.R.; O'Hara, R.B.; Simpson, G.L.; Solymos, P.; et al. Vegan: Community Ecology Package. R Package Version 2.4-5. 2017. Available online: https://CRAN.R-project.org/package=vegan (accessed on 3 March 2018).

43. McMurdie, P.J.; Holmes, S. Phyloseq: An R Package for Reproducible Interactive Analysis and Graphics of Microbiome Census Data. *PLoS ONE* **2013**, *8*. [CrossRef] [PubMed]

44. Parks, D.H.; Tyson, G.W.; Hugenholtz, P.; Beiko, R.G. STAMP: Statistical analysis of taxonomic and functional profiles. *Bioinformatics* **2014**, *30*, 3123–3124. [CrossRef] [PubMed]

45. Hobbs, R.J.; Huenneke, L.F. Disturbance, Diversity, and Invasion: Implications for Conservation. *Conserv. Biol.* **1992**, *6*, 324–337. [CrossRef]

46. Kaiser, C.; Koranda, M.; Kitzler, B.; Fuchslueger, L.; Schnecker, J.; Schweiger, P.; Rasche, F.; Zechmeister-Boltenstern, S.; Sessitsch, A.; Richter, A. Belowground carbon allocation by trees drives seasonal patterns of extracellular enzyme activities by altering microbial community composition in a beech forest soil. *New Phytol.* **2010**, *187*, 843–858. [CrossRef] [PubMed]

47. Urbanová, M.; Šnajdr, J.; Baldrian, P. Composition of fungal and bacterial communities in forest litter and soil is largely determined by dominant trees. *Soil Biol. Biochem.* **2015**, *84*, 53–64. [CrossRef]

48. Lejon, D.P.H.; Chaussod, R.; Ranger, J.; Ranjard, L. Microbial Community Structure and Density under Different Tree Species in an Acid Forest Soil (Morvan, France). *Microb. Ecol.* **2005**, *50*, 614–625. [CrossRef] [PubMed]

49. Ward, N.L.; Challacombe, J.F.; Janssen, P.H.; Henrissat, B.; Coutinho, P.M.; Wu, M.; Xie, G.; Haft, D.H.; Sait, M.; Badger, J.; et al. Three Genomes from the Phylum Acidobacteria Provide Insight into the Lifestyles of These Microorganisms in Soils. *Appl. Environ. Microbiol.* **2009**, *75*, 2046–2056. [CrossRef] [PubMed]

50. Burns, R.M.; Honkala, B.H. Fraxinus. In *Silvics of North America: Volume 2, Hardwoods*; Agriculture Handbook 654; United States Department of Agriculture, Forest Service: Washington, DC, USA, 1990; Volume 2, pp. 333–357, ISBN 1800553684.

51. Kielak, A.M.; Barreto, C.C.; Kowalchuk, G.A.; van Veen, J.A.; Kuramae, E.E. The ecology of Acidobacteria: Moving beyond genes and genomes. *Front. Microbiol.* **2016**, *7*, 1–16. [CrossRef] [PubMed]

52. Margalef, O.; Sardans, J.; Fernández-Martínez, M.; Molowny-Horas, R.; Janssens, I.A.; Ciais, P.; Goll, D.; Richter, A.; Obersteiner, M.; Asensio, D.; et al. Global patterns of phosphatase activity in natural soils. *Sci. Rep.* **2017**, *7*, 1–13. [CrossRef] [PubMed]

53. Turner, B.L.; Lambers, H.; Condron, L.M.; Cramer, M.D.; Leake, J.R.; Richardson, A.E.; Smith, S.E. Soil microbial biomass and the fate of phosphorus during long-term ecosystem development. *Plant Soil* **2013**, *367*, 225–234. [CrossRef]

54. Richardson, A.E.; Simpson, R.J. Soil Microorganisms Mediating Phosphorus Availability Update on Microbial Phosphorus. *Plant Physiol.* **2011**, *156*, 989–996. [CrossRef] [PubMed]

55. Ferrenberg, S.; Knelman, J.E.; Jones, J.M.; Beals, S.C.; Bowman, W.D.; Nemergut, D.R. Soil bacterial community structure remains stable over a 5-year chronosequence of insect-induced tree mortality. *Front. Microbiol.* **2014**, *5*, 1–11. [CrossRef] [PubMed]

forests

MDPI

Article

Downed Coarse Woody Debris Dynamics in Ash (*Fraxinus* spp.) Stands Invaded by Emerald Ash Borer (*Agrilus planipennis* Fairmaire)

Kayla I. Perry [1,*], Daniel A. Herms [1,2], Wendy S. Klooster [3], Annemarie Smith [1], Diane M. Hartzler [1], David R. Coyle [4,5] and Kamal J. K. Gandhi [4]

[1] Department of Entomology, 1680 Madison Avenue, Ohio Agricultural Research and Development Center, The Ohio State University, Wooster, OH 44691, USA; dan.herms@davey.com (D.A.H.); annemariesmith.bucki@gmail.com (A.S.); diane.hartzler@gmail.com (D.M.H.)

[2] Current Address: The Davey Tree Expert Company, 1500 Mantua Street, Kent, OH 44240, USA

[3] Department of Horticulture and Crop Science, 1680 Madison Avenue, Ohio Agricultural Research and Development Center, The Ohio State University, Wooster, OH 44691, USA; klooster.2@osu.edu

[4] Daniel B. Warnell School of Forestry and Natural Resources, University of Georgia, 180 E Green Street, Athens, GA 30602, USA; dcoyle@warnell.uga.edu (D.R.C.); kjgandhi@uga.edu (K.J.K.G.)

[5] Southern Regional Extension Forestry, University of Georgia, Athens, GA 30602, USA

* Correspondence: perry.1864@osu.edu

Received: 2 March 2018; Accepted: 5 April 2018; Published: 7 April 2018

Abstract: Emerald ash borer (EAB; *Agrilus planipennis* Fairmaire) has had major ecological impacts in forests of eastern North America. In 2008 and 2012, we characterized dynamics of downed coarse woody debris (DCWD) in southeastern Michigan, USA near the epicenter of the invasion, where the mortality of white (*Fraxinus americana* L.), green (*F. pennsylvanica* Marshall), and black (*F. nigra* Marshall) ash exceeded 99% by 2009. Percentage of fallen dead ash trees and volume of ash DCWD on the forest floor increased by 76% and 53%, respectively, from 2008 to 2012. Ash and non-ash fell non-randomly to the east and southeast, conforming to prevailing winds. More ash fell by snapping along the bole than by uprooting. By 2012, however, only 31% of ash snags had fallen, indicating that DCWD will increase substantially, especially if it accelerates from the rate of 3.5% per year documented during the study period. Decay of ash DCWD increased over time, with most categorized as minimally decayed (decay classes 1 and 2) in 2008 and more decayed (decay classes 2 and 3) in 2012. As the range of EAB expands, similar patterns of DCWD dynamics are expected in response to extensive ash mortality.

Keywords: *Agrilus planipennis*; ash; coarse woody debris; emerald ash borer; *Fraxinus*

1. Introduction

Above and belowground woody debris is a critical biological and structural component of forest ecosystems [1,2]. Through both standing material as snags and fallen material as logs, above-ground coarse woody debris (CWD) provides many ecological functions in terrestrial and aquatic ecosystems, including organic matter inputs; nutrient cycling; soil moisture retention; a habitat for vertebrate, invertebrate, and fungal species; micro-sites for plant regeneration; and altered fire behavior [3–9]. Depending on the disturbance rate, tree species, and rate of decomposition, CWD can impact forest ecosystem patterns and processes from decades to centuries [10,11].

Input and accumulation of CWD in forests is highly dynamic due to variation in the intensity and frequency of abiotic and biotic disturbances (e.g., fire, windstorms, ice-storms, and native and non-native insects and diseases) [12,13]. Consequently, tree death can occur on spatial scales that vary from individual to landscape levels, and time scales that vary from annually to 500 or more years [14–16]. Small-scale gap formation provides a slow but consistent input of CWD, whereas sudden large-scale ecosystem disturbances provide an infrequent but high input of CWD [14,17–19]. At the smallest spatial scale, tree death results in the formation of snags, with branches and stems breaking and falling at different rates [20]. CWD is also created when live trees fall, for example, during storms and/or under the weight of other falling trees. This results in a rich diversity of the type and volume of CWD inputs to the forest floor [20,21], which has major implications for nutrient cycling, successional pathways, and community dynamics, especially for organisms that depend upon dead and decaying wood for some part of their life cycles [6,9,22,23].

Alien insects and pathogens that kill trees provide a major input of CWD to forest ecosystems [24–26]. During the last century, alien insects and pathogens caused widespread tree mortality in North American forests, including hemlock woolly adelgid (*Adelges tsugae* Annand), as well as the chestnut blight (*Cryphonectria parasitica* (Murrill)), white pine blister rust (*Cronartium ribicola* J.C. Fisch.), Dutch elm disease (*Ophiostoma ulmi* (Buismann) Nannf.) pathogens, and beech bark disease (a complex of the scale insect *Cryptococcus fagisuga* Lindinger and *Neonectria* spp. fungi) [26–29]. Conversion of living trees to CWD by alien species at landscape scales within a relatively short time-frame represents a novel disturbance because these ecosystems have not previously experienced similar historical disturbance events. The impact on forests is patchy and confined to one or a few tree species in otherwise heterogeneous communities, and regeneration of the impacted species is typically limited by the alien invader. All of these factors interact to alter patterns of accumulation and decomposition of CWD. Despite the major ecological consequences of non-native insects and pathogens, only a few studies have quantified the effects of alien species on patterns of CWD accumulation and subsequent impacts on ecological processes [30,31].

Emerald ash borer (EAB; *Agrilus planipennis* Fairmaire) is the most damaging alien insect that has established in North American forests [32]. Introduced from eastern Asia, EAB attacks ash (*Fraxinus* spp.) [32], and to a lesser degree, white fringetree (*Chionanthus virginicus* L.) [33]. Since its first detection in North America in 2002 in southeast Michigan, USA, EAB has caused the widespread mortality of ash in invaded forests [34–37]. For example, Klooster et al. [35] observed >99% mortality of ash with stem diameters greater than 2.5 cm and a cessation of new seedling regeneration in southeastern Michigan near the epicenter of the invasion. Many North American ash tend to be abundant canopy trees in heterogeneous landscapes, and hence a major decline of ash, waves of tree mortality, and CWD formation will continue as EAB spreads across the landscape [31].

To better understand the dynamics of DCWD in response to tree mortality caused by alien species, we quantified the formation and accumulation of DCWD in forests in the Upper Huron River watershed in southeastern Michigan, where rates of ash decline and mortality were previously documented [35,38,39]. These forests are near the epicenter of the EAB invasion [40], and thus have been impacted longer than any other in North America. Our research objectives were to: (1) quantify the current volume (m^3 ha^{-1}) and species of DCWD (as fallen trees or logs); (2) assess whether the change in the rate of input of DCWD to the forest floor corresponded with the rate of change of ash mortality; (3) determine whether there were any differences in the pattern of accumulation of DCWD between ash and non-ash tree species; (4) assess the manner of treefall (broken or uprooted) and the height of broken stumps; and (5) assess the spatial patterns of DCWD accumulation (i.e., the general direction of treefall within these forest stands). Results from this study may contribute to the prediction and management of CWD dynamics (e.g., which ash species will fall first and when) in response to EAB-induced ash mortality in North America.

2. Materials and Methods

2.1. Study Sites

The study was conducted in a subset of 16 forested stands randomly selected from 38 stands previously established and characterized in the Livingston, Oakland, and Washtenaw counties in the

Upper Huron River watershed in southeastern Michigan [35,38,39,41]. The 16 stands represented a gradient of percentage ash mortality that decreased with the distance (25–45 km) from the presumed epicenter of EAB infestation in the township of Canton, Michigan, USA [38–40] and were located in the Highland, Island Lake, Pontiac, and Proud Lake State Recreation Areas, and Hudson Mills, Indian Springs, and Kensington Metro Parks (Table S1, Figure S1). The soils are Hapludalfs with loamy to clay and sandy texture [42]. Average annual temperature was 8.7–12.1 °C, average annual precipitation was 69–121 cm, and average annual snowfall was 66–241 cm from 1997–2017 near Detroit, Michigan, USA [43]. The forest overstory consisted of oak (*Quercus* spp.), maple (*Acer* spp.), basswood (*Tilia* spp.), cherry (*Prunus* spp.), elm (*Ulmus* spp.), and tamarack (*Larix* spp.) [38]. Ash was once abundant in these forests [38,39]. The density and basal area of ash ranged from 32.9 to 461.1 stems ha^{-1} and 2.8 to 14.5 m^2 ha^{-1}, respectively, in these stands in 2005 prior to significant treefall [39]. However, across all stands sampled, the mortality of ash with stem diameters greater than 2.5 cm exceeded 95% by 2008 and 99% by 2009 [35]. Detailed information about these stands, including overstory and understory species composition, soil conditions, and patterns of ash mortality, can be found in Smith [38], Klooster et al. [35], and Smith et al. [39].

Within each stand, we utilized a previously established transect that consisted of three 0.1 ha replicate circular plots, each with a radius of 18 m. When established, transects were oriented along a randomly selected compass heading between 0–90°, and plots contained at least two mature ash trees with stem diameters greater than 10 cm. Within a transect, the replicate plots were separated by ~80 m from their centers [35,38,39]. Each transect was previously classified according to soil moisture class as xeric, mesic, or hydric, in which white ash (*Fraxinus americana* L.), green ash (*Fraxinus pennsylvanica* Marsh.), or black ash (*Fraxinus nigra* Marsh.) was the most common ash species, respectively [35,38,39]. The formation and accumulation of DCWD was quantified in 48 circular plots within the 16 transects (three xeric, eight mesic, and five hydric).

2.2. Sampling of Downed Coarse Woody Debris

Percentage mortality of ash was monitored in each plot annually from 2004–2013 [35,38,39]. Sampling was conducted during the summers of 2008 and 2012 to assess temporal patterns of treefall and accumulation of DCWD. All ash trees with stem diameters ≥10 cm were cataloged in each plot as either standing or fallen. We sampled DCWD (i.e., logs which included boles and branches) rather than standing coarse woody debris (i.e., snags). To be considered DCWD, each piece had to meet the following criteria [as modified from the USDA Forest Service Forest Inventory Analysis (FIA) guidelines]: (1) diameter at the small end ≥7.6 cm; (2) length ≥ 1 m; and (3) distance above the forest floor ≤ 0.5 m (i.e., leaning snags and suspended woody debris were excluded) [44]. If a log had a diameter <7.6 cm at the small end, the diameter and length measurements (at least 1 m long) were taken from where the end diameter met the above requirements. If the bole was broken, each piece was independently assessed. Measurements of DCWD were taken using a dbh-tape for diameters and a logger tape measure for the length. Diameters and lengths were measured at the end of each intact side with the size criteria as described previously. Portions of DCWD that extended outside of the plot were excluded from the study (i.e., only the portions within the plots that met the above criteria were measured).

All sections of DCWD were categorized according to decay class using a 1–5 decay stage scale adapted from the USDA Forest Service FIA guidelines [44] as follows: Class 1—recently fallen CWD with intact bark, sound wood, and no decay; Class 2—mostly intact bark and wood (cannot be pulled apart by hand), but with softer sapwood and early signs of decay; Class 3—bark in advanced stage of decay (but present), sapwood can be pulled apart by hand, but heartwood is sound and can maintain its own weight; Class 4—bark and sapwood decayed, heartwood beginning to decay, the log cannot support its own weight, but maintains its shape; and Class 5—bark, sapwood, and heartwood has decayed, loss of structural integrity, and the log is incorporated into the forest floor. Based on these decay class descriptions, tree species were identified for DCWD in classes 1–3, but it was not possible to identify species for DCWD in classes 4–5. Therefore, DCWD was categorized as either known (to genus or species level) or unknown, and further as ash or non-ash. Fallen trees were categorized based on whether they uprooted or snapped along the trunk. Heights of broken stumps were measured at the highest point using a logger tape measure. Azimuth (degrees) was measured by pointing a compass towards either the top of the tree or smaller end of the piece of DCWD to assess the direction of fall.

2.3. Statistical Analyses

Percentage of fallen (versus standing) ash was calculated at the plot level for trees with stems $\geq$ 10 cm in diameter in 2008 and 2012. The year at which 90% ash mortality was reached at the transect level was identified, and the number of years since 90% ash mortality was determined for each transect in 2008 and 2012. Volume (m^3) of DCWD was calculated using the formula for a frustum cone:

$$V = \frac{\pi l}{3}\left(R^2 + Rr + r^2\right) \tag{1}$$

where r is the small radius, R is the large radius, and l is the length of the log [45,46]. Volumes of total DCWD, non-ash DCWD, and ash DCWD were calculated separately to assess patterns of woody debris accumulation from EAB-induced ash mortality.

Separate repeated measures analysis of variance (RMANOVA) tests were used to compare: (1) the percentage of ash trees that had fallen; (2) the volume of non-ash DCWD; (3) the volume of total DCWD; and (4) the volume of ash DCWD across the three habitat types based on soil moisture conditions (xeric, mesic, and hydric) from 2008 to 2012 using the package 'car' [47] in R version 3.7.1 [48]. All four response variables were calculated at the plot level and then averaged across the three replicate plots in each transect, which was the unit of replication in this study. Volume data were then multiplied by 10 to express values on a per hectare basis. All data were checked for statistical assumptions of normality and homogeneity of variance. Volume data for total DCWD and non-ash DCWD did not meet assumptions of normality and were rank transformed [49]. Predictor variables for the four ANOVA models were soil moisture class (xeric, mesic, or hydric) as a fixed factor, transect as a random factor, and year (2008 or 2012) as a repeated factor. Tukey's pairwise comparisons were used to evaluate mean separation following a significant F-test. Data are reported as mean $\pm$ standard error. At the stand level, the relationship between the percentage of ash trees that had fallen and the number of years since occurrence of 90% ash mortality was characterized by regression analysis.

Chi-square analyses were used to determine whether ash DCWD decay class or the method of treefall (broken or uprooted) were influenced by ash species (*F. americana*, *F. pennsylvanica*, or *F. nigra*), soil moisture class (xeric, mesic, or hydric), or year (2008 or 2012; assessed for decay class only). All percentage data met assumptions of equal variances. Only decay classes 1-3 were included in the analyses to ensure the accurate identification of ash species. Kruskal-Wallis tests were used to compare the heights of broken ash stumps (pooled across years) by species (*F. americana*, *F. pennsylvanica*, or *F. nigra*) and by soil moisture class (xeric, mesic, or hydric). Chi-square and Kruskal-Wallis analyses were completed using the package 'stats' in R version 3.7.1 [48].

For the treefall data, all analyses were performed at the transect level by taking an average of azimuth or angles of fallen trees from the three plots. Transects that had <3 fallen trees were excluded from the analyses [2008: N (transects) = 8 for ash and N = 12 for non-ash; 2012: N = 12 for ash and N = 12 for non-ash trees]. We performed analyses of uniformity to test the null hypothesis that there was no mean sample direction of treefall (based on the averages of angles of fallen trees) for ash and non-ash species. Rayleigh z analysis was used, which is a likelihood ratio test for uniformity within the von Mises distribution family [50,51]. Angular data were first transformed into polar coordinates (sine and cosine) to place them in a Cartesian space before conducting the Rayleigh z tests. Watson U^2 non-parametric tests were conducted to compare the directions of treefall between: (1) ash and non-ash trees within each year; and (2) across years for the same tree-type (2008 and 2012) [50]. Rose Diagrams were used to plot frequency distributions of the angles of individual fallen trees (rather than individual transects) in circular histograms [52].

3. Results

Across all stands, the overall mortality of ash with stems $\geq$ 10 cm in diameter was 99.5 $\pm$ 0.3% in 2008 and 99.6 $\pm$ 0.4% in 2012, with 90% mortality reached in three stands in 2006, in eight stands in 2007, and in the remaining five stands in 2008. The density of standing dead ash decreased from 123.8 $\pm$ 13.6 trees ha^{-1} in 2008 to 98.5 $\pm$ 12.3 trees ha^{-1} in 2012, while the density of fallen dead ash increased from 11.3 $\pm$ 2.1 to 39.2 $\pm$ 5.7 trees ha^{-1} in 2008 and 2012, respectively. The percentage of dead ash trees that had fallen increased from 7.6 $\pm$ 1.0% in 2008 to 30.7 $\pm$ 3.3% in 2012 ($F_{1,4}$ = 105.9; P < 0.001), with no differences among soil moisture classes ($F_{2,4}$ = 0.08; p = 0.776). At the stand (transect) level, the percentage of dead ash that fell increased 3.5% per year after ash mortality reached 90% (Figure 1; percentage of ash fallen = 2.81 + 3.53 $\times$ years; $F_{1,30}$ = 57.4; R^2 = 0.65; p < 0.001).

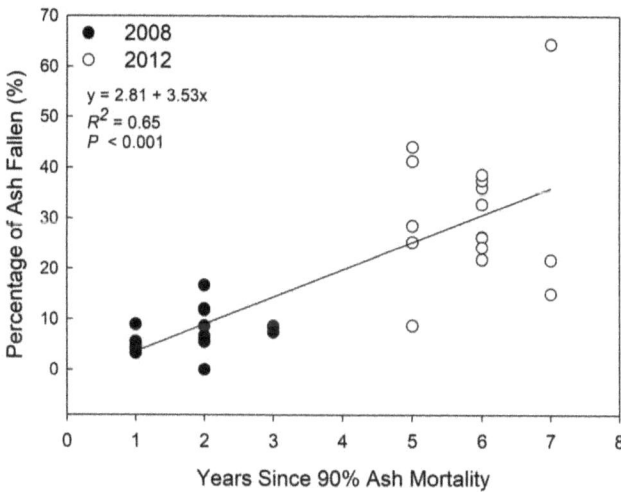

Figure 1. Relationship between the percentage of dead ash (*Fraxinus* spp.) that had fallen and the number of years since the occurrence of 90% ash mortality in 16 forest stands sampled in 2008 (open circles) and in 2012 (closed circles) in the Upper Huron River watershed in southeastern Michigan, USA. Stands reached 90% ash mortality in 2006, 2007, or 2008.

A higher percentage of ash trees fell to the forest floor by snapping along the bole (62.5 $\pm$ 4.4%) than by uprooting (37.4 $\pm$ 4.4%) (Figure 2A; χ^2 = 26.2; p = 0.010). A higher percentage of *F. americana* fell to the forest floor by breaking than by uprooting, as compared to *F. pennsylvanica* or *F. nigra* (Figure 2B; χ^2 = 17.1; P = 0.020). The method of ash treefall was not influenced by soil moisture class (χ^2 = 11.3; P = 0.788). Average stump height of broken ash trees was 2.5 $\pm$ 0.4 m and was not influenced by species (χ^2 = 5.50; P = 0.064) or by soil moisture class (χ^2 = 1.58; P = 0.452).

Ash and non-ash trees fell in a non-random manner in 2008 (ash: Rayleigh z = 4.602, 0.02 > P > 0.01; non-ash: z = 11.875, P < 0.001) and 2012 (ash: z = 9.063, P < 0.001; non-ash: z = 9.79, P < 0.001), indicating a distinct direction of treefall towards the east and southeast (Figure 3). Watson U^2 tests suggested that there were no differences in the direction of treefall between ash and non-ash trees in 2008 (U^2 = 0.06, P > 0.05) and 2012 (U^2 = 0.05, P > 0.05). Similarly, there were no differences in the direction of treefall between 2008 and 2012 for ash (U^2 = 0.102, P > 0.05) and non-ash (U^2 = 0.111, P > 0.05).

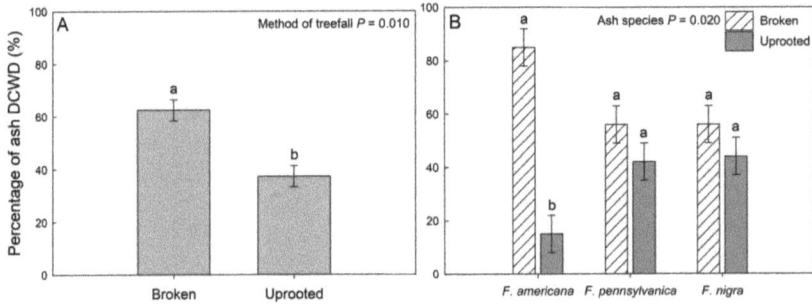

Figure 2. Average percentage (±SE) of downed coarse woody debris (DCWD) by method of treefall (broken or uprooted) (**A**) and by ash species (white ash, *Fraxinus americana* L.; green ash, *Fraxinus pennsylvanica* Marsh.; and black ash, *Fraxinus nigra* Marsh.) that fell to the forest floor by breaking (white bars) or uprooting (gray bars); (**B**) in 16 forest stands in the Upper Huron River watershed in southeastern Michigan, USA. Different letters indicate a significant difference at α = 0.05.

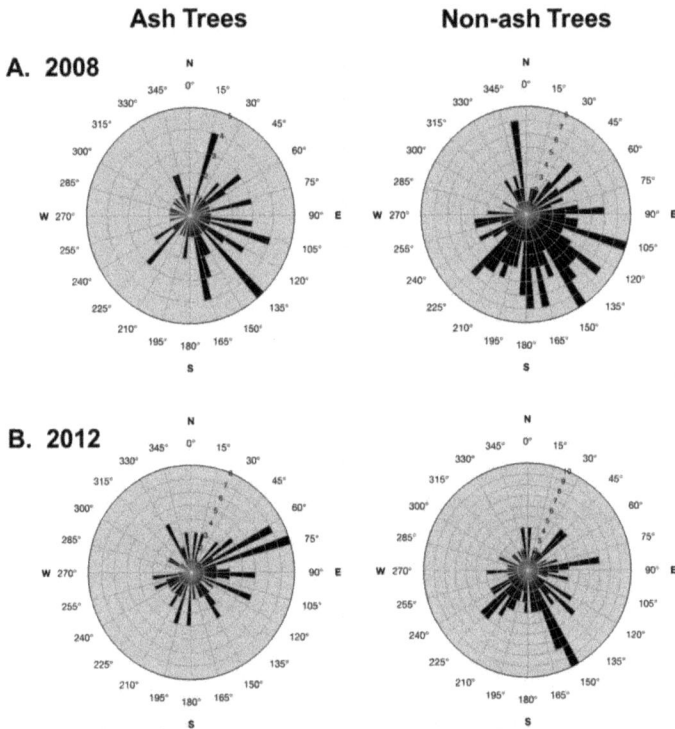

Figure 3. Average azimuth (angles) of the direction of fall for ash (left) and non-ash (right) trees in 2008 (**A**) and 2012 (**B**) in 16 forest stands in the Upper Huron River watershed in southeastern Michigan, USA. The numbers within the circles refer to frequency distribution of the angles of individual trees. Analyses were performed at the transect-level with transects that had <3 fallen trees excluded (2008: N (transects) = 8 for ash and N = 12 for non-ash; 2012: N = 12 for ash and N = 12 for non-ash trees).

Ash species comprised 17% and 27% of the total volume of DCWD sampled in 2008 and 2012, respectively (Table 1). Other tree species comprised 27% in 2008 and 11% in 2012, and DCWD classified as 'unknown' comprised 56% in 2008 and 62% in 2012 (Table 1). From 2008 to 2012, ash trees that had fallen increased the volume of ash DCWD by 53% ($F_{1,10}$ = 15.1; P = 0.003) (Table 2). Average volume of ash DCWD was 17.2 ± 3.3 m³ ha⁻¹ (range of 0.18–45.2 m³ ha⁻¹) in 2008 and 36.6 ± 4.1 m³ ha⁻¹ (range of 29.5–56.4 m³ ha⁻¹) in 2012 (Table 2). Volume of ash DCWD was higher in wetter (hydric and mesic) stands than in xeric stands ($F_{1,10}$ = 4.3; P = 0.043).

Table 1. Average ± SE volume (m³ ha⁻¹) and percentage of downed coarse woody debris (DCWD) by tree species in 2008 and 2012 in 16 forest stands in the Upper Huron River watershed in southeastern Michigan, USA.

Tree Species	2008		2012	
	Volume (m³ ha⁻¹)	Percentage of Total DCWD	Volume (m³ ha⁻¹)	Percentage of Total DCWD
Acer spp.	16.8 ± 16.1	16.5	4.5 ± 1.6	7.1
Fraxinus pennsylvanica Marsh.	9.3 ± 2.9	9.1	7.9 ± 2.1	11.4
Fraxinus nigra Marsh.	4.6 ± 1.7	4.5	7.1 ± 3.1	10.3
Fraxinus americana L.	3.5 ± 1.7	3.4	2.8 ± 1.7	5.5
Populus deltoides Bartr. ex Marsh.	5.8 ± 5.8	5.6	0.2 ± 0.1	0.3
Quercus spp.	2.5 ± 1.9	2.8	1.1 ± 0.6	1.7
Prunus serotina Ehrh.	0.5 ± 0.3	0.5	0.4 ± 0.3	0.5
Larix spp.	0.8 ± 0.8	0.8	0.0	0.0
Tilia americana L.	0.3 ± 0.1	0.3	0.3 ± 0.1	0.4
Ulmus spp.	0.2 ± 0.1	0.2	0.1 ± 0.1	0.1
Carya spp.	0.0	0.0	0.2 ± 0.1	0.3
Sassafras albidum (Nutt.) Nees	0.1 ± 0.1	0.05	0.1 ± 0.1	0.2
Fagus grandifolia Ehrh.	0.1 ± 0.1	0.1	0.0	0.0
Betula spp.	0.1 ± 0.1	0.05	0.0	0.0
Carpinus caroliniana Walt.	0.1 ± 0.1	0.3	0.0	0.0
Unknown	56.9 ± 10.3	55.8	39.8 ± 9.4	62.2

Table 2. Patterns of downed coarse woody debris (DCWD) sampled in 2008 and 2012 in 16 forest stands in the Upper Huron River watershed in southeastern Michigan, USA for the percentage of ash trees that had fallen, volume of ash DCWD (m³ ha⁻¹), volume of non-ash DCWD (m³ ha⁻¹), and volume of total (ash and non-ash) DCWD (m³ ha⁻¹) by year (2008 and 2012). Data are expressed as mean ± SE, with percentage of total DCWD in parentheses. Comparisons were made across years, with means followed by different letters being significantly different at α = 0.05.

	Year	
	2008	2012
Percentage of ash trees that had fallen	7.6 ± 1.0 a	30.7 ± 3.3 b
Volume of ash DCWD (m³ ha⁻¹)	17.2 ± 3.3 a (17.0%)	36.6 ± 4.1 b (27.2%)
Volume of non-ash DCWD (m³ ha⁻¹)	90.3 ± 24.2 a (83.0%)	68.3 ± 10.1 a (72.8%)
Volume of total DCWD (m³ ha⁻¹)	106.7 ± 26.0 a (100%)	93.5 ± 12.3 a (100%)

Volume of non-ash DCWD did not change from 2008 to 2012 (Table 2; $F_{1,12}$ = 0.15; P = 0.701), and did not differ across soil moisture classes ($F_{2,12}$ = 1.2; P = 0.310). Moreover, volume of total (ash and non-ash) DCWD also did not change over time (Table 2; $F_{1,12}$ = 2.2; P = 0.158), or differ across soil moisture classes ($F_{2,12}$ = 0.8; P = 0.437).

In 2008, higher percentages of ash DCWD were characterized as decay classes 1 and 2, but in 2012, higher percentages of ash DCWD were characterized as decay classes 2 and 3 (Figure 4; χ^2 = 83.1; P = 0.004). Decay classes of ash DCWD did not vary by ash species (P ranged from 0.143–0.837) or by soil moisture class (χ^2 = 58.3; P = 0.076).

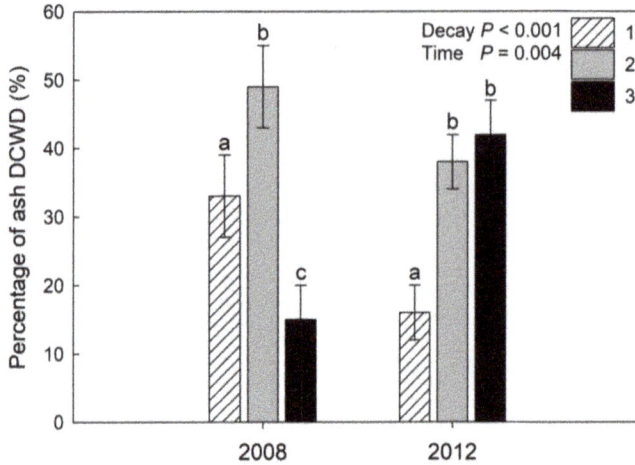

Figure 4. Average percentage (±SE) of downed coarse woody debris (DCWD) that was ash by year (2008 and 2012) for decay classes 1 (white bars), 2 (gray bars), and 3 (black bars) in 16 forest stands in the Upper Huron River watershed in southeastern Michigan, USA. Different letters indicate a significant difference at $\alpha = 0.05$.

4. Discussion

By 2009, ash mortality exceeded 99% in forests in the Upper Huron River watershed of SE Michigan near where EAB first established in North America [35]. By 2008, 8% of dead ash trees greater than 10 cm dbh had fallen, which increased to 31% by 2012. Temporal dynamics of ash and non-ash DCWD in these forests revealed the following patterns: (1) over the time-frame of our study, dead ash trees fell at a rate of 3.5% per year once stands reached 90% ash mortality; (2) dead ash trees fell primarily by snapping along the bole and to a lesser degree by uprooting; (3) ash and non-ash trees fell mostly towards the east and southeast in the direction of prevailing winds; (4) volume of ash DCWD increased nearly two-fold from 2008 to 2012, while volume of non-ash and total DCWD did not change; and (5) majority of ash DCWD was characterized as minimally decayed in 2008, with degree of decay increasing over time. To our knowledge, this is the first study that has investigated temporal dynamics of CWD accumulation following extensive tree mortality from an invasive insect.

Mortality of ash trees with stem diameters ≥ 10 cm reached 90% in 2006, 2007, or 2008 in the stands we sampled, but most dead trees were still standing or leaning as snags in 2008, with only 8% of trees having fallen. Once ash mortality reached 90% in a stand, the percentage of dead ash that fell and became DCWD increased at a rate of 3.5% per year through 2012. Volume of ash DCWD doubled from 2008 to 2012, increasing by 19 m^3 ha^{-1} in these stands. Volume of non-ash DCWD did not change over the four years of this study, which indicates that the increase in ash DCWD resulted from EAB-induced ash mortality, rather than another disturbance such as ice or wind that would have affected a broader diversity of tree species. By 2012, however, only 31% of ash snags had fallen to the forest floor, which indicates that the volume added to these forests by EAB-induced ash mortality will increase substantially in the years following this study. Higham et al. [31] observed faster rates of treefall in Ohio forests during earlier stages of EAB invasion, with 60–80% of the standing dead ash falling to the forest floor within six to seven years after ash mortality had reached only 25%. In the stands we sampled, ash mortality had already reached 40% by 2004 [35,39].

The magnitude of the pulse of ash DCWD accumulation that we observed from 2008 to 2012 was not enough to significantly increase the total volume of DCWD (combined ash and non-ash) at our study sites. A trend for a decline in the volume of non-ash species may have contributed to this pattern, although the difference between the two years sampled was not statistically significant.

As a basis of comparison, the volume of DCWD measured in southern Ohio forests prior to the EAB invasion averaged 42.0 ± 5.1 m^3 ha^{-1}, with ash species comprising 7.6 ± 2.5 m^3 ha^{-1} of the total [53]. This is less than half the total volume of DCWD that we observed in 2008 and 2012, but the percentage of DCWD that was ash (18%) in the Ohio forests was very similar to what we observed in Michigan in 2008 (17%). By 2012, the volume of ash DCWD in our study sites had more than doubled.

Only a few studies have investigated the accumulation of DCWD from tree mortality caused by alien insects. Higham et al. [31] found that the volume of DCWD in forests experiencing early stages of EAB-induced ash mortality ($\geq 25\%$) averaged 60.3 m^3 ha^{-1}, and was positively correlated with ash basal area. They did not differentiate between total volume of ash and non-ash DCWD, but ash comprised a higher proportion of the total volume of minimally decayed DCWD (decay class 1) in sites with a higher ash mortality. In forests of a similar composition [54], but much more advanced ash mortality, we observed that substantially lower percentages of dead ash had fallen, but higher volumes of total DCWD, when compared to Higham et al. [31].

Mortality of American beech (*Fagus grandifolia* Ehrh.) caused by beech bark disease resulted in DCWD accumulation of 13 and 38 m^3 ha^{-1} in maturing and old growth northern hardwood forests, respectively, in New York [46]. These values are comparable to the volume of ash DCWD observed in this study (17.2 m^3 ha^{-1} in 2008 and 36.6 m^3 ha^{-1} in 2012), although tree mortality from EAB has occurred faster and more recently than from beech bark disease, which has existed in North America since the early 1900s [55]. Although extensive mortality of hemlock (*Tsuga* spp.) caused by hemlock woolly adelgid (HWA) has occurred in eastern North America since its introduction in the 1950s, studies have generally reported percentage cover but not volume of DCWD on the forest floor [56–58]. A manipulative experiment designed to investigate the effects of HWA-induced tree mortality and preemptive management via logging [59] found that trees girdled to simulate early stages of decline and mortality by HWA were still standing after four years [60]. By year five, carbon inputs from hemlock DCWD in the girdled treatment were >three-fold higher than in the logging treatment, and >two-fold higher than in the treatment meant to simulate two decades post-HWA disturbance [61].

Ash species comprised 17% and 27% of the total volume of DCWD measured in 2008 and 2012, respectively, for this study, although most could not be confidently identified to species due to the loss of bark and decomposition. Nearly 25% of ash DCWD was determined to be decay class 1 and over 40% was categorized as decay class 2, indicating that this pool of downed wood was still intact with sound structural integrity (class 2 has early stages of decay, but heartwood is still sound), attached branches, and the absence of invading roots [44,62]. Based on ash decay class transition models [63], this suggests there was a pulse of ash DCWD accumulation on the forest floor approximately five years prior to this study. We observed a shift in the percentage of ash DCWD from decay classes 1 and 2 in 2008 to decay classes 2 and 3 in 2012, as decomposition progressed over time. Many biotic and abiotic factors contribute to the rate of woody debris decomposition, which can take up to several decades or longer depending on species-specific factors and site-specific conditions [1,62,64–66], although ash is considered to have a low resistance to decay [45]. We observed a similar progression of ash decay irrespective of ash species or soil moisture class.

Ash and non-ash trees fell primarily to the east and southeast, which conforms to prevailing winds, with no effects of tree species, year, or soil moisture class. Over 60% of ash trees fell to the forest floor by snapping along the bole at an average of 2.5 m above the ground with the rest uprooting. Understanding how ash trees fall and in which direction will contribute to the management of dead ash trees in urban and recreational landscapes where standing snags may pose safety hazards. Among species, this pattern of treefall was driven by *F. americana*, with over 80% of trees snapping, while similar percentages of *F. pennsylvanica* and *F. nigra* fell by snapping and uprooting. Method of treefall did not differ by soil moisture class, suggesting that other species-specific properties may have contributed to this pattern of treefall. Snags (i.e., standing dead trees) create habitat for a variety of forest species [1,67–71]. When they snap, the litter and soil layers are minimally disturbed beyond the point of impact. Trees that uproot, however, result in a patchy formation of pit-and-mound topography, which can have long-term effects on community dynamics by altering patterns of nutrient cycling, standing water, and the physical and chemical properties of the soil through mixing of the litter and soil layers, exposing the root mass, and adding fine and coarse woody debris to the forest floor [72–74].

Altered patterns of DCWD formation and accumulation from tree mortality caused by alien insects and pathogens indirectly affect plant and animal populations in forests. As a fundamental structural component, DCWD increases habitat complexity [75] and provides resources for wildlife such as food, habitat, and sites for sprouting, breeding, and overwintering [1,70,76]. DCWD acts as refugia for species including ground-dwelling invertebrates [77,78], amphibians and reptiles [79], and small mammals [80]. The accumulation and decomposition of ash DCWD caused by EAB-induced ash mortality impacted the abundance, diversity, and community composition of ground-dwelling invertebrates [81–84]. Because the distribution of ash species is widespread in North America [85], the pulse of ash DCWD resulting from ash mortality will have widespread ecological impacts on the flora and fauna of forests as EAB continues to expand its range.

Supplementary Materials: The following are available online at http://www.mdpi.com/1999-4907/9/4/191/s1, Figure S1: Location of transects within the Upper Huron River watershed in southeastern Michigan, USA. From bottom-left to top-right, parks include Hudson Mills MetroPark (red), Island Lake State Recreation Area (purple), Kensington MetroPark (blue), Proud Lake State Recreation Area (green), Highland State Recreation Area (pink), Pontiac State Recreation Area (brown), and Indian Springs MetroPark (orange), Table S1: GPS coordinates and additional information for transects within the Upper Huron River watershed in southeastern Michigan, USA. For this study, 16 transects were selected from 38 transects previously established and characterized by Smith [38], Klooster et al. [35], and Smith et al. [39].

Acknowledgments: We thank Sarah Rudawsky, Kara Taylor, Patty Verdesoto, and other members of the Herms Lab in the Department of Entomology, The Ohio State University, for their assistance in the field, and Derek Robertson, University of Georgia, for technical assistance and creating the map of transect locations. Two anonymous reviewers provided valuable comments that improved this manuscript. This project was supported by funding from the USDA Forest Service Northern Research Station, USDA National Institute for Food and Agriculture, state and federal funds appropriated to the Ohio Agricultural Research and Development Center, The Ohio State University, and the Daniel B. Warnell School of Forestry and Natural Resources, University of Georgia, Athens.

Author Contributions: K.J.K.G. and D.A.H. conceived and designed the experiments; K.I.P., W.S.K., A.S., D.M.H., D.R.C., and K.J.K.G performed the experiments; K.I.P. and K.J.K.G. analyzed the data; all authors contributed to the writing of the paper and gave final approval for publication.

Conflicts of Interest: The authors declare no conflict of interest.

References

1. Harmon, M.E.; Franklin, J.F.; Swanson, F.J.; Sollins, P.; Gregory, S.V.; Lattin, J.D.; Anderson, N.H.; Cline, S.P.; Aumen, N.G.; Sedell, J.R.; et al. Ecology of coarse woody debris in temperate ecosystems. In *Advances in Ecological Research*; MacFadyen, A., Ford, E.D., Eds.; Academic Press: Cambridge, MA, USA, 1986; Volume 15, pp. 133–302.
2. Stevens, V. *The Ecological Role of Coarse Woody Debris: An Overview of the Ecological Importance of CWD in BC Forests*; Working Paper 30; Forest Science Program, BC Ministry of Forests: Victoria, BC, Canada, 1997. Available online: https://www.for.gov.bc.ca/hfd/pubs/docs/wp/wp30.pdf (accessed on 9 January 2018).
3. Everett, R.A.; Ruiz, G.M. Coarse woody debris as a refuge from predation in aquatic communities. *Oecologia* **1993**, *93*, 475–486. [CrossRef] [PubMed]
4. Gurnell, A.M.; Gregory, K.J.; Petts, G.E. The role of coarse woody debris in forest aquatic habitats: Implications for management. *Aquat. Conserv. Mar. Freshw. Ecosyst.* **1995**, *5*, 143–166. [CrossRef]
5. Goodburn, J.M.; Lorimer, C.G. Cavity trees and coarse woody debris in old-growth and managed northern hardwood forests in Wisconsin and Michigan. *Can. J. For. Res.* **1998**, *28*, 427–438. [CrossRef]
6. Siitonen, J. Forest management, coarse woody debris and saproxylic organisms: Fennoscandian boreal forests as an example. *Ecol. Bull.* **2001**, *49*, 11–41.
7. Bowman, J.C.; Sleep, D.; Forbes, G.J.; Edwards, M. The association of small mammals with coarse woody debris at log and stand scales. *For. Ecol. Manag.* **2000**, *129*, 119–124. [CrossRef]

8. Brown, J.K.; Reinhardt, E.D.; Kramer, K.A. *Coarse Woody Debris: Managing Benefits and Fire Hazards in the Recovering Forest*; General Technical Report RMRS-GTR-105; Department of Agricultura, Forest Service, Rocky Mountain Research Station: Ogden, UT, USA, 2003; p. 16. Available online: http://openknowledge.nau.edu/2565/1/Brown_J_etal_2003_CoarseWoodyDebrisManagingBenefits.pdf (accessed on 9 January 2018).

9. Nordén, B.; Ryberg, M.; Götmark, F.; Olausson, B. Relative importance of coarse and fine woody debris for the diversity of wood-inhabiting fungi in temperate broadleaf forests. *Biol. Conserv.* **2004**, *117*, 1–10. [CrossRef]

10. Agee, J.K.; Huff, M.H. Fuel succession in a western hemlock/Douglas-fir forest. *Can. J. For. Res.* **1987**, *17*, 697–704. [CrossRef]

11. Muller, R.N.; Liu, Y. Coarse woody debris in an old-growth deciduous forest on the Cumberland Plateau, southeastern Kentucky. *Can. J. For. Res.* **1991**, *21*, 1567–1572. [CrossRef]

12. Tinker, D.B.; Knight, D.H. Coarse woody debris following fire and logging in Wyoming lodgepole pine forests. *Ecosystems* **2000**, *3*, 472–483. [CrossRef]

13. Pedlar, J.H.; Pearce, J.L.; Venier, L.A.; McKenney, D.W. Coarse woody debris in relation to disturbance and forest type in boreal Canada. *For. Ecol. Manag.* **2002**, *158*, 189–194. [CrossRef]

14. Spies, T.A.; Franklin, J.F.; Thomas, T.B. Coarse woody debris in Douglas-fir forests of western Oregon and Washington. *Ecology* **1988**, *69*, 1689–1702. [CrossRef]

15. Sturtevant, B.R.; Bissonette, J.A.; Long, J.N.; Roberts, D.W. Coarse woody debris as a function of age, stand structure, and disturbance in boreal Newfoundland. *Ecol. Appl.* **1997**, *7*, 702–712. [CrossRef]

16. Baker, T.R.; Honorio Coronado, E.N.; Phillips, O.L.; Martin, J.; van der Heijden, G.M.F.; Garcia, M.; Silva Espejo, J. Low stocks of coarse woody debris in a southwest Amazonian forest. *Oecologia* **2007**, *152*, 495–504. [CrossRef] [PubMed]

17. Spies, T.A.; Franklin, J.F.; Klopsch, M. Canopy gaps in Douglas-fir forests of the Cascade Mountains. *Can. J. For. Res.* **1990**, *20*, 649–658. [CrossRef]

18. Whigham, D.F.; Olmsted, I.; Cano, E.C.; Harmon, M.E. The impact of hurricane Gilbert on trees, litterfall, and woody debris in a dry tropical forest in the northeastern Yucatan Peninsula. *Biotropica* **1991**, *23*, 434–441. [CrossRef]

19. Hooper, M.C.; Arii, K.; Lechowicz, M.J. Impact of a major ice storm on an old-growth hardwood forest. *Can. J. Bot.* **2001**, *79*, 70–75.

20. Van Lear, D.H. Dynamics of coarse woody debris in southern forest ecosystems. In Proceedings of the Workshop on Coarse Woody Debris in Southern Forests: Effects on Biodiversity, Athens, GA, USA, 18–20 October 1993; McMinn, J.W., Crossley, D.A., Jr., Eds.; General Technical Report SE-94. United States Department of Agriculture, Southern Research Station: Athens, GA, USA, 1993.

21. Brassard, B.W.; Chen, H.Y.H. Effects of forest type and disturbance on diversity of coarse woody debris in boreal forest. *Ecosystems* **2008**, *11*, 1078–1090. [CrossRef]

22. Kruys, N.; Jonsson, B.G. Fine woody debris is important for species richness on logs in managed boreal spruce forests of northern Sweden. *Can. J. For. Res.* **1999**, *29*, 1295–1299. [CrossRef]

23. Perry, K.I.; Herms, D.A. Responses of ground-dwelling invertebrates to gap formation and accumulation of woody debris from invasive species, wind, and salvage logging. *Forests* **2017**, *8*, 1–13.

24. Ellison, A.M.; Bank, M.S.; Clinton, B.D.; Colburn, E.A.; Elliott, K.; Ford, C.R.; Foster, D.R.; Kloeppel, B.D.; Knoepp, J.D.; Lovett, G.M.; et al. Loss of foundation species: Consequences for the structure and dynamics of forested ecosystems. *Front. Ecol. Environ.* **2005**, *3*, 479–486. [CrossRef]

25. Klutsch, J.G.; Negrón, J.F.; Costello, S.L.; Rhoades, C.C.; West, D.R.; Popp, J.; Caissie, R. Stand characteristics and downed woody debris accumulations associated with a mountain pine beetle (*Dendroctonus ponderosae* Hopkins) outbreak in Colorado. *For. Ecol. Manag.* **2009**, *258*, 641–649. [CrossRef]

26. Gandhi, K.J.K.; Herms, D.A. Direct and indirect effects of alien insect herbivores on ecological processes and interactions in forests of eastern North America. *Biol. Invasions* **2010**, *12*, 389–405. [CrossRef]

27. Liebhold, A.M.; MacDonald, W.L.; Bergdahl, D.; Mastro, V.C. Invasion by exotic forest pests: A threat to forest ecosystems. *For. Sci.* **1995**, *41*, 1–49.

28. Lovett, G.M.; Canham, C.D.; Arthur, M.A.; Weathers, K.C.; Fitzhugh, R.D. Forest ecosystem responses to exotic pests and pathogens in eastern North America. *BioScience* **2006**, *56*, 395–405. [CrossRef]

29. Flower, C.E.; Gonzalez-Meler, M.A. Responses of temperate forest productivity to insect and pathogen disturbances. *Annu. Rev. Plant Biol.* **2015**, *66*, 547–569. [CrossRef] [PubMed]

30. Krasny, M.E.; DiGregorio, L.M. Gap dynamics in Allegheny northern hardwood forests in the presence of beech bark disease and gypsy moth disturbances. *For. Ecol. Manag.* **2001**, *144*, 265–274. [CrossRef]

31. Higham, M.; Hoven, B.M.; Gorchov, D.L.; Knight, K.S. Patterns of coarse woody debris in hardwood forests across a chronosequence of ash mortality due to the emerald ash borer (*Agrilus planipennis*). *Nat. Areas J.* **2017**, *37*, 406–411. [CrossRef]

32. Herms, D.A.; McCullough, D.G. Emerald ash borer invasion of North America: History, biology, ecology, impacts, and management. *Annu. Rev. Entomol.* **2014**, *59*, 13–30. [CrossRef] [PubMed]

33. Cipollini, D. White fringetree as a novel larval host for emerald ash borer. *J. Econ. Entomol.* **2015**, *108*, 370–375. [CrossRef] [PubMed]

34. Knight, K.S.; Brown, J.P.; Long, R.P. Factors affecting the survival of ash (*Fraxinus* spp.) trees infested by emerald ash borer (*Agrilus planipennis*). *Biol. Invasions* **2013**, *15*, 371–383. [CrossRef]

35. Klooster, W.S.; Herms, D.A.; Knight, K.S.; Herms, C.P.; McCullough, D.G.; Smith, A.; Gandhi, K.J.K.; Cardina, J. Ash (*Fraxinus* spp.) mortality, regeneration, and seed bank dynamics in mixed hardwood forests following invasion by emerald ash borer (*Agrilus planipennis*). *Biol. Invasions* **2014**, *16*, 859–873. [CrossRef]

36. Burr, S.J.; McCullough, D.G. Condition of green ash (*Fraxinus pennsylvanica*) overstory and regeneration at three stages of the emerald ash borer invasion wave. *Can. J. For. Res.* **2014**, *44*, 768–776. [CrossRef]

37. Kashian, D.M.; Witter, J.A. Assessing the potential for ash canopy tree replacement via current regeneration following emerald ash borer-caused mortality on southeastern Michigan landscapes. *For. Ecol. Manag.* **2011**, *261*, 480–488. [CrossRef]

38. Smith, A. Effects of Community Structure on Forest Susceptibility and Response to the Emerald Ash Borer Invasion of the Huron River Watershed in Southeastern Michigan. Master's Thesis, The Ohio State University, Columbus, OH, USA, 2006.

39. Smith, A.; Herms, D.A.; Long, R.P.; Gandhi, K.J.K. Community composition and structure had no effect on forest susceptibility to invasion by the emerald ash borer (Coleoptera: Buprestidae). *Can. Entomol.* **2015**, *147*, 318–328. [CrossRef]

40. Siegert, N.W.; McCullough, D.G.; Liebhold, A.M.; Telewski, F.W. Dendrochronological reconstruction of the epicentre and early spread of emerald ash borer in North America. *Divers. Distrib.* **2014**, *20*, 847–858. [CrossRef]

41. Gandhi, K.J.K.; Smith, A.; Hartzler, D.M.; Herms, D.A. Indirect effects of emerald ash borer-induced ash mortality and canopy gap formation on epigaeic beetles. *Environ. Entomol.* **2014**, *43*, 546–555. [CrossRef] [PubMed]

42. Albert, D.A. *Regional Landscape Ecosystems of Michigan, Minnesota and Wisconsin: A Working Map and Classification*; General Technical Report NC-178; U.S. Department of Agriculture, Forest Service, North Central Forest Experiment Station: St. Paul, MN, USA, 1995; p. 255.

43. National Oceanic and Atmospheric Administration (NOAA). *National Weather Service Forecast Office, NOAA Online Weather Data, (20 February)*; NOAA: Silver Spring, MD, USA, 2018.

44. Woodall, C.; Williams, M.S. *Sampling Protocol, Estimation, and Analysis Procedures for the Down Woody Materials Indicator of the FIA Program*; General Technical Report NC-256; United States Department of Agriculture, Forest Service, North Central Research Station: St. Paul, MN, USA, 2005.

45. Harmon, M.E.; Sexton, J. *Guidelines for Measurements of Woody Detritus in Forest Ecosystems*; Publication No. 20; U.S. LTER Network Office, University of Washington: Seattle, WA, USA, 1996; pp. 1–34.

46. McGee, G.G. The contribution of beech bark disease-induced mortality to coarse woody debris loads in northern hardwood stands of Adirondack Park, New York, USA. *Can. J. For. Res.* **2000**, *30*, 1453–1462. [CrossRef]

47. Fox, J.; Weisberg, S. *An {R} Companion to Applied Regression*, 2nd ed.; Sage: Thousand Oaks, CA, USA, 2011; Available online: http://socserv.socsci.mcmaster.ca/jfox/Books/Companion.

48. R Development Core Team. *R: A Language and Environment for Statistical Computing*; R Foundation for Statistical Computing: Vienna, Austria, 2017; Available online: www.r-project.org.

49. Quinn, G.P.; Keough, M.J. *Experimental Design and Data Analysis for Biologists*; Cambridge University Press: Cambridge, UK, 2002.

50. Zar, J.H. *Biostatistical Analysis*, 3rd ed.; Prentice-Hall, Inc.: Princeton, NJ, USA, 1996.

51. Mardia, K.V.; Jupp, P.E. *Directional Statistics*; John Wiley & Sons: London, UK, 2000.

52. Young Technology Inc. *GeoRose Version 0.5.1.1*; Young Technology Inc.: Edmonton, AB, Canada, 2014; Available online: Http://yongtechnology.Com/download/georose (accessed on 27 December 2017).

53. Rubino, D.L.; McCarthy, B.C. Evaluation of coarse woody debris and forest vegetation across topographic gradients in a southern Ohio forest. *For. Ecol. Manag.* **2003**, *183*, 221–238. [CrossRef]

54. Dyer, J.M. Revisiting the deciduous forests of eastern North America. *BioScience* **2006**, *56*, 341–352. [CrossRef]

55. Houston, D.R. Major new tree disease epidemics: Beech bark disease. *Annu. Rev. Phytopathol.* **1994**, *32*, 75–87. [CrossRef]

56. Kizlinski, M.L.; Orwig, D.A.; Cobb, R.C.; Foster, D.R. Direct and indirect ecosystem consequences of an invasive pest on forests dominated by eastern hemlock. *J. Biogeogr.* **2002**, *29*, 1489–1503. [CrossRef]

57. Eschtruth, A.K.; Cleavitt, N.L.; Battles, J.J.; Evans, R.A.; Fahey, T.J. Vegetation dynamics in declining eastern hemlock stands: 9 years of forest response to hemlock woolly adelgid infestation. *Can. J. For. Res.* **2006**, *36*, 1435–1450. [CrossRef]

58. Cleavitt, N.L.; Eschtruth, A.K.; Battles, J.J.; Fahey, T.J. Bryophyte response to eastern hemlock decline caused by hemlock woolly adelgid infestation. *J. Torrey Bot. Soc.* **2008**, *135*, 12–25. [CrossRef]

59. Ellison, A.M.; Barker-Plotkin, A.A.; Foster, D.R.; Orwig, D.A. Experimentally testing the role of foundation species in forests: The Harvard forest hemlock removal experiment. *Methods Ecol. Evol.* **2010**, *1*, 168–179. [CrossRef]

60. Orwig, D.A.; Barker Plotkin, A.A.; Davidson, E.A.; Lux, H.; Savage, K.E.; Ellison, A.M. Foundation species loss affects vegetation structure more than ecosystem function in a northeastern USA forest. *PeerJ* **2013**, *1*, e41. [CrossRef] [PubMed]

61. Raymer, P.C.L.; Orwig, D.A.; Finzi, A.C. Hemlock loss due to the hemlock woolly adelgid does not affect ecosystem C storage but alters its distribution. *Ecosphere* **2013**, *4*, 1–16. [CrossRef]

62. Stokland, J.N.; Siitonen, J. Mortality factors and decay succession. In *Biodiversity in Dead Wood*; Stokland, J.N., Siitonen, J., Jonsson, B.G., Eds.; Cambridge University Press: Cambridge, UK, 2012; pp. 110–149.

63. Russell, M.B.; Woodall, C.W.; Fraver, S.; D'Amato, A.W. Estimates of downed woody debris decay class transitions for forests across the eastern United States. *Ecol. Model.* **2013**, *251*, 22–31. [CrossRef]

64. Swift, M.J.; Heal, O.W.; Anderson, J.M. *Decomposition in Terrestrial Ecosystems*; University of California Press: Berkeley, CA, USA, 1979.

65. Freschet, G.T.; Weedon, J.T.; Aerts, R.; van Hal, J.R.; Cornelissen, J.H.C. Interspecific differences in wood decay rates: Insights from a new short-term method to study long-term wood decomposition. *J. Ecol.* **2012**, *100*, 161–170. [CrossRef]

66. Stokland, J.N. Wood decomposition. In *Biodiversity in Dead Wood*; Stokland, J.N., Siitonen, J., Jonsson, B.G., Eds.; Cambridge University Press: Cambridge, UK, 2012; pp. 10–28.

67. Raphael, M.G.; White, M. Use of snags by cavity-nesting birds in the Sierra Nevada. *Wildl. Monogr.* **1984**, *86*, 3–66.

68. Drapeau, P.; Nappi, A.; Imbeau, L.; Saint-Germain, M. Standing deadwood for keystone bird species in the eastern boreal forest: Managing for snag dynamics. *For. Chron.* **2009**, *85*, 227–234. [CrossRef]

69. McComb, W.; Lindenmayer, D.B. Dying, dead, and downed trees. In *Maintaining Biodiversity in Forest Ecosystems*; Hunter, M.L., Jr., Ed.; Cambridge University Press: Cambridge, UK, 1999; pp. 335–372.

70. Siitonen, J. Microhabitats. In *Biodiversity in Dead Wood*; Stokland, J.N., Siitonen, J., Jonsson, B.G., Eds.; Cambridge University Press: Cambridge, UK, 2012; pp. 150–182.

71. Siitonen, J.; Jonsson, B.G. Other associations with dead woody material. In *Biodiversity in Dead Wood*; Stokland, J.N., Siitonen, J., Jonsson, B.G., Eds.; Cambridge University Press: Cambridge, UK, 2012; pp. 58–81.

72. Schaetzl, R.J.; Burns, S.F.; Johnson, D.L.; Small, T.W. Tree uprooting: Review of impacts on forest ecology. *Vegetatio* **1988**, *79*, 165–176. [CrossRef]

73. Liechty, H.O.; Jurgensen, M.F.; Mroz, G.D.; Gale, M.R. Pit and mound topography and its influence on storage of carbon, nitrogen, and organic matter within an old-growth forest. *Can. J. For. Res.* **1997**, *27*, 1992–1997. [CrossRef]

74. Clinton, B.D.; Baker, C.R. Catastrophic windthrow in the southern Appalachians: Characteristics of pits and mounds and initial vegetation responses. *For. Ecol. Manag.* **2000**, *126*, 51–60. [CrossRef]

75. McElhinny, C.; Gibbons, P.; Brack, C.; Bauhus, J. Forest and woodland stand structural complexity: Its definition and measurement. *For. Ecol. Manag.* **2005**, *218*, 1–24. [CrossRef]

76. Tews, J.; Brose, U.; Grimm, V.; Tielborger, K.; Wichmann, M.C.; Schwager, M.; Jeltsch, F. Animal species diversity driven by habitat heterogeneity/diversity: The importance of keystone structures. *J. Biogeogr.* **2004**, *31*, 79–92. [CrossRef]

77. Evans, A.M.; Clinton, P.W.; Allen, R.B.; Frampton, C.M. The influence of logs on the spatial distribution of litter-dwelling invertebrates and forest floor processes in New Zealand forests. *For. Ecol. Manag.* **2003**, *184*, 251–262. [CrossRef]

78. Ulyshen, M.D.; Hanula, J.L. Litter-dwelling arthropod abundance peaks near coarse woody debris in loblolly pine forests of the southeastern United States. *Fla. Entomol.* **2009**, *92*, 163–164. [CrossRef]

79. Whiles, M.R.; Grubaugh, J.W. Importance of Coarse Woody Debris to Southern Forest Herpetofauna. In Proceedings of the Biodiversity and Coarse Woody Debris in Southern Forests: Effects on Biodiversity, Athens, GA, USA, 18–20 October 1993; McMinn, J.W., Crossley, D.A., Jr., Eds.; USDA Forest Service: Athens, GA, USA, 1993.

80. Fauteux, D.; Imbeau, L.; Drapeau, P.; Mazerolle, M.J. Small mammal responses to coarse woody debris distribution at different spatial scales in managed and unmanaged boreal forests. *For. Ecol. Manag.* **2012**, *266*, 194–205. [CrossRef]

81. Ulyshen, M.D.; Klooster, W.S.; Barrington, W.T.; Herms, D.A. Impacts of emerald ash borer-induced tree mortality on leaf litter arthropods and exotic earthworms. *Pedobiologia* **2011**, *54*, 261–265. [CrossRef]

82. Perry, K.I.; Herms, D.A. Short-term responses of ground beetles to forest changes caused by early stages of emerald ash borer (Coleoptera: Buprestidae)-induced ash mortality. *Environ. Entomol.* **2016**, *45*, 616–626. [CrossRef] [PubMed]

83. Perry, K.I.; Herms, D.A. Response of the forest floor invertebrate community to canopy gap formation caused by early stages of emerald ash borer-induced ash mortality. *For. Ecol. Manag.* **2016**, *375*, 259–267. [CrossRef]

84. Perry, K.I.; Herms, D.A. Effects of late stages of emerald ash borer (Coleoptera: Buprestidae)-induced ash mortality on forest floor invertebrate communities. *J. Insect Sci.* **2017**, *17*, 1–10. [CrossRef]

85. MacFarlane, D.W.; Meyer, S.P. Characteristics and distribution of potential ash tree hosts for emerald ash borer. *For. Ecol. Manag.* **2005**, *213*, 15–24. [CrossRef]

forests

MDPI

Article

Optimizing Conservation Strategies for a Threatened Tree Species: In Situ Conservation of White Ash (*Fraxinus americana* L.) Genetic Diversity through Insecticide Treatment

Charles E. Flower [1,2,*], Jeremie B. Fant [3], Sean Hoban [4], Kathleen S. Knight [1], Laura Steger [3], Elijah Aubihl [5], Miquel A. Gonzalez-Meler [2], Stephen Forry [6], Andrea Hille [6] and Alejandro A. Royo [7]

[1] USDA Forest Service, Northern Research Station, 359 Main Rd., Delaware, OH 43015, USA; ksknight@fs.fed.us

[2] Biological Sciences, Department of Ecology and Evolution, University of Illinois at Chicago, 845 W, Taylor St, Chicago, IL 60607, USA; mmeler@uic.edu

[3] Chicago Botanic Gardens, Department of Plant Science, 1000 Lake Cook Rd, Glencoe, IL 60022, USA; jfant@chicagobotanic.org (J.B.F.); ldsteger@gmail.com (L.S.)

[4] The Morton Arboretum, 4100 Illinois Route 53, Lisle, IL 60532, USA; shoban@mortonarb.org

[5] Department of Biology, Miami University, 316 Pearson Hall, Oxford, OH 45056, USA; aubihled@miamioh.edu

[6] USDA Forest Service, Allegheny National Forest, 4 Farm Colony Dr., Warren, PA 16365, USA; sforry@fs.fed.us (S.F.); ahille@fs.fed.us (A.H.)

[7] The USDA Forest Service, Northern Research Station, P.O. Box 267, Irvine, PA 16329, USA; aroyo@fs.fed.us

* Correspondence: charlesflower@fs.fed.us; Tel.: +1-740-368-0068

Received: 20 February 2018; Accepted: 10 April 2018; Published: 13 April 2018

Abstract: Forest resources face numerous threats that require costly management. Hence, there is an increasing need for data-informed strategies to guide conservation practices. The introduction of the emerald ash borer to North America has caused rapid declines in ash populations (*Fraxinus* spp. L.). Natural resource managers are faced with a choice of either allowing ash trees to die, risking forest degradation and reduced functional resilience, or investing in conserving trees to preserve ecosystem structure and standing genetic diversity. The information needed to guide these decisions is not always readily available. Therefore, to address this concern, we used eight microsatellites to genotype 352 white ash trees (*Fraxinus americana* L.) across 17 populations in the Allegheny National Forest; a subset of individuals sampled are part of an insecticide treatment regimen. Genetic diversity (number of alleles and He) was equivalent in treated and untreated trees, with little evidence of differentiation or inbreeding, suggesting current insecticidal treatment is conserving local, neutral genetic diversity. Using simulations, we demonstrated that best practice is treating more populations rather than more trees in fewer populations. Furthermore, through genetic screening, conservation practitioners can select highly diverse and unique populations to maximize diversity and reduce expenditures (by up to 21%). These findings will help practitioners develop cost-effective strategies to conserve genetic diversity.

Keywords: emerald ash borer (*Agrilus planipennis* Fairmaire); in situ conservation; forest pest; disturbance; *Fraxinus*; integrated pest management

---

1. Introduction

The intentional and unintentional spread of non-native species is contributing to the global reduction of plant diversity and the homogenization of plant communities [1,2]. Non-native insect pests and pathogens represent major destructive forces in forests, impacting productivity, biogeochemical cycling, hydrology and forest successional dynamics [3–5]. Conserving the genetic diversity of imperiled forest tree species is essential for maintaining the long-term sustainability and resilience of forest ecosystems. Genetic diversity provides the basis for adaptation to environmental and anthropogenic change [6], increases ecosystem stability and resilience [7,8], and promotes species diversity [9]. Particularly influential is the genetic diversity of foundational species, such as large trees. Although it is well established that conserving genetic diversity is crucial in preventing the extinction of populations and species, very few management plans incorporate genetic monitoring in their action plans [10].

Conserving genetic diversity typically relies on both an ex situ approach, in which a species is preserved outside of its natural habitat, and an in situ approach, in which viable, reproducing populations are maintained within the species' natural habitat. Ex situ conservation can involve either maintenance of living trees, or storing seeds in long term storage such as a seed bank. Although ex situ approaches have the potential to save a majority of a species' diversity, they also have their limitations. Seed banks only work for species with orthodox seeds, and accessions may be compromised either during the collection (selection bias) [11] or storage (e.g., seed death, mutation accumulation). Likewise, trees maintained in living collections require large planting areas, diversity can be lost through cycles of regeneration, and there is potential for adaptation to cultivated conditions [12,13]. The value of ex situ collections may also be limited if it is not associated with any in situ strategies. Reintroductions, with ex situ material, may be hampered by the degradation of underlying ecosystem properties associated with the initial disturbance and extirpation and loss of in situ mutualists, like pollinators, seed dispersers, and fungi. Recent evidence suggests that *Fraxinus* are associated with unique soil microbial assemblages which may be altered if ash were to be lost from a forest ecosystem [14]. Additionally, black ash (*Fraxinus nigra* Marshall) has been shown to help to maintain the hydraulic balance of sensitive wetland forests, and its loss from these systems because of the invasive emerald ash borer (*Agrilus planipennis* Fairmaire, EAB) may prevent the future establishment of black ash and other tree species [15]. Therefore, in situ conservation is often the preferred means of conserving a species, especially when used in conjunction with ex situ conservation.

Refinement of in situ approaches for maintaining imperiled species is required if managers are to maximize the genetic diversity conserved while minimizing costs. Developing such tools is increasingly necessary as outbreaks of forest pest and pathogen become more prevalent and result in widespread disturbances [16]. EAB, a non-native forest pest accidentally introduced into North America in the 1990's, has spread from south eastern Michigan to over 31 states, resulting in a dramatic decline in numbers of ash trees [17]. There are 16 native species of ash trees (*Fraxinus* spp.) across North America [18], which collectively represent ~2.5% of the trees in the United States, with the greatest density and abundance in the Great Lakes Region [19]. EAB larvae, which feed predominately on phloem and cambial tissue, create serpentine galleries that effectively girdle host trees [20,21] and result in >99% tree mortality (for black, white and green ash see [22,23], relative to ~71% survival for blue ash see [24]). Due to this infestation, five ash species are now listed as critically endangered on the International Union for Conservation of Nature red list (*F. pennsylvanica* Marshall, *F. americana*, *F. nigra*, *F. profunda* (Bush) Bush, and *F. quadrangulata* Michx.) and one is listed as endangered (*F. caroliniana* Mill.) [25]. The loss of ash trees has been estimated to cost upwards of $60 billion USD, not including replacement costs [26], which will be significant in cities like Chicago, where there were significant numbers of ash trees (>600,000). Because of the lethal effects and speed with which EAB is spreading, managers need new tools to mitigate the impact.

The threat to ash trees has led to the development of a combined ex situ and in situ conservation approach. Ex situ methods include seed collection and storage, which have yielded collections from 1982 ash trees from five species (US National Plant Germplasm System). Meanwhile, the in situ methods include insecticide treatment of standing trees to protect them from future EAB attacks. In situ preservation will help maintain breeding populations of large trees in the landscape, thereby protecting ecosystems services, helping to ensure future ash reintroduction, and allowing continued natural selection and adaptation. Evidence from early in situ *Fraxinus* conservation efforts suggest a degree of associational protection, in which insecticide treatment of a small number of ash trees in a stand can promote the health of untreated trees [27]. However, the optimal treatment densities to achieve maximal associational protection remain unknown and are under investigation [28]. While insecticides have been used frequently for protection of urban street trees and yard trees, these cultivated ash are not representative of, and do not help to preserve, natural levels of genetic diversity. To address this concern, insecticide treatment has been expanded to multiple, naturally-occurring populations, managed by a diverse assemblage of landowners. Multiple insecticide formulations are available, but all require repeated treatments (yearly or every 3 years, depending on the insecticide used). The cost of emamectin benzoate injections, which have been demonstrated to be most effective at controlling EAB [29,30], range from $50 to $150 USD per tree, depending on tree diameter. Multiple entities have expressed interest in expanding these efforts; however, practical information is needed to help managers maximize the benefits of insecticide treatments while minimizing resource use.

As treating all trees in an area is usually not an option, a subset of trees will need to be chosen for treatment. If the objective is to maintain local standing genetic diversity, population models suggest that treating more trees will maximize the chance of conserving a substantial portion of the local alleles [31,32]. However, the relationship between allelic diversity and sample size is non-linear, hence there is a point at which additional individuals provide minimal gains in diversity. Ex situ seed collections protocols suggest that in order to maximize the local diversity collected, you need to sample a minimum of 50 unrelated maternal trees per population, and repeat for as many populations as possible [33]. However, the appropriate design of a conservation collection depends on various aspects of the target species, including population size, reproductive biology, recent population history, and connectivity among populations [34–39]. Given that it is nearly impossible to preserve all genetic material, determining the optimum number and spatial arrangement of treated trees will be important in efforts to conserve the majority of local genetic variation.

Here we investigated the genetic diversity of white ash trees (*F. americana*) across the Allegheny National Forest (ANF), and utilized the results to optimize a regional in situ conservation plan. Our goals were specifically to: (1) determine the genetic diversity and structure of white ash populations across the ANF; (2) quantify how varying levels of treatment (number of trees and populations) will affect the preservation of ash genetic diversity; and (3) formulate optimal strategies, including identifying which populations preserve the greatest proportion of genetic diversity. To achieve this, we conducted a genetic survey of white ash trees, then used computer algorithms to subsample the dataset with different possible sampling strategies (100 plants, 20 plants, etc.). Finally, we discuss implications for white ash, as well as caveats to this approach.

2. Materials and Methods

2.1. Site Selection

Our study was conducted in the Allegheny National Forest (ANF) in northwestern Pennsylvania, USA, which covers over 2075 km^2 (Figure 1). The forest sits atop an unglaciated portion of the Allegheny plateau, and exhibits considerable variability in topographic relief. The forest ecosystem is comprised largely of secondary successional mixed deciduous forests, which regenerated after extensive anthropogenic disturbances in the late 1800's and early 1900's. The forest is managed intensively for timber and is dominated by black cherry, maples, and other hardwoods; *Fraxinus* spp. comprise nearly 3% of trees across the ANF, comparable to densities across the eastern US, but not as high as densities near the epicenter of the EAB outbreak.

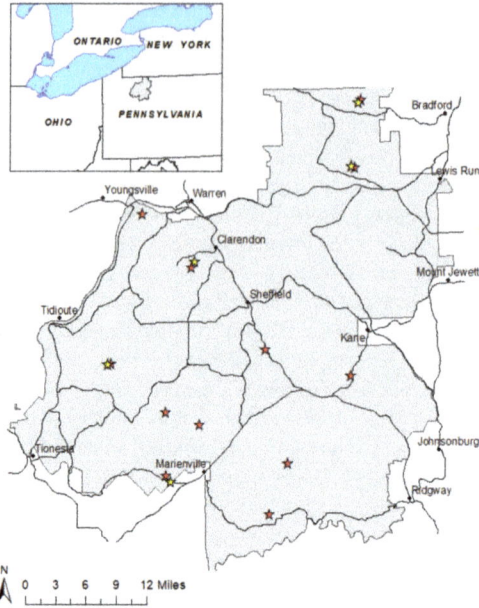

Figure 1. Map of the Allegheny National Forest with white ash foliar sampling locations denoted with stars (red stars denote low elevation and yellow sites upper elevation sites). Inset map denotes the position of the ANF in the greater region.

2.2. Sample Collection

During 2015, 27 (100 m radius, 3.14 ha) permanent research plots were established across the ANF for in situ conservation efforts. These plots represent a portion of a larger network of plots across the ANF which was established to track the arrival and spread of EAB, as well as the decline of ash in the region. For the insecticide conservation effort, plots widely distributed across the ANF containing at least 20 ash trees were selected (Figure 1). These plots contained anywhere from 21 to 201 specimens. In May 2015, to protect these trees from EAB, a random subset of 20 ash trees within each site was treated with 0.157 g a.i./cm diameter at breast height of the systemic insecticide emamectin benzoate (Tree-äge®; Arborjet, Woburn, MA, USA) using the Arborjet Quick-Jet system (as directed by the manufacturer: Arborjet, Woburn, MA, USA). As such, in these plots, between ~10% and 95% of the ash were inoculated with insecticide. In July 2016, foliar samples were collected from 17 plots distributed across the ANF (see Figure 1; Table 1); these plots represent a subset of those assigned to the insecticide trial, as well as several other plots that did not receive insecticide treatment that comprise a wider geographic footprint. Furthermore, because many of the insecticide treated plots are paired (high and low elevation), sampling each of the paired plots would not be necessary to assess population genetics across the ANF, and would represent similar local alleles. Although it would be unexpected to see genetic differentiation between groups of individuals in close vicinity, it is possible that trees may exhibit phenological differences in flowering across systems with considerable topographic complexity as in this system. As such, we incorporated five of these areas into the study (denoted in Figure 1 below). In each plot, foliar samples were collected from ~30 trees (if available) which were randomly selected. Foliar samples were stored on ice in the field, and frozen until analysis, for a total of 352 trees sampled (consisting of 274 inoculated trees and 78 non-inoculated trees). Obviously, because of randomly selecting trees within the treatment plots, foliar samples were sourced from both insecticide and non-insecticide treated trees.

Table 1. Averages for genetic parameters by population, including population name (PopID), whether trees were treated with insecticide (Y = yes, N = no), topographic position of the site (upland or lowland), number of trees sampled (n), average number of alleles (Na), average effective number of alleles (Ne) corrected for sample size, gene diversity (He), number of Private alleles (P) and Inbreeding coefficient (Fis).

PopID	Insecticide Treated	Topography	n	Na	Ne	He	P	Fis
104	N	Lowland	16	4.88	2.89	0.60	0	0.00
150	N	Lowland	23	5.88	3.13	0.64	0	0.13
162	N	Lowland	3	2.63	2.11	0.48	0	−0.09
186	N	Lowland	2	1.63	1.58	0.23	0	−0.64
200	N	Lowland	23	5.00	2.88	0.61	2	0.04
103	N	Upland	7	4.13	2.83	0.60	0	0.16
149	N	Upland	1	1.38	1.38	0.25	0	−1.00
6	Y	Lowland	30	8.25	3.40	0.62	7	0.08
26	Y	Lowland	29	7.50	3.14	0.61	7	0.07
88	Y	Lowland	30	7.50	3.32	0.65	0	0.07
126	Y	Lowland	24	6.13	2.89	0.61	1	0.11
142	Y	Lowland	22	5.38	2.76	0.59	1	0.09
158	Y	Lowland	23	5.75	2.99	0.63	0	0.02
166	Y	Lowland	30	7.50	3.45	0.64	2	0.08
5	Y	Upland	29	7.13	3.26	0.65	1	0.08
25	Y	Upland	30	6.63	3.10	0.63	2	0.08
87	Y	Upland	30	7.25	3.58	0.64	3	0.17

2.3. Genetic Analysis

Genomic DNA was extracted from ~1 cm^2 of silica dried leaf tissues using a modified Cetyl trimethylammonium bromide (CTAB) method, developed by Doyle and Doyle [40]. Following DNA extraction and purification, the quality and concentration of DNA were evaluated using spectrophotometry (Nanodrop 2000, ThermoScientific, Waltham, MA, USA). A subset of samples was screened using 23 microsatellites primers pairs previously designed for *F. excelsior* L. [41–44], and 15 expressed sequence-simple sequence repeats (EST-SSR) designed from *F. americana* [41]. From these, 10 primer pairs amplified reliably and were polymorphic in our samples. The remainder produced no bands (9), were monomorphic (6), or did not amplify reliably (13). The final primer sets used in this study included FEMSATL 11 & FEMSATL 16 [41], FR639485 [44], M230 [45] and Fp21068, Fp20239, Fp19681; Fp12378; Fp14665; and Fp17710 [46].

To visualize the alleles, each forward primer was modified with the addition of an M13 sequence to the 5′ end (5′-CACGACGTTGTAAAACGAC-3′), to allow post-polymerase chain reaction (PCR) labeling with fluorescent dyes [47]. An initial PCR was conducted in a 10 µL reaction mixture containing 5 ng template DNA, 25 nM of forward and reverse primer, and 5 µL of MyTaqTM Master mix (Bioline, Taunton, MA, USA). This PCR mix was run for 2 min initial annealing (94 °C), then 15 cycles of 94 °C for 40 s, 57 °C for 40 s, and 72 °C for 90 s, and a final extension (72 °C) for 5 min. Once the initial PCR product was generated, it was labeled through a second PCR through the addition of 2.5 µL of MyTaqTM Master mix (Bioline, Taunton, MA, USA), 2.25 µL of DNA grade water and 0.25 µL M13 primer labeled with either WellRed Black (D2), Green (D3) or Blue (D4) fluorescent dye (Sigma-Proligo, The Woodlands, TX, USA-Sigma-Aldrich, St. Louis, MO, USA). The labeling PCR's were conducted at 94 °C for 2 min, 27 cycles of 94 °C for 30 s, 55 °C for 30 s, and 72 °C for 1 min, and an extension of 72 °C for 10 min. PCR products were analyzed on a CEQ 8000 Genetic Analysis System with GenomeLab 400 internal size standard (ABSCIEX, Chicago, IL, USA).

2.4. Data Analysis

The program GENALEX [48] was used to generate descriptive parameters, including common metrics of genetic diversity such as mean number of alleles per locus (A$_p$) effective number of alleles per loci (A_e), expected heterozygosity (H_e), and number of private alleles (P), as well as measures of inbreeding, using Weir and Cockerham's [49] estimates of Wright's inbreeding co-efficient (F_{IS}). The average pairwise relatedness of each population was calculated using the Queller and Goodnight [50] estimator in GENALEX. A generalized linear model was used to test for differences between treated and untreated trees for all genetic parameters using the lme4 statistical package in the statistical program R version 3.3.1 [51] (R Development Core Team, 2009). Pairwise genetic distances (*Fst*) were calculated using GENALEX [48]. The Bayesian clustering analysis software STRUCTURE [52] was used to determine if there was any geographic structure of genotypes within the ANF. We used the parameters: ploidy level two, length of burnin period 100,000, and the number of Markov chain Monte Carlo reps after burn 100,000 for the admixture model. To identify the optimal value of K, 20 replicates of each value of K was used from 1 to 20, which is 3 more than a total number of populations used [53]. Structure Harvester [54] was used to choose the most likely K.

2.4.1. Assessing the Success of In Situ Treatments

To determine the degree to which currently treated ash trees represent the genetic diversity of the entire ash population, we counted alleles present in the entire genotyped sample (A_{Total}), as well as alleles present only in the insecticide-treated sample ($A_{Treated}$). $A_{Treated}/A_{Total}$ is the proportion of genetic diversity currently protected. We calculated this proportion for each of several categories of alleles: all, very common (overall frequency > 0.10), common (overall frequency > 0.05), low frequency (overall frequency < 0.10 and > 0.01), rare (overall frequency < 0.01), and "locally common" (present in only one population at frequency > 0.15 and in all other populations at frequency < 0.05). "Locally common" is included because this is the pattern reminiscent of local adaptation (high frequency in one local population only).

2.4.2. Optimizing In Situ Treatments

The following methods aim to test possible conservation strategies, in terms of number of trees and populations to treat. To test other possible treatment strategies, we used a resampling technique in which we repeatedly selected (from the genotyped dataset) at random a given number of trees from a given number of local populations for "treatment" (i.e., in situ conservation). For each sampling strategy, we calculated the proportion of genetic diversity protected in terms of all alleles (as above), and the identity of the selected populations was recorded. We tested various possible strategies, i.e., all combinations of 2 to 20 randomly chosen trees per population and 1 to 10 randomly chosen populations (thus there are 190 combinations of trees and trees per population), without replacement. As some populations are small (see Table 1), if a sampling strategy was attempted that exceeded population size, all trees available were sampled. Each of the 190 sampling strategies was performed 100,000 times (replicates), using the adegenet package [55] and a custom written set of functions in R version 3.3.1 [51] (see the R code file in Supplemental Materials). For each possible treatment strategy, we calculated the mean and standard deviation of alleles captured over all replicates.

It is possible that treating certain populations will be more effective than treating others, as some have more unique allelic diversity than others. As such, treating certain sets of populations can provide complementarity to cover all the unique alleles. Thus, an optimal set has the highest number of alleles. To identify the "best" or optimal sets of populations to treat, the 100,000 replicates (additional runs of 200,000 replicates showed identical results) were sorted by genetic diversity preserved for each of the 190 sampling combinations. The identity of the particular combinations of populations resulting in the highest number of alleles were recorded as the optimal set. For each of the 190 optimal sets, these populations were resampled an additional 100,000 times with different individuals. This resampling is designed to determine the distribution of allelic diversity from treating these populations. Lastly, the improvement due to sampling the optimal set was calculated as the difference between sampling a random set of populations and the optimal set of populations, assuming the random choice of trees within populations, for the 190 combinations, and for each allele category.

3. Results

3.1. White Ash Genetic Variability across the ANF

Across all sites, the average number of alleles per locus (Na) ranged from 1.4 to 8.2, the effective number of alleles ranged from 1.4 to 3.6, and gene diversity (He) ranged from 0.23 to 0.65 (Table 1). A total of 26 private alleles (P) were detected from 8 microsatellites across all populations. The numbers by population ranged from 0 to 7, with all being found in treated populations except for 2, which were located in one lowland population of untreated trees (PopID 200). For a few untreated sites, the death rate was so high that our sample size was reduced to only a handful of trees. As a consequence, the genetic diversity for these sites was the lowest. As genetic diversity metrics can be limited at small sample sizes, these populations were excluded for the statistical comparisons between treated and untreated plots. Once these plots were removed we found a significant difference between treated and untreated populations for number of alleles per loci ($t_{1,11}$ = 2.85, p = 0.02), with treated populations being higher. There were no significant differences in the number of effective alleles ($t_{1,11}$ = 1.82, p = 0.10) or gene diversity ($t_{1,11}$ = 0.94, p = 0.37). Inbreeding coefficient ranged from −1.00 to 0.16, although for populations which had larger samples sizes (n > 16), the range was restricted from 0.00 to 0.14, suggesting low to moderate levels of inbreeding. The result was supported by a comparison of the degree of relatedness between trees within populations, which was close to zero, suggesting most trees sampled were unrelated (Figure 2). There was no significant difference between treated and untreated trees for inbreeding ($t_{1,11}$ = 0.31, p = 0.76). Pairwise genetic distances (Fst) showed low (0.01) to moderate (0.09) genetic differentiation between populations, with no obvious differences for treated and untreated sites. This was supported by the Structure analysis which showed a gradual drop in maximum likelihood from its peak at K = 1, which suggests little to no structure. This was supported by visualization with all populations being comprised of both genetic groups for K = 2.

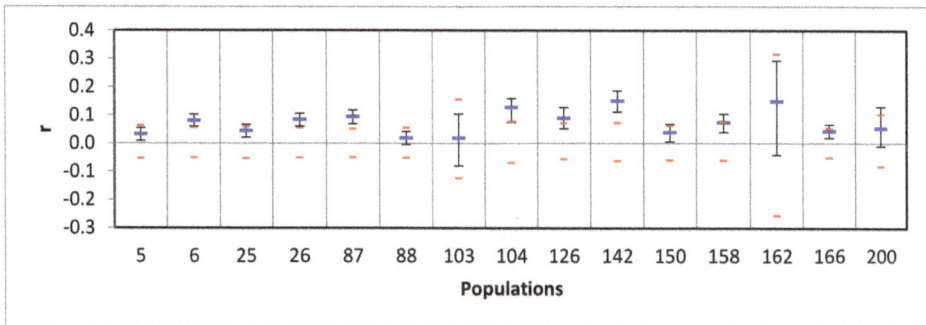

Figure 2. Mean relatedness (r) within the population, box plots (blue) represent average relatedness and standard error. Red bars represent 95% confidence intervals about zero, which is determined by sample size.

3.2. Assessing the Success of In Situ Efforts

Overall, there were 131 alleles (28 very common, 38 commons, 45 low frequency, 58 rare, 8 locally common and 68 locally rare). Assuming the genotyped trees are representative of the ash meta-population across the ANF (see Discussion), the 536 trees that were treated in the larger scale treatment regimen capture a substantial proportion of the neutral genetic variation present. Specifically, the proportion of all alleles captured at least once in the genotyped dataset is 97.7%. The proportion of all other categories is 100%, except with rare alleles (94.8%).

A forest manager is often faced with decision making about how to spend funds. We first aimed to examine the consequences of potential alternative choices. These simple choices are arbitrary of course, but illustrate major lessons about gene conservation. We modeled different treatment scenarios to determine the genetic diversity that would be preserved with less expensive alternatives. Protecting the same number of trees in fewer populations is one management option, i.e., treating 4 populations as opposed to 10 populations. Our results show that this decrease would result in a lower proportion of genetic diversity preserved for the same number of trees (Table 2). This decrease is most noticeable in the low frequency and locally common alleles (about 6% and 13% respectively). It is worth noting for this comparison that a maximum of 61 trees was used, which is on average the maximum number of trees that can be sampled from 4 randomly chosen populations in our dataset. This is illustrative of the difference that can be expected for other numbers of trees.

Table 2. Proportion of genetic diversity preserved in 61 trees, whether these trees are chosen from 4 (option 1) or 10 (option 2) populations, randomly.

Management Option	No. of Populations	All	Very Common	Common	Low Freq	Rare	Locally Common
Option 1	4	64.4%	~100%	99.4%	84.8%	31.4%	75.5%
Option 2	10	66.6%	~100%	99.9%	91.1%	31.4%	88.4%
Difference		2.2%	<0.1%	0.5%	6.3%	<0.1%	12.9%

Another option is protecting the same number of populations but fewer trees. This would also substantially reduce the genetic variation protected. Reducing the number of currently protected trees approximately by half (in our dataset, from 274 to 140), albeit with trees still spread across 10 populations, would decrease the proportion of alleles saved from 97.7% to 81.1%.

We then examined all possible combinations of the number of trees per population and number of populations. In general, our results show that it is better to protect more populations (Figure 3). The slightly higher slope and the narrower distance between lines on the left panels suggest that typically, more genetic variation is gained by adding a new population than a new individual. It can also be observed that the lines for adding more trees look to be reaching a plateau sooner than is the case when more populations are added. A specific example of the choice of more trees or more populations is that the same proportion of locally common alleles can be protected by treating 77 trees across 10 populations or 102 trees across 7 populations.

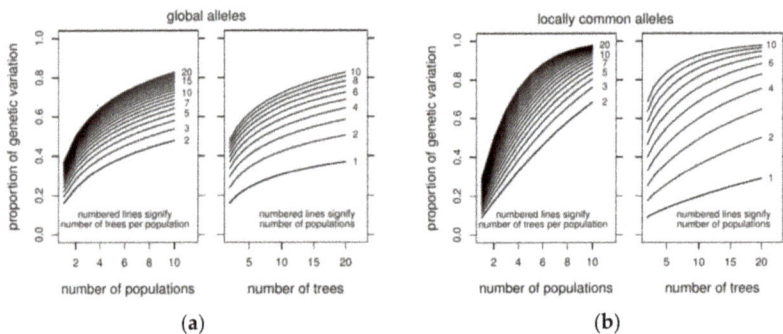

Figure 3. Accumulation of genetic variation as more populations or trees are added. The two plots in panel (**a**) represent "global alleles" (i.e., all alleles) and the two in panel (**b**) "locally common" alleles. In the graphs with number of populations on the X axis, the successive lines are additional numbers of trees per population; likewise, in the graphs with number of trees on the X axis, the lines are additional numbers of populations those trees are sampled from.

3.3. Optimizing Insecticide Treatment

An optimal set of populations to sample was recorded for each of the 190 treatment combinations. Our iterative approach consistently identified several populations across the span of minimum numbers of populations. For 10 treated populations, the optimal set is 5, 6, 25, 26, 87, 88, 142, 166, 200, and one of either 103 or 149 (Figure 4). Note that this is the ranking when choosing sets of populations to maximize preservation of all alleles. If the optimality "criteria" is to maximize a given category of the allele, the ranking of populations by their optimality does change; for example, to maximize locally common alleles the top populations are 5, 88, 149, and 166. Note that for pairs of plots that are close together, i.e., 25 and 26 (one is upland, and one is lowland), typically only one of these pairs shows up as optimal, suggesting that maintaining a minimum distance between treated populations would be prudent for maximizing diversity.

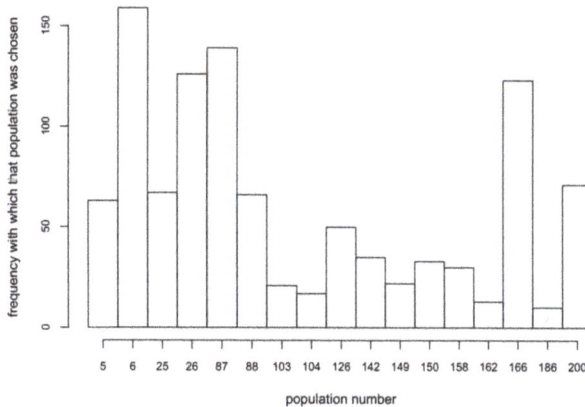

Figure 4. Frequency with which each population was observed in the 190 optimal sets of populations when optimality is defined as capturing all alleles.

The optimal populations are often, but not exclusively, those with a larger number of genotyped trees in our study. For example, populations 6, 26, 87, and 166 were frequently chosen and all have 30 trees genotyped; population 200, however, is frequently chosen and only 22 trees were sampled. In addition, we find that when treatment strategies are limited to three or fewer trees per population, nearly the same populations are chosen as optimal, suggesting they are not optimal merely because of the larger sample size. To further investigate whether the number of trees genotyped influenced the inclusion of a population as optimal, we plotted the genetic diversity captured compared to the number of trees treated (Figure 5), which shows an improvement on allele preservation using the optimal set.

On average, treating an optimal set of populations increased the proportion of all alleles preserved by 7.7% and a maximum of 16.1%, with similar gains observed for the low frequency, rare, and "locally common" alleles. Again, this result is based on using all alleles as the ranking criteria, although examining plots for other allele categories reveals similar substantial improvement. On average, the optimal populations captured more variation than randomly chosen ones, even for the same number of trees (Figure 5), suggesting some populations do indeed have an optimal genetic diversity content. Optimal populations tend to have higher allelic diversity and heterozygosity, although not consistently (populations 150 and 158). The fact that all populations are chosen at least a few times (Figure 4), suggests that there can be multiple combinations of populations that can conserve the same level of diversity. As such, there is not just one optimal set for managers to use.

These results show that if a conservation practitioner chose populations to treat at random, they would have to treat more trees in order to preserve the same amount of genetic diversity as an optimal set. For example, two strategies—treating a total of 153 from 10 *random* populations and treating a total of 121 trees from 10 *optimal* populations—both capture about 83% of all alleles. This equates to about 21% fewer trees requiring treatment, under the optimal plan, to preserve the same genetic variation. If this percentage is transferred to the cost of treating 536 trees as opposed to 675 trees (21% fewer), these savings would be 139 fewer trees and thus nearly $13,900 USD per year that the treatment is applied (assuming $100 USD/tree treatment costs). Costs of genotyping may be between $5 USD and $10 USD per tree, which would total $6750 ($10/tree × 675 trees). Thus, in year one, the savings of treating an optimal set more than offsets the genotyping costs. From then on, the savings will only accumulate with every treatment cycle. The exact cost savings will depend on the number and distribution of trees of course, and the criteria for optimality, but it is likely that the cost of sampling and genetic analysis of trees before embarking on an insecticide treatment plan could lead to greater savings in treatment costs by allowing managers to optimize treatment strategies.

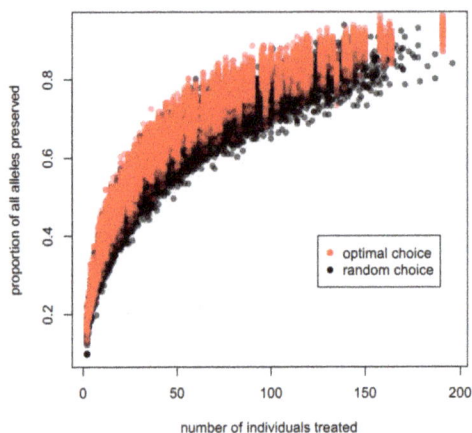

Figure 5. Proportion of genetic variation preserved for a given number of individuals treated depending on whether optimal populations or random populations are selected.

4. Discussion

In this study, we have found that the optimal tree sampling strategy identified here, i.e., treating more populations, rather than more trees per population, maximizes conservation of genetic diversity of ash while reducing management costs by >20%. The *F. americana* populations of the Allegheny National Forest showed minimal genetic structure (in terms of low Fst), suggesting that genetic diversity within white ash is evenly distributed across the region. Efficiencies may be even higher in forests with more structured populations. The cost savings associated with reduced treatment expenses can easily offset the costs of genetic analyses. The low genetic structure and high genetic diversity observed herein compliments findings in other ash species (i.e., *F. excelsior* [56,57]), and is consistent with other common, wind-pollinated tree species (oaks [58,59] such as pine [60,61] and beech [62]). This likely reflects adequate gene flow between populations, common in species with wind-dispersed pollen and seeds. Hence, we suggest that our findings on landscape population patterns and management may be transferable to conservation efforts of other wind dispersed tree species. However, in complex landscapes with high levels of environmental variability, a high degree of adaptation to local environments may occur.

A number of studies have examined how many trees and populations to sample for ex situ conservation, both empirically [34,36–38], and theoretically [33,63–65]. These studies have shown that appropriate sampling strategies depend on the species and the situation. They have also provided several consistent messages: (1) a relatively small number of well-chosen samples can preserve most of the genetic diversity; (2) many samples are needed to preserve all of the genetic diversity; and (3) diminishing returns are observed as more samples are added. More importantly, protecting multiple populations is almost always necessary to capture most of the genetic diversity. These results are consistent with these commonalities from previous studies. We found that even for a species with moderately low genetic differentiation across populations, protecting more populations is more valuable for maintaining a maximum level of genetic diversity. Our study helps address the paucity of specific information aimed at designing an optimal in situ or ex situ strategy for conserving the genetic variation in plants (though see [66,67]), and provides the first gene conservation strategy for designing in situ pest management approaches.

We show that for *F. americana* in the Allegheny region, the current strategy of treating 536 trees from 27 plots (of which 10 plots were in our analyzed data) will likely protect the majority of the known genetic variation, based on our samples. The alleles that are not protected are typically rare (frequency less than 0.01), or are restricted to, but common in, one population (locally common). We demonstrate that protection of a larger number of populations with insecticide treatment is more useful and cost effective than a treatment applied to more trees per population. Protecting numerous populations has the additional benefit of ensuring against disturbances caused by fire, windstorms, or other local environmental change. However, the number of trees protected per population should still be substantial (20 or more if possible) so that future seed production will remain highly heterozygous. To test whether it is possible to reduce costs without sacrificing genetic diversity, we examined alternative strategies for treating fewer trees (thus reducing the amount of insecticide needed), or fewer populations (reducing travel time and transportation costs for treatment). Our results show that protecting the same number of trees but in fewer populations would result in a small loss (2%) of diversity, but a larger loss (13%) in locally common diversity. An attempt to reduce cost expenditures further might consider reducing the number of individuals treated by half; this would result in a larger loss (16%) of diversity (primarily rare alleles). Lastly, we demonstrate that an optimal set of populations can be identified that could result in greater genetic preservation and cost reductions. This is likely due to the principle of complementarity; an optimal set of populations has minimal overlap in alleles, and thus each population complements the others [34]. Considering that it was possible to identify optimal strategies for *F. americana*, a common and wind-pollinated species with low genetic structure, this method should provide even greater benefit for species with higher genetic structure.

While we consider our results to be robust in these main findings, we need to be cautious as we used a small number of neutral microsatellite markers and a small sample size of untreated trees. While neutral markers are good at detecting effective population sizes and migration rates, their connection to fitness and "adaptive potential" is more tenuous. A number of studies have demonstrated that with low differentiation there can still be extensive between-population differences in adaptive traits [60,61,68]. Thus, there may be adaptive genetic differences that were not captured in our study, nor might they be captured in a treatment program guided by neutral markers; indeed, capturing them is a difficult task, regardless of methods used [69]. Despite this limitation, neutral markers are the most cost-effective tools available, and alleles are the only "currency" we have for measuring the gain in genetic diversity from a conservation action. In addition, sparse densities of ash as well as EAB induced mortality in the region limited the number of untreated trees available for the study. The exact estimates of the 'proportion of genetic variation' preserved for a given number of trees would decrease as more of the untreated trees were genotyped, and future studies may benefit from collecting additional samples before EAB expands to those areas (Figure 6). An accumulation curve of resampling our datasets, adding individuals one by one, supports this supposition; there are likely many more rare alleles in existence, but the other categories have likely reached an asymptote (Figure 6). It is important to note that the broad conclusions about sampling more trees or more populations derived from our approach is unlikely to be changed by using other marker approaches or increasing sampling size.

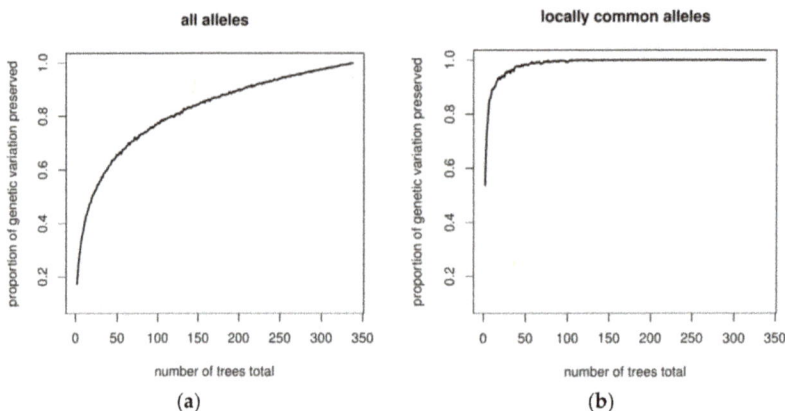

Figure 6. Accumulation analysis which shows an asymptote of low frequency (**a**) and rare (**b**) alleles as more individuals trees are selected across all samples, demonstrating an asymptote is likely reached for locally common but not for all alleles (due to rare alleles).

Natural demographic loss is an important consideration when determining an optimal number of populations and trees for treatment. With ex situ seed collections, the minimum number of seeds needed to preserve genetic variation is increased to account for germination and recruitment failures [70]. For our in situ treatment, the main dangers include the loss of populations due to abiotic disturbances (e.g., fire or windstorms), or the lack of treatment efficacy [30]. A recent estimate of disturbance levels gleaned from Forest Inventory and Analysis Data suggests that in the eastern US, approximately 3% of all trees are lost across a 5-year period [71]. Although these disturbance levels are relatively low, we suggest factoring disturbance into treatment plans, in order to help ensure that the 27 treated populations persist long term (i.e., 100 years). This may be achieved through a moderate increase in the number of populations treated (5 to 10 additional populations), and an increase in individuals per population (to perhaps 25), as well as the wide spatial distribution of populations. A more detailed regional analysis of mortality frequency from past catastrophic events may be useful to determine the likelihood that sufficient treated populations and trees will survive over the long term. Insecticide treatment efficacy should also be considered when deciding how many individuals to treat within a population. Because treatment efficacy decreases in less healthy trees [30], more trees must be treated in strains that are already in decline, to ensure the survival of the target number of trees for genetic conservation. In sum, our analysis represents a minimum number of white ash trees and populations to treat based on standing genetic diversity; as some trees will be lost to disturbance or insecticide failure, this minimum number could be increased to ensure that a minimum persists into the future.

A more complete genotyping of the populations containing fewer samples would help fully verify these results. Sequencing a large number of genes with a known function would provide an important complement to this work, by providing information on functional genetic variation [72]. Additionally, layering in a biological control strategy may help mitigate EAB pressures while contributing to the maintenance of ecosystem stability [73]. As such, efforts should be made to incorporate insecticide treatments into broader in situ or combined in situ and ex situ conservation program(s) [74]. Lastly, including information on the soil, vegetation, slope and other environmental variables may be useful to ensure the preservation of ecologically relevant adaptations, by demonstrating that treated trees are found through the full environmental range of the species.

5. Conclusions

In this study, we used genetic data from microsatellite markers to determine the effectiveness of current insecticide treatments for protecting the neutral genetic variation of white ash in situ in the Allegheny National Forest. We determined that treating a smaller number of trees across more populations is more effective than more trees in fewer populations, especially for locally common alleles. Furthermore, our iterative approach for identifying optimal sets of populations can be used to increase the efficiency of treatment, resulting in cost savings that would essentially offset the cost of the genetic study within one to two years.

Supplementary Materials: The source code for R scripts is available online at http://www.mdpi.com/1999-4907/9/4/202/s1.

Acknowledgments: This research was supported by the U.S.D.A. Forest Service Northern Research Station. We appreciate field assistance from M. Pickar and lab assistance from C. Courtney. Additionally, we thank the three reviewers for their thoughtful comments/suggestions.

Author Contributions: C.E.F., M.A.G.-M. and J.B.F. conceived and designed the experiment; K.S.K., C.E.F. and A.A.R. conceived and deployed the larger experiment; C.E.F. managed field work; E.A., S.F., and A.H. collected foliar samples; J.B.F., L.S. and S.H. analyzed the data; C.E.F., S.H. and J.B.F. wrote the paper.

Conflicts of Interest: The authors declare no conflict of interest.

References

1. McKinney, M.L. Measuring floristic homogenization by non-native plants in North America. *Glob. Ecol. Biogeogr.* **2004**, *13*, 47–53. [CrossRef]
2. Corvalan, C.; Hales, S.; McMichael, A. Ecosystems and human well-being: Biodiversity synthesis. In *Millenium Ecosystem Assessment*; World Resources Institute: Washington, DC, USA, 2005.
3. Lovett, G.M.; Canham, C.D.; Arthur, M.A.; Weathers, K.C.; Fitzhugh, R.D. Forest ecosystem responses to exotic pests and pathogens in eastern North America. *Bioscience* **2006**, *56*, 395–405. [CrossRef]
4. Flower, C.E.; Gonzalez-Meler, M.A. Responses of temperate forest productivity to insect and pathogen disturbances. *Annu. Rev. Plant Biol.* **2015**, *66*, 547–569. [CrossRef] [PubMed]
5. Liebhold, A.M.; Brockerhoff, E.G.; Kalisz, S.; Nuñez, M.A.; Wardle, D.A.; Wingfield, M.J. Biological invasions in forest ecosystems. *Biol. Invasions* **2017**, *19*, 3437–3458. [CrossRef]
6. Schaberg, P.G.; DeHayes, D.H.; Hawley, G.J.; Nijensohn, S.E. Anthropogenic alterations of genetic diversity within tree populations: Implications for forest ecosystem resilience. *For. Ecol. Manag.* **2008**, *256*, 855–862. [CrossRef]
7. Hughes, A.R.; Inouye, B.D.; Johnson, M.T.J.; Underwood, N.; Vellend, M. Ecological consequences of genetic diversity. *Ecol. Lett.* **2008**, *11*, 609–623. [CrossRef] [PubMed]
8. Hughes, A.R.; Stachowicz, J.J. Genetic diversity enhances the resistance of a seagrass ecosystem to disturbance. *Proc. Natl. Acad. Sci. USA* **2004**, *101*, 8998–9002. [CrossRef] [PubMed]
9. Clark, J.S. Individuals and the variation needed for high species diversity in forest trees. *Science* **2010**, *327*, 1129–1132. [CrossRef] [PubMed]
10. Laikre, L.; Allendorf, F.W.; Aroner, L.C.; Baker, C.S.; Gregovich, D.P.; Hansen, M.M.; Jackson, J.A.; Kendall, K.C.; Mckelvey, K.; Neel, M.C.; et al. Neglect of genetic diversity in implementation of the convention on biological diversity. *Conserv. Biol.* **2009**, *24*, 86–88. [CrossRef] [PubMed]
11. Schoen, D.J.; Brown, A.H. The Conservation of Wild Plant Species in Seed Banks: Attention to both taxonomic coverage and population biology will improve the role of seed banks as conservation tools. *Bioscience* **2001**, *51*, 960–966. [CrossRef]
12. Fant, J.; Havens, K.; Kramer, A.; Walsh, S.; Callicrate, T.; Lacy, R.; Maunder, M.; Meyer, A.; Smith, P. What to do when we can't bank on seeds: What botanic gardens can learn from the zoo community about conserving plants in living collections. *Am. J. Bot.* **2016**, *103*, 1541–1543. [CrossRef] [PubMed]
13. Basey, A.C.; Fant, J.B.; Kramer, A.T. Producing native plant materials for restoration: 10 rules to collect and maintain genetic diversity. *Nativ. Plants J.* **2015**, *16*, 37–53. [CrossRef]
14. Ricketts, M.P.; Flower, C.E.; Knight, K.S.; Gonzalez-Meler, M.A. Evidence of ash tree (*Fraxinus* spp.) association with soil bacterial community structure and function. *Forests* **2018**, *9*, 187. [CrossRef]

15. Slesak, R.A.; Lenhart, C.F.; Brooks, K.N.; D'Amato, A.W.; Palik, B.J. Water table response to harvesting and simulated emerald ash borer mortality in black ash wetlands in Minnesota, USA. *Can. J. For. Res.* **2014**, *44*, 961–968. [CrossRef]

16. Fady, B.; Aravanopoulos, F.A.; Alizoti, P.; Mátyás, C.; Von Wühlisch, G.; Westergren, M.; Belletti, P.; Cvjetkovic, B.; Ducci, F.; Huber, G.; et al. Evolution-based approach needed for the conservation and silviculture of peripheral forest tree populations. *For. Ecol. Manag.* **2016**, *375*, 66–75. [CrossRef]

17. Herms, D.A.; McCullough, D.G. Emerald ash borer invasion of North America: History, biology, ecology, impacts, and management. *Annu. Rev. Entomol.* **2014**, *59*, 13–30. [CrossRef] [PubMed]

18. MacFarlane, D.W.; Meyer, S.P. Characteristics and distribution of potential ash tree hosts for emerald ash borer. *For. Ecol. Manag.* **2005**, *213*, 15–24. [CrossRef]

19. Flower, C.E.; Knight, K.S.; Gonzalez-Meler, M.A. Impacts of the emerald ash borer (*Agrilus planipennis* Fairmaire) induced ash (*Fraxinus* spp.) mortality on forest carbon cycling and successional dynamics in the eastern United States. *Biol. Invasions* **2013**, *15*, 931–944. [CrossRef]

20. Flower, C.E.; Knight, K.S.; Rebbeck, J.; Gonzalez-Meler, M.A. The relationship between the emerald ash borer (*Agrilus planipennis*) and ash (*Fraxinus* spp.) tree decline: Using visual canopy condition assessments and leaf isotope measurements to assess pest damage. *For. Ecol. Manag.* **2013**, *303*, 143–147. [CrossRef]

21. Flower, C.E.; Lynch, D.J.; Knight, K.S.; Gonzalez-Meler, M.A. Biotic and abiotic drivers of sap flux in mature green ash trees (*Fraxinus pennsylvanica*) experiencing varying levels of emerald ash borer (*Agrilus planipennis*) infestation. *Forests* **2018**, in press.

22. Knight, K.S.; Brown, J.P.; Long, R.P. Factors affecting the survival of ash trees (*Fraxinus* spp.) infested by emerald ash borer (*Agrilus planipennis*). *Biol. Invasions* **2013**, *15*, 371–383. [CrossRef]

23. Klooster, W.S.; Herms, D.A.; Knight, K.S.; Herms, C.P.; McCullough, D.G.; Smith, A.; Gandhi, K.J.K.; Cardina, J. Ash (*Fraxinus* spp.) mortality, regeneration, and seed bank dynamics in mixed hardwood forests following invasion by emerald ash borer (*Agrilus planipennis*). *Biol. Invasions* **2013**, *16*, 859–873. [CrossRef]

24. McCullough, D.G.; Tanis, S.R.; Robinett, M.; Limback, C.; Poland, T.M. White ash—Is EAB always a death sentence? In Proceedings of the 2014 Emerald Ash Borer National Research and Technology Development Meeting, Wooster, OH, USA, 15–16 October 2014; Buck, J., Parra, G., Lance, D., Reardon, R., Binion, D., Eds.; U.S. Department of Agriculture, Forest Service: Washington, DC, USA, 2015; pp. 10–11.

25. The IUCN Red List of Threatened Species. Available online: http://www.iucnredlist.org/ (accessed on 11 April 2018).

26. Cappaert, D.; Mccullough, D.G.; Poland, T.M.; Siegert, N.W. Emerald ash borer in North America: A research and regulatory challenge. *Am. Entomol.* **2005**, *51*, 152–165. [CrossRef]

27. O'Brien, E.M. *Conserving Ash (Fraxinus) Populations and Genetic Variation in Forests Invaded by Emerald Ash Borer Using Large-Scale Insecticide Applications*; Ohio State University: Columbus, OH, USA, 2017.

28. Flower, C.E.; Aubihl, E.; Fant, J.; Forry, S.; Hille, A.; Knight, K.S.; Oldland, W.K.; Royo, A.A.; Richard, M. In-situ Genetic Conservation of White Ash (*Fraxinus americana*) at the Allegheny National Forest. In Proceedings of the Workshop on Gene Conservation of Tree Species-Banking on the Future, Chicago, IL, USA, 16–19 May 2016; Sniezko, R.A., Man, G., Hipkins, V., Woeste, K., Gwaze, D., Kliejunas, J.T., McTeague, B.A., Eds.; U.S. Department of Agriculture, Forest Service: Washington, DC, USA; Pacific Northwest Research Station: Portland, OR, USA, 2016; pp. 165–169.

29. Herms, D.A.; Mccullough, D.G.; Smitley, D.R.; Sadof, C.S.; Cranshaw, W. *Insecticide Options for Protecting Ash Trees from Emerald Ash Borer*, 2nd ed.; North Central IPM Center Bulletin: Champaign, IL, USA, 2014; 16p.

30. Flower, C.E.; Dalton, J.E.; Knight, K.S.; Brikha, M.; Gonzalez-Meler, M.A. To treat or not to treat: Diminishing effectiveness of emamectin benzoate tree injections in ash trees heavily infested by emerald ash borer. *Urban For. Urban Green.* **2015**, *14*. [CrossRef]

31. Gregorious, H.-R. The probability of losing an allele when diploid genotypes are sampled. *Bioinformatics* **1980**, *36*, 643–652. [CrossRef]

32. Bashalkhanov, S.; Pandley, M.; Rahjora, O.P. A simple method for estimating genetic diversity in large populations from finite sample sizes. *BMC Genet.* **2009**, *10*, 84. [CrossRef] [PubMed]

33. Brown, A.; Marshall, D. A basic sampling strategy: Theory and practice. In *Collecting Plant Genetic Diversity: Technical Guidelines*; Guarino, L., Rao, V.R., Reid, R., Eds.; CAB International: Wallingford, UK, 1995; pp. 75–91.

34. Richards, C.M.; Antolin, M.F.; Reilley, A.; Poole, J.; Walters, C. Capturing genetic diversity of wild populations for ex situ conservation: Texas wild rice (*Zizania texana*) as a model. *Genet. Resour. Crop Evol.* **2007**, *54*, 837–848. [CrossRef]

35. Hoban, S.; Strand, A. Ex situ seed collections will benefit from considering spatial sampling design and species' reproductive biology. *Biol. Conserv.* **2015**, *187*, 182–191. [CrossRef]

36. Griffith, M.P.; Calonje, M.; Meerow, A.W.; Francisco-Ortega, J.; Knowles, L.; Aguilar, R.; Tut, F.; Sánchez, V.; Meyer, A.; Noblick, L.R.; et al. Will the same ex situ protocols give similar results for closely related species? *Biodivers. Conserv.* **2017**, *26*, 2951–2966. [CrossRef]

37. McGlaughlin, M.E.; Riley, L.; Brandsrud, M.; Arcibal, E.; Helenurm, M.K.; Helenurm, K. How much is enough? Minimum sampling intensity required to capture extant genetic diversity in ex situ seed collections: Examples from the endangered plant *Sibara filifolia* (Brassicaceae). *Conserv. Genet.* **2015**, *16*, 253–266. [CrossRef]

38. Kashimshetty, Y.; Pelikan, S.; Rogstad, S.H. Effective seed harvesting strategies for the ex situ genetic diversity conservation of rare tropical tree populations. *Biodivers. Conserv.* **2017**, *26*, 1311–1331. [CrossRef]

39. Schoen, D.J.; Brown, A.H.D. Intraspecific variation in population gene diversity and effective population size correlates with the mating system in plants. *Proc. Natl. Acad. Sci. USA* **1991**, *88*, 4494–4497. [CrossRef] [PubMed]

40. Doyle, J.J.; Doyle, J.L. A rapid DNA isolation procedure for small quantities of fresh leaf tissue. *Phytochem. Bull.* **1987**, *19*, 11–15.

41. Lefort, F.; Brachet, S.; Frascaria-Lacoste, N.; Edwards, K.J.; Douglas, G. Identification and characterization of microsatellite loci in ash (*Fraxinus excelsior* L.) and their conservation in the olive family (Oleaceae). *Mol. Ecol.* **1999**, *8*, 1088–1089. [CrossRef]

42. Harbourne, M.E.; Douglas, C.G.; Waldren, S.; Hodkinson, T.R. Characterization and primer development for amplification of chloroplast microsatellite regions of *Fraxinus excelsior*. *J. Plant Res.* **2005**, *118*, 339–341. [CrossRef] [PubMed]

43. Sutherland, B.G.; Belaj, A.; Nier, S.; Cottrell, J.E.; P Vaughan, S.; Hubert, J.; Russell, K. Molecular biodiversity and population structure in common ash (*Fraxinus excelsior* L.) in Britain: Implications for conservation. *Mol. Ecol.* **2010**, *19*, 2196–2211. [CrossRef] [PubMed]

44. Beatty, G.E.; Brown, J.A.; Cassidy, E.M.; Finlay, C.M.V.; McKendrick, L.; Montgomery, W.I.; Reid, N.; Tosh, D.G.; Provan, J. Lack of genetic structure and evidence for long-distance dispersal in ash (*Fraxinus excelsior*) populations under threat from an emergent fungal pathogen: Implications for restorative planting. *Tree Genet. Genomes* **2015**, *11*, 1–13. [CrossRef]

45. Brachet, S.; Jubier, M.; Richard, M.; Jung-Muller, B.; Frascaria-Lacoste, N. Rapid identification of microsatellite loci using 5′ anchored PCR in the common ash *Fraxinus excelsior*. *Microb. Ecol.* **1999**, *8*, 160–163.

46. Noakes, A.G.; Best, T.; Staton, M.E.; Koch, J.; Romero-Severson, J. Cross amplification of 15 EST-SSR markers in the genus *Fraxinus*. *Conserv. Genet. Resour.* **2014**, *6*, 969–970. [CrossRef]

47. Schuelke, M. An economic method for the fluorescent labeling of PCR fragments. *Nat. Biotechnol.* **2000**, *18*, 233–234. [CrossRef] [PubMed]

48. Peakall, R.; Smouse, P. GenAlEx 6: Genetic analysis in Excel. Population genetic software for teaching and research. *Mol. Ecol. Notes* **2006**, *6*, 288–295. [CrossRef]

49. Weir, B.; Cockerham, C. Estimating F-statistics for the analysis of population structure. *Evolution* **1984**, *38*, 1358–1370. [PubMed]

50. Queller, D.C.; Goodnight, K.F. Estimating relatedness using genetic markers. *Evolution* **1989**, *43*, 258. [CrossRef] [PubMed]

51. Bates, D.; Maechler, M.; Bolker, B.; Walker, S. Fitting linear mixed-effects models using lme4. *J. Stat. Softw.* **2015**, *67*, 1–48. [CrossRef]

52. Pritchard, J.; Stephens, M.; Donnelly, P. Inference of population structure using multilocus genotype data. *Genetics* **2000**, *155*, 945–959. [PubMed]

53. Evanno, G.; Regnaut, S.; Goudet, J. Detecting the number of clusters of individuals using the software STRUCTURE: A simulation study. *Mol. Ecol.* **2005**, *14*, 2611–2620. [CrossRef] [PubMed]

54. Earl, D.A.; von Holdt, B.M. STRUCTURE HARVESTER: A website and program for visualizing STRUCTURE output and implementing the Evanno method. *Conserv. Genet. Resour.* **2012**, *4*, 359–361. [CrossRef]

55. Jombart, T. Adegenet: A R package for the multivariate analysis of genetic markers. *Bioinformatics* **2008**, *24*, 1403–1405. [CrossRef] [PubMed]

56. Heuertz, M.; Hausman, J.F.; Tsvetkov, I.; Frascaria-Lacoste, N.; Vekemans, X. Assessment of genetic structure within and among Bulgarian populations of the common ash (*Fraxinus excelsior* L.). *Mol. Ecol.* **2001**, *10*, 1615–1623. [CrossRef] [PubMed]

57. Heuertz, M.; Hausman, J.F.; Hardy, O.; Vendramin, G.G.; Frascaria-Lacoste, N.; Vekemans, X. Nuclear microsatellites reveal contrasting patterns of genetic structure between western and southeastern European populations of the common ash (*Fraxinus excelsior* L.). *Evolution* **2004**, *58*, 976–988. [PubMed]

58. Craft, K.J.; Ashley, M.V. Landscape genetic structure of bur oak (*Quercus macrocarpa*) savannas in Illinois. *For. Ecol. Manag.* **2007**, *239*, 13–20. [CrossRef]

59. Streiff, R.; Ducousso, A.; Lexer, C.; Steinkellner, H.; Gloessl, J.; Kremer, A. Pollen dispersal inferred from paternity analysis in a mixed oak stand of *Quercus robur* L. and *Q. petraea* (Matt.) Liebl. *Mol. Ecol.* **1999**, *8*, 831–841. [CrossRef]

60. Nadeau, S.; Meirmans, P.G.; Aitken, S.N.; Ritland, K.; Isabel, N. The challenge of separating signatures of local adaptation from those of isolation by distance and colonization history: The case of two white pines. *Ecol. Evol.* **2016**, *6*, 8649–8664. [CrossRef] [PubMed]

61. Bower, A.D.; Aitken, S.N. Ecological genetics and seed transfer guidelines for *Pinus albicaulis* (Pinaceae). *Am. J. Bot.* **2008**, *95*, 66–76. [CrossRef] [PubMed]

62. Jump, A.S.; Penuelas, J. Genetic effects of chronic habitat fragmentation in a wind-pollinated tree. *Proc. Natl. Acad. Sci. USA* **2006**, *103*, 8096–8100. [CrossRef] [PubMed]

63. Brown, A.H.D. Isozymes, plant population genetic structure and genetic conservation. *Theor. Appl. Genet.* **1978**, *52*, 145–157. [CrossRef] [PubMed]

64. Hoban, S.; Schlarbaum, S. Optimal sampling of seeds from plant populations for ex-situ conservation of genetic biodiversity, considering realistic population structure. *Biol. Conserv.* **2014**, *177*, 90–99. [CrossRef]

65. Guerrant, E.O.; Fiedler, P.L.; Havens, K.; Maunder, M. Revised genetic sampling guidelines for conservation collections of rare and endangered plants. In *Ex Situ Plant Conservation: Supporting Species Survival in the Wild*; Island Press: Washington, DC, USA, 2004; pp. 419–448.

66. Vinceti, B.; Loo, J.; Gaisberger, H.; van Zonneveld, M.J.; Schueler, S.; Konrad, H.; Kadu, C.A.C.; Geburek, T. Conservation priorities for prunus Africana defined with the aid of spatial analysis of genetic data and climatic variables. *PLoS ONE* **2013**, *8*, e59987. [CrossRef] [PubMed]

67. Schlottfeldt, S.; Walter, M.E.M.T.; de Carvalho, A.C.P.L.F.; Soares, T.N.; Telles, M.P.C.; Loyola, R.D.; Diniz-Filho, J.A.F. Multi-objective optimization for plant germplasm collection conservation of genetic resources based on molecular variability. *Tree Genet. Genomes* **2015**, *11*, 16. [CrossRef]

68. Jump, A.S.; Hunt, J.M.; Martínez-Izquierdo, J.; Peñuelas, J. Natural selection and climate change: Temperature-Linked spatial and temporal trends in gene frequency in *Fagus sylvatica*. *Mol. Ecol.* **2006**, *15*, 3469–3480. [CrossRef] [PubMed]

69. Hoban, S.; Kelley, J.L.; Lotterhos, K.E.; Antolin, M.F.; Bradburd, G.; Lowry, D.B.; Poss, M.L.; Reed, L.K.; Storfer, A.; Whitlock, M.C. Finding the genomic basis of local adaptation: pitfalls, practical solutions, and future directions. *Am. Nat.* **2016**, *188*, 379–397. [CrossRef] [PubMed]

70. Way, M.J. Collecting seed from non-domesticated plants for long-term conservation. In *Seed Conservation: Turning Science into Practice*; Smith, R., Dickie, J., Linington, S., Pritchard, H., Probert, R., Eds.; Royal Botanic Gardens Kew: London, UK, 2003; pp. 165–201.

71. Vanderwel, M.C.; Coomes, D.A.; Purves, D.W. Quantifying variation in forest disturbance, and its effects on aboveground biomass dynamics, across the eastern United States. *Glob. Chang. Biol.* **2013**, *19*, 1504–1517. [CrossRef] [PubMed]

72. Flanagan, S.P.; Forester, B.R.; Latch, E.K.; Aitken, S.N.; Hoban, S. Guidelines for planning genomic assessment and monitoring of locally adaptive variation to inform species conservation. *Evol. Appl.* **2017**, 1–18. [CrossRef]

73. Margulies, E.; Bauer, L.; Ibáñez, I. Buying time: Preliminary assessment of biocontrol in the recovery of native forest vegetation in the aftermath of the invasive emerald ash borer. *Forests* **2017**, *8*, 369. [CrossRef]

74. Schoettle, A.W.; Klutsch, J.G.; Sniezko, R.A. Integrating regeneration, genetic resistance, and timing of intervention for the long-term sustainability of ecosystems challenged by non-native Pests—A novel proactive approach. In Proceedings of the 4th International Workshop on the Genetics of Host-Parasite Interactions in Forestry: Disease and Insect Resistance in Forest Trees, Eugene, OR, USA, 31 July–5 August 2011; Sniezko, R.A., Yanchuk, A.D., Kliejunas, J.T., Palmieri, K.M., Alexander, J.M., Frankel, S.J., Eds.; Pacific Southwest Research Station: Redding, CA, USA; Forest Service, U.S. Department of Agriculture: Washington, DC, USA, 2012; pp. 112–123.

forests

MDPI

Review

Ecological Impacts of Emerald Ash Borer in Forests at the Epicenter of the Invasion in North America

Wendy S. Klooster [1,*], Kamal J. K. Gandhi [2], Lawrence C. Long [3], Kayla I. Perry [4], Kevin B. Rice [5] and Daniel A. Herms [6]

1   Department of Entomology, The Ohio State University, Columbus, OH 43210, USA
2   Daniel B. Warnell School of Forestry and Natural Resources, University of Georgia, Athens, GA 30602, USA; kgandhi@warnell.uga.edu
3   Department of Entomology and Plant Pathology, North Carolina State University, Raleigh, NC 27695, USA; lclong2@ncsu.edu
4   Department of Entomology, The Ohio State University, Ohio Agricultural Research and Development Center, Wooster, OH 44691, USA; perry.1864@osu.edu
5   Division of Plant Sciences, University of Missouri, Columbia, MO 65211, USA; ricekev@missouri.edu
6   The Davey Tree Expert Company, Kent, OH 44240, USA; dan.herms@davey.com
*   Correspondence: klooster.2@osu.edu; Tel.: +1-614-292-2764

Received: 31 March 2018; Accepted: 3 May 2018; Published: 5 May 2018

Abstract: We review research on ecological impacts of emerald ash borer (EAB)-induced ash mortality in the Upper Huron River watershed in southeast Michigan near the epicenter of the invasion of North America, where forests have been impacted longer than any others in North America. By 2009, mortality of green, white, and black ash exceeded 99%, and ash seed production and regeneration had ceased. This left an orphaned cohort of saplings too small to be infested, the fate of which may depend on the ability of natural enemies to regulate EAB populations at low densities. There was no relationship between patterns of ash mortality and ash density, ash importance, or community composition. Most trees died over a five-year period, resulting in relatively simultaneous, widespread gap formation. Disturbance resulting from gap formation and accumulation of coarse woody debris caused by ash mortality had cascading impacts on forest communities, including successional trajectories, growth of non-native invasive plants, soil dwelling and herbivorous arthropod communities, and bird foraging behavior, abundance, and community composition. These and other impacts on forest ecosystems are likely to be experienced elsewhere as EAB continues to spread.

Keywords: Invasive species; *Fraxinus* spp.; *Agrilus planipennis* Fairmaire; disturbance; gap ecology; coarse woody debris; non-target impacts; forest succession; soil arthropods; tri-trophic interactions

1. Introduction

Alien phytophagous insects, including emerald ash borer (EAB, *Agrilus planipennis* Fairmaire (Coleoptera: Buprestidae)), have altered forest composition, structure, and function throughout much of North America [1–3]. EAB was first detected in North America in 2002 in southeast Michigan and neighboring Ontario [4–6]. Subsequent analyses of dendrochronological data indicated that the beetle was established and killing trees by the 1990s [7]. Since its introduction to North America, EAB has caused extensive mortality of ash (*Fraxinus* spp.) [8–15], and to a lesser degree white fringetree (*Chionanthus virginicus* L.) [16,17]. Since its initial detection, numerous studies have examined the biology, ecology, and management of EAB [5,18–20].

The objective of this paper is to review research on the direct and indirect ecological impacts of the EAB invasion on the flora and fauna of forests in the Upper Huron River watershed, which extends across western Oakland County, southeastern Livingston County, and north central Washtenaw County in southeast Michigan. These forests are near the presumed epicenter of the EAB invasion in Canton Township, Michigan [7], and thus have been impacted by EAB longer than others in North America. Prior to the EAB invasion, black (*F. nigra* Marshall), green (*F. pennsylvanica* Marshall), and white (*F. americana* L.) ash were the most common ash species on hydric swamps, mesic lowlands and flood plains, and xeric upland sites, respectively [14,21]. As EAB continues to expand its distribution in North America, the results of these studies provide insights into the ways EAB may impact other ecosystems, which are predicted to be substantial at multiple scales [22,23]. Furthermore, EAB is also causing extensive mortality of European ash (*F. excelsior* L.) in eastern Europe [24,25] where it may have ecological impacts comparable to those in North America.

2. Timing and Patterns of Ash Mortality

EAB has caused extensive ash mortality in the Upper Huron River watershed [8,11–14]. Dendrochronological analyses revealed that EAB-induced ash mortality occurred in this watershed as early as 1994 (L. Becker, D.A. Herms, and G.C. Wiles, unpublished data), and overall mortality of ash with stem diameters >2.5 cm had reached 40% by 2005 [14,21]. Initially, decline of black ash slightly exceeded that of green and white ash [14]. By 2008, however, mortality of all three species was greater than 95%, and peaked at 99.7% in 2009 [13]. Hence, following a long lag period since the onset of mortality, nearly 60% of trees died over a five-year period from 2005–2009, resulting in nearly simultaneous, widespread gap formation (Figure 1). The extremely high mortality of these North American ash species has been attributed to their low resistance to EAB relative to coevolved Asian ash hosts [26]. As EAB continues to spread in "defense free space" [3], white, green, and black ash may experience functional extirpation (sensu [27]) in which their populations decline to the point that they no longer provide significant ecosystem function and services [22].

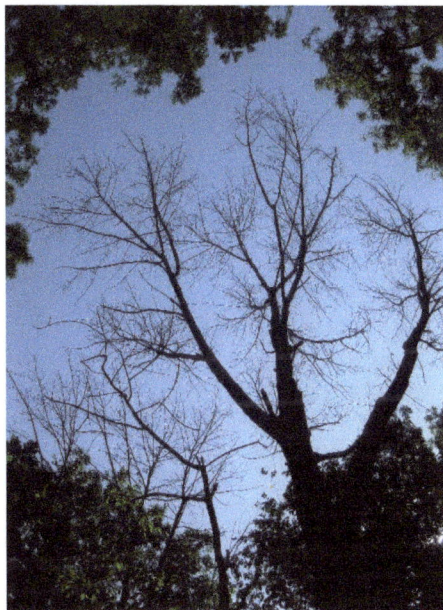

Figure 1. Widespread formation of canopy gaps occurred throughout forests of the Upper Huron River watershed in southeast Michigan as mortality of ash increased from 40% to >99% between 2005 and 2009.

The relationships between host density or tree species diversity and population and impact of alien phytophagous insects have been documented [28–30]. However, Smith et al. [14] found no relationship between EAB-induced ash mortality and ash density, nor any other measure of community composition including ash basal area, ash importance, total stand density, total stand basal area, or any indices of tree diversity. Similarly, Knight et al. [10] observed no relationship between ash density and percentage ash mortality in Ohio, although ash mortality proceeded faster in stands with lower density of ash. These studies, conducted across an ash density gradient from low to very high and across a broad spatial area, suggest the potential to prevent ash mortality via silvicultural management is extremely limited [10,14].

From 2004–2006, there was a negative relationship between percentage ash mortality in the Upper Huron River watershed and distance from the presumed epicenter of the invasion in Canton Township, Michigan [7], with mortality decreasing 2% per km from the epicenter [14]. By 2007, however, this relationship plateaued as ash mortality exceeded 90% across the entire watershed [14]. Decreasing ash decline and mortality with increasing distance from the invasion epicenter was also documented by other studies conducted at various spatial scales [8,31].

3. Ash Recruitment and Regeneration

3.1. Ash Seed Bank, Seedling Regeneration and Basal Sprouting

Where mature ash trees are present, their regeneration is generally substantial [32]. This was the case in the Upper Huron River watershed, where ash recruitment and regeneration have been assessed in several studies in response to the near complete mortality of reproductively mature trees [8,13,33]. Klooster et al. [13] conducted extensive soil sampling from 2005–2008 to characterize changes in the ash seed bank. The soil seed bank depleted quickly as ash mortality approached 95%, and the number of viable ash seeds declined until none were detected in 2007 or 2008. Rapid depletion of the seed bank was confirmed by the lack of newly germinated ash seedlings (with cotyledons), which were not detected after 2008 despite extensive sampling of the seedling layer [13]. These data from both soil samples and forest floor surveys suggest that new ash regeneration ceased completely as mortality of ash trees exceeded 95%. Kashian and Witter [8] also observed steep declines in the density of ash seedlings in the Upper Huron River watershed.

Epicormic basal sprouting can contribute to ash regeneration [34] and is a common response of ash trees that have had their canopies killed by EAB [33,35], especially for open-grown trees (Figure 2). However, no such regeneration was observed by Klooster et al. [13] in the closed-canopy mixed deciduous forests of the Upper Huron River watershed, where basal sprouts exhibited low vigor and died with the canopy or soon thereafter, perhaps due to strong interspecific competition for light and other resources in the understory of these diverse forests [14]. Conversely, Kashian [33] observed significant regeneration from basal sprouts (with some producing seed) in small, nearly pure stands of green ash where interspecific competition would not have been a factor. In addition, the 58% ash mortality documented by Kashian [33] would have generated larger canopy gaps than observed by Klooster et al. [13], where ash was a significantly lower component of more diverse forest stands [14]. In southeastern Ontario, Aubin et al. [35] also observed substantial ash regeneration from basal sprouting. However, inter- and intraspecific competition experienced by regenerating ash would have been limited there as well, because the amount of pre-EAB ash basal area in the sampled stands was greater than twice that of all other species combined, and more than 99% ash basal area died following EAB establishment [35].

Figure 2. Vigorous epicormic basal sprouting often occurs in response to canopy decline in open-grown ash infested with emerald ash borer (EAB) but was not observed by Klooster et al. [13] in closed canopy, mixed deciduous forests of the Upper Huron River watershed in southeast Michigan.

3.2. The Orphaned Cohort: Demography of Regenerating Ash

Prior to the EAB invasion, ash recruitment and regeneration were substantial in the Upper Huron River watershed, as *Fraxinus* was the most common genus in the understory and seedling layers of the stands sampled by Smith et al. [14] (Figure 3). As ash mortality exceeded 99%, the ash seed bank became depleted and ash seedling recruitment ceased, leaving only an orphaned cohort of previously established ash seedlings and saplings too small to be colonized by EAB, where they may persist for many years (Figure 4). The EAB population also continued to persist in the region at low levels despite the precipitous decline in its carrying capacity [36]. Each year, a proportion of ash saplings grows large enough to be colonized by EAB, and in aftermath forests in southeastern Ontario, EAB was found to be colonizing 19% of regenerating stems as small as 2.0 cm in diameter [35]. The fate of ash in the Upper Huron River watershed will depend on the degree that the orphaned cohort of regenerating saplings can survive and reproduce in the presence of low-density EAB populations [13].

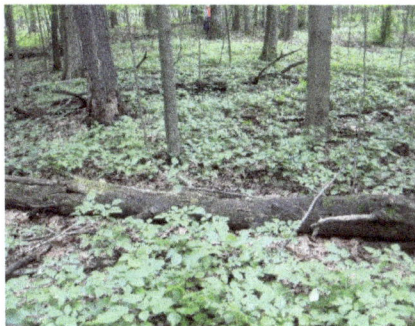

Figure 3. Regenerating ash seedlings and saplings too small to be infested by EAB were the most common woody species in the forest understory in the Upper Huron River watershed in southeast Michigan.

Figure 4. Ash seedlings and small saplings can persist in the understory for long periods with little growth, as evidenced by this plant that grew less that one cm between 2009 when it was tagged and 2016 when it was remeasured.

3.3. Biological Control and the Fate of the Orphaned Cohort

The degree to which ash survive to reproduce may be dependent in large part on whether natural enemies can regulate EAB populations at low levels [32,37]. Woodpeckers are the most important predators of EAB and are capable of causing high mortality on individual trees [38–42]. Predation rates by woodpeckers, however, were highly variable across sites and from tree-to-tree [38,40,42]. Woodpeckers caused limited mortality of EAB in saplings [43] and have been observed to decrease parasitoid populations by preying on parasitized EAB larvae, which may interfere with biological control [41]. In another study, however, woodpeckers did not affect rates of EAB parasitism [44].

Native and introduced parasitoids can also be important sources of EAB mortality [39,40]. Braconid wasps (*Atanycolus* spp.) native to North America parasitize EAB in Michigan, but with variable effects on EAB populations [43,45]. In a classical biological control program, several EAB parasitoids native to Asia have been introduced to North America [46]. Although *Spathius agrili* Yang (Hymenoptera: Braconidae) has had little success becoming established in the northern United States, *Oobius agrili* Zhang and Huang (Hymenoptera: Encyrtidae) has contributed to EAB mortality, and *Tetrastichus planipennisi* Yang (Hymenoptera: Eulophidae) has become the dominant biotic factor causing EAB mortality in southeastern Michigan [37,43,47]. Based on life table analyses, Duan et al. [43] concluded that *T. planipennisi* decreased the growth rate of EAB populations in saplings by more than 50%, and Margulies et al. [48] found more live ash saplings where higher numbers of parasitoids had been released. However, given the long residence time of ash seedlings and saplings in the understory, this may reflect their density when parasitoids were initially released, which was not reported.

If biological control agents and other natural enemies can regulate EAB at low levels, perhaps ash can regenerate at densities sufficient to restore significant ecosystem services lost during the EAB invasion [13,33,37]. However, it remains to be demonstrated that parasitoids and other mortality agents can exert temporal density dependent effects powerful enough to regulate EAB at low densities. Parasitism rates by *T. planipennisi* declined substantially in trees with stem diameters >12 cm due to the inability of their short ovipositors to penetrate thicker bark [37,40,43]. Furthermore, North American ash species planted in Asia have experienced high mortality from EAB [49,50], even in the presence of coevolved natural enemies.

4. Impacts on Other Flora and Fauna

Widespread and relatively simultaneous mortality of ash has been predicted to have substantial direct and indirect ecological impacts on forest structure, function, and community composition via gap formation as trees die, as well as accumulation of coarse woody debris as dead trees fall [3,51]. This disturbance can alter soil microbial communities [52], hydrology [53,54], and carbon and nutrient cycling [22,52,54], ultimately leading to community-level effects on successional trajectories [55], facilitation of the establishment and spread of exotic plants [56], and impacts on native fauna [57,58]. Some effects of ash mortality will dissipate relatively quickly as canopy gaps close via regeneration in the understory and growth of dominant and subdominant trees [59,60]. For example, the effects of increased light availability on soil moisture and the foliar chemistry of understory plants will be more ephemeral than the ecological impacts of the accumulation and decomposition of coarse woody debris, and the persistent legacy of altered succession.

4.1. Successional Trajectories Following Ash Mortality

EAB-induced ash mortality is likely to alter successional trajectories, as other overstory and understory species respond to widespread, relatively simultaneous gap formation [14,22,61]. As ash mortality in the Upper Huron River watershed reached a peak, the most common genera in the overstory were oak (*Quercus*) and maple (*Acer*), which thus appear likely to benefit from released competition, at least in the short term [14]. Conversely, oaks were underrepresented in the understory [14], perhaps due to limited recruitment and/or deer browsing (e.g., [62,63]), while maple and basswood (*Tilia*) species were the most common taxa in the understory (other than ash), suggesting that their dominance could increase over time [14]. Elm (*Ulmus*) was underrepresented in the overstory relative to the understory [14], probably due to the impact of Dutch elm disease [64].

The effects of ash mortality and gap formation on radial trunk growth varied by species [65]. Of 11 taxa sampled, all of which are native to the study site, the majority of species that exhibited positive correlations between ash importance value (prior to EAB-induced mortality) and diameter growth (increased ring width) were shade-tolerant (sugar maple, *A. saccharum* Marshall; red maple, *A. rubrum* L.) or intermediate (hickory, *Carya*; white oak, *Q. alba* L.; red oak, *Q. rubra* L.) tree species. Diameter growth of most shade-intolerant species (black cherry, *Prunus serotina* Ehrh.; poplar, *Populus*; larch, *Larix*; tulip tree, *Liriodendron tulipifera* L.) was not correlated with ash importance value, with the exception of walnut (*Juglans*). At sites in Ohio, the radial growth of maples and elm increased following EAB-induced ash mortality [22,61].

4.2. Facilitation of Invasive Plants

Some invasive plants are more vigorous and reproductive in forest gaps than under closed canopies where light is limited (e.g., [56,66,67]). EAB may trigger an "invasional meltdown" [68] if widespread gap formation caused by ash mortality facilitates the establishment and spread of invasive plants by increasing light availability and/or relaxing interspecific competition for other resources [3]. Consistent with this hypothesis, Klooster [69] found that in the Upper Huron River watershed the growth rate of alien woody shrubs—specifically multiflora rose (*Rosa multiflora* Thunb.), Amur honeysuckle (*Lonicera maackii* (Rupr.) Herder), and autumn olive (*Elaeagnus umbellata* Thunb.)—increased to a much greater degree in canopy gaps created by ash mortality than did the growth rate of native understory plants, such as ash seedlings, spicebush (*Lindera benzoin* (L.) Blume), American hornbeam (*Carpinus caroliniana* Walter), and American hophornbeam (*Ostrya virginiana* (Mill.) K. Koch). Hoven et al. [56] observed a similar pattern in Ohio forests where radial growth of Amur honeysuckle was directly related to the degree of ash mortality. These patterns are consistent with the species' adaptions to light availability. The dominant species of alien flora are adapted to respond to increased light availability, while the native shrubs consisted largely of shade-adapted, understory species, which typically exhibit lower phenotypic plasticity in response to variation in light availability [70,71]. However, Klooster [69] found no effect of EAB-induced gap formation on the density of alien plants, perhaps because not enough time had lapsed since the onset of ash mortality to impact their population dynamics.

4.3. Arthropod Herbivores of Ash

The decline and mortality of ash trees are expected to directly impact phytophagous arthropods that use ash as a host for at least part of their life cycle [72,73]. In a review of published literature, Gandhi and Herms [72] found host records for 281 arthropod herbivores of ash in six taxa (Arachnida: Acari; Hexapoda: Coleoptera, Diptera, Hemiptera, Hymenoptera, and Lepidoptera), including folivores, sap feeders, phloem/xylem feeders, gall formers, and seed predators. Most species (208) were polyphagous and thus were considered to face a low risk of population decline in response to ash mortality due to the prevalence of alternative host plants. However, 43 native and one alien species were reported to be specialist herbivores of ash, and thus were considered to face a high risk of local extirpation [72]. Wagner and Todd [73] conducted an appraisal of published and unpublished host records for specialist invertebrate herbivores of ash based on expert assessment by taxonomic authorities and concluded that 98 species may be imperiled by the EAB invasion.

In the short term, populations of some wood-borers and bark beetles that colonize declining and dead ash trees may increase in parallel with availability of suitable hosts [72]. However, their populations are predicted to eventually decline as snags fall and subsequently decay (e.g., [74]). For example, in a study conducted in the Upper Huron River watershed, Ulyshen et al. [75] reared 18 species of saproxylic beetles from ash limbs that had been suspended in the canopy or placed on the ground. The highly polyphagous cerambycid, *Neoclytus acuminatus* Fabricius, was the most common species collected. The buprestid *Agrilus subcinctus* Gory and the curculionid *Hylesinus aculeatus* Say, were also collected and face greater threat of local extirpation because they are largely or entirely restricted to ash [75]. Population declines of arthropod species that utilize ash as a host will likely have cascading impacts on biota with which they interact (e.g., symbionts and natural enemies), and the impacts may reverberate across the food web [72,73].

4.4. Ground-Dwelling Invertebrates

Widespread tree mortality caused by alien insects may also have indirect effects on invertebrate populations and communities [3]. Perry and Herms [76] proposed a model of dynamic temporal effects of disturbance caused by tree-killing invasive insects, including gap formation and accumulation of coarse woody debris (CWD) (Figure 5), on ground-dwelling invertebrate populations and communities. The model predicts the magnitude of effects of gap formation and accumulation of CWD will transition over time in opposing ways as ash mortality in the stand progresses from early to late stages. The formation of canopy gaps is predicted to have the greatest impact on ground-dwelling invertebrate diversity and abundance during early stages of ash mortality when gaps are presumably at their maximum size after tree death, with impacts diminishing over time as gaps close. Impacts of CWD, in contrast, are predicted to increase over time [76] as ash trees die, standing snags fall, and CWD accumulates and decomposes on the forest floor. For example, Higham et al. [74] observed rapid accumulation of CWD across a chronosequence of ash mortality in Ohio, and in the Upper Huron River watershed, the number of fallen ash trees increased by 76% from 2008–2012, and volume of ash CWD increased by 53% [77].

Experimental tests have been broadly consistent with these predictions. In a study conducted in stands experiencing early stages of ash mortality in northern Ohio, gap formation decreased the abundance of ground beetles (Carabidae) and other ground-dwelling arthropod taxa, as well as species richness and diversity, while the effects of CWD were less substantial [78,79]. Similarly, during early stages of ash mortality in the Upper Huron River watershed in southeast Michigan, ground beetle abundance and diversity decreased as ash mortality and gap size increased [80]. At the same sites during late stages of ash mortality, the effects of gaps—which by then were smaller—on ground-dwelling invertebrate communities were minimal, while the abundance, evenness, and diversity of soil arthropods and exotic earthworms were highest adjacent to decomposing ash CWD [81,82].

Figure 5. Coarse woody debris (CWD) accumulates steadily on the forest floor as dead ash trees fall.

4.5. Tri-Trophic Impacts on Swallowtail Butterflies

As ash mortality generates canopy gaps, insect herbivores of understory plants may be impacted indirectly by the effects of increased light availability on the quality of their host plants. For example, foliar concentrations of secondary metabolites are often higher in plants in the sun than in the same species growing in shade [83,84]. Common prickly ash (*Zanthoxylum americanum* Mill.), a native understory shrub in southeast Michigan, is the only host in the Upper Huron River watershed for giant swallowtail butterfly (*Papilio cresphontes* Cramer) larvae (Figure 6). The foliage of prickly ash contains furanocoumarins [85], which are photoactivated secondary metabolites that become more bioactive and toxic to herbivores when exposed to ultraviolet light [86]. Rice [87] found that prickly ash growing in canopy gaps created by ash mortality contained higher foliar concentrations of furanocoumarins than conspecifics in the shaded understory. Although giant swallowtail butterfly larvae are capable of detoxifying furanocoumarins [88], larvae still grew more slowly on plants in canopy gaps [87].

Figure 6. Prickly ash is the only host for giant swallowtail larvae in the Upper Huron River watershed.

The slow growth–high mortality hypothesis predicts that slower growing larvae will experience greater mortality because of their longer exposure to natural enemies [89,90]. Average daily probability of mortality from natural enemies (15%) was equivalent for larvae feeding on plants in gaps and shade [87]. Hence, if the lower growth rate of larvae feeding on plants in canopy gaps delays completion of the larval stage, mortality from natural enemies should increase as indirect effects of EAB-induced ash mortality and gap formation cascade across trophic levels [87].

4.6. Effects on Bird Behavior and Communities

EAB-induced ash mortality may also affect bird behavior and communities indirectly by altering the availability of food resources and nesting habitat. Woodpeckers and other insectivorous birds that forage primarily on bark or dead wood may be ecologically primed to benefit from the EAB invasion, at least temporarily, as a dramatic pulse of food from the EAB outbreak leads to increased reproduction and population growth, followed by a sharp population decline caused by resource depletion as ash trees die and the EAB population crashes (e.g., [5,91]). For example, data from the citizen science program Project FeederWatch revealed a signature of the EAB invasion near the epicenter in southeast Michigan that was not detected elsewhere, as Red-bellied Woodpecker (*Melanerpes carolinus* L.) and White-breasted Nuthatch (*Sitta canadensis* L.) numbers initially increased, while those of Downy Woodpecker (*Picoides pubescens* L.) and Hairy Woodpecker (*Picoides villosus* L.) initially declined and then increased several years later [92].

Long [93] monitored bird communities and foraging behavior during the winter across a gradient of EAB impact ranging from near complete ash mortality in southeast Michigan to early stages of EAB invasion in southwestern Ohio. He found that Downy, Hairy, Red-bellied, and Pileated (*Dryocopus pileatus* L.) Woodpecker all spent more time foraging on ash trees in stands with active EAB infestations, and that these stands had higher numbers of Downy Woodpecker. Red-bellied Woodpecker was significantly less abundant in stands in which the EAB outbreak had run its course.

Forest stands with high ash mortality had more diverse bird assemblages than did stands experiencing low ash mortality. Stands with high ash mortality had greater herbaceous groundcover, shrubby regeneration, and canopy fragmentation relative to stands with low ash mortality, which created nesting habitat and resulted in a shift in the breeding bird community to species more typical of open habitats [93].

5. Summary and Conclusions

It is clear from this review that EAB already has substantially impacted forests near the epicenter of the invasion of North America. In the Upper Huron River watershed in southeast Michigan, mortality of black, green, and white ash exceeded 99% by 2009, with nearly 60% occurring over a five-year period. As would be expected when mortality is so comprehensive, there were no relationships between ash mortality and ash density, species diversity, or any other measure of stand composition. New ash recruitment ceased as the ash seedbank was depleted and no new seedlings were detected, leaving only an orphaned cohort of previously established ash seedlings and saplings too small to be infested by EAB. The degree to which ecosystem services provided by ash can be restored may depend in large part on whether introduced biological control agents and other natural enemies can regulate EAB populations at densities low enough to facilitate significant ash regeneration.

The relatively simultaneous, widespread canopy gap formation followed by a steady accumulation of downed coarse woody debris has triggered a cascade of direct and indirect effects on plant and animal communities. Forest successional trajectories have been altered, growth rates of exotic plants have increased, specialist herbivores of ash are threatened with local extirpation, and the abundance and diversity of ground-dwelling invertebrates have been impacted, as have behavior and abundance of overwintering and breeding birds.

While these studies have increased our understanding of the ecological impacts of EAB, future research may focus on elucidating rates and patterns of gap closure and successional trajectories in different forest types; whether ash mortality and accumulation of CWD alter nutrient cycling and hydrological processes; long-term impacts of gap formation on alien and native understory flora; and impacts of ash mortality on ash herbivores and biodiversity at the landscape-level. Such studies will inform efforts focused on increasing resilience and restoration of ash ecosystems as the EAB invasion of North America proceeds.

Author Contributions: W.S.K. and D.A.H. conceived of the review, wrote many of the sections, and organized all contributions into the final version. All other co-authors contributed equally and substantially and are listed in alphabetical order.

Acknowledgments: Annemarie Smith and Diane Hartzler conducted the field-work to establish the long-term monitoring plots within the Upper Huron River watershed where much of the research reviewed in this paper was conducted from 2004–2014 with cooperation and approval from Paul Muelle of the Huron-Clinton MetroParks and Glenn Palmgren of the Michigan Department of Natural Resources. This research was funded by grants from the United States Department of Agriculture (USDA) Forest Service Northeastern Research Station's Research on Biological Invasions of Northeastern Forests program; USDA National Research Initiative Biology of Weedy and Invasive Species in Agroecosystems competitive grants program; the National Institute of Food and Agriculture; Cooperative Agreements with the USDA Forest Service Northern Research Station, Delaware, OH; and state and federal funds appropriated to the Ohio Agricultural Research and Development Center and The Ohio State University. We thank Catherine Herms and John Cardina for reviews of earlier drafts, which greatly improved this manuscript.

Conflicts of Interest: The authors declare no conflict of interest.

References

1. Liebhold, A.M.; MacDonald, W.L.; Bergdahl, D.; Mastro, V.C. Invasion by exotic forest pests: A threat to forest ecosystems. *For. Sci. Monograph.* **1995**, *41*, 1–49.

2. Lovett, G.M.; Canham, C.D.; Arthur, M.A.; Weathers, K.C.; Fitzhugh, R.D. Forest ecosystem responses to exotic pests and pathogens in eastern North America. *BioScience* **2006**, *56*, 395–405. [CrossRef]

3. Gandhi, J.K.J.; Herms, D.A. Direct and indirect effects of alien insect herbivores on ecological processes and interactions in forests of eastern North America. *Biol. Invasions* **2010**, *12*, 389–405. [CrossRef]

4. Haack, R.A.; Jendak, E.; Houping, L.; Marchant, K.R.; Petrice, T.R.; Poland, T.M.; Ye, H. The emerald ash borer: A new exotic pest in North America. *Newsl. Mich. Entomol. Soc.* **2002**, *47*, 1–5.

5. Cappaert, D.; McCullough, D.G.; Poland, T.M.; Siegert, N.W. Emerald ash borer in North America: A research and regulatory challenge. *Am. Entomol.* **2005**, *51*, 152–163. [CrossRef]

6. Poland, T.M.; McCullough, D.G. Emerald ash borer: Invasion of the urban forest and the threat to North America's ash resource. *J. For.* **2006**, *104*, 118–124.

7. Siegert, N.W.; McCullough, D.G.; Liebhold, A.M.; Telewski, F.W. Dendrochronological reconstruction of the epicentre and early spread of emerald ash borer in North America. *Divers. Distrib.* **2014**, *20*, 847–858. [CrossRef]

8. Kashian, D.M.; Witter, J.A. Assessing the potential for ash canopy tree replacement via current regeneration following emerald ash borer-caused mortality on southeastern Michigan landscapes. *For. Ecol. Manag.* **2011**, *261*, 480–488. [CrossRef]

9. Pugh, S.A.; Liebhold, A.M.; Morin, R.S. Changes in ash tree demography associated with emerald ash borer invasion, indicated by regional forest inventory data from the Great Lakes States. *Can. J. For. Res.* **2011**, *41*, 2165–2175. [CrossRef]

10. Knight, K.S.; Brown, J.P.; Long, R.P. Factors affecting the survival of ash (*Fraxinus* spp.) trees infested by emerald ash borer (*Agrilus planipennis*). *Biol. Invasions* **2013**, *15*, 371–383. [CrossRef]

11. Marshall, J.M.; Smith, E.L.; Mech, R.; Storer, A.J. Estimates of *Agrilus planipennis* infestation rates and potential survival of ash. *Am. Midl. Nat.* **2013**, *169*, 179–193. [CrossRef]

12. Burr, S.J.; McCullough, D.G. Condition of green ash (*Fraxinus pennsylvanica*) overstory and regeneration at three stages of the emerald ash borer invasion wave. *Can. J. For. Res.* **2014**, *44*, 768–776. [CrossRef]

13. Klooster, W.S.; Herms, D.A.; Knight, K.S.; Herms, C.P.; McCullough, D.G.; Smith, A.M.; Gandhi, K.J.K.; Cardina, J. Ash (*Fraxinus* spp.) mortality, regeneration, and seed bank dynamics in mixed hardwood forests following invasion by emerald ash borer (*Agrilus planipennis*). *Biol. Invasions* **2014**, *16*, 859–873. [CrossRef]

14. Smith, A.; Herms, D.A.; Long, R.P.; Gandhi, K.J.K. Community composition and structure had no effect on forest susceptibility to invasion by the emerald ash borer (Coleoptera: Buprestidae). *Can. Entomol.* **2015**, *147*, 318–328. [CrossRef]

15. Morin, R.S.; Leibhold, A.M.; Pugh, S.A.; Crocker, S.J. Regional assessment of emerald ash borer, *Agrilus planipennis*, impacts in forests of the Eastern United States. *Biol. Invasions* **2017**, *19*, 703–711. [CrossRef]

16. Cipollini, D.; Rigsby, C.M. Incidence of infestation and larval success of emerald ash borer (*Agrilus planipennis*) on white fringetree (*Chionanthus virginicus*), and devilwood (*Osmanthus americanus*). *Environ. Entomol.* **2015**, *44*, 1375–1383. [CrossRef] [PubMed]

17. Peterson, D.L.; Cipollini, D. Distribution, predictors, and impacts of emerald ash borer (*Agrilus planipennis*) (Coleoptera: Buprestidae) infestation of white fringetree (*Chionanthus virginicus*). *Environ. Entomol.* **2017**, *46*, 50–57. [CrossRef] [PubMed]

18. Herms, D.A.; McCullough, D.G. Emerald ash borer invasion of North America: History, biology, ecology, impacts, and management. *Annu. Rev. Entomol.* **2014**, *59*, 13–30. [CrossRef] [PubMed]

19. Van Driesche, R.G.; Reardon, R. (Eds.) *Biology and Control of Emerald Ash Borer*; Technical Bulletin FHTET 2014-09; USDA Forest Service: Morgantown, WV, USA, 2015; p. 180.

20. Liu, H. Under siege: Ash management in the wake of the emerald ash borer. *J. Integr. Pest. Manag.* **2017**, *9*, 5. [CrossRef]

21. Smith, A. Effects of Community Structure on Forest Susceptibility and Response to the Emerald Ash Borer Invasion of the Huron River Watershed in Southeast Michigan. Master's Thesis, The Ohio State University, Columbus, OH, USA, 2006.

22. Flower, C.E.; Knight, K.S.; Gonzalez-Meler, M.A. Impacts of the emerald ash borer (*Agrilus Planipennis* Fairmaire) induced ash (*Fraxinus* spp.) mortality on forest carbon cycling and successional dynamics in the eastern United States. *Biol. Invasions* **2013**, *15*, 931–944. [CrossRef]

23. Nisbet, D.; Kreutzweiser, D.; Sibley, P.; Scarr, T. Ecological risks posed by emerald ash borer to riparian forest habitats: A review and problem formulation with management implications. *For. Ecol. Manag.* **2015**, *358*, 165–173. [CrossRef]

24. Baranchikov, Y.; Mozolevskaya, E.; Yurchenko, G.; Kenis, M. Occurrence of the emerald ash borer, *Agrilus planipennis* in Russia and its potential impact on European forestry. *EPPO Bull.* **2008**, *38*, 233–238. [CrossRef]

25. Orlova-Bienkowskaja, M.J. Ashes in Europe are in danger: The invasive range of *Agrilus planipennis* in European Russia is expanding. *Biol. Invasions* **2014**, *16*, 1345–1349. [CrossRef]

26. Villari, C.; Herms, D.A.; Whitehill, J.G.A.; Cipollini, D.; Bonello, P. Progress and gaps in understanding mechanisms of ash tree resistance to emerald ash borer, a model for wood boring insects that kill angiosperm trees. *New Phytol.* **2016**, *209*, 63–79. [CrossRef] [PubMed]

27. Valiente-Banuet, A.; Aizen, M.A.; Alcántara, J.M.; Arroyo, J.; Cocucci, A.; Galetti, M.; García, M.B.; García, D.; Gómez, J.M.; Jordano, P.; et al. Beyond species loss: The extinction of ecological interactions in a changing world. *Funct. Ecol.* **2015**, *29*, 299–307. [CrossRef]

28. Brockerhoff, E.G.; Liebhold, A.M.; Jactel, H. The ecology of forest insect invasions and advances in their management. *Can. J. For. Res.* **2006**, *36*, 263–268. [CrossRef]

29. Jactel, H.; Menassieu, P.; Vetillard, F.; Gaulier, A.; Samalens, J.C.; Brockerhoff, E.G. Tree species diversity reduces the invasibility of maritime pine stands by the bast scale, *Matsucoccus feytaudi* (Homoptera: Margarodidae). *Can. J. For. Res.* **2006**, *36*, 314–323. [CrossRef]

30. Guyot, V.; Castagneyrol, B.; Vialatte, A.; Deconchat, M.; Selvi, F.; Bussotti, F.; Jactel, H. Tree diversity limits the impact of an invasive forest pest. *PLoS ONE* **2015**, *10*, e0136469. [CrossRef] [PubMed]

31. Smitley, D.; Tavis, T.; Rebek, E. Progression of ash canopy thinning and dieback outward from the initial infestation of emerald ash borer (Coleoptera: Buprestidae) in southeastern Michigan. *J. Econ. Entomol.* **2008**, *101*, 1643–1650. [CrossRef] [PubMed]

32. Granger, J.J.; Zobel, J.M.; Buckley, D.S. Potential for regenerating major and minor ash species (*Fraxinus* spp.) following EAB infestation in the eastern United States. *For. Ecol. Manag.* **2017**, *389*, 296–305. [CrossRef]

33. Kashian, D.M. Sprouting and seed production may promote persistence of green ash in the presence of the emerald ash borer. *Ecosphere* **2016**, *7*, e01332. [CrossRef]

34. Dietze, M.C.; Clarke, J.S. Changing the gap dynamics paradigm: Vegetative regeneration control on forest response to disturbance. *Ecol. Monograph.* **2008**, *78*, 331–347. [CrossRef]

35. Aubin, I.; Cardou, F.; Ryall, K.; Kreutzweiser, D.; Scarr, T. Ash regeneration capacity after emerald ash borer (EAB) outbreaks: Some early results. *For. Chron.* **2015**, *91*, 291–298. [CrossRef]

36. Burr, S.J.; McCullough, D.G.; Poland, T.M. Density of emerald ash borer (Coleoptera: Buprestidae) adults and larvae at three stages of the invasion wave. *Environ. Entomol.* **2018**, *47*, 121–132. [CrossRef] [PubMed]

37. Duan, J.J.; Van Driesche, R.G.; Bauer, L.S.; Kashian, D.M.; Herms, D.A. Risk to ash from emerald ash borer: Can biological control prevent the loss of ash stands? In *Biology and Control of Emerald Ash Borer*; Technical Bulletin FHTET 2014-09; Van Driesche, R.G., Reardon, R., Eds.; USDA Forest Service: Morgantown, WV, USA, 2015; pp. 65–73.

38. Lindell, C.A.; McCullough, D.G.; Cappaert, D.; Apostolou, N.M.; Roth, M.B. Factors influencing woodpecker predation on emerald ash borer. *Am. Midl. Nat.* **2008**, *159*, 434–444. [CrossRef]

39. Duan, J.J.; Ulyshen, M.D.; Bauer, L.S.; Gould, J.; Van Driesche, R.G. Measuring the impact of biotic factors on populations of immature emerald ash borers (Coleoptera: Buprestidae). *Environ. Entomol.* **2010**, *39*, 1513–1522. [CrossRef] [PubMed]

40. Jennings, D.E.; Gould, J.R.; Vandenberg, J.D.; Duan, J.J.; Shrewsbury, P.M. Quantifying the impact of woodpecker predation on population dynamics of the emerald ash borer (*Agrilus planipennis*). *PLoS ONE* **2013**, *8*, e83491. [CrossRef] [PubMed]

41. Jennings, D.E.; Duan, J.J.; Shrewsbury, P.M. Biotic mortality factors affecting emerald ash borer (*Agrilus planipennis*) are highly dependent on life stage and host tree condition. *Bull. Entomol. Res.* **2015**, *105*, 598–606. [CrossRef] [PubMed]

42. Flower, C.E.; Long, L.L.; Knight, K.S.; Rebbeck, J.; Brown, J.S.; Gonzalez-Meler, M.A.; Whelan, C.J. Native bark-foraging birds preferentially forage in infected ash (*Fraxinus* spp.) and prove effective predators of the invasive emerald ash borer (*Agrilus planipennis* Fairmaire). *For. Ecol. Manag.* **2014**, *313*, 300–306. [CrossRef]

43. Duan, J.J.; Bauer, L.S.; Van Driesche, R.G. Emerald ash borer biocontrol in ash saplings: The potential for early stage recovery of North American ash trees. *For. Ecol. Manag.* **2017**, *394*, 64–72. [CrossRef]

44. Murphy, T.C.; Gould, J.R.; Van Driesche, R.G.; Elkinton, J.S. Interactions between woodpecker attack and parasitism by introduced parasitoids of the emerald ash borer. *Biol. Control* **2018**, *122*, 109–117. [CrossRef]

45. Cappaert, D.; McCullough, D.G. Occurrence and seasonal abundance of *Atanycolus cappaerti* (Hymenoptera: Braconidae) a native parasitoid of emerald ash borer, *Agrilus planipennis* (Coleoptera: Buprestidae). *Great Lakes Entomol.* **2009**, *42*, 16–29.

46. Bauer, L.S.; Duan, J.J.; Gould, J.G.; Van Driesche, R.G. Progress in the classical biological control of *Agrilus planipennis* Fairmaire (Coleoptera: Buprestidae). *Can. Entomol.* **2015**, *147*, 300–317. [CrossRef]

47. Abell, K.J.; Bauer, L.S.; Duan, J.J.; Van Driesche, R.G. Long-term monitoring of the introduced emerald ash borer (Coleoptera: Buprestidae) egg parasitoid, *Oobius agrili* (Hymenoptera: Encyrtidae), in Michigan, USA and evaluation of a newly developed monitoring technique. *Biol. Control* **2014**, *79*, 36–42. [CrossRef]

48. Margulies, E.; Bauer, L.S.; Ibáñez, I. Buying time: Preliminary assessment of biocontrol in the recovery of native forest vegetation in the aftermath of the invasive emerald ash borer. *Forests* **2017**, *8*, 369. [CrossRef]

49. Wei, X.; Reardon, R.; Wu, Y.; Sun, J.-H. Emerald ash borer, *Agrilus planipennis* Fairmaire (Coleoptera: Buprestidae), in China: A review and distribution survey. *Acta Entomol. Sin.* **2004**, *47*, 679–685.

50. Wei, X.; Wu, Y.; Reardon, R.; Sun, T.H.; Lu, M.; Sun, J.-H. Biology and damage traits of emerald ash borer (*Agrilus planipennis* Fairmaire) in China. *J. Insect Sci.* **2007**, *14*, 367–373. [CrossRef]

51. Flower, C.E.; Gonzalez-Meler, M.A. Responses of temperate forest productivity to insect and pathogen disturbances. *Annu. Rev. Plant Biol.* **2015**, *66*, 547–569. [CrossRef] [PubMed]

52. Ricketts, M.P.; Flower, C.E.; Knight, K.S.; Gonzalez-Meler, M.A. Evidence of ash tree (*Fraxinus* spp.) specific associations with soil bacterial community structure and functional capacity. *Forests* **2018**, *9*, 187. [CrossRef]

53. Van Grinsven, M.J.; Shannon, J.P.; Davis, J.C.; Bolton, N.W.; Wagenbrenner, J.W.; Kolka, R.K.; Pypker, T.G. Source water contributions and hydrologic responses to simulated emerald ash borer infestations in depressional black ash wetlands. *Ecohydrology* **2017**, *10*, e1862. [CrossRef]

54. Kolka, R.K.; D'Amato, A.W.; Wagenbrenner, J.W.; Slesak, R.A.; Pypker, T.G.; Youngquist, M.B.; Grinde, A.R.; Palik, B.J. Review of ecosystem level impacts of emerald ash borer on black ash wetlands: What does the future hold? *Forests* **2018**, *9*, 179. [CrossRef]

55. Bowen, A.K.M.; Stevens, M.H.H. Predicting the effects of emerald ash borer on hardwood swamp forest structure and composition in southern Michigan. *J. Torrey Bot. Soc.* **2018**, *145*, 41–54. [CrossRef]

56. Hoven, B.M.; Gorchov, D.L.; Knight, K.S.; Peters, V.E. The effect of emerald ash borer-caused tree mortality on the invasive shrub Amur honeysuckle and their combined effects on tree and shrub seedlings. *Biol. Invasions* **2017**, *19*, 2813–2836. [CrossRef]

57. Jennings, D.E.; Duan, J.J.; Bean, D.; Rice, K.A.; Williams, G.L.; Bell, S.K.; Shurtleff, A.S.; Shrewsbury, P.M. Effects of emerald ash borer invasion on the community composition of arthropods associated with ash tree boles in Maryland, USA. *Agric. For. Entomol.* **2017**, *19*, 122–129. [CrossRef]

58. Savage, M.B.; Rieske, L.K. Coleopteran communities associated with forests invaded by emerald ash borer. *Forests* **2018**, *9*, 69. [CrossRef]

59. Whitmore, T.C. Canopy gaps and the two major groups of forest trees. *Ecology* **1989**, *70*, 536–538. [CrossRef]

60. Valverde, T.; Silvertown, J. Canopy closure rate and forest structure. *Ecology* **1997**, *78*, 1555–1562. [CrossRef]

61. Costilow, K.C.; Knight, K.S.; Flower, C.E. Disturbance severity and canopy position control the radial growth response of maple trees (*Acer* spp.) in forests of northwest Ohio impacted by emerald ash borer (*Agrilus planipennis*). *Ann. For. Sci.* **2017**, *74*, 10. [CrossRef]

62. Rooney, T.P.; Waller, D.M. Direct and indirect effects of white-tailed deer in forest ecosystems. *For. Ecol. Manag.* **2003**, *181*, 165–176. [CrossRef]

63. Leonardsson, J.; Löf, M.; Götmark, F. Exclosures can favour natural regeneration of oak after conservation-oriented thinning in mixed forests in Sweden: A 10-year study. *For. Ecol. Manag.* **2015**, *354*, 1–9. [CrossRef]

64. Barnes, B.V. Succession in deciduous swamp communities of southern Michigan formerly dominated by American elm. *Can. J. Bot* **1976**, *54*, 19–24. [CrossRef]

65. Klooster, W.S.; Goebel, P.C.; Herms, D.A. Forest responses following emerald ash borer-induced ash mortality in southeastern Michigan. In Proceedings of the 2016 Emerald Ash Borer National Research and Technology Development Meeting, Wooster, OH, USA, 19–20 October 2017; Buck, P., Lance, R., Binion, Eds.; USDA Forest Service: Morgantown, WV, USA, 2017; pp. 40–41.

66. Burnham, K.M.; Lee, T.D. Canopy gaps facilitate establishment, growth, and reproduction of invasive *Frangula alnus* in a *Tsuga canadensis* dominated forest. *Biol. Invasions* **2010**, *12*, 1509–1520. [CrossRef]

67. Driscoll, A.G.; Angeli, N.F.; Gorchov, D.L.; Jiang, Z.; Zhang, J.; Freeman, C. The effect of treefall gaps on the spatial distribution of three invasive plants in a mature upland forest in Maryland. *J. Torrey Botan. Soc.* **2016**, *143*, 349–358. [CrossRef]

68. Simberloff, D.; Von Holle, B. Positive interactions of nonindigenous species: Invasional meltdown? *Biol. Invasions* **1999**, *1*, 21–32. [CrossRef]

69. Klooster, W.S. Forest Responses to Emerald Ash Borer-Induced Ash Mortality. Ph.D. Dissertation, The Ohio State University, Columbus, OH, USA, 2012.

70. Valladares, F.; Niinemets, U. Shade tolerance, a key plant feature of complex nature and consequences. *Ann. Rev. Ecol. Evol. Syst.* **2008**, *39*, 237–257. [CrossRef]

71. Heberling, M.; Fridley, J.D. Resource-use strategies of native and invasive plants in Eastern North American forests. *New Phytol.* **2013**, *200*, 523–533. [CrossRef] [PubMed]

72. Gandhi, K.J.K.; Herms, D.A. North American arthropods at risk due to widespread *Fraxinus* mortality caused by the alien emerald ash borer. *Biol. Invasions* **2010**, *12*, 1839–1846. [CrossRef]

73. Wagner, D.L.; Todd, K. New ecological assessment for the emerald ash borer: A cautionary tale about unvetted host-plant literature. *Am. Entomol.* **2016**, *62*, 26–35. [CrossRef]

74. Higham, M.; Hoven, B.M.; Gorchov, D.L.; Knight, K.S. Patterns of coarse woody debris in hardwood forests across a chronosequence of ash mortality due to the emerald ash borer (*Agrilus planipennis*). *Nat. Area. J.* **2017**, *37*, 406–411. [CrossRef]

75. Ulyshen, M.D.; Barrington, W.T.; Hoebeke, R.; Herms, D.A. Vertically stratified ash-limb beetle fauna in northern Ohio. *Psyche* **2012**, *2012*, 215891. [CrossRef]

76. Perry, K.I.; Herms, D.A. Responses of ground-dwelling invertebrates to gap formation and accumulation of woody debris from invasive species, wind, and salvage logging. *Forests* **2017**, *8*, 174. [CrossRef]

77. Perry, K.I.; Herms, D.A.; Klooster, W.S.; Smith, A.; Hartzler, D.M.; Coyle, D.R.; Gandhi, K.J.K. Downed coarse woody debris dynamics in ash (*Fraxinus* spp.) stands invaded by emerald ash borer (*Agrilus planipennis* Fairmaire). *Forests* **2018**, *9*, 191. [CrossRef]

78. Perry, K.I.; Herms, D.A. Response of the forest floor invertebrate community to canopy gap formation caused by early stages of emerald ash borer-induced ash mortality. *For. Ecol. Manag.* **2016**, *375*, 259–267. [CrossRef]

79. Perry, K.I.; Herms, D.A. Short-term responses of ground beetles to forest changes caused by early stages of emerald ash borer (Coleoptera: Buprestidae)-induced ash mortality. *Environ. Entomol.* **2016**, *45*, 616–626. [CrossRef] [PubMed]

80. Gandhi, K.J.K.; Smith, A.M.; Hartzler, D.M.; Herms, D.A. Indirect effects of emerald ash borer-induced ash mortality and canopy gap formation on epigaeic beetles. *Environ. Entomol.* **2014**, *43*, 546–555. [CrossRef] [PubMed]

81. Ulyshen, M.D.; Klooster, W.S.; Barrington, W.T.; Herms, D.A. Impacts of emerald ash borer-induced tree mortality on leaf litter arthropods and exotic earthworms. *Pedobiologia* **2011**, *54*, 261–265. [CrossRef]

82. Perry, K.I.; Herms, D.A. Effects of late stages of emerald ash borer (Coleoptera: Buprestidae)-induced ash mortality on forest floor invertebrate communities. *J. Insect Sci.* **2017**, *17*, 119. [CrossRef]

83. Herms, D.A.; Mattson, W.J. The dilemma of plants: To grow or defend. *Q. Rev. Biol.* **1992**, *67*, 282–335. [CrossRef]

84. Koricheva, J.; Larsson, S.; Haukioja, E.; Keinanen, M. Regulation of woody plant secondary metabolism by resource availability: Hypothesis testing by means of meta-analysis. *Oikos* **1998**, *83*, 212–226. [CrossRef]

85. Bafi-Yeboa, N.; Arnason, J.; Baker, J.; Smith, M. Antifungal constituents of Northern prickly ash, *Zanthoxylum americanum* Mill. *Phytomedicine* **2005**, *12*, 370–377. [CrossRef] [PubMed]

86. Berenbaum, M.; Scriber, J.M.; Tsubaki, Y.; Lederhouse, R.C. Chemistry and oligophagy in the Papilionidae. In *Swallowtail Butterflies: Their Ecology and Evolutionary Biology*; Scriber, J.M., Tsubaki, Y., Lederhouse, R.C., Eds.; Scientific Publishers: Gainesville, FL, USA, 1995; pp. 27–38.

87. Rice, K.B. Cascading Ecological Impacts of Emerald Ash Borer: Tritrophic Interactions between Prickly Ash, Giant Swallowtail Butterfly Larvae, and Larval Predators. Ph.D. Dissertation, The Ohio State University, Columbus, OH, USA, 2013.

88. Lee, K.; Berenbaum, M.R. Ecological aspects of antioxidant enzymes and glutathione-S-transferases in three *Papilio* species. *Biochem. Syst. Ecol.* **1992**, *20*, 197–207. [CrossRef]

89. Price, P.W.; Bouton, C.E.; Gross, P.; McPheron, B.A.; Thompson, J.N.; Weis, A.E. Interactions among three trophic levels: Influence of plants on interactions between insect herbivores and natural enemies. *Annu. Rev. Ecol. Evol. Syst.* **1980**, *11*, 41–65. [CrossRef]

90. Clancy, K.M.; Price, P.W. Rapid herbivore growth enhances enemy attack: Sublethal plant defenses remain a paradox. *Ecology* **1987**, *68*, 733–737. [CrossRef]

91. Ostfeld, R.S.; Keesing, F. Pulsed resources and community dynamics of consumers in terrestrial ecosystems. *Trends Ecol. Evol.* **2000**, *15*, 232–237. [CrossRef]

92. Koenig, W.D.; Liebhold, A.M.; Bonter, D.N.; Hochachka, W.M.; Dickinson, J.L. Effects of the emerald ash borer invasion on four species of birds. *Biol. Invasions.* **2013**, *15*, 2095–2103. [CrossRef]

93. Long, L.C. Direct and Indirect Impacts of Emerald Ash Borer on Forest Bird Communities. Master's Thesis, The Ohio State University, Columbus, OH, USA, 2013.

MDPI

Article

Potential Impacts of Emerald Ash Borer Biocontrol on Ash Health and Recovery in Southern Michigan

Daniel M. Kashian [1,*], Leah S. Bauer [2], Benjamin A. Spei [3], Jian J. Duan [4] and Juli R. Gould [5]

1   Department of Biological Sciences, Wayne State University, Detroit, MI 48202, USA
2   United States Department of Agriculture, Forest Service, Northern Research Station, Lansing, MI 48910, USA; lbauer@fs.fed.us
3   Department of Biological Sciences, Wayne State University, Detroit, MI 48202, USA; spei.ben@gmail.com
4   United States Department of Agriculture, Agricultural Research Service, Beneficial Insects Introduction Research Unit, Newark, DE 19713, USA; jian.duan@ars.usda.gov
5   United States of Department of Agriculture, Animal and Plant Health Inspection Service, Plant Protection and Quarantine, Science and Technology, Buzzards Bay, MA 02542, USA; juli.r.gould@aphis.usda.gov
*   Correspondence: dkash@wayne.edu; Tel.: +1-313-577-9093

Received: 30 March 2018; Accepted: 23 May 2018; Published: 25 May 2018

Abstract: Emerald ash borer (EAB) is an invasive beetle that kills native North American ash species, threatening their persistence. A classical biological control program for EAB was initiated in 2007 with the release of three specialized EAB parasitoids. Monitoring changes in the health and regeneration of ash where EAB biocontrol agents have been released is critical for assessing the success of EAB biocontrol and predicting future changes to the ash component of North American forests. We sampled release and control plots across southern Michigan over a three-year period to measure ash health and recruitment to begin assessing the long-term impact of EAB biological control on ash populations. We noted a reduced mortality of larger trees between 2012 and 2015 in release plots compared to control plots and increases in ash diameter, but our results were otherwise inconsistent. Ash regeneration was generally higher in release plots compared to control plots but highly variable among sites, suggesting some protection of ash saplings from EAB by parasitoids. We conclude that EAB biocontrol is likely to have a positive effect on ash populations, but that the study duration was not long enough to definitively deduce the long-term success of the biocontrol program in this region.

Keywords: *Agrilus planipennis*; disturbance; forest recovery; *Fraxinus* spp.; invasive species; *Tetrastichus planipennisi*; *Oobius agrili*; *Spathius galinae*; Michigan; tree regeneration

---

1. Introduction

Introduced insects and pathogens are some of the most significant threats to forest growth and diversity in North America [1]. Most genera of woody plants today are associated with one or more species of phytophagous invasive insect [2–5]. Although containment and eradication strategies may exist for some invasive species [6,7], the United States continues to experience establishment events by destructive, non-indigenous forest pests [4,8]. Increased rates of new introductions are facilitated by increases and advances in global trade and travel [4,8,9]. Moreover, a linear increase in the establishment of new insect species since 1860 has been documented, including an average of 2.5 nonindigenous forest insects established per year over the last 150 years, 14% of which have caused significant damage to native trees [4]. In particular, phloem- and wood-boring insects have dominated new establishments of nonindigenous insects into forests of the United States since 1980.

The emerald ash borer (EAB), *Agrilus planipennis* (Coleoptera: Buprestidae), is an invasive phloem-feeding beetle native to Asia that attacks and kills native North American species of ash (*Fraxinus* spp.) trees [10]. Unlike Asian ash species that co-evolved varying levels of resistance to EAB,

North American ash species show little resistance to infestation and consequently are more vulnerable to EAB attack, phloem damage, and subsequent tree death [10–12]. Emerald ash borer was discovered in southeastern Michigan, USA and nearby Ontario, Canada in 2002 [13]. Despite early eradication efforts and ongoing quarantines, EAB has now spread to 33 states, the District of Columbia, and three Canadian provinces as of May 2018 [14]. Emerald ash borer attacks ash trees with a diameter at breast height (DBH) >2.5 cm [13,15] and has killed hundreds of millions of ash in urban and natural areas across North America at billions of dollars of estimated costs [16,17]. The loss or reduction of ash in North America will permanently alter forest ecosystems wherever ash represents a significant fraction of tree community composition, and many species that are dependent on ash are likely to be affected. For example, at least 44 species of arthropods within the current range of EAB are known to feed only on ash, including 24 lepidopterans and a number of coleopterans, and are now considered at risk should ash become rare or locally extinct [18–20].

In North America, the mortality rates of EAB caused by native insect natural enemies are relatively low compared to those attacking native species of *Agrilus* and EAB in Asia [20–23]. Consequently, EAB became a candidate for management using classical biological control–the importation and introduction of specialized natural enemies from the pest's native range with the goal of permanent control [24–26]. In northeast China, researchers found three specialized hymenopteran parasitoids attacking EAB: *Tetrastichus planipennisi* Yang (Eulophidae), a larval endoparasitoid; *Spathius agrili* Yang (Braconidae), a larval ectoparasitoid; and *Oobius agrili* Zhang and Huang (Encyrtidae), an egg parasitoid [21,27]. After several years of research in quarantine laboratories in the United States and field studies in China, these parasitoids were found to have narrow host ranges and to reduce EAB populations in China [22,23]. The three parasitoids were first approved for release in Michigan in 2007 [28]. Similar research was completed for another EAB parasitoid, *Spathius galinae* Belokobylskij (Braconidae) from the Russian Far East, and its release was approved in 2015 [29]. To conserve native *Fraxinus* species, the United States Department of Agriculture (USDA) began an EAB biocontrol program in which approved EAB parasitoids are provided to researchers and land managers for release in EAB-infested areas of North America [30]. These cooperators are currently introducing the EAB biocontrol agents in most states and provinces invaded by EAB [26] with all parasitoid release and recovery data maintained in a geospatial database for long-term monitoring [31].

The establishment and spread of *O. agrili* and *T. planipennisi* have been studied at permanent research plots in southern Michigan since releases first began in 2007. The establishment of *S. agrili*, however, has not been confirmed [32,33]. The egg parasitoid *O. agrili*, which kills EAB before any phloem damage occurs regardless of ash diameter, is establishing and slowly expanding its distribution, but its impact on EAB population dynamics has been difficult to quantify [34]. The most dominant of the introduced parasitoids, *T. planipennisi*, is spreading quickly and was found to protect ash saplings and small ash trees up to ~15 cm DBH from EAB damage in forests recovering from the EAB invasion [32,33,35,36]. Releases of *S. galinae* began in Michigan and several other states in 2015. Although the reproduction and spread of *S. galinae* has been documented in Michigan and several other northern states [26], it is too early to confirm sustained establishment. In other regions of North America where early EAB parasitoid-release and recovery data are available, the establishment and spread of *T. planipennisi* and *O. agrili*, and recoveries of *S. galinae*, are also documented [31].

Monitoring changes in ash condition following the invasion of forests by EAB is critical for predicting future changes to the ash component. Initial efforts to assess the biological control impact on ash health, based on the decline rate and survival of ash trees and saplings >5 cm DBH, have detected a significant reduction in EAB density in saplings due to the released biocontrol agent *T. planipennisi*, but have failed to demonstrate the direct effect of the biocontrol agent on ash health and survival [36]. However, higher densities of ash saplings (<5 cm DBH) were found in proximity to EAB biocontrol release plots, suggesting that *T. planipennisi* may be protecting ash saplings from EAB and potentially favoring the recruitment of native hardwood regeneration over invasive plant species [37]. Nevertheless, data collected from long-term monitoring plots are necessary to evaluate the effects of

EAB biological control, especially because ash trees survive for about five years once infested [38]. Results from other biological control programs have shown that there is a considerable lag time for the establishment and population increase of biocontrol agents [39,40]. In particular, baseline information on ash health and regeneration (specifically, monitoring the growth, density, and survival of seedlings, saplings, basal sprouts, some larger survivors, and seed production/germination) is needed to predict the long-term impact that parasitoids will have on EAB, and the potential outcome for ash species in North America.

In this study, we evaluated the impacts of the EAB biological control on ash health and recruitment in southern Lower Michigan in 2012 and again in 2015. This work was done at long-term biocontrol study sites, each comprised of paired biocontrol-release and non-release control plots. Our objectives were to (1) examine and characterize ash health, mortality, and regeneration in these EAB biocontrol study sites; and (2) quantify the initial impacts of EAB biocontrol on ash health conditions and regeneration and provide a framework for long-term comparisons of biocontrol-release and non-release plots. We predicted that plots with parasitoid releases should feature larger and healthier ash trees and higher ash regeneration densities compared to control plots.

2. Materials and Methods

2.1. Study Sites and Field Sampling

We examined ash health and regeneration at nine study sites, with a total of 17 sample plots, in two areas of southern Michigan (Figure 1, Appendixs A and B). Sampling methods for ash health and plot design approximated those described in the Emerald Ash Borer Biological Control Release and Recovery Guidelines [30]. Forest composition was mixed and variable from site to site; ash sampled at study sites was either green ash (*F. pennsylvanica* Marsh.) or white ash (*F. americana* L.) and represented a small fraction of the overstory at the time of sampling (following EAB impact; Appendix A).

In central Michigan, we collected ash health data at five study sites, each consisting of a paired release and non-release control plot. Parasitoids were released between 2007 and 2010; EAB densities and overstory ash mortality peaked in 2010–2011 in this region (Appendix B) [32,34,36]. Results of annual sampling of infested ash trees at the central Michigan study sites since 2008 revealed that larval parasitism by *T. planipennisi* increased on average from ~1% to ~21% in release plots from 2009 to 2012, and from 0.2 to ~13% in control plots during this same period [32]. More recently, larval parasitism by *T. planipennisi* has spread more evenly across the control and release plots, although prevalence varies from year to year [33,36] [LSB,JJD unpublished data]. Egg parasitism by *O. agrili* was also detected in 2008 in release plots one year after first release, but not until 2012 in control plots. In 2012, egg parasitism averaged ~20% and ~5% in the release and control plots, respectively, and was variable between trees and sampling years [34,41] [LSB,JJD unpublished data].

In southeastern Michigan, ash health was evaluated at four release and three control plots (Figure 1) where the mortality of overstory ash had peaked in about 2007 across this region [42–44]. The EAB parasitoids were released in 2011 onto young ash trees, saplings, and basal sprouts, although EAB densities appeared low. In the winter of 2014–2015, ash trees were sampled for egg and larval parasitism, and the establishment and spread of *T. planipennisi* at both the release and control plots was confirmed [LSB, JRG unpublished data]. Larval parasitism by *T. planipennisi* was highly variable between sites and averaged ~26% and ~4% at release and control plots, respectively. Egg parasitism by *O. agrili* was confirmed at one release plot (Lake Erie Metropark) at a rate of ~6% [LSB, JRG unpublished data].

At each site, ash health and regeneration data were collected in 400 × 400 m release and control plots, each subdivided into 50-m^2 grid cells, which allowed for a spatial reference of sampling in relation to parasitoid-release epicenter trees [30]. In 2012, 25 live ash trees >4 cm DBH (hereafter called "large trees") were randomly selected beginning near the center of the 400 m^2 grid. Trees were measured, permanently numbered with aluminum tags, and georeferenced with a GPS unit for re-measurement and assessment in subsequent sampling years. Live ash trees were identified to species and their DBH measured. Evidence of EAB attack in live ash trees was assessed using the identification of external symptoms, including EAB exit holes, epicormic sprouting, bark splitting, and woodpecker feeding [30,45]. Crown dieback and crown class were evaluated using a five-class categorical scale where 1 is a healthy tree and 5 is a dead tree [45]. This process was repeated for 25 sub-canopy trees and saplings 2.5–4.0 cm DBH (hereafter called "small trees"). Ash regeneration (all living ash <2.5 cm DBH or shorter than breast height) was sampled along two perpendicular belt transects that intersected the center point of the grid; transect length varied across sample plots but included a minimum of 300 m of total transect length at all plots. Ash regeneration was tallied into three height classes (<0.5 m, 0.5–1 m, >1 m) that served as a proxy for seedling/sapling age and potential for recruitment into the canopy [42]. Sampling was conducted at all sites in 2012 and then repeated in 2015.

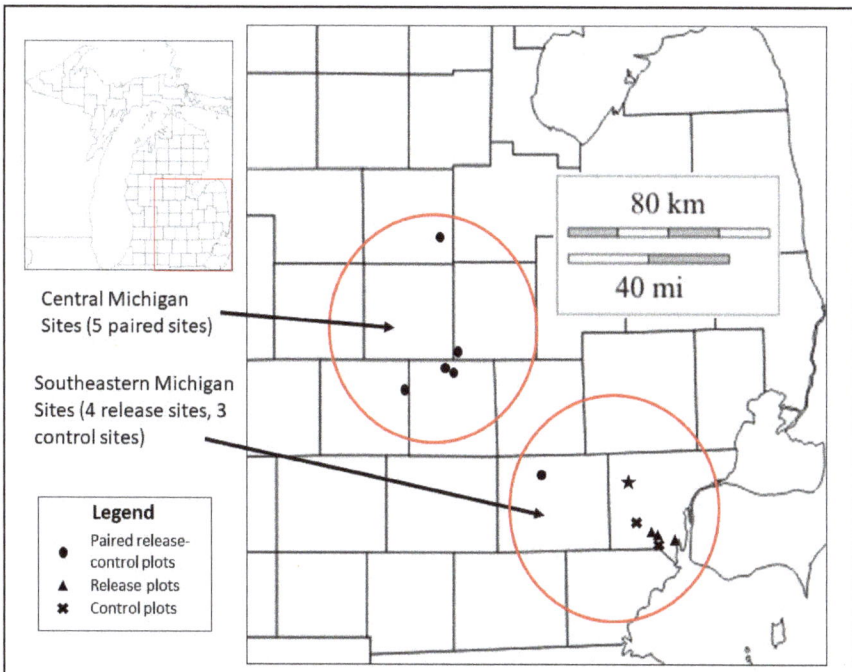

Figure 1. Location of study sites in two regions of southern Michigan where ash health data collection occurred in 2012 and 2015. Central Michigan sites include five paired release and control plots with parasitoid releases from 2007 to 2010; southeastern Michigan sites include four release and three control plots with parasitoid releases in 2011 (Table 1, Appendixs A and B). The star indicates the approximate location of the EAB introduction point in Wayne County, Michigan [44].

2.2. Data Analysis

All data were pooled by region (central Michigan and southeastern Michigan, Figure 1) for analysis to coincide with parasitoid release dates and presumed effects on ash tree health and regeneration. Site was the experimental unit within each region ($n = 5$ in central Michigan; $n = 4$ in southeastern Michigan). Four of the five central Michigan study sites had parasitoid releases beginning in 2007, 2008, or 2009, three to five years prior to sampling; parasitoid releases occurred in 2010 in the fifth central Michigan site (Rose Lake; Table 1). Parasitoid establishment was confirmed at all central Michigan sites by 2012, and the spread of both *T. planipennisi* and *O. agrili* into control plots was observed by 2015 [32,33]. The southeastern Michigan sites included three control plots and four release plots (Lower Huron Metropark served as the control plot for both Willow Metropark North and Willow Metropark South release plots). Parasitoids were released at all southeastern Michigan plots in 2011, the year prior to initial sampling, and thus data sampled in southeastern Michigan in 2012 are probably more representative of pre-release conditions because parasitoid establishment was likely low or absent. This experimental design allowed us to make relevant comparisons between release and control plots in each year of sampling (2012 and 2015), but also within release and control plots across sample years (e.g., release plots in 2012 vs. 2015, control plots in 2012 vs. 2015). Therefore, we were able to directly examine the effect of the treatment (parasitoid release and establishment) each sample year. We were also able to examine changes in ash health and regeneration over a three-year period using time as the effect.

Table 1. EAB biological control study sites in southern Lower Michigan.

Site	Location	Treatment	Year Parasitoid Releases Began
Central Michigan			
Central Park	Okemos, MI	Paired	2007
Legg Park	Okemos, MI	Paired	2008
Burchfield Park	Holt, MI	Paired	2008
Gratiot-Saginaw West	Ashley, MI	Paired	2009
Rose Lake	Lansing, MI	Paired	2010
Southeastern Michigan			
Lake Erie Metropark *	Brownstown, MI	Release	2011
Oakwoods Metropark *	Flat Rock, MI	Control	
Pinckney Recreation Area	Pinckney, MI	Paired	2011
Willow Metropark North **	New Boston, MI	Release	2011
Willow Metropark South **	New Boston, MI	Release	2011
Lower Huron Metropark **	Belleville, MI	Control	

* paired release and control plots. ** two release plots paired with one control plot.

Statistical analyses were conducted using IBM SPSS statistics version 23 [46]. All data were examined for normality and equal variances and analyzed using $\alpha = 0.05$. Diameters of large trees (>4.0 cm DBH) and small trees (2.5–4.0 cm DBH) could not be normalized; differences in tree diameters between release and control plots within and among sample years were therefore analyzed with the nonparametric Wilcoxon ranked sign test. Counts of trees dying between sample years were compared for release and control plots using a chi-square test for heterogeneity. We compared changes in the health of large and small trees using crown condition class ratings and the presence of other EAB symptoms. Crown class ratings were ordinal data and were thus analyzed with a nonparametric Kruskal-Wallis test. When significant differences were found, pairwise comparisons were made with the Mann Whitney U test with a Bonferroni correction. We analyzed differences in the proportion of sampled living large and small trees exhibiting signs of EAB infestation using a chi-square test for heterogeneity. Seedling counts within belt transects were converted to density (stems/ha) and analyzed separately for each height class. Density values for seedlings <0.5 m were transformed using the formula [square root (x + 0.5)] to correct for normality and unequal variances and analyzed with a paired sample t-test. Density values for seedlings 0.5–1 m and >1 m could not be normalized, and were analyzed using a Wilcoxon signed rank test.

3. Results

3.1. Ash Tree Diameters and Mortality

In the central Michigan sites, where parasitoids were released from 2007 to 2010, the mean diameter of living large trees did not differ between release and control plots in either 2012 or 2015; large tree diameter also did not differ within release or control plots among years (Figure 2). However, a higher proportion of large trees in control plots (20.8%) died between 2012 and 2015 than in release plots (3.2%; $\chi^2 = 18.3$, $p < 0.001$) (Figure 3). Mean diameter of small trees (2.5–4.0 cm DBH) was higher in control plots (3.1 cm) in 2012 compared to release plots (2.9 cm; $Z = 2886.5$, $p < 0.001$). The mean diameter of small trees was higher in 2015 than 2012 for both release plots (3.6 vs. 2.9 cm; $Z = 3371$, $p < 0.001$) and for control plots (3.8 vs. 3.1 cm; $Z = 3257.5$, $p < 0.001$), consistent with expected small tree growth over the three-year sampling period (Figure 2). The proportion of small trees that died between 2012 and 2015 was higher in control plots (Figure 3), and this difference was nearly significant ($p = 0.058$).

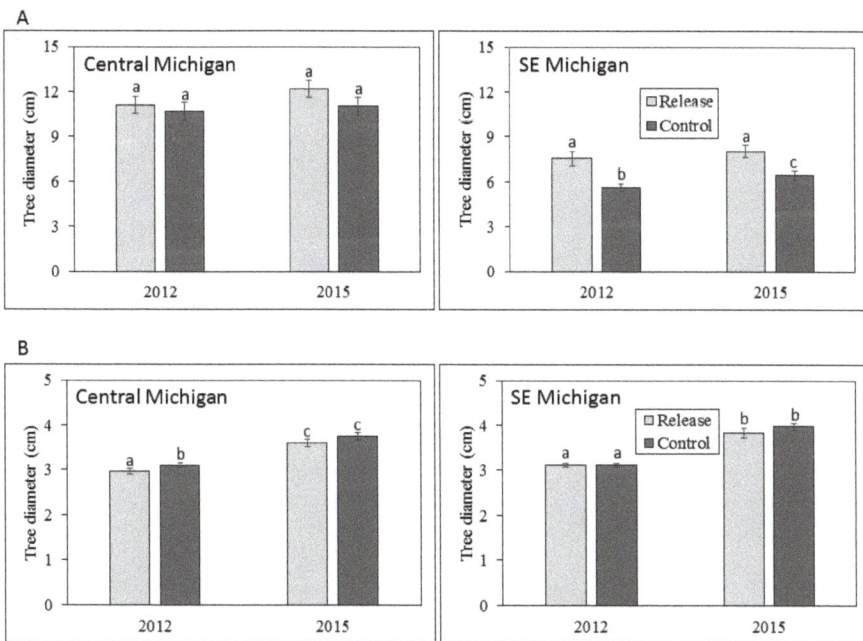

Figure 2. Mean tree diameters at breast height (DBH) for (**A**) large trees >4.0 cm DBH and (**B**) small trees 2.5–4.0 cm DBH in central and southeastern Michigan in 2012 and 2015. Error bars represent ± 1 SE. Mean values with the same letter are not significantly different.

In southeastern Michigan, where parasitoid releases occurred in 2011, mean large tree diameter was higher in release sites compared to control sites in both 2012 (7.6 vs. 5.6 cm; $Z = 1165.5$, $p = 0.016$) and in 2015 (8.0 vs. 6.4 cm; $Z = 431.5$, $p = 0.019$). In addition, mean large tree diameter was higher in 2015 than in 2012 for control sites (6.4 vs. 5.6 cm; $Z = 63$, $p < 0.001$) but not release sites. A much greater proportion of large trees died between 2012 and 2015 in control plots (72%) vs. release plots (3%; $\chi^2 = 92.9$, $p < 0.001$) (Figure 3). Mean diameter of small trees did not differ between release and control sites, but similar to that of large trees, mean diameter was higher in 2015 than in 2012 for both release sites (3.8 vs. 3.1 cm; $Z = 287.5$, $p < 0.001$) and control sites (4.0 vs. 3.1 cm; $Z = 2239$, $p < 0.001$; Figure 2). The proportion of small trees that died between 2012 and 2015 did not differ between release and control plots (Figure 3).

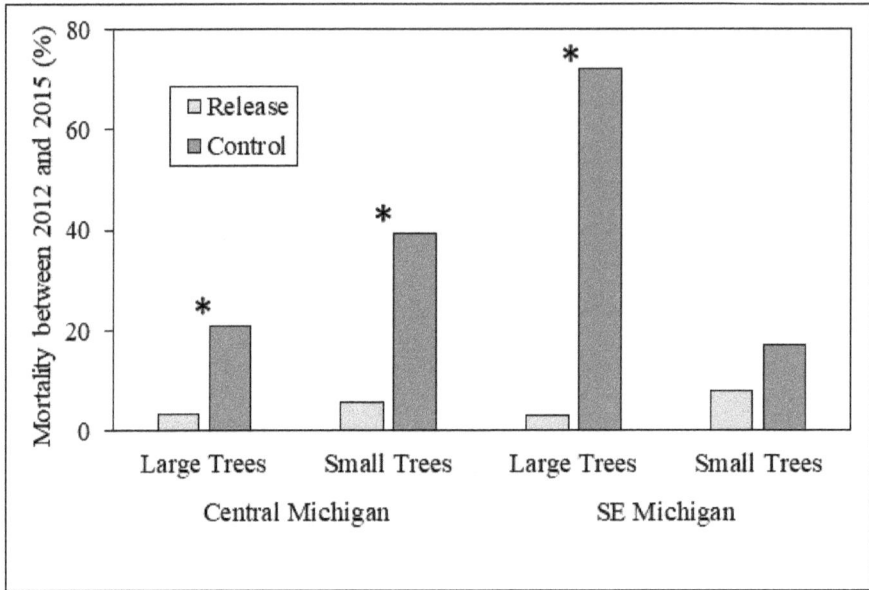

Figure 3. Tree mortality between 2012 and 2015 in release and control plots in central and southeastern Michigan study sites. Differences between release and control plots were significant for large trees in both central and southeastern Michigan ($p < 0.001$ for both), and were nearly significant ($p = 0.058$) in central Michigan for small trees. Asterisks indicate significantly different comparisons between release and control plots.

3.2. Ash Tree Health

In central Michigan, large tree crown class ratings did not differ between release and control sites in either 2012 or 2015 (Table 2). However, the crown condition for large trees worsened significantly from 2012 to 2015 in both release (2.6 to 3.5; $U = 3017$, $p < 0.001$) and control plots (2.3 to 3.3; $U = 2625$, $p < 0.001$), despite the recently established parasitoids.

Table 2. Mean (± 1 SE) crown condition ratings for large and small trees in release and control plots in central and southeastern Michigan in 2012 and 2015. Crown condition ratings range from "1" for a healthy tree to "5" for a dead tree.

Central Michigan	Release	Control
Large Trees 2012	2.60 (0.11) [a]	2.34 (0.14) [a]
Large Trees 2015	3.49 (0.15) [b]	3.27 (0.21) [b]
Small Trees 2012	1.48 (0.08) [1]	1.42 (0.07) [1]
Small Trees 2015	2.30 (0.12) [2]	2.14 (0.12) [2]
Southeastern Michigan	Release	Control
Large Trees 2012	1.83 (0.08) [a]	2.05 (0.16) [a]
Large Trees 2015	3.46 (0.15) [b]	3.53 (0.15) [b]
Small Trees 2012	1.80 (0.09) [1]	1.85 (0.12) [1]
Small Trees 2015	3.20 (0.14) [2]	2.68 (0.16) [3]

Differing superscripts (letters for large trees and numbers for small trees) represent significant differences at $\alpha = 0.05$.

Other symptoms of EAB infestation (EAB exit holes, epicormic sprouting, bark splitting, and/or woodpecker feeding) were found on a higher proportion of large trees in the release plots compared to the control plots in 2012 (86.4 vs. 72.8%; $\chi^2 = 7.12$, $p = 0.007$) but not in 2015 (Table 3). EAB symptoms were found on a higher percentage of large trees in 2015 compared to 2012 in both release plots (97.5 vs. 86.4%; $\chi^2 = 10.18$, $p = 0.001$) and control plots (94.9 vs. 72.8%; $\chi^2 = 18.8$, $p < 0.001$), consistent with our expectation that EAB infestation would continue over the three-year study period. Similar to that of large trees, the crown class rating of small trees did not differ between release and control plots in 2012 or 2015 (Table 2).

Table 3. Mean proportions of trees exhibiting symptoms of EAB infestation (see text for explanation) for large and small trees in release and control plots in central and southeastern Michigan in 2012 and 2015. Differing superscripts (letters for large trees and numbers for small trees) represent significant differences at $\alpha = 0.05$.

Central Michigan	Release	Control
Large Trees 2012	86.4 [a]	72.8 [b]
Large Trees 2015	97.5 [c]	94.9 [c]
Small Trees 2012	48.8 [1]	37.6 [1]
Small Trees 2015	74.6 [2]	89.5 [3]
Southeastern Michigan	Release	Control
Large Trees 2012	85.0 [a]	70.7 [b]
Large Trees 2015	90.7 [a]	98.1 [a]
Small Trees 2012	61.0 [1]	50.7 [1]
Small Trees 2015	84.3 [2]	85.5 [2]

Small tree crown class rating worsened between 2012 and 2015 for both release (1.5 to 2.3; $U = 4252.5$, $p < 0.001$) and control plots (1.4 to 2.1; $U = 3292.5$, $p < 0.001$). The proportion of small trees exhibiting EAB symptoms did not differ between release and control plots in central Michigan in 2012, but was higher in control plots in 2015 (89.5 vs. 74.6, $\chi^2 = 6.5$, $p = 0.01$; Table 3). The proportion of infested trees increased between 2012 and 2015 for both release plots (48.8 to 74.6%; $\chi^2 = 17.0$, $p < 0.001$) and control plots (37.6 to 89.5%; $\chi^2 = 52.0$, $p < 0.001$).

In southeastern Michigan, trends in ash tree health were similar to those in central Michigan. Large tree crown class ratings did not differ between release and control plots in southeastern Michigan in either 2012 or 2015 (Table 2). Large tree crown condition declined from 2012 to 2015 in both release (1.8 to 3.5, $U = 2209.5$, $p < 0.001$) and control plots (1.9 to 2.7, $U = 894$, $p < 0.001$). Symptoms of EAB infestation were found on a higher proportion of large trees in the release plots (85%) compared to the control plots (70.7%) in 2012 ($\chi^2 = 5.28$, $p = 0.02$), but not in 2015 (Table 3). EAB symptoms were found on a higher proportion of large trees in 2015 compared to 2012 in control plots ($\chi^2 = 16.2$; $p < 0.001$), but not in release plots. Small tree crown class rating was lower in release plots (3.2) compared to control plots (2.7) in 2015 ($U = 2223.5$, $p = 0.15$) but did not differ in 2012 (Table 2). Similar to large trees, crown class rating for small trees declined from 2012 to 2015 for both control plots (1.9 to 2.7, $U = 1635$, $p < 0.001$) and release plots (1.8 to 3.2, $U = 1671.5$, $p < 0.001$). The proportion of small trees with EAB symptoms did not differ between release and control plots in 2012 or 2015 (Table 3), but the proportion of infested trees increased between 2012 and 2015 for both release plots (61 to 84.3%; $\chi^2 = 12.1$, $p < 0.001$) and control plots (50.7 to 85.5%; $\chi^2 = 19.8$, $p < 0.001$).

3.3. Ash Regeneration

We found no differences in the densities of ash seedlings and saplings of any height class between release and control sites in either year of sampling, or among sample years. Extremely variable seedling densities among sites and years created high standard errors that led to a lack of statistical significance given the sample sizes available within our experimental design. Notably, however, seedling densities were consistently higher in release plots compared to the control plots, with two exceptions in the twelve comparisons (seedlings <0.5 m tall in central Michigan, and seedlings >1 m tall in southeastern Michigan; Figure 4). Seedlings were less abundant in successively taller size height classes in both central and southeastern Michigan, but did not show consistent trends across sample years (Figure 4).

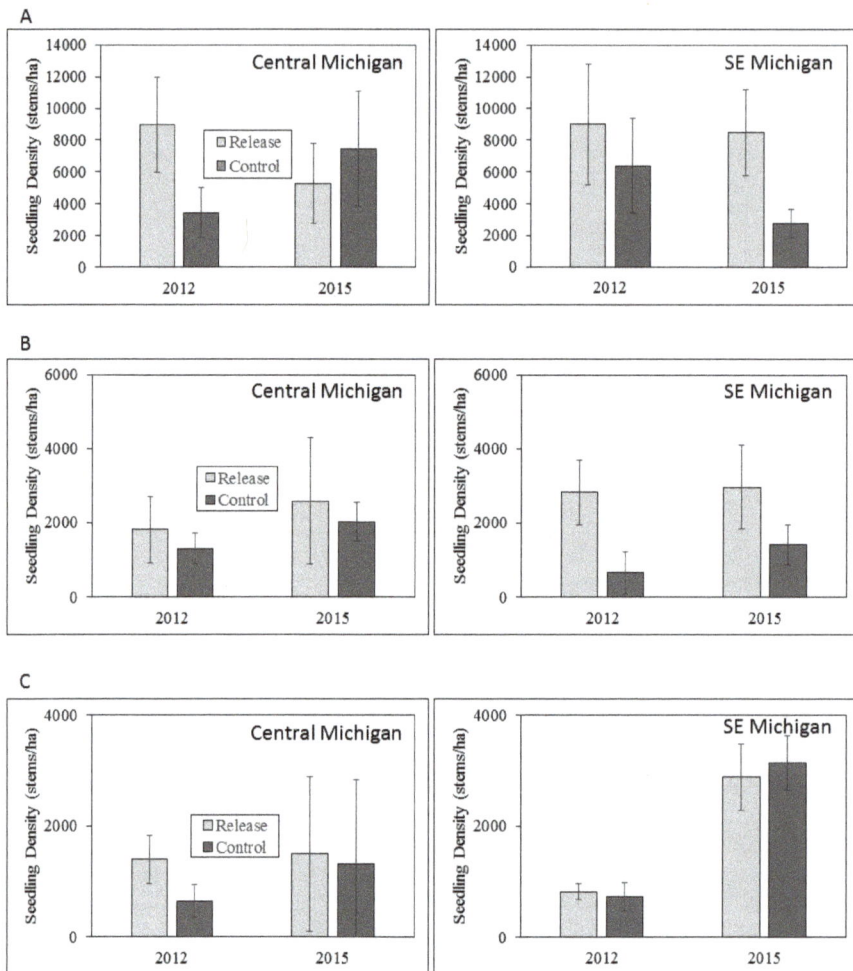

Figure 4. Mean seedling densities for (**A**) seedlings <0.5 m tall; (**B**) seedlings 0.5 – 1.0 m tall; and (**C**) >1 m tall in central and southeastern Michigan in 2012 and 2015. Error bars represent ± 1 SE. Note differences in scale of Y axis among seedling height classes.

4. Discussion

The introduction and establishment of insect biocontrol agents requires monitoring and documenting impacts of the new natural enemy on the target pest. However, a comprehensive understanding of the subsequent changes in the impacts of the target pest on the host plant is critical for determining the future viability of the vegetation [24]. In the case of EAB, ash trees survive for about five years after infestation [38], whereas the lag time for parasitoid establishment and population increase is considerably longer. Thus it remains difficult to precisely assess the impacts of EAB biocontrol on ash populations. Moreover, up to 99% of the overstory ash trees were killed by high EAB population densities in southeastern Michigan by 2009 [43]. This widespread loss of most EAB host ash trees, as well as predation of EAB larvae by woodpeckers and some native parasitoids [23,33], resulted in the collapse of EAB populations below a density threshold that is currently favoring the survival and growth of small ash trees, basal sprouts, saplings, and seedlings [33,43,47]. Parasitoids introduced as part of biological control programs should therefore be effective in supporting the survival of sub-canopy and regenerating ash [36,37]. As such, long-term studies that assess and monitor the survival, growth, and density of ash at all its life stages are necessary to understand the potential for ash in North America to persist in the presence of EAB.

4.1. Ash Diameter and Health

Our work provided very mixed results in terms of the potential impacts of the EAB biocontrol on ash diameter and health. Optimally, impacts of biological control would result in the survival of both larger and smaller trees, and maintain the health of living ash trees (this would have been indicated in our study by the relative conditions of tree crowns and the presence or absence of EAB symptoms). At the central Michigan sites, large tree diameter did not differ between release and control plots and did not appear to increase over the three-year sampling period. These data suggest that the growth of larger trees was minimal, possibly due to EAB infestation. Small trees in central Michigan grew to larger diameters over the three years of the study, but were larger in control plots than release plots. Moreover, crown condition declined on large trees between 2012 and 2015 in both release and control plots, and parasitoid releases did not appear to reduce the evidence of EAB infestation.

Results were more similar to our expectations in southeastern Michigan, where both large and small trees exhibited significant diameter growth over the three-year period, and large tree diameter was higher in release compared to control plots in both 2012 and 2015 (Figure 2). Stronger evidence of biocontrol effects in southeastern Michigan was somewhat unexpected because parasitoid releases occurred in 2011, only the year before initial sampling. These observed differences in southeastern Michigan by 2015 may be explained by the high prevalence of *T. planipennisi* in the release plots compared to the control plots, and *O. agrili* had not spread beyond the release plots. In central Michigan, however, differences in ash growth between release to the control plots were not found because both *T. planipennisi* and *O. agrili* had spread throughout the area by 2012 [26,32,34,41]. Perhaps the best evidence for positive biocontrol effects in our study is that fewer large trees died between 2012 and 2015 in release plots compared to control plots in both central and southeastern Michigan. Our sampling was not done on an annual basis, and thus we are unable to document the timing of large tree death between 2012 and 2015 or to ascertain whether parasitoid establishment was directly correlated to reduced tree mortality, but the three-year trends are promising. Although tree diameter impacts are difficult to ascertain in a three-year study, trends in biocontrol impacts on ash diameters appear to be mixed at best.

4.2. Ash Regeneration

Our experimental design was insufficient to detect statistical differences in ash regeneration, but we expected that the survival and improved health of mature ash trees due to EAB biocontrol, if present, should increase ash seed rain such that biocontrol should indirectly increase the densities of seedlings and saplings [47]. We highlight that despite high standard errors, seedling densities were substantially higher in release plots compared to controls for most height classes in both central and southeastern Michigan. Given the lack of significant improvement of ash tree health or survival at our study sites, these trends support the conclusions of other researchers that ash seedlings and saplings are likely protected by the establishment of *T. planipennisi* [36,37]. In this context, EAB biocontrol agents may have a direct positive effect on ash sapling and seedling survival. Assuming that at least some of the ash regeneration and small trees at our study sites will grow to diameters >16 cm, when their bark becomes too thick for *T. planipennisi* to reach EAB larvae feeding in the phloem [35], expanded releases of *S. galinae* are warranted because they can attack EAB larvae in ash trees up to ~57 cm DBH [48]. We also note that browsing pressure by deer in both regions is likely to have a confounding effect on ash regeneration data in this and other studies [49]. Deer browsing appears to be especially prevalent in central Michigan (LSB, pers. observ) compared to southeastern Michigan [42], and could have important implications for ash persistence [49].

4.3. Future Research

The premise of EAB biocontrol is that the introduced parasitoids, once established, will work together with other biotic and abiotic mortality factors to maintain the EAB population below a threshold level such that at least some small ash trees and regeneration will be able to recruit to the overstory, thus maintaining a seed source and allowing ash to persist. Several studies have documented the persistence of blue ash (*F. quadrangulata* Michx.) [11,12], green ash [47], and white ash [12] in small populations in southeastern Michigan, not necessarily in the proximity of EAB parasitoid releases. Thus if even only partially successful, EAB biocontrol may have an important impact on ash persistence in an area where EAB populations have peaked and declined. Studies that document the degree of success of EAB biocontrol, therefore, are critical for predicting longer-term EAB impacts in areas where EAB populations have not yet peaked. To our knowledge, ours is the only empirical study to monitor ash health at all ash life stages as an attempt to directly assess the effects of classic biological control on the ash component of forests within the current range of EAB.

Despite the presence of some strong trends that suggest a positive impact of biocontrol on ash persistence both in the literature [36,37] and in our study, we suggest that it remains too soon to assess the effect of biocontrol on ash health and regeneration. The temporal scale of forest monitoring studies in general is longer than three years, given the life span of trees and their rates of growth and mortality. Particularly with ash responses to EAB, for example, a three-year monitoring study would be unable to capture trends in basal sprouting from killed ash trees, which has been shown to be an important vector of post-EAB ash persistence following a significant lag time before sprouts are observable [47]. Monitoring studies should also be long enough to capture the temporal dynamics of established parasitoids and EAB populations in the presence of the parasitoids, which a three-year study is unlikely to do. Moreover, we caution that our study lacks the pre-release vegetation data necessary to make short-term conclusions about the effects of EAB biocontrol on ash health and regeneration. Long-term studies of vegetation responses to biocontrol programs would reduce the importance of pre-release data once trends in the data become evident. Until that time and without pre-release data for comparison, our short-term results should be considered preliminary.

Finally, we emphasize that despite the susceptibility of all North American ash species to EAB, studies that document and monitor the response of ash to a biocontrol should be careful to examine ash response on a species-by-species basis. Our study examines green and white ash, which themselves have very different life history traits and are likely to respond quite differently to EAB [47] and thus EAB biocontrol in the long term. Other species of ash in the region, including black ash (*F. nigra* Marsh.) and pumpkin ash (*F. profunda* (Bush) Bush), are rarer in southern Michigan and were not sampled. Black ash is of significant concern, both for its cultural significance [50] and its dominance in other geographical regions [51], but both its site-species relationships and its life history traits differ from the other species to the extent that the effect of EAB biocontrol may also differ, and thus future EAB biocontrol research and monitoring should target this important species of *Fraxinus*.

5. Conclusions

Our data suggest that EAB biocontrol in southern Michigan is likely to have a positive effect on ash populations. Trends in our data that suggest the establishment of EAB parasitoids may reduce the death of large ash trees are promising, though inconsistent. Similarly, higher densities of ash regeneration in most release plots suggests that EAB biocontrol and its presumed protection of ash seedlings and saplings may provide an important avenue for future ash recruitment into the canopy.

In either case, however, the study duration is currently too short to definitively deduce the long-term success of EAB biocontrol in this region, and additional data collection over a longer time period is therefore critical for EAB and ash management in other regions.

Author Contributions: D.M.K. and L.S.B. conceived and designed the study; D.M.K., L.S.B., B.A.S., J.J.D., and J.R.G. collected and analyzed the data; D.M.K. and L.S.B. wrote the paper.

Acknowledgments: This work was supported by USDA Forest Service Joint Venture Agreement 12-JV-11242303-053 and the Department of Biological Sciences at Wayne State University. We gratefully acknowledge the hard work by our field crews in 2012 and 2015, which included Jake Dombrowski, Dana Sugar, Clare Surmont, and Vanessa Verstraete; and we especially wish to acknowledge Doug Putt, who was instrumental in leading the 2015 field crew and entering all of the data (Wayne State University, Detroit, MI, USA). We also thank Deborah Miller (USDA FS NRS, Lansing, MI, USA) and Jonathan Lelito (USDA APHIS PPQ, Brighton, MI, USA) for providing many of the EAB parasitoids released at these study sites; Benjamin Sorensen (USDA APHIS CPHST, Brighton, MI USA) for help with parasitoid recovery in southeast Michigan; Kim Hoelmer (USDA ARS, Newark, DE, USA) and Noah Koller (USDA ARS, E. Lansing, MI, USA) for helpful reviews of this manuscript; and lastly, the tremendous encouragement and support provided by the late Kurt Gottschalk, USDA FS NRS Research Leader. We also thank the Michigan Department of Natural Resources, Meridian Township, Ingham County, and Clinton-Huron Metroparks for long-term use of the lands used in this study.

Conflicts of Interest: The authors declare no conflict of interest. The founding sponsors had no role in the design of the study; in the collection, analyses, or interpretation of data; in the writing of the manuscript, and in the decision to publish the results.

Appendix A.

Table A1. Dominant overstory vegetation at study sites in central and southeastern Michigan. Ash relative dominance was measured at the time of sampling (i.e., after EAB-caused ash mortality occurred).

Site Name *	Release-Plot Codes **	Treatment ***	Ash Relative Dominance (%)	Dominant Overstory Tree Species
Central Michigan				
Central Park	CPNMLT	Release	3.7	American elm, boxelder
		Control	8.3	American elm
Legg Park	LPRFLT	Release	2.3	Silver maple, American elm, hackberry
		Control	1.0	Silver maple, sycamore, American elm
Burchfield	BFPKLT	Release	3.8	Silver maple, American elm
		Control	3.8	Silver maple, American elm
Gratiot-Saginaw	GSW	Release	2.1	Red maple, red oak
		Control	1.9	Red maple, red oak
Rose Lake	RL	Release	1.7	Red maple, basswood, American elm
		Control	1.1	Red maple, black cherry, American elm
Southeastern Michigan				
Lake Erie	LKEMP	Release	0.1	American elm, silver maple
Oakwoods		Control	0.1	Red maple, American elm, shagbark hickory
Pinckney	PNKSL	Release	0.1	Red oak, red maple
		Control	0.1	Red oak, red maple
Willow North	WMPKN	Release	0.1	American elm
Willow South	WMPKS	Release	4.0	Eastern cottonwood
Lower Huron		Control	0.1	Silver maple, American elm

* See Table 1 and Appendix B for more information on locations. ** Codes used for release plots in Mapbiocontrol.org [31]. *** See Appendix B for more information on parasitoid releases.

Appendix B.

Table A2. Summary of adult *O. agrili* and *T. planipennisi* females released at the study release plots in central and southeastern Michigan.

Site Codes	Year	O. agrili Released			T. planipennisi Released		
		Month(s)	Total N (Females)	No. of Releases	Month(s)	Total N (Females)	No. of Releases
Central Michigan *							
CPNMLT	2007	August	700	6	September–October	870	3
	2008	June–August	330	4	June	150	1
	2009	June	300	1	June–October	3000	9
LPRFLT	2008	July	200	2	September–October	200	5
	2009	June	300	1	May–September	3250	10
BFPKLT	2008	July–August	200	2	July	110	1
	2009	June	300	1	May–September	3200	9
GSW	2009	August	375	1	August–September	700	2
	2010	June–July	1110	2	June–September	3290	6
RL	2010	July	1160	2	June–September	3880	7
SE Michigan **							
LKEMP	2011	June	240	1	June	1740	3
PNKSL	2011	July–Aug	300	3	July	710	2
WMPKN	2011	June–July	180	2	June–July	1830	2
WMPKS	2011	June–July	160	2	June–July	400	2

* Site codes for data reported in MapBiocontrol.org [31] and earlier publications [32–37,41]: CPNMLT (Central-Nancy Moore Parks) and LPRFLT (Legg Park-River Front Parks), Meridian Township Parks, Ingham County, Okemos, MI; BFPKLT (Burchfield Park), Ingham County Parks, Holt, MI; GSW (Gratiot-Saginaw State Game Area), Michigan Department of Natural Resources, Gratiot County, Ashley, MI; RL (Rose Lake State Wildlife Area), Michigan Department of Natural Resources, Shiawassee County, Bath, MI. ** Site codes for data in MapBiocontrol.org [31]: LKEMP (Lake Erie Metropark), Wayne County, Brownstown, MI; PNKSL (Pinckney Recreation Area), Washtenaw County, Pinckney, MI; WMPKN and WMPKS (Willow Metropark), Wayne County, New Boston, MI.

References

1. Mattson, W.; Vanhanen, H.; Veteli, T.; Sivonen, S.; Niemelä, P. Few immigrant phytophagous insects on woody plants in Europe: Legacy of the European crucible? *Biol. Invasions* **2007**, *9*, 957–974. [CrossRef]
2. Liebhold, A.M.; MacDonald, W.L.; Bergdahl, D.; Mastro, V.C. Invasion by exotic forest pests: A threat to forest ecosystems. *For. Sci. Monogr.* **1995**, *41*, 1–49. [CrossRef]
3. Simberloff, D. Nonindigenous species: A global threat to biodiversity and stability. In *Nature and Human Society: The Quest for a Sustainable World*; Raven, P., Williams, T., Eds.; National Academy Press: Washington, DC, USA, 2000; pp. 325–336.
4. Aukema, J.E.; McCullough, D.G.; Von Holle, B.; Liebhold, A.M.; Britton, K.; Frankel, S.J. Historical accumulation of non-indigenous forest pests in the continental United States. *BioScience* **2010**, *60*, 886–897. [CrossRef]
5. Lovett, G.M.; Weiss, M.; Liebhold, A.M.; Holmes, T.; Leung, B.; Lambert, K.F.; Orwig, D.A.; Campbell, F.T.; Rosenthal, J.; McCullough, D.G.; et al. Nonnative forest insects and pathogens in the United States: Impacts and policy options. *Ecol. Soc. Am.* **2016**, *26*, 1437–1455. [CrossRef] [PubMed]
6. Kenis, M.; Hurley, B.P.; Hajek, A.E.; Cock, M.J.W. Classical biological control of insect pests of trees: Facts and figures. *Biol. Invasions* **2017**, *19*, 3401–3417. [CrossRef]
7. Sadof, C.S.; Hughes, G.P.; Witte, A.R.; Peterson, D.J.; Ginzel, M.D. Tools for staging and managing emerald ash borer in the urban forest. *Arboric. Urban For.* **2017**, *43*, 15–26.
8. Liebhold, A.M.; McCullough, D.G.; Blackburn, L.M.; Frankel, S.J.; Von Holle, B.; Aukema, J.E. A highly aggregated geographical distribution of forest pest invasions in the USA. *Divers. Distrib.* **2013**, *19*, 1208–1216. [CrossRef]
9. Levine, J.M.; D'Antonio, C.M. Forecasting biological invasions with increasing international trade. *Conserv. Biol.* **2003**, *17*, 322–326. [CrossRef]
10. Rebek, E.J.; Herms, D.A.; Smitley, D.R. Interspecific variation in resistance to emerald ash borer (Coleoptera: Buprestidae) among North American and Asian ash (*Fraxinus* spp.). *Environ. Entomol.* **2008**, *37*, 242–246. [CrossRef]

11. Tanis, S.R.; McCullough, D.G. Differential persistence of blue ash (*Fraxinus quadrangulata*) and white ash (*Fraxinus americana*) following emerald ash borer (*Agrilus planipennis*) invasion. *Can. J. For. Res.* **2012**, *42*, 1542–1550. [CrossRef]

12. Spei, B.A.; Kashian, D.M. Potential for persistence of blue ash in the presence of emerald ash borer in southeastern Michigan. *For. Ecol. Manag.* **2017**, *392*, 137–143. [CrossRef]

13. Cappaert, D.; McCullough, D.G.; Poland, T.M.; Siegert, N.W. Emerald ash borer in North America: A research and regulatory challenge. *Am. Entomol.* **2005**, *51*, 152–165. [CrossRef]

14. USDA–APHIS. Initial County EAB Detection Map. 2018. Available online: https://www.aphis.usda.gov/plant_health/plant_pest_info/emerald_ash_b/downloads/MultiState.pdf (accessed on 15 May 2018).

15. Marshall, J.M.; Smith, E.L.; Mech, R. Estimates of *Agrilus planipennis* infestation rates and potential survival of ash. *Am. Midland Nat.* **2013**, *169*, 179–193. [CrossRef]

16. Nowak, D.; Crane, D.; Stevens, J.; Walton, J. *Potential Damage from Emerald Ash Borer*; United States Department of Agriculture, Forest Service, Northern Research Station: Syracuse, NY, USA, 2003; pp. 1–5. Available online: https://www.nrs.fs.fed.us/disturbance/invasive_species/eab/local-resources/downloads/EAB_potential.pdf (accessed on 20 March 2018).

17. Herms, D.A.; McCullough, D.G. Emerald ash borer invasion of North America: History, biology, ecology, impacts, and management. *Ann. Rev. Entomol.* **2014**, *59*, 13–30. [CrossRef] [PubMed]

18. Wagner, D.L.; Todd, K.J. New ecological assessment for the emerald ash borer: A cautionary tale about unvetted host-plant literature. *Am. Entomol.* **2016**, *62*, 26–35. [CrossRef]

19. Jennings, D.E.; Duan, J.J.; Bean, D.; Kimberly, A.R.; Williams, G.L.; Bells, S.K.; Shurtleff, A.S.; Shrewsbury, P.M. Effects of the emerald ash borer invasion on the community composition of arthropods associated with ash tree boles in Maryland, U.S.A. *Agric. For. Entomol.* **2017**, *19*, 122–129. [CrossRef]

20. Taylor, P.B.; Duan, J.J.; Fuester, R.W.; Hoddle, M.; Van Driesche, R.G. Parasitoid Guilds of *Agrilus* Woodborers Coleoptera: Buprestidae: Their Diversity and Potential for Use in Biological Control. *Psyche* **2012**, *2012*, 813929.

21. Liu, H.; Bauer, L.S.; Gao, R.; Zhao, T.; Petrice, T.R.; Haack, R.A. Exploratory survey for the emerald ash borer, *Agrilus planipennis* (Coleoptera: Buprestidae), and its natural enemies in China. *Great Lakes Entomol.* **2003**, *36*, 191–204.

22. Liu, H.; Bauer, L.S.; Miller, D.L.; Zhao, T.; Gao, R.; Song, L.; Luan, Q.; Jin, R.; Gao, C. Seasonal abundance of *Agrilus planipennis* (Coleoptera: Buprestidae) and its natural enemies *Oobius agrili* (Hymenoptera: Encyrtidae) and *Tetrastichus planipennisi* (Hymenoptera: Eulophidae) in China. *Biol. Control* **2007**, *42*, 61–71. [CrossRef]

23. Bauer, L.S.; Duan, J.J.; Gould, J.R. Emerald ash borer *Agrilus planipennis* Fairmaire Coleoptera: Buprestidae. In *The Use of Classical Biological Control to Preserve Forests in North America*; FHTET-2013-2; Van Driesche, R., Reardon, R., Eds.; United States Department of Agriculture, Forest Service, Forest Health and Technology Enterprise Team: Morgantown, WV, USA, 2014; pp. 189–209. Available online: https://www.nrs.fs.fed.us/pubs/48051 (accessed on 20 March 2018).

24. Van Driesche, R.; Hoddle, M.; Center, T. *Control of Pests and Weeds by Natural Enemies*; Blackwell Publishing: Malden, MA, USA, 2008.

25. Bauer, L.S.; Duan, J.J.; Gould, J.R.; Van Driesche, R.G. Progress in the classical biological control of *Agrilus planipennis* Fairmaire (Coleoptera: Buprestidae) in North America. *Can. Entomol.* **2015**, *147*, 300–317. [CrossRef]

26. Duan, J.J.; Bauer, L.S.; Van Driesche, R.G.; Gould, J.G. Progress and challenges of protecting North American ash trees from the emerald ash borer using biological control. *Forests* **2018**, *9*, 142. [CrossRef]

27. Wang, X.Y.; Cao, L.M.; Yang, Z.Q.; Duan, J.J.; Gould, J.R.; Bauer, L.S. Natural enemies of emerald ash borer (Coleoptera: Buprestidae) in northeast China, with notes on two species of parasitic Coleoptera. *Can. Entomol.* **2016**, *148*, 329–342. [CrossRef]

28. Federal Register. Availability of an environmental assessment for the proposed release of three parasitoids for the biological control of the emerald ash borer *Agrilus planipennis* in the Continental United States. *Fed. Regist.* **2007**, *72*, 28947–28948. Available online: http://www.regulations.gov/#!documentDetail;D=APHIS-2007-0060-0043 (accessed on 20 March 2018).

29. Federal Register. Availability of an environmental assessment for field release of the parasitoid *Spathius galinae* for the biological control of the emerald ash borer (*Agrilus planipennis*) in the contiguous United States. *Fed. Regist.* **2015**, *80*, 7827–7828. Available online: https://www.regulations.gov/docket?D=APHIS-2014-0094 (accessed on 20 March 2018).

30. USDA–APHIS/ARS/FS. USDA Animal Plant Health Inspection Service/Agricultural Research Service/Forest Service. Emerald Ash Borer Biological Control Release and Recovery Guidelines. 2017. Available online: https://www.aphis.usda.gov/plant_health/plant_pest_info/emerald_ash_b/downloads/EAB-FieldRelease-Guidelines.pdf (accessed on 20 March 2018).

31. MapBioControl.org. Agent Release Tracking and Data Management for Federal, State, and Researchers Releasing Biocontrol Agents for Management of the Emerald Ash Borer. 2018. Available online: http://www.mapbiocontrol.org/ (accessed on 20 March 2018).

32. Duan, J.J.; Bauer, L.S.; Abell, K.J.; Lelito, J.P.; Van Driesche, R.G. Establishment and abundance of *Tetrastichus planipennisi* (Hymenoptera: Eulophidae) in Michigan: Potential for success in classical biocontrol of the invasive emerald ash borer (Coleoptera: Buprestidae). *J. Econ. Entomol.* **2013**, *106*, 1145–1154. [CrossRef] [PubMed]

33. Duan, J.J.; Bauer, L.S.; Abell, K.J.; Ulyshen, M.D.; Van Driesche, R.G. Population dynamics of an invasive forest insect and associated natural enemies in the aftermath of invasion: Implications for biological control. *J. Appl. Ecol.* **2015**, *52*, 1246–1254. [CrossRef]

34. Abell, K.J.; Bauer, L.S.; Duan, J.J.; Van Driesche, R.G. Long-term monitoring of the introduced emerald ash borer (Coleoptera: Buprestidae) egg parasitoid, *Oobius agrili* (Hymenoptera: Encyrtidae), in Michigan, USA and evaluation of a newly developed monitoring technique. *Biol. Control* **2014**, *79*, 36–42. [CrossRef]

35. Abell, K.J.; Duan, J.J.; Bauer, L.S.; Lelito, J.P.; Van Driesche, R.G. The effect of bark thickness on the effectiveness of *Tetrastichus planipennisi* (Hymen: Eulophidae) and *Atanycolus* spp. (Hymen: Braconidae) two parasitoids of emerald ash borer (Coleop: Buprestidae). *Biol. Control* **2012**, *63*, 320–325. [CrossRef]

36. Duan, J.J.; Bauer, L.S.; Van Driesche, R.G. Emerald ash borer biocontrol in ash saplings: The potential for early stage recovery of North American ash trees. *For. Ecol. Manag.* **2017**, *394*, 64–72. [CrossRef]

37. Margulies, E.; Bauer, L.; Ibanez, I. Buying time: Preliminary assessment of biocontrol in the recovery of native forest vegetation in the aftermath of the invasive emerald ash borer. *Forests* **2017**, *8*, 369. [CrossRef]

38. Knight, K.S.; Brown, J.P.; Long, R.P. Factors affecting the survival of ash (*Fraxinus* spp.) trees infested by emerald ash borer (*Agrilus planipennis*). *Biol. Invasions* **2013**, *15*, 371–383. [CrossRef]

39. Vercken, E.; Vincent, F.; Mailleret, L.; Ris, N.; Tabone, E.; Fauvergue, X. Time-lag in extinction dynamics in experimental populations: Evidence for a genetic Allee effect? *J. Anim. Ecol.* **2013**, *82*, 621–631. [CrossRef] [PubMed]

40. Vercken, E.; Fauvergue, X.; Ris, N.; Crochard, D.; Mailleret, L. Temporal autocorrelation in host density increases establishment success of parasitoids in an experimental system. *Ecol. Evol.* **2015**, *5*, 2684–2693. [CrossRef] [PubMed]

41. Abell, K.J.; Bauer, L.S.; Miller, D.L.; Duan, J.J.; Van Driesche, R.G. Monitoring the establishment and flight phenology of parasitoids of emerald ash borer (Coleoptera: Buprestidae) in Michigan by using sentinel eggs and larvae. *Fla. Entomol.* **2016**, *99*, 667–672. [CrossRef]

42. Kashian, D.M.; Witter, J.A. Assessing the potential for ash canopy tree replacement via current regeneration following emerald ash borer-caused mortality on southeastern Michigan landscapes. *For. Ecol. Manag.* **2011**, *261*, 480–488. [CrossRef]

43. Klooster, W.S.; Herms, D.A.; Knight, K.S.; Herms, C.P.; McCullough, D.G.; Smith, A.S.; Gandhi, K.J.K.; Cardina, J. Ash (*Fraxinus* spp.) mortality, regeneration, and seed bank dynamics in mixed hardwood forests following invasion by emerald ash borer (*Agrilus planipennis*). *Biol. Invasions* **2014**, *16*, 859–873. [CrossRef]

44. Siegert, N.W.; McCullough, D.G.; Liebhold, A.M.; Telewski, F.W. Dendrochronological reconstruction of the epicentre and early spread of emerald ash borer in North America. *Divers. Distrib.* **2014**, *20*, 847–858. [CrossRef]

45. Smith, A. Effects of community structure on forest susceptibility and response to the emerald ash borer invasion of the Huron River Watershed in southeastern Michigan. Master's Thesis, Ohio State University, Columbus, OH, USA, 2006.

46. IBM Corp. *IBM SPSS Statistics for Windows*, version 23.0; Released; IBM Corp.: Armonk, NY, USA, 2015.

47. Kashian, D.M. Sprouting and seed production may promote persistence of green ash in the presence of the emerald ash borer. *Ecosphere* **2016**, *7*, 1–15. [CrossRef]

48. Murphy, T.C.; Van Dreische, R.G.; Gould, J.R.; Elkinton, J. Can *Spathius galinae* attack emerald ash borer larvae feeding in large ash trees? *Biol. Control* **2017**, *114*, 8–14. [CrossRef]

49. Hausman, C.E. The Ecological Impacts of the Emerald Ash Borer (*Agrilus plannipennis*): Identification of Conservation and Forest Management Strategies. Ph.D. Thesis, Kent State University, Kent, OH, USA, 2010; p. 174.
50. Costanza, K.K.L.; Livingston, W.H.; Kashian, D.M.; Slesak, R.A.; Tardif, J.C.; Dech, J.P.; Diamond, A.K.; Daigle, J.J.; Ranco, D.J.; Siegert, N.W.; et al. The precarious state of a cultural keystone species: Biological and tribal assessments of the role and future of black ash. *J. For.* **2017**, *115*, 435–446. [CrossRef]
51. Palik, B.J.; Ostry, M.E.; Venette, R.C.; Abdela, E. Tree regeneration in black ash (*Fraxinus nigra*) stands exhibiting dieback in northern Minnesota. *For. Ecol. Manag.* **2012**, *269*, 26–30. [CrossRef]

forests

MDPI

Article

Biotic and Abiotic Drivers of Sap Flux in Mature Green Ash Trees (*Fraxinus pennsylvanica*) Experiencing Varying Levels of Emerald Ash Borer (*Agrilus planipennis*) Infestation

Charles E. Flower [1,2,*], Douglas J. Lynch [2,3], Kathleen S. Knight [1] and Miquel A. Gonzalez-Meler [2]

[1]   USDA Forest Service, Northern Research Station, 359 Main Rd., Delaware, OH 43015, USA; ksknight@fs.fed.us
[2]   Department of Biological Sciences, University of Illinois at Chicago, 845 W. Taylor St., Chicago, IL 43015, USA; doug.lynch@licor.com (D.J.L.); mmeler@uic.edu (M.A.G.-M.)
[3]   LI-COR Biosciences, 4421 Superior Street, Lincoln, NE 68504, USA
*   Correspondence: charlesflower@fs.fed.us; Tel.: +1-740-368-0038

Received: 16 April 2018; Accepted: 24 May 2018; Published: 28 May 2018

Abstract: While the relationship between abiotic drivers of sap flux are well established, the role of biotic disturbances on sap flux remain understudied. The invasion of the emerald ash borer (*Agrilus planipennis* Fairmaire, EAB) into North America in the 1990s represents a significant threat to ash trees (*Fraxinus* spp.), which are a substantial component of temperate forests. Serpentine feeding galleries excavated by EAB larvae in the cambial and phloem tissue are linked to rapid tree mortality. To assess how varying levels of EAB infestation impact the plant water status and stress levels of mature green ash (*Fraxinus pennsylvanica* Marshall) trees, we combined tree-level sap flux measurements with leaf-level gas exchange, isotopes, morphology and labile carbohydrate measurements. Results show sap flux and whole tree water use are reduced by as much as 80% as EAB damage increases. Heavily EAB impacted trees exhibited reduced leaf area and leaf mass, but maintained constant levels of specific leaf area relative to lightly EAB-impacted trees. Altered foliar gas exchange (reduced light saturated assimilation, internal CO_2 concentrations) paired with depleted foliar $\delta^{13}C$ values of heavily EAB impacted trees point to chronic water stress at the canopy level, indicative of xylem damage. Reduced photosynthetic rates in trees more impacted by EAB likely contributed to the lack of nonstructural carbohydrate (soluble sugars and starch) accumulation in leaf tissue, further supporting the notion that EAB damages not only phloem, but xylem tissue as well, resulting in reduced water availability. These findings can be incorporated into modeling efforts to untangle post disturbance shifts in ecosystem hydrology.

Keywords: emerald ash borer (*Agrilus planipennis*); invasive species; *Fraxinus*; forest disturbance; sap flux; tree water use; thermal dissipation probe

1. Introduction

Patterns of sap flux in trees are used in estimates of whole-tree water use, tree-level transpiration, and are even scaled to ecosystem transpiration, as such precise measurements of water exchanges are essential for coupled biosphere-atmosphere models. The variables driving water fluxes through the soil-plant continuum are generally well understood [1–4]. However, the impacts of biotic factors such as tree boring forest pests on host water use are largely understudied (but see [5]), despite the considerable impacts they have on forest systems [6]. Abiotic variables including light levels, vapor pressure deficit (VPD), and soil moisture have long been shown to drive sap flux, generally with higher irradiance, higher VPD, and lower soil moisture driving higher rates of sap flux [2,7,8]. From a biotic standpoint, substantial variability has been shown to occur among species [9–11], between trees of the same species which differ in size [2], and based on tree canopy position (i.e., dominant, codominant or suppressed) [12]. Several studies have demonstrated reduced sap flux and water use associated with the disease progression and susceptibility of host trees to fungal pathogens (e.g., [13–16]) as well as forest response to defoliation (e.g., [17,18]). To date, we could identify only a single study describes tree-level sap flux declines following a tree boring beetle attack (i.e., mountain pine beetle (*Dendroctonous poderosae* Hoplins) infestation of lodgepole pine (*Pinus contorta* Douglas) [5]. When assessing the impacts of biotic forest disturbances on tree sap fluxes however, potentially unique characteristics of a given disturbance agent may result in a variety of responses necessitating further investigation.

Emerald ash borer (*Agrilus planipennis* Fairmaire, EAB) is a tree-boring beetle native to Asia, which was inadvertently imported into the United States in the 1990s [19–23]. Emerald ash borers feed almost exclusively on trees in the genus *Fraxinus* [24,25], but see [26] whose constituent species are widely distributed across xeric (e.g., *F. Americana* L., *F. quadrangulata* Michx.), mesic (e.g., *F. pennsylvanica* Marshall) and hydric (e.g., *F. nigra* Marshall, *F. profunda* Bush) forests in the continental US [27–29] and southern Canada [30]. In forests of the Great Lakes Region of the U.S., ash occupies an ecologically important niche in riparian and wetland systems and accounts for ~5–9% of aboveground live carbon mass [28]. Unlike most native tree boring beetles which typically attack trees experiencing physiological stress, in its invaded range EAB attacks healthy and stressed trees, with preference for damaged trees [31]. EAB causes progressive canopy decline [32] that culminates with ash tree mortality in 2–5 years and almost complete ash tree mortality at the forest level within 6 years [33]. It is generally recognized that EAB larvae cause rapid tree mortality via their feeding in the cambial and phloem tissue, which creates serpentine galleries that sever sap transport between shoots and roots [22,34,35]. While simulated girdling studies have been conducted to assess the decline of black ash in wetlands (*F. nigra*) [36], the effects of EAB on tree-level water uptake and sap flux remain undocumented. Thus, direct evidence of the physiological mechanisms altered by EAB infestation are lacking. Changes in water relations caused by EAB will not only accelerate the rate of tree mortality, but also may have local consequences for surface hydrology and the energy balance of associated ecosystems [6,37].

Forest pests that damage cambial tissue can impact tree water use and canopy processes by at least three non-exclusive mechanisms. First, forest pests can impact tree water use via direct feedback mechanisms that regulate photosynthesis [38]. Specifically, leaf stomatal closure may be caused by direct loss of turgor pressure in guard cells associated with changes in the hydraulic conductance along the soil-leaf continuum [39–41] or by increased concentrations of free abscisic acid [42]. Second, the sink limitation imposed by the forest pest's girdling behavior may result in the accumulation of photosynthate in source leaves [43] causing down regulation of photosynthesis, photoinhibition and related effects [44]. Third, via persistent and chronic direct mechanical damage to vascular tissue over a period of years, impacted trees may exhibit reduced leaf area at both canopy and single leaf levels, resulting in concomitant reductions in water use [45,46]. Understanding the underlying mechanisms involved in ash tree responses to EAB will facilitate a better understanding of the associated hydrological responses to forest pests [36]. Studying these effects on EAB impacted forests is of particular importance because our understanding of the hydrological responses to forest pest disturbances predominantly rely largely on girdling studies or inoculations of fungal pathogens, whose damage symptom progressions differ at the tree level. Specifically, sap flux measurements studying mortality from fungal pathogen inoculations can occur in as few as 1 year [14,15], girdling in as quickly as 2 years [47], and EAB induced mortality from 2–5 years [33].

The objective of this study was to examine the impacts of the invasive EAB on ash foliar morphology and tree water use within an ash dominated forest in central Ohio, U.S. The even-aged forest stand utilized in this study is comprised of mature overstory green ash (*F. pennsylvanica*

Marshall) trees displaying all stages of EAB infestation [35], thus providing a unique opportunity to investigate biotic (i.e., EAB) and abiotic drivers of tree water use. Using Granier type sap flux probes (described by Lu et al. [8]), we measured sap flux density as a proxy for direct water use in nine green ash which comprised the full range of EAB impacted conditions. We also measured the bulk leaf carbon (C) isotopic composition, $\delta^{13}C_{leaf}$, of leaf tissue from the focal trees as a proxy for relative changes in intrinsic water use efficiency (the ratio of water used per carbon gain at the leaf level) [48]. Foliar nonstructural carbohydrate concentrations were measured to assess sink limitations [43]. Finally, we conducted leaf level gas exchange measurements of canopy leaves and collected leaf tissue from the canopy of ash trees across a gradient of EAB induced mortality to assess morphological characteristics (surface area (SA), dry mass (DM), and specific leaf area (SLA)). These parameters are used to elucidate the plant hydrological consequences of EAB infestation and to assess the role of water on mortality of infected ash.

2. Materials and Methods

2.1. Site Description and Tree Selection

This study was conducted in 2010 within a 35-year old (as determined dendrochronologically) abandoned tree plantation established on an agricultural field in central Ohio located at the USDA Forest Service Northern Research Station (40° 23' N, 83° 02' W) previously described by Flower et al. [49]. Following abandonment ~30 years prior to the study, succession occurred unimpeded. The forest is characterized by silty loam soils, is at an elevation of 290 m above sea level, and its average annual precipitation from 2006–2010 was 92.04 cm ± 3.71 cm. The forest canopy is dominated by *F. pennsylvanica* (green ash) and *F. americana* L. (white ash), which represent ~80% of tree basal area, as well as *Ulmus americana* L. (American elm), and *Tilia americana* L. (basswood).

Green ash trees were selected for inclusion in the sap flux study by first assessing the diameter at breast height (DBH, ~1.37 m from ground) of each tree in the forest stand and if an ash tree, rating the ash canopy condition (see below for description of AC classes). Additionally, the height of each tree was measured prior to the study with a hypsometer (Vertex IV; Haglöf, Sweden). Nine trees of comparable DBH were randomly selected within the plot across the range of AC classes (see Table 1 for more details on study trees). Investigative sap flux studies have been successfully conducted with low replication when the effect sizes for the variables of interest are large [13,50]. Only co-dominant overstory ash trees were selected for sap flux measurements, since canopy position has been shown to impact sap flow rates [51]. Selected trees were located >10 m from the forest edge to minimize the microclimatic variability and differences in turbulence and momentum fluxes near forest edges which can impact canopy level boundary conditions, transpiration and sap flux rates [52–55]. Measurements of solar radiation and VPD were derived from a meteorological station located at the USDA FS station <0.5 km from the study site.

Table 1. Characteristics of the experimental *Fraxinus pensylvanica* sampled for sap flux density.

Tree No.	AC	DBH [a] (cm)	Tree Height (m)	Avg. Daily Sap Flux (kg/day ± se)	Leaf $\delta^{13}C$	% EAB Gallery Cover
1	1	18.8	11.9	13.69 ± 0.33	−28.12	0
2	1	22.4	13.2	13.13 ± 0.10	−28.11	5
3	1	17.8	11.2	13.01 ± 0.31	−27.39	5
4	2	20.1	12.1	4.54 ± 0.30	−26.97	10
5	2	19.8	12.3	3.78 ± 0.28	−25.94	17.5
6	3	19.3	11.6	3.74 ± 0.16	−26.49	25
7	4	17.0	10.5	2.20 ± 0.15	−25.91	45
8	4	18.2	11.7	1.17 ± 0.22	−25.94	50
9	4	18.8	11.5	1.48 ± 0.10	−25.92	55

[a] Diameter at breast height.

Each live co-dominant ash tree within the interior of the stand was evaluated for EAB infestation using a visual assessment of canopy health called ash canopy condition (AC). This visual assessment has been shown to exhibit a significant relationship with tree-level EAB densities (larvae/adult) and EAB gallery cover and thus it represents a solid visual proxy for EAB infestation levels [35,56]. Briefly, ash canopy condition is graded on a 1 (healthy) to 5 (dead) scale: (1) healthy/full canopy—a healthy ash canopy is full and exhibits no defoliation; (2) thinning canopy—slight reduction in leaf mass, all top branches exposed to sunlight have leaves; (3) dieback—canopy is thinning and some top branches exposed to sunlight are defoliated, lower branches which exhibit natural thinning are not considered; (4) >50% dieback—the canopy has less than 50% of the leaves of a class 1 tree and/or over half of the top branches are defoliated; and (5) dead canopy—no leaves remain in the canopy portion of the tree, regardless of epicormic sprouts [57]. Canopy condition was determined visually in June after leaf expansion was complete. Additionally, EAB larval gallery cover was quantified on 22 August 2018 following termination of sap flux measurements on two debarked 10 × 10 cm windows (on the north and south side of each tree) at 1.5 m height as described by Flower et al. [35].

2.2. Sap flux Measurements

Granier [58] type systems consisting of thermal dissipation probes (TDP, manufactured by PlantSensors, Nakara, Australia) were deployed to measure sap flux in nine ash trees along a gradient of EAB infestation. Each system is a pair of TDP probes (each 2 mm in diameter) inserted radially into the sapwood of the bole to a depth of 2 cm and separated vertically, with the upper probe heated and the lower probe an unheated reference. Two TDP systems were deployed per tree at a height of 1.37 m, one on the north-facing side and one on the south facing-side of each tree to account for flux variability along the stem circumference. In all trees, the depth of the sapwood exceeded the depth of the TDP. The heated and reference probes were placed ~10 cm apart (so the reference measurements are not affected by the heating) and the heating power was adjusted to 0.2 W and induced a maximum temperature difference of 8–10 °C under zero flux conditions as described by Do and Rocheteau [59]. In order to reduce the effect of natural temperature gradients which can impact measurement accuracy, we utilized a heating/cooling cycle of 15 min/15 min as described by Do and Rocheteau [59]. Additionally, the TDP systems were surrounded by closed cell foam and wrapped in a reflective heat shield (a foil coated bubble sheet ~0.6 m in height, wrapped around the TDP system and tree) to minimize temperature gradients associated with direct radiation. Silicone caulking was used at the top and bottom of the heat shield to exclude rain and stem flow from interfering with the TDP systems. Sap flux probes were deployed on 17 June 2010 and data was recorded through 1 August, because of interference with the sap flux probes from raccoons we will only report results from before this interference (17–20 June). Temperature differentials were measured every 5 min between the heated and reference thermocouple junctions on each probe and 30 min means were recorded on a Campbell CR10 datalogger (Campbell Scientific, Logan, UT, USA). Temperature differentials between probes were influenced by sap flux density surrounding the heated probe. Sap flux density (SFD, m s^{-1}) for each sensor was calculated according to Granier [12,58] using the following equations:

$$SFD = \alpha K^{\beta},\tag{1}$$

where $\beta = 1.231$ and $\alpha = 119 \times 10^{-6}$ m s^{-1} [59]. The flux index (K, a dimensionless value) is defined as:

$$K = (\Delta T_0 - \Delta T_u)/\Delta T_u,\tag{2}$$

where ΔT_0 is the maximum temperature difference obtained under zero flux conditions and ΔT_u is the measured temperature difference at a given flux density. It was assumed that no sap flux occurred at pre-dawn. Tree level water use (TWU, kg h^{-1}) was calculated as:

$$TWU = SFD \times S_A,\tag{3}$$

where S_A is the cross-sectional area of the sapwood at the level of the heating probe [12]. The cross-sectional area of sapwood was estimated visually, by wood color as described by Meadows and Hodges [60], from increment cores removed following sap flux measurements. Daily tree level water use (TWU_D, kg^{-1} day^{-1}) was calculated as the 24-h daily sum of TWU. In summary, as sap flows past the unheated (lower) thermocouple, it records the ambient sap temperature and the upper heated needle is cooled. Rapid sap flux results in low differences in the thermocouple temperatures as the upper (heated) thermocouple is cooled.

2.3. Leaf Tissue Sampling

In late June 2010, leaf tissue was excised from the upper sun exposed canopy of each focal (sap flux) ash tree during mid-day using a shotgun. Collected leaf tissue from each experimental ash tree was immediately oven-dried at 60 °C for ~48 h to constant mass to remove any moisture prior to isotopic analysis. The bulk leaf carbon isotopic composition, $\delta^{13}C$, represents an unbiased integrator of a plant water relations over the longevity of the leaf tissue and can be used as a surrogate for intrinsic water use efficiency. The intrinsic water use efficiency is non-linearly proportional to the difference between ambient CO_2 concentrations (C_a) and internal leaf CO_2 concentrations (C_i) [48]. Thus, under constant C_a, and constant $\delta^{13}C$ of atmospheric CO_2, $\delta^{13}C$ reflects a relative and integrated measure of intrinsic water use efficiency among trees of different AC.

On 11–12 July 2013, at least 3 replicate leaf samples were collected from the upper sun exposed canopy of 20 ash trees across a gradient of AC classes to assess differences in leaf morphology. Leaf tissue was collected from 09:00–10:30 and the surface area determined immediately following collection using a LI-COR 3100 leaf area meter (LI-COR, Lincoln, NE, USA). Samples were then dried in a forced-air convection oven at 60 °C for 48 h or until no change in mass was detected and analyzed for % carbon (%C), % nitrogen (%N), and $\delta^{13}C$ as described below. Replicates were averaged prior to statistical analyses.

2.4. Foliar Gas-Exchange Measurements

During June 2015, sun-exposed branches were collected between 09:30–11:30 from the upper canopy of green ash trees across the AC gradient (n = 12 trees). Branches were re-cut under deionized water to reduce embolism and gas-exchange measurements were conducted immediately thereafter. An open-system portable IRGA (LI-6400 equipped with the 6400-02 LED light source attachment, LI-COR, Inc., Lincoln, NE, USA) was used to measure the light saturated photosynthetic rate (A_{sat} @ 1600 PAR; μmol CO_2 m^2 s^{-1}), internal CO_2 concentrations (C_i; ppm CO_2), and stomatal conductance (gs; mmol m^{-2} s^{-1}) for 3 fully expanded leaves/branch. Before each measurement the leaf chamber was checked for leaks and stability in flow (500 μmol s^{-1}), CO_2 (400 ppm), and H_2O (<80%).

2.5. Elemental and Isotopic Analysis of Leaf Tissue

Dried plant tissue was finely ground using a ball mill prior to analysis. The carbon content of leaf tissue was made via flash combustion using a Costech Elemental Analyzer (Valencia, CA, USA) with a zero-blank autosampler and electronic actuator to eliminate air contamination in the samples. Stable isotope analyses of C were conducted on the same sample using an IRMS (isotope ratio mass spectrometer, Finnigan Deltaplus XL, Bremen, Germany) operated in continuous flow mode and calibrated to the ^{13}C Vienna PeeDee Belemnite (VPDB) scale using USGS 40 with a precision of 0.05‰. Elemental and solid international and secondary isotope standards were used for instrument calibration and sample conversion to δ values. Isotopic values are reported in delta notation relative to the standard VPDB for carbon, (δ_{sample} = 1000 (($R_{sample}/R_{standard}$) − 1), R = $^{13}C/^{12}C$). All measurements were done at the stable isotope laboratory at the University of Illinois at Chicago.

2.6. Plant Tissue Non-Structural Carbohydrate Analysis

In 2010, leaf tissue (as previously described) from sap flux trees was analyzed for nonstructural carbohydrates (starch and soluble sugars; NSC). Briefly, we quantified soluble sugars and starch in woody tissues using standard methods [61]. Leaf tissue was dried in a forced convection oven immediately following collection until no mass loss and ground with a ball mill for 3 min. Soluble sugars (sucrose, glucose, and fructose) were extracted from 25 mg of tissue with 80% ethanol (5 mL) at 80 °C for 5 min. Extracts were centrifuged and the supernatants pooled; a 2 mL aliquot was removed and dried using a vacuum evaporator. Dried extract was resuspended with 3 mL deionized water and 40 mg polyvinylpolypyrrolidone and spun down using a centrifuge. A 0.5 mL aliquot was colorimetrically assayed for soluble sugars [61] and modified for use on deciduous tissues according to Curtis et al. [62]. Soluble sugar recovery was >95%.

Starch was quantified from the soluble sugar extracted tissue pellets by resuspension using 1 mL of 0.2 N KOH and incubated at 80 °C for 25 min. KOH was neutralized by adding 0.2 mL of 1 N acetic acid. Starch was hydrolyzed to glucose with α-amyloglucosidase solution (pH 7.05) at 55 °C for 1.5 h and assayed according to Jones et al. [61]. Starch recovery was >95%. The total nonstructural carbohydrate concentration (TNC) was the sum of soluble sugars and starch.

2.7. Statistical Analysis

The effect of AC on ash tree sap flux density rates (SFD) were investigated using a repeated measures analysis of variance (RM ANOVA) test, with the hourly SFD measurements as the repeated measure. Sap flux density rates were compared between AC using Tukey's HSD, α = 0.05. Because of lack of replication, trees in AC 2 and 3 were pooled prior to analysis and thus represent a moderate level of EAB decline. Separately, we tested the role of AC and abiotic factors on SFD using ANOVA, in which AC was treated as a main effect and the environmental parameters (radiation and VPD) were treated as covariates. Additionally, the effect of AC on average TWU_D were tested using ANOVA, daily tree level water use was compared between AC using Tukey's HSD, α = 0.05. The relationship between daily sap flux rates and the carbon isotopic composition of leaf tissue ($\delta^{13}C_{Leaf}$) was tested using Spearman's Rank Correlation Analysis, due to the non-linear relationship between daily sap flux rates and the carbon isotopic composition of leaf tissue ($\delta^{13}C_{Leaf}$). This test provides a distribution free test of independence between the variables. The effect of ash condition classes (AC) on morphological leaf measurements (surface area, dry mass, specific leaf area), gas exchange parameters (A_{sat}, g_s, and C_i) and leaf carbon isotopic composition were investigated using independent analysis of variance (ANOVA) tests. Variables were compared between AC classes using Tukey's HSD, α = 0.05. The effect of AC on leaf starch, soluble sugar, and TNC concentrations was assessed with ANOVA, trees in AC 2 and 3 were pooled prior to analysis. All statistical analyses were conducted using SYSTAT statistical software (v13, SYSTAT 2007).

3. Results

3.1. Sap Flux Density Rates by Ash Canopy Condition Classes and Abiotic Drivers

Sap flux densities ranged from 0.00 to 1.49 m s^{-1} and exhibited significant variability attributed to ash canopy condition classes (ANOVA; $F_{(2624)}$ = 566.3, p < 0.001). Sap flux densities in healthy trees (AC 1; 0.583 m^3 m^{-2}s^{-1} ± 0.045 SE) were significantly higher than those of moderately (AC 2 or 3; 0.144 m^3 m^{-2} s^{-1} ± 0.011 SE) and heavily (AC 4; 0.112 m^3 m^{-2} s^{-1} ± 0.009 SE) infested ash trees (Tukey's HSD p < 0.001 for all AC values; Figure 1). Sap flux density also exhibited significant temporal variability across the sampling period (ANOVA; $F_{(103,624)}$ = 105.018, p < 0.001). Finally, we observed a significant difference in the interaction term (AC × time), highlighting increased responsiveness in sap flux in the healthy trees over time (Figure 1; ANOVA; $F_{(206,624)}$ = 36.251, p < 0.001).

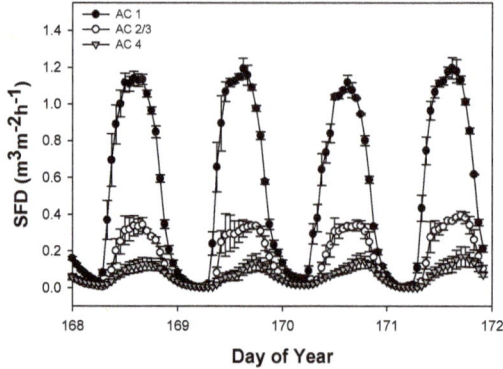

Figure 1. Sap flux density (SFD) of *Fraxinus pennsylvanica* trees during 26–30 June 2010 (AC1 = low infestation levels, AC 2 & 3 moderate infestation levels, and AC4 = high infestation levels). Error bars denote ±1 SE.

Although substantial variation existed between AC classes and over time, SFD exhibited considerable diurnal fluctuations within trees generally peaking during mid-day when the evaporative demand was greatest (Figure 1; Table 2). The SFD was most strongly driven by incoming solar radiation and VPD (Table 2). Higher order interactions of AC*VPD and AC*radiation were also significant, indicating the differential impacts that the abiotic factors have on the SFD of trees across different AC classes (Table 2). Specifically, SFD was less responsive to increased radiation as EAB induced canopy decline manifested (Figure 2).

Table 2. Analysis of variance statistics for the main effects of ash canopy condition class, vapor pressure deficit (VPD), and radiation on ash tree sap flux (SFD) density rates.

Source	Type III SS	df	MS	F-Ratio	*p*-Value
AC	0.215	2	0.108	7.34	0.001
VPD	0.904	1	0.904	61.65	<0.001
Radiation	11.867	1	11.867	809.37	<0.001
AC × VPD	0.426	2	0.213	14.53	<0.001
AC × Radiation	7.441	2	3.721	253.76	<0.001
AC × VPD × Radiation	1.562	2	0.781	53.284	<0.001
Error	13.562	925	0.015		

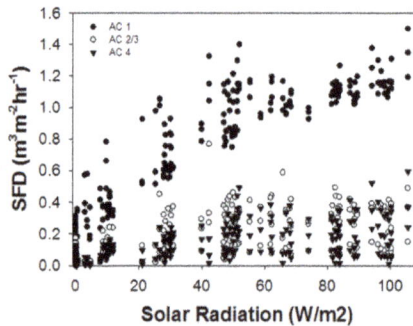

Figure 2. Half hourly sap flux density (SFD) of *Fraxinus* trees in relation to solar radiation (W/m^2) during 26–30 June 2010 (AC1 = low, AC2/3 = moderate, AC 4 = high infestation levels).

3.2. Daily Sap Flux

Reduced tree level SFD resulted in significant differences in daily sap flux rates (TWU$_D$) between lightly and heavily infested trees (Figure 3, ANOVA; F$_{(2,6)}$ = 554.014; p < 0.001). Values of TWU$_D$ tended to be greater for trees which were lightly or not infested with EAB (AC 1 = 13.27 kg day^{-1} ± 0.25 SE) relative to moderately (AC 2 & 3 = 4.02 kg day^{-1} ± 0.31 SE) and heavily infested trees (1.62 kg day^{-1} ± 0.37 SE, Figure 3). Furthermore, a Spearman's rank correlation revealed that TWU$_D$ was negatively correlated with the $\delta^{13}C_{Leaf}$ (r$_s$ = −0.9372, p < 0.05).

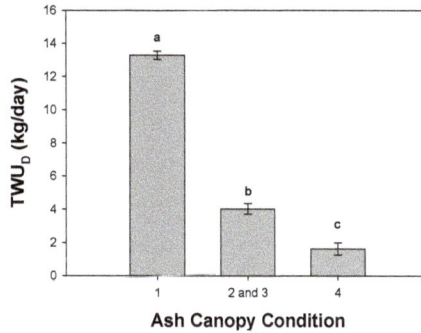

Figure 3. Mean daily sap flux (TWU$_D$) across trees of different ash condition classes. Letters represent significance levels determined by Tukey's HSD pairwise comparisons (p < 0.05), error bars denote ±1 SE.

3.3. Leaf Traits

Individual leaf surface area varied from 7.1 to 41.2 cm^2 and was significantly larger in healthy trees (AC 1 and AC 2) relative to trees exhibiting EAB induced canopy decline (AC 3 and AC4) (Figure 4A; ANOVA; F$_{(3,16)}$ = 10.731, p < 0.001). Leaf dry mass was also significantly larger in healthy trees (AC 1 and AC 2) relative to trees exhibiting canopy decline (AC 3 and AC4) (Figure 4B; ANOVA; F$_{(3,16)}$ = 9.026, p = 0.001). As such, leaf dry mass exhibited a linear positive relationship with leaf surface area (Figure S1; Adj r^2 = 0.98). Because of the conserved relationship between leaf dry mass and leaf surface area, no significant differences were observed in the specific leaf area (data not presented; p = 0.472).

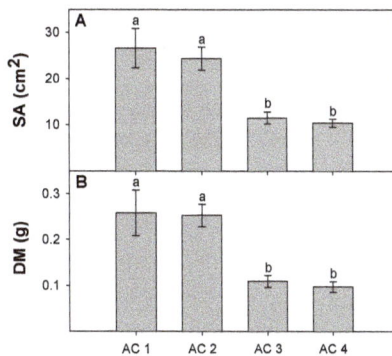

Figure 4. Leaf surface area (SA) (**A**) and leaf dry mass (DM) (**B**) in relation to ash tree canopy condition class (AC). Significance levels depict results of Tukeys HSD pairwise comparisons (p < 0.05).

3.4. Foliar Chemistry

Leaf carbon isotopic signatures exhibited significant enrichment along with progressive canopy decline (Figure 5A; ANOVA; $F_{(3,16)}$ = 9.08, $p < 0.001$), while no significant differences were observed in the foliar nitrogen isotopic signatures (Figure 5B; ANOVA; $F_{(3,16)}$ = 2.152, $p = 0.134$). Bulk leaf carbon content generally increased as ash canopy condition increases. The carbon content of leaves in AC 1 was significantly lower than that of trees in AC 2 and AC 4 (Posthoc Tukey's HSD $p < 0.05$; data not shown). Bulk leaf nitrogen content also exhibited variability across AC classes, although differences were less pronounced, and AC 1 was only different than AC 2 and AC3 (Posthoc Tukey's HSD $p < 0.05$; data not shown). High foliar bulk N and low bulk C in the foliar tissue of trees in AC 1 largely led to the significant differences in leaf C:N ratios between AC classes (Figure 5C; ANOVA; $F_{(3,16)}$ = 6.241, $p = 0.005$).

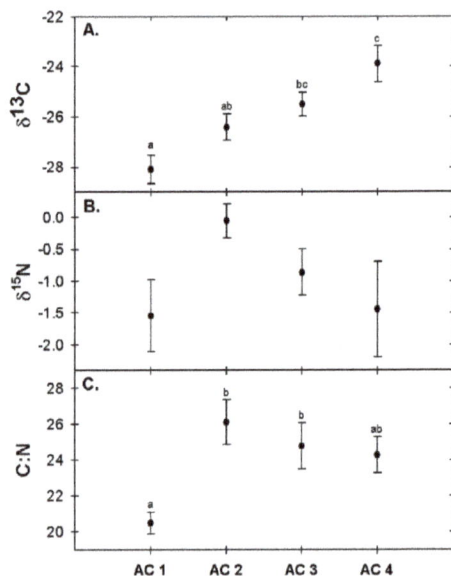

Figure 5. Relationship between green ash foliar 13C and ash canopy decline (**A**), foliar 15N (**B**), and foliar carbon to nitrogen ratio (**C**). Significance levels depict results of Tukeys HSD pairwise comparisons ($p < 0.05$).

Soluble sugar concentrations ranged from 13 to 51 mg g^{-1} and starch concentrations ranged from 1.2 to 22 mg g^{-1} (see Table 3 for means). In general, foliar soluble sugar concentrations were greater than starch concentrations (Table 3). No significant differences in soluble sugar ($p = 0.13$), starch ($p = 0.361$), or TNC concentrations ($p = 0.66$) were observed between AC classes.

Table 3. Foliar soluble sugar, starch concentrations, and total nonstructural carbohydrates (TNC) (mg/g) between ash canopy condition classes. Values represent mean and (SE). No significant differences across AC classes were observed between soluble sugars, starch, or TNC concentrations were revealed via one-way ANOVA.

AC	Soluble Sugars	Starch	TNC
1	38.0 (0.9)	4.9 (1.5)	42.9 (2.3)
2/3	46.1 (4.0)	4.7 (2.3)	50.8 (2.1)
4	28.3 (7.9)	12.0 (5.9)	40.3 (13.8)

3.5. Foliar Gas Exchange

Foliar gas exchange parameters were tightly linked to EAB induced canopy deterioration. Specifically, A_{sat} declined progressively from a maximum of 14.5 µmol m^2 s^{-1} (±0.39) to 6.69 (±0.35) as ash trees became increasingly impacted by EAB (Figure 6A). Additionally, internal CO_2 concentrations (C_i) increased in leaves of ash canopies experiencing significant EAB damage (Figure 6B).

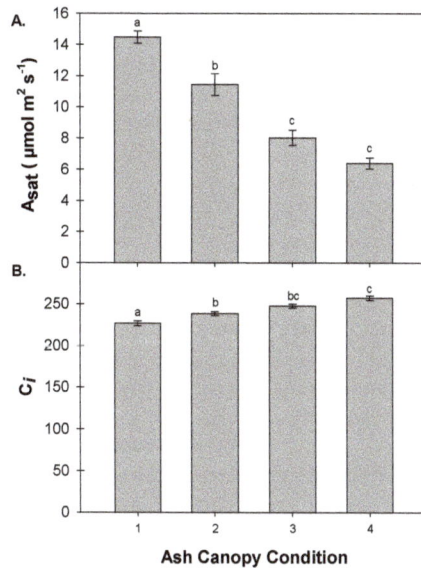

Figure 6. Mean light saturated photosynthetic rate (A_{sat}; at 1600 PAR; panel (**A**)) and internal leaf CO_2 concentration (Ci; panel (**B**)) across foliage from ash trees of different ash condition classes. Letters represent significance levels determined by Tukey's HSD pairwise comparisons ($p < 0.05$), error bars denote ±1 SE.

4. Discussion

The results presented herein indicate that EAB larvae damage ash tree cambial tissue resulting in drastic changes to plant water relations leading to chronic water stress. As EAB symptom progression increased, SFD was reduced, inhibiting A_{sat} and reducing leaf size and mass. Increasing EAB larval feeding gallery cover or emergence holes within the stem have previously been linked to shifts in leaf and canopy level traits [24,32,35]. Although the experiment presented herein was not designed to investigate progressive canopy thinning as a mechanism by which ash trees can re-assimilate nutrients for later use as a means to cope with stress [45], our results lend support to this notion. Specifically, EAB damage resulted in significant reductions in leaf morphological characteristics (SA and DM; Figure 4a,b), consistent with leaf responses to water stress, yet SLA was conserved (Figure S1). The reductions in SA and dry mass seen in this study were comparable to those of Possen et al. [63]. However, the lack of a SLA response as EAB impacted trees contrasts the findings of Possen et al. [63] who observed a 5–10% increase in the SLA of *Betula* and *Populus* associated with wet and dry treatments. The $\delta^{13}C$ isotopic enrichment of leaf tissue as ash trees progressed from healthy (AC 1) to heavily EAB impacted (AC 4) further supports the notion of EAB-induced chronic water stress in ash trees of this forest (Figure 5A) and is consistent with the previous reports of shifts in foliar carbon isotopic composition with differences in plant water use efficiency [64,65].

At the tree-level, sap flux rates were impacted by both biotic (EAB larvae) and abiotic (solar radiation and VPD) drivers (Table 2, Figure 1). The diurnal patterns of SFD of healthy trees reported within this study are consistent with those of other deciduous species and are indicative of being driven by mostly abiotic factors [1,11,66–68]. Specifically, diel patterns of SFD are positively correlated with incoming solar radiation, as observed by Bovard et al. [1]. In our experiments EAB infestation altered the magnitude of the response to solar radiation as this response was larger on lightly EAB infested trees relative to moderately/heavily EAB infested ash trees (Figure 2). This result indicates the interaction of biotic factors with abiotic factors affecting ecosystem level fluxes of forest stands. Additionally, SFD was more positively related to VPD in non-infested or lightly EAB impacted trees relative to moderately and heavily EAB infested trees (Table 2, Figure 3).

Results suggest that EAB larvae cause significant damage to the cambial tissue of infested ash trees reducing not only the expected mass flow of nutrients between roots and shoots, but also the water flux. Larval damage reduced SFD in moderately and heavily EAB impacted trees relative to healthier trees (Table 2; Figure 3). These measurements reveal that moderately and heavily EAB impacted trees did not exhibit a compensatory increase in SFD to counter EAB activity, i.e., heavily impacted trees did not exhibit increased SFD in functioning xylem tissue to maintain canopy water relations. The consequence of xylem damage by EAB is the overall reduced water use. Reduced SFD in heavily EAB impacted trees paralleled reduced TWU_D, whereby lightly EAB impacted trees used 8 times more water per day compared to heavily EAB impacted trees (Figure 3). Daily water use of healthy ash trees observed in this study (13.4 kg day^{-1}) were lower than previously described for *F. excelsior* (27 kg day^{-1}), although focal trees in this study were nearly half the diameter and contained reduced sapwood area perhaps explaining this discrepancy [9]. We measured reduction of sap flux below the expected 1:1 relationship between relative daily sap flux and relative canopy cover as well as relative daily sap flux (a proxy for cambial damage [35]; see Figure S2). This relationship indicates that reductions in daily water use are not proportional to reductions in leaf area and therefore, the proportion of functional cambial tissue which governs mass flow is not solely responsible for the reductions seen in daily sap flux. Reductions in daily sap flux occur before canopy decline becomes apparent; indicating active down-regulation of photosynthesis or stomatal closure associated with EAB induced water stress (enriched foliar $\delta^{13}C$ further supports the notion of direct EAB induced stomatal closure) and/or direct xylem tissue damage. Increased tree-level water stress has previously been linked to increased EAB larval numbers and mean larval mass [69], these results highlight how rapid alterations to plant water status may facilitate forest pest invasions.

The relationship between source activity and sink metabolism has been thoroughly demonstrated, and changes in sink demand (because of the likely EAB altered phloem transport) may be met with reduced photosynthetic rates and/or the accumulation of photosynthate (TNC) at the source tissues [70–72]. The lack of foliar TNC accumulation in heavily impacted trees relative to lightly infested trees does not support the occurrence of sink limitation (Table 3). The lack of a shift in foliar TNC associated with EAB is not consistent with girdling studies which indicate a regulation of photosynthesis by end-product accumulation [73–75]. The lack of end-product accumulation demonstrated here paired with the reduced leaf SA, enriched foliar $\delta^{13}C$, and reduced sap flux indicates broad damage to the vascular tissue caused by EAB larvae.

Other studies investigating different pests or diseases have found similar results. In a study of lodgepole pine inoculated with the blue stain fungi (*O. clavigerum*; associated with the mountain pine beetle, a common forest pest in the western US) it was observed that inoculated trees exhibited significantly reduced sap flow relative to control trees [76]. Similarly, Kirisits and Offenthaler [15] witnessed reductions in sap flow of Norway spruce (*Picea abies* (L.) H. Karst) following inoculation with *Ceratocystis polonica* (a fungal associate of the spruce bark beetle, *Ips typographyus* L.). Additionally, Hultine et al. [18] observed reduced sap flux in tamarisk (*Tamarix* spp.) trees during a defoliation event by the saltcedar leaf beetle (*Diorhabda carinulata* Desbrochers). Results presented herein indicate that the EAB larval induced girdling of ash trees disrupts sapwood function, reducing water availability to leaves and altering the isotopic composition of leaf tissue (Table 1; Figure 5). Excavation of prepupal galleries that penetrate the xylem tissue may contribute to the EAB induced water stress [77].

The negative correlation observed between TWU_D and $\delta^{13}C_{Leaf}$ indicates that the carbon isotopic composition of leaf tissue may be used as a proxy for whole plant water use ($r_s = -0.9372$, $p < 0.05$). The carbon isotopic signature of a leaf has been shown to be affected by the plants photosynthetic pathway, the diffusion of $^{13}CO_2$ into the leaf, and the partial pressure of CO_2 inside the leaf relative to the atmosphere [78]. In the current study, the shift in $\delta^{13}C$ composition of leaf tissue resulted from high stomatal resistance in heavily infested trees (because of leaf level water stress), thus resulting in the reduced discrimination against the isotopically enriched ^{13}C (and hence enriched ^{13}C content in the photosynthetic tissue) (Table 1). While enriched foliar $\delta^{13}C$ values of moderately and highly EAB impacted trees are consistent with several of our proposed mechanisms (stem girdling and foliar carbohydrate accumulation), when paired with the reduction in TWU_D, these results support the hypothesis that inhibited water and nutrient movement between roots and shoots associated with EAB infestation is due to the larval feeding which damages cambial tissue. Although the down regulation of photosynthesis by accumulation of carbohydrates will also result in photoinhibition, the absence of foliar TNC loading in ash trees with moderate to high levels of EAB-induced canopy decline suggests carbohydrate accumulation is not playing a significant role.

In temperate deciduous forests with a high leaf area index, transpiration can represent a substantial portion of the water budget, thus tree mortality can greatly impact forest hydrology [79]. We calculated TWU_D in healthy ash trees at ~13.5 kg day^{-1} compared to ~2 kg day^{-1} in heavily infested trees. Scaling those rates to a forest where ash represents ~80% of the basal area, such a disturbance could result in a significant alteration of the hydrological cycle, and the energy balance of ecosystems with potential impacts to the regional climate system [37]. The reduced sap flux and TWU_D could either increase available soil moisture or reduce soil moisture content (via canopy thinning which could increase soil temperature and evaporation or from reduced hydraulic lift). While nocturnal hydraulic lift has been widely observed in arid systems, it remains uncertain how widespread the phenomenon of hydraulic lift is in deciduous forest systems [80–83].

As a result of resource limitations the number of replicates included in the study was limited, as such trees in categories AC 2 and AC 3 were pooled (despite the recognition that underlying stress levels may differ). As such, caution should be utilized when extrapolating these findings to trees exhibiting intermediate decline patterns. Furthermore, due to the fact that SFD rates of the ash trees in this study were not measured across the entire growing season and that SFD rates of non-ash species were not measured, the authors felt that it was inappropriate to scale EAB induced ash tree decline to annual forest ecosystem water use. However, ash represents an ecologically significant component of riparian forests across the eastern United States and its exclusion from the landscape may result in considerable change to the ecosystem functioning of these sensitive systems [6,28,57].

5. Conclusions

This study reveals the negative impacts tree boring insects on tree water use and highlights the canopy decline, leaf morphological and chemical shifts that accompany tree-level EAB infestation. Our results demonstrate that as ash trees succumb to increased levels of EAB stress, they exhibit reduced canopy cover (via ash canopy condition), reduced leaf area, and reduced leaf dry mass, but no shift in SLA or foliar NSC concentrations (Figure 4 and Figure S2, Table 3). Our results support the theory that the serpentine galleries created by EAB larval feeding results in tree-level water stress as indicated by significant reductions in SFD and subsequently TWU_D (Figures 1–3). Reductions in relative TWU_D were greater than would have been expected if TWU_D was directly proportional to relative canopy cover or the proportion of intact cambium, suggesting active down-regulation of photosynthesis (Figures S1 and S2). This theory is also consistent with the depleted (more negative) carbon isotopic composition of leaves from trees in AC 1 relative to AC 4 (Table 1; Figure 5A). The significant reduction in TWU_D caused by EAB, paired with the rapid decline of infested trees, highlights the potentially severe impacts of this non-native forest pest on individual trees and forest ecosystems. Rapid alterations in the water table caused by reductions in tree water use could have consequences for successional dynamics.

Supplementary Materials: The following are available online at http://www.mdpi.com/1999-4907/9/6/301/s1, Figure S1: Relationship between leaf dry mass and surface area, Figure S2: Relationship between relative daily water use and relative canopy cover.

Author Contributions: C.E.F., D.J.L., and M.A.G.-M. conceived and designed the experiment; K.S.K. provided plot access; C.E.F. conducted the field work, sampling and analysis; C.E.F. and D.J.L. analyzed the data; C.E.F. and D.J.L. wrote the manuscript; M.A.G.-M. edited the manuscript.

Acknowledgments: This research was supported in part by the National Science Foundation Grant DGE-0549245 (C.E.F.), the University of Illinois at Chicago, and the United States Department of Agriculture Northern Research Station. Additionally, we thank J. Rucks for her assistance with labile carbohydrate analyses.

Conflicts of Interest: The authors declare no conflict of interest.

References

1. Bovard, B.D.; Curtis, P.S.; Vogel, C.S.; Su, H.-B.; Schmid, H.P. Environmental controls on sap flow in a northern hardwood forest. *Tree Physiol.* **2005**, *25*, 31–38. [CrossRef] [PubMed]

2. Oren, R.; Phillips, N.; Ewers, B.E.; Pataki, D.E.; Megonigal, J.P. Sap-Flux-Scaled Transpiration Responses To Light, Vapor Pressure Deficit, and Leaf Area Reduction in a flooded *Taxodium distichum* forest. *Tree Physiol.* **1999**, *19*, 337–347. [CrossRef] [PubMed]

3. Lösch, R. Plant Water Relations. In *Progress in Botany: Genetics Cell Biology and Physiology Systematics and Comparative Morphology Ecology and Vegetation Science*; Esser, K., Kadereit, J.W., Lüttge, U., Runge, M., Eds.; Springer: Berlin/Heidelberg, Germany, 1999; pp. 193–233. ISBN 978-3-642-59940-8.

4. Wullschleger, S.D.; Meinzer, F.C.; Vertessy, R.A. A Review of Whole-Plant Water Use Studies in Tree. *Tree Physiol.* **1998**, *18*, 499–512. [CrossRef] [PubMed]

5. Hubbard, R.M.; Rhoades, C.C.; Elder, K.; Negron, J. Changes in transpiration and foliage growth in lodgepole pine trees following mountain pine beetle attack and mechanical girdling. *For. Ecol. Manag.* **2013**, *289*, 312–317. [CrossRef]

6. Flower, C.E.; Gonzalez-Meler, M.A. Responses of temperate forest productivity to insect and pathogen disturbances. *Annu. Rev. Plant Biol.* **2015**, *66*, 547–569. [CrossRef] [PubMed]

7. Lagergren, F.; Lindroth, A. Transpiration response to soil moisture in pine and spruce trees in Sweden. *Agric. For. Meteorol.* **2002**, *112*, 67–85. [CrossRef]

8. Lu, P.; Urban, L.; Zhao, P. Granier's thermal dissipation probe (TDP) method for measuring Sap Flow in trees: theory and practice. *Acta Bot. Sin.* **2004**, *46*, 631–646.

9. Hölscher, D.; Koch, O.; Korn, S.; Leuschner, C. Sap flux of five co-occurring tree species in a temperate broad-leaved forest during seasonal soil drought. *Trees* **2005**, *19*, 628–637. [CrossRef]

10. Gebauer, T.; Horna, V.; Leuschner, C. Variability in radial sap flux density patterns and sapwood area among seven co-occurring temperate broad-leaved tree species. *Tree Physiol.* **2008**, *28*, 1821–1830. [CrossRef] [PubMed]

11. Oren, R.; Pataki, D.E. Transpiration in response to variation in microclimate and soil moisture in southeastern deciduous forests. *Oecologia* **2001**, *127*, 549–559. [CrossRef] [PubMed]

12. Granier, A. Evaluation of transpiration in a Douglas-fir stand by means of sap flow measurements. *Tree Physiol.* **1987**, *3*, 309–320. [CrossRef] [PubMed]

13. Ploetz, R.C.; Schaffer, B.; Vargas, A.I.; Konkol, J.L.; Salvatierra, J.; Wideman, R. Impact of Laurel Wilt, Caused by *Raffaelea lauricola*, on Leaf Gas Exchange and Xylem Sap Flow in Avocado, *Persea americana*. *Phytopathology* **2015**, *105*, 433–440. [CrossRef] [PubMed]

14. Urban, J.; Dvořák, M. Sap flow-based quantitative indication of progression of Dutch elm disease after inoculation with *Ophiostoma novo-ulmi*. *Trees Struct. Funct.* **2014**, *28*, 1599–1605. [CrossRef]

15. Kirisits, T.; Offenthaler, I. Xylem sap flow of Norway spruce after inoculation with the blue-stain fungus *Ceratocystis polonica*. *Plant Pathol.* **2002**, *51*, 359–364. [CrossRef]

16. Park, J.-H.; Juzwik, J.; Cavender-Bares, J. Multiple *Ceratocystis smalleyi* infections associated with reduced stem water transport in bitternut hickory. *Phytopathology* **2013**, *103*, 565–574. [CrossRef] [PubMed]

17. Schäfer, K.V.R.; Clark, K.L.; Skowronski, N.; Hamerlynck, E.P. Impact of insect defoliation on forest carbon balance as assessed with a canopy assimilation model. *Glob. Chang. Biol.* **2010**, *16*, 546–560. [CrossRef]

18. Hultine, K.R.; Nagler, P.L.; Morino, K.; Bush, S.E.; Burtch, K.G.; Dennison, P.E.; Glenn, E.P.; Ehleringer, J.R. Sap flux-scaled transpiration by tamarisk (*Tamarix* spp.) before, during and after episodic defoliation by the saltcedar leaf beetle (*Diorhabda carinulata*). *Agric. For. Meteorol.* **2010**, *150*, 1467–1475. [CrossRef]

19. Bauer, L.; Haack, R.A.; Miller, D.; Petrice, T.; Liu, H. Emerald ash borer life cycle. In Proceedings of the Emerald Ash Borer Research and Technology Development Meeting, Port Huron, MI, USA, 30 September–1 October 2003; FHTET-2004-02. USDA Forest Service, Forest Health Technology Enterprise Team: Morgantown, WV, USA, 2004; p. 8.

20. Siegert, N.W.; McCullough, D.G.; Liebhold, A.M.; Telewski, F.W. Dendrochronological reconstruction of the epicentre and early spread of emerald ash borer in North America. *Divers. Distrib.* **2014**, *20*, 847–858. [CrossRef]

21. Wang, X.; Yang, Z.; Gould, J.; Zhang, Y.; Liu, G.; Liu, E. The biology and ecology of the emerald ash borer, Agrilus planipennis, in China. *J. Insect Sci.* **2010**, *10*, 128. [CrossRef] [PubMed]

22. Cappaert, D.; Mccullough, D.G.; Poland, T.M.; Siegert, N.W. Emerald ash borer in north america a research and regulatory challenge. *Am. Entomol.* **2005**, *51*, 152–165. [CrossRef]

23. Herms, D.A.; McCullough, D.G. Emerald Ash Borer Invasion of North America: History, Biology, Ecology, Impacts, and Management. *Annu. Rev. Entomol.* **2014**, *59*, 13–30. [CrossRef] [PubMed]

24. Rebek, E.J.; Herms, D.A.; Smitley, D.R. Interspecific variation in resistance to emerald ash borer (Coleoptera: Buprestidae) among North American and Asian ash (*Fraxinus* spp.). *Environ. Entomol.* **2008**, *37*, 242–246. [CrossRef]

25. Anulewicz, A.C.; McCullough, D.G.; Cappaert, D.L.; Poland, T.M. Host range of the emerald ash borer (*Agrilus planipennis* Fairmaire) (Coleoptera: Buprestidae) in North America: results of multiple-choice field experiments. *Environ. Entomol.* **2008**, *37*, 230–241. [CrossRef]

26. Cipollini, D. White fringetree as a novel larval host for emerald ash borer. *J. Econ. Entomol.* **2015**, *108*, 370–375. [CrossRef] [PubMed]

27. Pugh, S.A.; Liebhold, A.M.; Morin, R.S. Changes in ash tree demography associated with emerald ash borer invasion, indicated by regional forest inventory data from the Great Lakes States. *Can. J. For. Res.* **2011**, *41*, 2165–2175. [CrossRef]

28. Flower, C.E.; Knight, K.S.; Gonzalez-Meler, M.A. Impacts of the emerald ash borer (*Agrilus planipennis* Fairmaire) induced ash (*Fraxinus* spp.) mortality on forest carbon cycling and successional dynamics in the eastern United States. *Biol. Invasions* **2013**, *15*, 931–944. [CrossRef]

29. MacFarlane, D.W.; Meyer, S.P. Characteristics and distribution of potential ash tree hosts for emerald ash borer. *For. Ecol. Manag.* **2005**, *213*, 15–24. [CrossRef]

30. Burns, R.; Honkala, B. *Silvics of North America, Vol. 2: Hardwoods. Agriculture Handbook 654*; United States Department of Agriculture Forest Service: Washington, DC, USA, 1990.

31. McCullough, D.G.; Poland, T.M.; Cappaert, D. Attraction of the emerald ash borer to ash trees stressed by girdling, herbicide treatment, or wounding. *Can. J. For. Res.* **2009**, *39*, 1331–1345. [CrossRef]

32. Smitley, D.; Davis, T.; Rebek, E. Progression of ash canopy thinning and dieback outward from the initial infestation of emerald ash borer (Coleoptera: Buprestidae) in southeastern Michigan. *J. Econ. Entomol.* **2008**, *101*, 1643–1650. [CrossRef] [PubMed]

33. Knight, K.S.; Brown, J.P.; Long, R.P. Factors affecting the survival of ash trees (*Fraxinus* spp.) infested by emerald ash borer (Agrilus planipennis). *Biol. Invasions* **2013**, *15*, 371–383. [CrossRef]

34. Flower, C.E.; Lynch, D.J.; Knight, K.S.; González-Meler, M.A. EAB induced tree mortality impacts tree water use and ecosystem respiration in an experimental forest. In Proceedings of the Emerald Ash Borer National Research and Technology Development Meeting, Wooster, OH, USA, 12–13 October 2011; FHTET-2011-06. pp. 115–116.

35. Flower, C.E.; Knight, K.S.; Rebbeck, J.; Gonzalez-Meler, M.A. The relationship between the emerald ash borer (*Agrilus planipennis*) and ash (*Fraxinus* spp.) tree decline: Using visual canopy condition assessments and leaf isotope measurements to assess pest damage. *For. Ecol. Manag.* **2013**, *303*, 143–147. [CrossRef]

36. Telander, A.C.; Slesak, R.A.; D'Amato, A.W.; Palik, B.J.; Brooks, K.N.; Lenhart, C.F. Sap flow of black ash in wetland forests of northern Minnesota, USA: Hydrologic implications of tree mortality due to emerald ash borer. *Agric. For. Meteorol.* **2015**, *206*, 4–11. [CrossRef]

37. Drewniak, B.; Gonzalez-Meler, M.A. Earth system model needs for including the interactive representation of nitrogen deposition and drought effects on forested ecosystems. *Forests* **2017**, *8*, 267. [CrossRef]

38. Paul, M.J.; Pellny, T.K. Carbon metabolite feedback regulation of leaf photosynthesis and development. *J. Exp. Bot.* **2003**, *54*, 539–547. [CrossRef] [PubMed]

39. Rashke, K. Stomatal action. *Annu. Rev. Plant Physiol.* **1975**, *26*, 309–340. [CrossRef]

40. Collatz, G.J.; Ball, J.T.; Grivet, C.; Berry, J.A. Physiological and environmental regulation of stomatal conductance, photosynthesis and transpiration—A model that includes a laminar boundary-layer. *Agric. For. Meteorol.* **1991**, *54*, 107–136. [CrossRef]

41. Sperry, J.S. Hydraulic constraints on plant gas exchange. *Agric. For. Meteorol.* **2000**, *104*, 13–23. [CrossRef]

42. Setter, T.L.; Brun, W.A.; Brenner, M.L. Effect of obstructed translocation on leaf abscisic Acid, and associated stomatal closure and photosynthesis decline. *Plant Physiol.* **1980**, *65*, 1111–1115. [CrossRef] [PubMed]

43. Cheng, J.; Fan, P.; Liang, Z.; Wang, Y.; Niu, N.; Li, W.; Li, S. Accumulation of End Products in Source Leaves Affects Photosynthetic Rate in Peach via Alteration of Stomatal Conductance and Photosynthetic Efficiency. *J. Am. Soc. Hortic. Sci.* **2009**, *134*, 667–676.

44. Azcón-Bieto, J.; Osmond, C.B. Relationship between photosynthesis and respiration—The effect of carbohydrate status on the rate of CO_2 production by respiration in darkened and illuminated wheat leaves. *Plant Physiol.* **1983**, *71*, 574–581. [CrossRef] [PubMed]

45. Munné-Bosch, S.; Alegre, L. Die and let live: Leaf senescence contributes to plant survival under drought stress. *Funct. Plant Biol.* **2004**, *31*, 203–216. [CrossRef]

46. Otieno, D.O.; Schmidt, M.W.T.; Adiku, S.; Tenhunen, J. Physiological and morphological responses to water stress in two Acacia species from contrasting habitats. *Tree Physiol.* **2005**, *25*, 361–371. [CrossRef] [PubMed]

47. Slesak, R.A.; Lenhart, C.F.; Brooks, K.N.; D'Amato, A.W.; Palik, B.J. Water table response to harvesting and simulated emerald ash borer mortality in black ash wetlands in Minnesota, USA. *Can. J. For. Res.* **2014**, *44*, 961–968. [CrossRef]

48. Saurer, M.; Siegwolf, R.T.W.; Schweingruber, F.H. Carbon isotope discrimination indicates improving water-use efficiency of trees in northern Eurasia over the last 100 years. *Glob. Chang. Biol.* **2004**, *10*, 2109–2120. [CrossRef]

49. Flower, C.E.; Dalton, J.E.; Knight, K.S.; Brikha, M.; Gonzalez-Meler, M.A. To treat or not to treat: Diminishing effectiveness of emamectin benzoate tree injections in ash trees heavily infested by emerald ash borer. *Urban For. Urban Green.* **2015**, *14*. [CrossRef]

50. Ford, C.R.; McGuire, M.A.; Mitchell, R.; Teskey, R.O. Assensing variations in the radial profile of sap flow density in *Pinus* species and its effects on daily water use. *Tree Physiol.* **2004**, *24*, 241–249. [CrossRef] [PubMed]

51. Kelliher, F.M.; Kostner, B.M.M.; Hollinger, D.Y.; Byers, J.N.; Hunt, J.E.; McSeveny, T.M.; Meserth, R.; Weir, P.L.; Schulze, E.D. Evaporation, Xylem Sap Flow, and Tree Transpiration in a New-Zealand Broad-Leaved Forest. *Agric. For. Meteorol.* **1992**, *62*, 53–73. [CrossRef]

52. Young, A.; Mitchell, N. Microclimate and vegetation edge effects in a fragmented podocarp-broadleaf forest in New Zealand. *Biol. Conserv.* **1994**, *67*, 63–72. [CrossRef]

53. Veen, A.W.L.; Klaassen, W.; Kruijt, B.; Hutjes, R.W.A. Forest edges and the soil-vegetation-atmosphere interaction at the landscape scale: The state of affairs. *Prog. Phys. Geogr.* **1996**, *20*, 292–310. [CrossRef]

54. Davies-Colley, R.J.; Payne, G.W.; Van Elswijk, M. Microclimate gradients across a forest edge. *N. Z. J. Ecol.* **2000**, *24*, 111–121.

55. Herbst, M.; Roberts, J.M.; Rosier, P.T.W.; Taylor, M.E.; Gowing, D.J. Edge effects and forest water use: A field study in a mixed deciduous woodland. *For. Ecol. Manag.* **2007**, *250*, 176–186. [CrossRef]

56. Flower, C.E.; Knight, K.S.; González-Meler, M.A. Using stable isotopes as a tool to investigate the impacts of EAB on tree physiology and EAB spread. In Proceedings of the Emerald Ash Borer Research and Technology Development Meeting, Pittsburgh, PA, USA, 20–21 October 2009; FHTET-2010-01. USDA Forest Service, Forest Health Technology Enterprise Team: Morgantown, WV, USA, 2010; pp. 54–55.

57. Smith, A. Effects of Community Structure on Forest Susceptibility and Response to the Emerald Ash Borer iNvasion of the Huron River Watershed in Southeast Michigan. Master's Thesis, The Ohio State University, Columbus, OH, USA, 2006.

58. Granier, A. A new method of sap flow measurement in tree stems. *Ann. Des. Sci. For.* **1985**, *42*, 193–200. [CrossRef]

59. Do, F.; Rocheteau, A. Influence of natural temperature gradients on measurements of xylem sap flow with thermal dissipation probes. 2. Advantages and calibration of a noncontinuous heating system. *Tree Physiol.* **2002**, *22*, 649–654. [CrossRef] [PubMed]

60. Meadows, J.S.; Hodges, J.D. Sapwood area as an estomator of leaf area and oliar weight in cherrybark oak and green ash. *For. Sci.* **2001**, *48*, 69–76.

61. Jones, M.G.; Outlaw, W.H.; Lowry, O.H. Enzymic assay of 10^{-7} to 10^{-14} moles of sucrose in plant tissues. *Plant Physiol.* **1977**, *60*, 379–383. [CrossRef] [PubMed]

62. Curtis, P.S.; Vogel, C.S.; Wang, X.Z.; Pregitzer, K.S.; Zak, D.R.; Lussenhop, J.; Kubiske, M.; Teeri, J.A. Gas exchange, leaf nitrogen, and growth efficiency of Populus tremuloides in a CO_2-enriched atmosphere. *Ecol. Appl.* **2000**, *10*, 3–17.

63. Possen, B.J.H.M.; Oksanen, E.; Rousi, M.; Ruhanen, H.; Ahonen, V.; Tervahauta, A.; Heinonen, J.; Heiskanen, J.; Kärenlampi, S.; Vapaavuori, E. Adaptability of birch (*Betula pendula* Roth) and aspen (*Populus tremula* L.) genotypes to different soil moisture conditions. *For. Ecol. Manag.* **2011**, *262*, 1387–1399. [CrossRef]

64. Farquhar, G.D.; Ball, M.C.; Roksandic, Z.; City, C. Effect of salinity and humidity on δ13C value of halophytes-evidence for diffusional isotope fractionation determined by the ratio of intercellular/atmospheric partial pressure of CO_2 under different environmental conditions. *Oecologia* **1982**, *52*, 121–124. [CrossRef] [PubMed]

65. Farquhar, G.D.; Richards, R.A. Isotopic composition of plant carbon correlates with water-use efficiency of wheat genotypes. *Aust. J. Plant Physiol.* **1984**, *11*, 539–552. [CrossRef]

66. Granier, A.; Anfodillo, T.; Sabatti, M.; Cochard, H.; Dreyer, E.; Tomasi, M.; Valentini, R.; Bréda, N. Axial and radial water flow in the trunk of oak trees: a quantitative and qualitative analysis. *Tree Physiol.* **1994**, *14*, 1383–1396. [CrossRef] [PubMed]

67. Lu, P.; Biron, P.; Breda, N.; Granier, A. Water relations of adult Norway spruce (*Picea abies* (L) Karst) under soil drought in the Vosges mountains: water potential, stomatal conductance and transpiration. *Ann. For. Sci.* **1995**, *52*, 117–129. [CrossRef]

68. Stöhr, A.; Lösch, R. Xylem sap flow and drought stress of *Fraxinus excelsior* saplings. *Tree Physiol.* **2004**, *24*, 169–180. [CrossRef] [PubMed]

69. Chakraborty, S.; Whitehill, J.G.A.; Hill, A.L.; Opiyo, S.O.; Cipollini, D.; Herms, D.A.; Bonello, P. Effects of water availability on emerald ash borer larval performance and phloem phenolics of Manchurian and black ash. *Plant Cell Environ.* **2014**, *37*, 1009–1021. [CrossRef] [PubMed]

70. Azcón-Bieto, J. Inhibition of photosynthesis by carbohydrates in wheat leaves. *Plant Physiol.* **1983**, *73*, 681–686. [CrossRef] [PubMed]

71. Paul, M.J.; Foyer, C.H. Sink regulation of photosynthesis. *J. Exp. Bot.* **2001**, *52*, 1383–1400. [CrossRef] [PubMed]

72. Iglesias, D.J.; Lliso, I.; Tandeo, F.R.; Talon, M. Regulation of photosynthesis through source: sink imbalance in citrus mediated by carbohydrate content in leaves. *Physiol. Plant.* **2002**, *116*, 563–572. [CrossRef]

73. Goldschmidt, E.E.; Huber, S.C. Regulation of photosynthesis by end-product accumulation in leaves of plants storing starch, sucrose, and hexose sugars. *Plant Physiol.* **1992**, *99*, 1443–1448. [CrossRef] [PubMed]

74. Jordan, M.-O.; Habib, R. Mobilizable carbon reserves in young peach trees as evidenced by trunk girdling experiments. *J. Exp. Bot.* **1996**, *47*, 79–87. [CrossRef]

75. Myers, D.; Thomas, R.; DeLucia, E. Photosynthetic responses of loblolly pine (*Pinus taeda*) needles to experimental reduction in sink demand. *Tree Physiol.* **1999**, *19*, 235–242. [CrossRef] [PubMed]

76. Yamaoka, Y.; Swanson, R.; Hiratsuka, Y. Inoculation of lodgepole pine with four blue-stain fungi associated with mountain pine beetle, monitored by a heat pulse velocity (HPV) insturment. *Can. J. For. Res.* **1990**, *20*, 31–36. [CrossRef]

77. Mack, R. Characterization of emerald ash borer (EAB) pupal chamber location in ash log sections. In *2014 Emerald Ash Borer National Research and Technology Development Meeting*; Buck, J., Parra, G., Lance, D., Reardon, R., Binion, D., Eds.; USDA Forest Service, Forest Health Technology Enterprise Team: Morgantown, WV, USA, 2015; p. 52.

78. Farquhar, G.D.; O'Leary, M.H.; Berry, J.A. On the relationship between carbon isotope discrimination and the intercellular carbon dioxide concentration in leaves. *Aust. J. Plant Physiol.* **1982**, *9*, 121–137. [CrossRef]

79. Running, S.W.; Coughlan, J.C. A general-model of forest ecosystem processes for regional applications. I. Hydrologic balance, canopy gas-exchange and primary production processes. *Ecol. Model.* **1988**, *42*, 125–154. [CrossRef]
80. Richards, J.; Caldwell, M. Hydraulic lift: Substantial nocturnal water transport between soil layers by *Artemisia tridentata* roots. *Oecologia* **1987**, *73*, 486–489. [CrossRef] [PubMed]
81. Caldwell, M.M.; Richards, J.H. Hydraulic lift: water efflux from upper roots improves effectiveness of water uptake by deep roots. *Oecologia* **1989**, *79*, 1–5. [CrossRef] [PubMed]
82. Dawson, T.E. Determining water use by trees and forests from isotopic, energy balance and transpiration analyses: the roles of tree size and hydraulic lift. *Tree Physiol.* **1996**, *16*, 263–272. [CrossRef] [PubMed]
83. Dawson, T.E. Hydraulic lift and water use by plants: implications for water balance, performance and plant-plant interactions. *Oecologia* **1993**, *95*, 565–574. [CrossRef] [PubMed]

forests

MDPI

Article

Response of Black Ash Wetland Gaseous Soil Carbon Fluxes to a Simulated Emerald Ash Borer Infestation

Matthew Van Grinsven [1,2,*], Joseph Shannon [2], Nicholas Bolton [2], Joshua Davis [2], Nam Jin Noh [2,3], Joseph Wagenbrenner [2,4], Randall Kolka [5] and Thomas Pypker [6]

[1] Department of Earth, Environmental and Geographical Sciences, Northern Michigan University, Marquette, MI 49855, USA
[2] School of Forest Resources & Environmental Science, Michigan Technological University, Houghton, MI 49931, USA; jpshanno@mtu.edu (J.S.); nwbolton@mtu.edu (N.B.); joshuad@mtu.edu (J.D.); n.noh@westernsydney.edu.au (N.J.N.); jwagenbrenner@fs.fed.us (J.W.)
[3] Hawkesbury Institute for the Environment, Western Sydney University, Penrith, NSW 2753, Australia
[4] USDA Forest Service, Pacific Southwest Research Station, Arcata, CA 95521, USA
[5] USDA Forest Service, Northern Research Station, Grand Rapids, MN 55744, USA; rkolka@fs.fed.us
[6] Department of Natural Resource Sciences, Thompson Rivers University, Kamloops, BC V2C 0C8, Canada; TPypker@tru.ca
* Correspondence: mvangrin@nmu.edu

Received: 31 March 2018; Accepted: 23 May 2018; Published: 4 June 2018

Abstract: The rapid and extensive expansion of emerald ash borer (EAB) in North America since 2002 may eliminate most existing ash stands, likely affecting critical ecosystem services associated with water and carbon cycling. To our knowledge, no studies have evaluated the coupled response of black ash (*Fraxinus nigra* Marsh.) wetland water tables, soil temperatures, and soil gas fluxes to an EAB infestation. Water table position, soil temperature, and soil CO_2 and CH_4 fluxes were monitored in nine depressional headwater black ash wetlands in northern Michigan. An EAB disturbance was simulated by girdling (girdle) or felling (ash-cut) all black ash trees with diameters greater than 2.5 cm within treated wetlands (n = 3 per treatment). Soil gas fluxes were sensitive to water table position, temperature, and disturbance. Soil CO_2 fluxes were significantly higher, and high soil CH_4 fluxes occurred more frequently in disturbed sites. Soil CH_4 fluxes in ash-cut were marginally significantly higher than girdle during post-treatment, yet both were similar to control sites. The strong connection between depressional black ash wetland study sites and groundwater likely buffered the magnitude of disturbance-related impact on water tables and carbon cycling.

Keywords: forested wetlands; *Fraxinus nigra*; invasive pest disturbance; greenhouse gas fluxes; soil carbon; biogeosciences; EAB

---

1. Introduction

Disturbance events are known to impact the natural function of wetland ecosystems by altering nutrient, carbon, energy, and hydrologic fluxes [1,2]. Recently, emerald ash borer (EAB) (Coleoptera: Buprestidae, *Agrilus planipennis* Fairmaire) was introduced as an invasive pest to ash (*Fraxinus* spp.) forests in North America [3]. Hundreds of millions of ash trees were estimated to have been killed in North America within the decade following the initial detection of EAB in southeastern Michigan in July 2002 [4,5]. The expansion of EAB results in significant economic costs [6,7] and causes significant perturbation to forest ecosystems [5,8,9], and will likely continue to decimate existing ash stands throughout North America.

Forested wetlands sequester large quantities of carbon from the atmosphere [10], and their carbon storage capacity is largely a result of prolonged periods of elevated water table position that limits decomposition [11,12]. Forest and soil gaseous-carbon fluxes are sensitive to disturbance of overstory trees [13,14], and fluxes from wetland soils are also regulated by water table position and soil temperature [15,16]. Disturbance to the dominant living woody component of a forested wetland has been shown to affect wetland water table positions [17,18] and may affect soil temperatures due to the decreased amount of transpiration and increased amount of incoming solar radiation. Ecohydrological responses to a simulated EAB infestation were evaluated concurrently with gaseous soil carbon fluxes in depressional black ash wetland study sites examined in this study. Significantly lower site transpiration, smaller rates of water table drawdown during the growing season, and elevated summertime water tables were detected following a simulated EAB infestation treatment in depressional black ash wetland study sites [17]. Similarly, mineral soil black ash wetlands in northern Minnesota treated with a simulated EAB infestation had significantly elevated water tables when compared to control sites [18]. Ultimately, wetland soil carbon pools and fluxes will be concomitantly influenced by overstory disturbance and by the response of water table position and soil temperature.

Most research on wetland carbon storage and fluxes has focused on boreal or sub-boreal peatland ecosystems dominated by *Sphagnum* species and sparse coniferous tree species [19,20], yet large stores of carbon also exist in mixed deciduous forested wetlands with histic, histic-mineral, and mineral soils [21,22]. Histic-mineral and mineral soil wetlands are five times more abundant on earth's surface than histosol wetlands, and comprise an estimated 22% of the global wetland carbon pool [10,23]. Freshwater mineral soil wetlands may contain significant organic soil deposits (20 to 40 cm depth), and have been inconsistently classified as either peatlands or fresh water mineral soil wetlands in regional- or global-scale wetland carbon cycle studies [24]. To our knowledge, no studies have specifically examined carbon cycling processes in deciduous forested peatlands.

Biogeochemophysical conditions such as temperature, pH, water table position, soil moisture, redox potential, and soil microbial communities affect the production and fluxes of gaseous soil carbon constituents (CO_2, CH_4) [19]. There are complex relationships between soil microbial community composition, physical and chemical conditions, the quality of available organic matter, and rates of soil decomposition [25]. As a result, it is often difficult to separate the influences of individual biogeochemophysical factors from one another. Peatland decomposition processes have consistently been shown to be sensitive to both soil temperature and redox conditions [26,27].

Carbon dioxide fluxes from wetlands generally increase following episodic short-term water table fluctuations; however, the magnitude of the response can be highly variable [26]. Specifically, CO_2 production and flux may increase following episodic water table fluctuations due to redox reaction–induced chemical breakdown and activation of exo-enzymes following short-term exposure to oxygen [11,28,29]. In contrast, the quality of organic substrate and presence of certain microbial communities can inhibit decomposition rates following episodic oxidation-state modifying events [19]. It is unclear how EAB-induced alteration of water tables may affect oxidation–reduction processes and ultimately decomposition rates in black ash wetlands.

Temperature controls on CO_2 production from wetlands have been well documented in both field and laboratory settings [15,30]. In laboratory studies, a 10 °C temperature increase resulted in a positive change in CO_2 production (Q_{10}) of two- to threefold [15]. However, in situ Q_{10} values have been inconsistent across vertical and lateral gradients within peatlands [30,31]. The uncertainty associated with the response of CO_2 production rates to experimental temperature manipulations in peatlands has likely resulted from the complexity of in situ biogeochemical conditions. Additionally, the need to characterize soil warming impacts on peatland ecosystems remains critical for the development of accurate global carbon cycle models [32]. Removal of the overstory following an EAB infestation may increase insolation, increase wetland soil temperatures, and ultimately alter gaseous carbon fluxes.

Similar to CO_2 production and fluxes, there are high degrees of spatial and temporal variability associated with CH_4 production and fluxes from peatland soils [33]. Because both fermentation and CO_2 reduction processes occur under anoxic conditions, the location of production and fate of

CH_4 once produced are highly sensitive to water table position [34]. Specific physical and chemical conditions need to be present for methane production to occur. Among the most important of these is the combined depletion of oxygen and absence of other, more efficient electron acceptors (e.g., NO_3^-, Fe^{3+}, Mn^{4+}, and SO_4^{2-}) that must be present before methanogenesis and CO_2 reduction can proceed [28,35]. The initiation or rate of methane production has also been shown to be sensitive to pH [36], temperature [15], and substrate quality and availability [19,37].

Due to the co-occurrence of methanogens and methanotrophs in peatland soils, water table position is the single most influential physical parameter that regulates the quantity of CH_4 fluxes from the soil surface. Several studies have evaluated the response of CH_4 fluxes in peatlands to temperature and water table conditions [33,38–40]. Yet, to our knowledge, no study has examined the relationship between water table, soil temperature, and soil CO_2 and CH_4 fluxes in deciduous forested peatlands, nor has any study evaluated the response of gaseous soil carbon fluxes to a simulated invasive pest disturbance in deciduous forested peatlands. Temperature-dependent incubation experiments of CH_4 production in cores extracted from hummock, lawn, and hollow locations in a peatland in southwestern Scotland revealed that CH_4 production was similar among sites, yet in situ CH_4 fluxes were highly variable and closely associated with the position of the water table [33]. The response of water tables and soil temperatures to an EAB-induced disturbance in black ash (*Fraxinus nigra* Marsh.) wetlands will likely alter CH_4 production processes, and ultimately the amount of CH_4 emitted from the soil surface.

This study examines the relationship between, and the coupled response of, water table, soil temperature, and gaseous soil carbon fluxes within the context of the previously established altered hydrologic regime in depressional black ash wetlands. The specific objectives of this study were to (1) characterize the relationships between water table position, soil temperature, and soil CO_2 and CH_4 fluxes; and (2) evaluate the responses of soil temperature and gaseous soil carbon fluxes to disturbances in black ash wetlands. Ultimately, this study will augment the currently limited knowledge of carbon cycling processes in deciduous forested peatlands and examine the ecosystem-level implications of a looming emerald ash borer infestation. We hypothesize that (1) water table position and soil temperature will be major influences on soil CO_2 and CH_4 fluxes from these soils; (2) soil temperature will increase because of increased insolation in disturbed sites; (3) soil CO_2 fluxes will increase with the increased soil temperature in disturbed sites; and (4) soil CH_4 fluxes will be higher because of the expected increased temperatures and elevated water tables detected in response to a simulated EAB infestation.

2. Materials and Methods

2.1. Study Area

The Ottawa National Forest (ONF) is located in the western portion of the upper peninsula of Michigan, USA (Figure 1). The ONF contains approximately 525 km^2 of wetlands within 4000 km^2 of mixed-hardwood forested area. Based upon US Department of Agriculture Forest Service Forest Inventory Analysis data, it is estimated that 70% of the 180 km^2 deciduous forested wetlands within the ONF are dominated by black ash [41].

Black ash wetlands in the ONF commonly occur as landform depressions surrounded by mixed-hardwood upland tree species such as sugar maple (*Acer saccharum* L.) and white pine (*Pinus strobus* L.). Other tree species, such as red maple (*Acer rubrum* L.), yellow birch (*Betula alleghaniensis* Britt.), balsam fir (*Abies balsamea* L. (Mill)), and white cedar (*Thuja occidentalis* L.), are common, though typically at low abundance within these black ash–dominated wetlands [42]. Shrub, fern, and sedge species are abundant but unevenly distributed in the understory.

Black ash wetland soils in this study were composed of woody peats. The study site soils were, on average, 75% organic matter and 47% organic carbon [17]. Peat thicknesses in the study sites ranged from 40 cm to greater than 690 cm, with an average peat depth of 140 cm. A clay lens or a poorly sorted clay loam was commonly detected at the bottom of the peat layer. Black ash wetlands in the ONF are also characterized by seasonally inundated conditions and commonly have discernible surface drainage outlet channels. Most soils in the ONF originate from glacial landforms and may exist as thin horizons overlying near-surface bedrock [43].

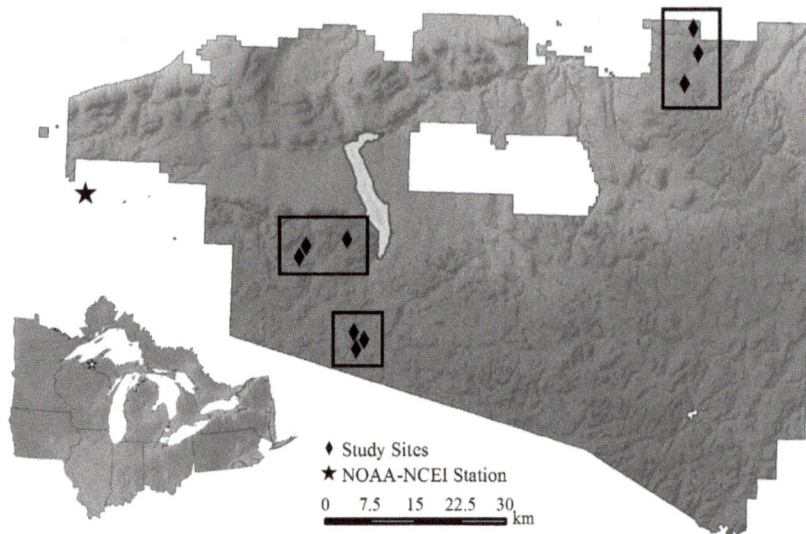

Figure 1. Map of Ottawa National Forest and black ash wetland study site locations within the Great Lakes region of North America (lower-left inset). Three geographically distinct locations (squares) were each subdivided into three treatment sites (diamonds). Regional precipitation data were obtained from the National Oceanic and Atmospheric Administration's National Centers for Environmental Information (NOAA-NCEI) [44] station in Ironwood, Michigan (star).

2.2. Experimental Study Design

Nine study sites within the Ottawa National Forest, ranging in size from 0.25 to 1.25 ha, were established in 2011 in isolated landform depressions within first-order watersheds [17] (Figure 1). Black ash basal area ranged from 26% to 85% of total basal area among study sites (mean = 66%), and black ash was always the most abundant overstory tree [42]. Treatments were assigned randomly to 3 sites in 3 geographically distinct locations (*n* = 3 per treatment). Each distinct location included 1 untreated control, 1 girdle treatment, and 1 ash-cut treatment. A 1-year baseline data-collection period (2012) preceded the 2-year posttreatment monitoring period (2013–2014) of the study. The girdle treatment was used to simulate an active EAB infestation, wherein near-complete stand mortality could take 3–4 years [45]. In the girdled sites, all black ash trees greater than 2.5 cm in diameter at a height of 1.37 m had the bark, cambium, and phloem removed in a 15–30 cm tall circumferential band. The ash-cut treatment was used to simulate post-disturbance ecosystem conditions that might occur following complete loss of black ash from the overstory. In the ash-cut sites, all black ash trees greater than 2.5 cm at a height of 1.37 m were cut with chainsaws within 1.2 m of the soil surface during February 2013 and left on site. Treatments were applied between November 2012 and February 2013.

2.3. Instrumentation and Monitoring

Three random locations within each wetland site were determined with ArcGIS (Esri, Redlands, CA, USA), and were located at least 8 m from the wetland boundary. Two 13 cm tall gas collars constructed of 25.4 cm diameter PVC pipe were co-located 6 m apart and driven into the soil to a depth of 5 cm at each random location (n = 6 per site). Portable static dark chambers were used to measure both CO_2 and CH_4 fluxes [46,47], and were constructed of 25.4 cm ID PVC slip caps. A rim of closed cell foam was adhered to a PVC ring mounted within the chamber to ensure that there was an airtight seal between the gas collar and the chamber. The dark static chambers used to measure CH_4 fluxes were equipped with a small fan to mix the chamber headspace [40]. When water table elevations exceeded 12 cm above the ground surface (6.3% of samples), high-density polyethylene extensions 50 cm long and 28 cm in diameter were used to connect the soil gas collar with the dark chamber.

CO_2 and CH_4 fluxes were measured at approximately monthly intervals during snow-free seasons (June–October) from 2012–2014. All measurements were collected on calm days between 11:00 and 17:30 h (EDT) to reduce insolation variability. Measurements were not collected if more than 6.4 mm of precipitation fell within the previous 8 h. All 18 gas collars within 1 geographically distinct location were measured on the same day, and all sites were measured within a 4-day period each month.

An infrared gas analyzer (IRGA) (PP Systems EGM-4, PP Systems International Inc., Amesbury, MA, USA) was used with a static dark chamber, equipped with a small fan to mix the chamber headspace, to measure CO_2 fluxes [39,40]. CO_2 concentrations were logged by the IRGA at 2–3 s intervals for a total of 124 s. The slope of CO_2 concentration versus time (t, s) was estimated by linear regression [48]. Portions of nonlinear responses that occurred within the first 80 s of the measurement period were manually removed prior to calculating the slope. CO_2 fluxes were then estimated according to Equation (1):

$$CO_2 Flux = \frac{dCO_2}{dt} \times \frac{PV}{ART} \tag{1}$$

where $\frac{dCO_2}{dt}$ is the slope from the regression, P is the atmospheric pressure (Pascal), V is the volume of the headspace within the chamber (m^3), A is the area of the soil encircled by the gas collar (m^2), R is the gas constant (J K^{-1} mol^{-1}), and T is the air temperature (K). All dark chamber headspace volume estimates used in Equation (1) were corrected for water table position and extension headspace volume when necessary.

Gas-tight syringes were used to remove five 20 mL gas samples at consecutive 10 min intervals from each gas chamber [46]. Gas samples were immediately injected into pre-evacuated 6 mL EXETAINER® glass vials (Labco Ltd., Lampeter, Wales, UK), and the septa were coated with 100% silicone for added leakage protection. Gas samples were analyzed within 30 days using gas chromatography (Varian 3800, Agilent Technologies Inc., Santa Clara, CA, USA) at Michigan Technological University's Hydrology and Wetlands Laboratory. As a quality-control measure, concentration standards were injected into EXETAINER® vials in the laboratory, transported to study sites during sampling excursions, and analyzed alongside field samples. The slope of the CH_4 concentration versus time response was calculated using least-squares regression as described for CO_2 above, and the slopes of the CH_4 responses ($\frac{dCH}{dt}$) were used in place of $\frac{dCO_2}{dt}$ in Equation (1) to estimate CH_4 fluxes [15]. Gaseous soil carbon fluxes were reported as mg m^{-2} day^{-1} of CO_2 and CH_4 molecules. Approximately 8% of samples with nonlinear slopes were flagged as possible human-induced ebullition events and were not included in the analysis.

Water levels were manually measured during monthly collection intervals and were recorded at 15 min intervals (Levelogger Junior M5, Solinst Canada Ltd., Georgetown, ON, Canada) in a steel monitoring well in each wetland [17]. The water level relative to the wetland soil surface (WL_{mw}, (cm)) was calculated for each gas collar (WL_{col}, (cm)) using total station surveyed elevations [17] at gas collars and metal monitoring wells, respectively. Soil temperature was measured within the gas collar 5 cm below the ground surface (ST_{5cm}, °C) immediately following each CO_2 flux measurement. Soil temperature probes (107, Campbell Scientific, Logan, UT, USA) connected to on-site data loggers (CR800, Campbell Scientific) were installed to record every 15 min in all sites in 2011, but repeat damage by animals resulted in large amounts of missing data during 2012 and 2013. Temperature loggers (iButtons, Embedded Data Systems LLC, Lawrenceburg, KY, USA) set to record at 2 h intervals were subsequently installed 5 cm beneath the soil surface and within PVC solar shields located 1 m above the soil surface near the wetland monitoring well from June through November of 2014.

2.4. Statistical Analysis

Linear mixed-effects models (LMMs) were fit with the restricted maximum likelihood procedure in R [48,49] to assess the impact of water level, temperature, and treatment on soil CO_2 and CH_4 fluxes. Zero-concentration data ($n = 400$) from gas chromatograph calibration curves were used to determine the 0.36 ppm instrument detection limit for CH_4 [50]. Based upon the 40 min collection period, the lower detectable soil CH_4 flux limit was determined to be 1.25 (mg m^{-2} day^{-1}). Approximately 20% of soil CH_4 flux estimates were below the lower detection limit. Distribution parameters of the natural-log transformed CH_4 flux data were calculated using maximum likelihood estimation to adjust for the censored values. Estimates for censored values were then imputed using random quantiles below the detection limit drawn from the fitted distribution as replacement values [51,52]. Following natural log-transformation of estimates of CO_2 and CH_4 from Equation (1), zero-mean assumptions of normality and homogeneity of variance were determined to be acceptable based on visual examination of residual plots for all linear mixed-effects models [48,49]. LMM methods were used to determine significance [53], the Tukey method was used to adjust p-values for all pairwise comparisons [54], and degrees of freedom were calculated according to the Kenward–Roger approximation [55]. Significance was reported where $p < 0.05$, and marginal significance was reported where $p < 0.1$. Following the transformation of CO_2 and CH_4 fluxes, the back-transformed regression equations took the form of Equation (2):

$$C_{Flux} = \beta_\circ e^{\beta_1 T_{soil}} \tag{2}$$

where $\beta_\circ$ is the back-transformed model intercept coefficient, and β_1 is the model slope coefficient.

Logged temperature and water level data from 2014 were used to test the response of maximum daily soil temperatures (ST_{max}) to maximum daily air temperatures (AT_{max}) and WL_{mw}. The influence of WL_{mw} on the relationship between ST_{max} and AT_{max} was evaluated by comparing a two-level LMM ($AT_{max} \cdot WL_{mw}$) with a single-level LMM (AT_{max}), where AT_{max} and WL_{mw} were used as fixed effects, and site and day were used as crossed random effects. Additionally, the response of ST_{max} to AT_{max}, WL_{mw}, and treatment during the posttreatment study period was tested with a three-level LMM where AT_{max}, WL_{mw}, and treatment were used as fixed effects, and site and day were used as crossed random effects.

LMM selection criteria, including the conditional variance (r_c^2) and marginal variance (r_m^2), the Akaike information criterion (AIC), the residual proportion change in variance (PCV), and residual standard error (RSE) were used to evaluate model performance [53,56]. Multivariate LMM response surfaces were predicted [57] by generating slopes for four water level positions: 10 cm above, 10 cm below, 30 cm below, and 50 cm below the wetland soil surface. Observations were binned with equidistant spacing between the four water level positions and used for visual interpretation only.

The influence of WL_{col} on the relationship between soil gas flux and ST_{5cm} was evaluated by comparing a two-level LMM ($ST_{5cm} \cdot WL_{col}$) with a single-level LMM (ST_{5cm}) for CO_2 and CH_4, respectively (Table 1). Both ST_{5cm} and WL_{col} were used as fixed effects, and correlated random intercepts and slopes were generated for each collar nested within site. Posttreatment CO_2 and CH_4 fluxes and responses to ST_{5cm} and WL_{col} were modeled with a three-level LMM using ST_{5cm}, WL_{col}, and treatment as interacting fixed effects with a correlated random intercept and slope for each collar nested within site. Annual differences in CO_2 and CH_4 soil fluxes by treatment were examined using a four-level LMM with ST_{5cm}, WL_{col}, treatment, and year as interacting fixed-effects terms, with a correlated random intercept and slope for each collar nested within site. A four-level LMM was constructed to compare CO_2 fluxes between treatments and study periods using ST_{5cm}, WL_{col}, treatment, and study period (pre- or posttreatment) as interacting fixed effects, with a correlated random intercept and slope for each collar nested within site. Finally, mean untransformed soil gas fluxes were calculated for each site during each sampling event, and Kruskal–Wallis test was used to compare between-treatment means within each year.

Table 1. Model evaluation statistics for soil temperature, CO_2 flux, and CH_4 flux linear mixed models (LMMs; Figure 2). Residual standard error (RSE) was reported for soil temperature for soil CO_2 and CH_4 flux LMMs. AIC, Akaike information criterion; PCV, proportion change in variance.

LMM	Predictors	r_c^2	r_m^2	AIC	PCV	RSE
Soil Temp.	Null: $x = AT_{max}$	0.97	0.73	1763	–	0.62
	Full: $x = AT_{max} \cdot WL_{mw}$	0.98	0.73	1225	0.6%	0.39
CO_2 Flux	Null: $x = ST_{5cm}$	0.34	0.14	1564	–	2.48
	Full: $x = ST_{5cm} \cdot WL_{col}$	0.51	0.31	1489	0.2%	2.27
CH_4 Flux	Null: $x = ST_{5cm}$	0.49	0.05	621	–	0.86
	Full: $x = ST_{5cm} \cdot WL_{col}$	0.50	0.10	653	0.1%	0.85

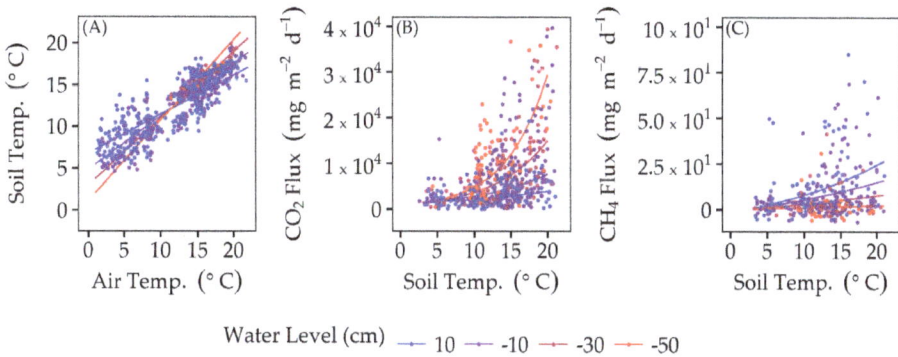

Figure 2. (**A**) Responses of daily maximum soil temperatures to daily maximum air temperatures and water levels; (**B**) Responses of soil CO_2 fluxes to soil temperatures and water levels; (**C**) Responses of soil CH_4 fluxes to soil temperatures and water levels. The lines in each panel represent the predicted [57] responses for four water level positions relative to the wetland soil surface. Observations (points) were color coordinated using equidistant binning for the four water level positions, and are used for visual interpretation only.

3. Results

3.1. Regional Climate and Soil Temperature

Precipitation data for Ironwood, Michigan [44], indicated that the 83 cm of precipitation received in the ONF during 2012 was below the 30-year average of 89 cm, whereas the 127 cm and 111 cm received during 2013 and 2014, respectively, were above the 30-year average. The ONF received approximately 150% more snow-water equivalent (SWE) precipitation during winter 2012–2013 and 100% more SWE during winter 2013–2014 compared to winter 2011–2012. In addition, the ONF received approximately 25% more rain during 2013 and 15% more rain during 2014 compared to 2012.

A comparison of model selection criteria indicated that the full LMM ($AT_{max} \cdot WL_{mw}$) for ST_{max} performed better than the null model (AT_{max} only) by improving the relative quality (AIC), increasing the residual proportion change in variance (PCV), and decreasing the residual standard error (RSE). The full LMM explained 0.98 of the conditional variance (r_c^2) and 0.73 of the marginal variance (r_m^2) (Table 1). The 0.88 (°C °C^{-1}) slope predicted for the ST_{max} response to AT_{max} when WL_{mw} was farthest beneath the soil surface was larger than the 0.39 (°C °C^{-1}) slope predicted when WL_{mw} was above the soil surface (Figure 2). The mean daily soil temperatures of 13.2 and 13.6 °C in the girdle and ash-cut sites, respectively, were both greater than the 12.7 °C soil temperature detected in the controls, although these differences were not significant (Table 2).

Table 2. Summary statistics, including degree of freedom (df), standard error (SE), Tukey honest significant difference (Mean Test), and Kruskal–Wallis rank means comparisons (KW Test), for each soil gas (CO_2 and CH_4), study period (pre- and posttreatment), and treatment.

Gas	Study Period	Treatment	df	Mean Flux (mg m^{-2} d^{-1})	SE (mg m^{-2} d^{-1})	Mean Test	n	Rank Mean	KW Test
CO_2	Pre	Control	18.9	3.0×10^3	1.26	a	10	13.5	a
		Girdle	16.5	4.0×10^3	1.25	a	10	16.5	a
		Ash-Cut	30.9	5.1×10^3	1.32	a	10	16.5	a
	Post	Control	6.2	3.1×10^3	1.18	a	24	30.8	a
		Girdle	7.5	5.7×10^3	1.19	b'	24	37.4	ab
		Ash-Cut	8.7	6.7×10^3	1.24	b	24	41.3	b'
CH_4	Post	Control	5.3	6.6	1.10	ab	24	33.3	a
		Girdle	5.5	2.4	1.11	a	24	26.6	a
		Ash-Cut	5.9	11.9	1.11	b'	24	47.6	b

Significant between-treatment mean differences are indicated by (a) and (b) at $p < 0.05$, where (') indicates detection of marginally significant between-treatment mean differences at $p < 0.1$.

3.2. Soil CO_2 and CH_4 Fluxes

As with soil temperature, a comparison of model statistics indicated that the full LMM ($ST_{5cm} \cdot WL_{col}$) performed better than the null LMM (ST_{5cm} *only*) for CO_2 and CH_4, respectively, by increasing the r_c^2 and r_m^2, increasing the PCV, and decreasing the RSE (Table 1). The predicted slope of the CO_2 flux response to soil temperature was 0.18 (mg m^{-2} day^{-1} °C^{-1}) when WL_{col} was farthest beneath the soil surface, and this value was larger than the 0.06 (mg m^{-2} day^{-1} °C^{-1}) response slope predicted when WL_{col} was above the soil surface (Figure 2) [57]. Conversely, the predicted slope of the CH_4 flux response to soil temperature was 0.05 (mg m^{-2} day^{-1} °C^{-1}) when WL_{col} was above the soil surface, and this value was larger than the 0.006 (mg m^{-2} day^{-1} °C^{-1}) response slope predicted when WL_{col} was farthest below the soil surface (Figure 2).

Soil temperature (ST_{5cm}) and water level (WL_{col}) were used to predict soil CO_2 and CH_4 fluxes for each treatment during the posttreatment study period (Figure 3). The CO_2 flux LMM had a 0.52 r_c^2 and 0.35 r_m^2 and a residual standard error of 2.1 mg m^{-2} day^{-1}, whereas the CH_4 flux LMM had a 0.37 r_c^2 and 0.24 r_m^2 and a residual standard error of 0.9 mg m^{-2} day^{-1}. Due to the limited number of CO_2 flux observations collected when soil temperatures were less than 10 °C during the pretreatment study period and the absence of pretreatment CH_4 flux observations, the pretreatment response slope estimates were either not reliable or nonexistent and therefore were excluded from the three-level LMM. The girdle, ash-cut, and control response slopes for soil CO_2 flux (0.10, 0.08, and 0.10 mg m^{-2} day^{-1} °C^{-1}) and soil CH_4 flux (0.02, 0.04, and 0.06 mg m^{-2} day^{-1} °C^{-1}) were not significantly different from one another during the posttreatment study period. The mean CO_2 flux detected in ash-cut sites was marginally significantly higher than the controls ($p = 0.094$), and the mean CH_4 fluxes detected in the girdle and ash-cut sites were not significantly different ($p > 0.1$) than the control sites during the posttreatment study period (Figure 3).

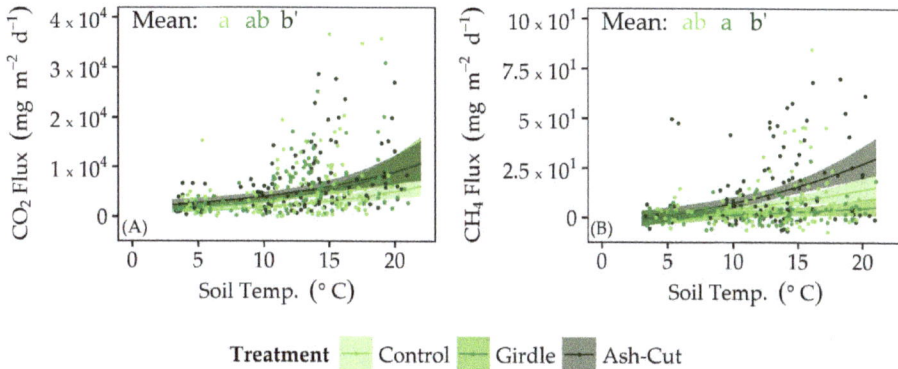

Figure 3. (A) Predicted soil CO_2 flux response to soil temperature and water level (WL_{col}) for each treatment, with shaded 95% confidence intervals [57]; (B) Predicted soil CH_4 flux response to soil temperature and water level (WL_{col}) for each treatment, with shaded 95% confidence intervals. WL_{col} was embedded, using a correlated random intercept and slope, within each LMM as an interacting fixed effect. Significant differences between treatment means are indicated by (a) and (b) at $p < 0.05$, where (') indicates detection of marginally significant between-treatment means differences at $p < 0.1$.

There were no significant differences in mean soil CO_2 flux detected among treatments during 2012 (pretreatment), whereas the 6599 mg m^{-2} day^{-1} ash-cut mean soil CO_2 flux was marginally significantly higher ($p = 0.076$) than the 3193 mg m^{-2} day^{-1} control mean soil CO_2 flux during 2013, and the 6722 and 6835 mean soil CO_2 fluxes in the girdle ($p = 0.042$) and ash-cut ($p = 0.042$), respectively, were significantly higher than the 2979 mg m^{-2} day^{-1} control mean soil CO_2 flux during 2014 (Figure 4). Neither the girdle nor the ash-cut soil CH_4 fluxes were significantly different than the control during 2013 and 2014 (Figure 4). When the entire pre- and posttreatment periods were considered, the 5625 and 6545 mg m^{-2} day^{-1} mean soil CO_2 fluxes detected in the girdle ($p = 0.096$) and ash-cut sites ($p = 0.046$), respectively, were significantly higher than the 3038 mg m^{-2} day^{-1} detected in the control sites during the posttreatment study period (Table 2). The nonparametric Kruskal–Wallis test shows that higher CO_2 fluxes occurred more frequently in the ash-cut than the controls ($p = 0.085$) (Table 2). The 11.9 mg m^{-2} day^{-1} ash-cut mean soil CH_4 flux was almost twice as high as the 6.6 mg m^{-2} day^{-1} control mean soil CH_4 flux, and marginally significantly ($p = 0.079$) higher than the 2.4 mg m^{-2} day^{-1} girdle mean soil CH_4 flux during the posttreatment study period (Table 2). While the mean CH_4 fluxes detected in girdle and ash-cut sites were similar to controls during the posttreatment period, the nonparametric Kruskal–Wallis test shows that higher CH_4 fluxes occurred more frequently in the ash-cut sites than the controls ($p = 0.026$) (Table 2). Within-treatment soil CO_2 flux comparisons were not considered due to the limited number of flux observations collected when soil temperatures were less than 10 °C during the pretreatment study period, and no pretreatment CH_4 flux data were available.

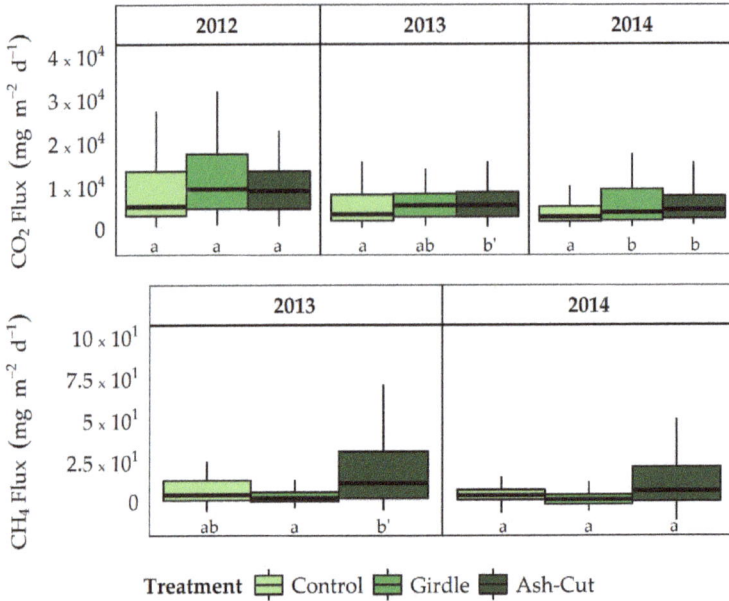

Figure 4. (A) Soil CO_2 flux for each treatment during 2012, 2013, and 2014; **(B)** Soil CH_4 flux for each treatment during 2013 and 2014. Significant between-treatment differences in mean flux at $p > 0.05$ are indicated by (a) and (b), where (′) indicates detection of marginally significant between-treatment mean differences at $p < 0.1$. Boxplot lines represent median, 25%, and 75% quantiles, while whiskers represent quantiles $\pm 1.5 \cdot$ interquartile range, respectively [58].

4. Discussion

4.1. Soil Temperature and Water Level

Wetland water level (WL_{mw}) significantly altered the response of soil temperature (ST_{max}) to air temperature (AT_{max}; Figure 2). When WL_{mw} was lower and the soil presumably drier, ST_{max} more closely paralleled AT_{max}. In contrast, when WL_{mw} was higher (near or above surface), ST_{max} was less correlated to AT_{max}. The combined influence of the higher specific heat capacity of water and the consistent groundwater discharge into black ash wetlands [17] likely decreased the sensitivity of ST_{max} to AT_{max} when soils were saturated. Furthermore, soil temperature response to treatment was likely mitigated by the consistent influx of groundwater because of the expected increase in magnitude of latent heat exchange and reduction in magnitude of sensible heat exchange between soil and atmosphere [59].

The wetter conditions (Table 2) and higher wetland water levels during the posttreatment period [17] may have buffered the magnitude of ST_{max} response to disturbance. Because of the wetter-than-average posttreatment water years and the resultant elevated water tables, the average soil temperatures in girdle and ash-cut sites were not significantly greater than the controls, despite the presumed increase in canopy openness and insolation. We theorized that the increased amount of insolation likely received by soils in the disturbed study sites would have a greater influence on the response of ST_{max} to AT_{max} during drier periods. It is likely that average ST_{max} differences would have been larger during the posttreatment study period under drier conditions.

4.2. Responses of Soil CO_2 and CH_4 Fluxes to Disturbance

The water level position relative to the wetland soil surface (WL_{col}) regulated the response of soil CO_2 and CH_4 fluxes to soil temperature (ST_{5cm}; Figure 2), and co-varied soil CO_2 and CH_4 flux responses to ST_{5cm} and WL_{col} have been observed in numerous wetland types [15,19,60]. Soil CO_2 fluxes tended to be higher and increase more rapidly in response to increased temperature when WL_{col} was below the wetland soil surface, whereas soil CH_4 fluxes tended to be higher and increase more rapidly in response to increased temperature when WL_{col} was at or above the wetland soil surface. The magnitude of observed soil CO_2 and CH_4 fluxes from depressional black ash wetlands was similar to fluxes observed in wetlands under field and laboratory conditions [15,61–63].

Given that both the magnitude of soil CO_2 and CH_4 fluxes and the correlation of ST_{max} to AT_{max} were controlled by wetland water levels, it is likely that interannual climatic conditions also affect the relative magnitude of mean annual soil CO_2 and CH_4 fluxes. Specifically, mean annual soil CO_2 fluxes may be higher during drier periods and lower during wetter periods. Conversely, soil CH_4 fluxes would be expected to be higher during wetter periods and lower during drier periods. Therefore, it is possible that the soil CO_2 flux response to disturbance was buffered and the soil CH_4 flux response was enhanced because of the wetter-than-average conditions during the posttreatment study period [17]. Black ash wetland water levels tended to be higher during the two wetter water years [17], and soil temperature tended to be less responsive to air temperature when water levels were higher (Figure 2). Moreover, wetland soil CO_2 fluxes are typically lower when water tables are above the surface compared to fluxes measured during lower water table positions [19,62] Therefore, soil CO_2 fluxes from disturbed sites may have been higher and possibly more dissimilar than control sites if drier climatic conditions had occurred during the two posttreatment monitoring years.

The magnitude of soil CO_2 fluxes was significantly higher in disturbed sites, and high soil CH_4 fluxes occurred more frequently in the ash-cut than the control during the posttreatment study period (Figures 3 and 4; Table 2). As evidenced by the similarity among slopes, the sensitivity of soil CO_2 fluxes to environmental variables (ST_{5cm} and WL_{col}) did not change following disturbance (Figure 3). Therefore, the higher soil CO_2 fluxes from disturbed sites were not caused by an altered sensitivity to environmental variables. Similarly, soil CH_4 response slopes for the three treatments were not significantly different from one another during the posttreatment study period (Figure 3), and the significantly more frequent occurrence of high CH_4 fluxes in ash-cut sites may not have been caused by an altered sensitivity to environmental variables.

It is possible that the increased rate of heterotrophic respiration from disturbed soils stemmed from an alteration in the organic chemical composition of peatland soil [64]. Carbon mineralization rates were positively correlated with labile carbon compounds and negatively correlated with recalcitrant carbon compounds in a northern Minnesota peatland [65]. Increased rates of CH_4 production were detected following the addition of glucose to lignin-rich peat in northern North American peatlands matter [66], and CO_2 and CH_4 production was significantly affected by pH and C:N ratios in Dutch peat grasslands [67]. Fine root respiration may also account for a substantial proportion of soil CO_2 fluxes, and increased soil CO_2 fluxes have been attributed to turnover and decomposition of fine roots [68–70]. Black ash trees have a shallow, fibrous root system [71], and the majority of this fine root system was likely available for decomposition following disturbance. Unlike the rapid infilling from co-occurring tree species that occurred following EAB infestation in green ash (*Fraxinus pennsylvanica* Marsh.) stands in northwest Ohio [72], neither overstory or understory stems of co-occurring tree species within these black ash wetland study sites responded positively following treatment [42]. Therefore, the increased magnitude of CO_2 fluxes (Figure 4) from disturbed black ash wetland soils may be at least partially attributed to the transient large influx of decomposable fine roots and an associated response in microbial activity and biomass [63,73], and not likely the result of increased autotrophic respiration from the release of co-occurring tree species. Inversely, the loss of autotrophic root respiration from mature individuals may lead to a decrease in soil CO_2 flux following disturbance [74,75]. A lack of response of inorganic soil nitrogen availability following treatment in these sites during this study period [76] supports the explanation that increased microbial activity drove the observed CO_2 response.

The study sites were consistently connected to mineral-rich groundwater during the study period [17], and this connection likely influenced carbon mineralization dynamics within depressional black ash wetland soils. The relative standard error of soil CH_4 fluxes (18.0%) was considerably larger than the relative standard error of soil CO_2 fluxes (<0.1%; Table 2). Because the highest rates of methane oxidation are expected near oxic/anoxic interfaces [28], the large variation of soil CH_4 fluxes that occurred when wetlands were saturated and/or inundated (Figure 3) was not likely caused by increased rates of methanotrophy within the peat profile. This further suggests that methanogenesis may have been constrained by the availability of electron acceptors [77], inhibited by chemical compounds [28] or low pH [78], or transported to the atmosphere through vegetation with arenchyma [79,80].

When compared to control sites, and because of the warmer and wetter conditions observed in the disturbed sites after treatment, we expected to observe higher posttreatment mean soil CH_4 fluxes in ash-cut and girdle sites. However, the persistent connection with mineral-rich groundwater during the wetter posttreatment study period may, in fact, have buffered the relative magnitude of CH_4 flux response to disturbance by inhibiting methanogenesis. Specifically, it is likely that study sites were consistently being replenished with more efficient electron-accepting redox species such as Fe^{2+}, Mn^{4+}, or SO_4^{2-} during the wetter posttreatment period, which effectively inhibited methanogenesis. It is also possible that methane production within the soil increased following disturbance, but increased CH_4 fluxes were not detected from the soil surface due to plant-mediated transport.

The large increase of herbaceous cover following disturbance in these sites may have limited natural woody regeneration recruitment to larger seedling size classes or saplings [42], and canopy removal has been shown to convert black ash–dominated wetlands to herbaceous or scrub-shrub–dominated wetlands [81]. As a result, canopy replacement through natural regeneration may take several decades [82], where planting mitigation efforts will influence the tree species composition, likely decrease the duration of canopy recovery, and help maintain ecosystem function following disturbance [83–85]. Though the hydrogeomorphic setting is distinctly different from the black ash wetlands considered here, these findings may have important implications for potential effects in other ash-dominated wetlands such as those dominated by green and pumpkin ash (*Fraxinus profunda* Bush). The contribution of herbaceous vegetation to soil respiration following disturbance may be substantial [86], and the contribution of the herbaceous community to autotrophic soil respiration may have increased, given the increased amount of herbaceous cover detected in these sites post treatment [42]. In addition, the increased magnitude of heterotrophic respiration associated with the elevated soil temperatures and transient large influx of decomposable fine roots likely contributed to the significantly higher girdle and ash-cut mean soil CO_2 fluxes during the posttreatment period when compared to controls. Warmer and wetter conditions are expected in the upper Great Lakes region under elevated atmospheric CO_2 conditions, which may also affect soil CH_4 fluxes in EAB-affected wetlands [87]. Therefore, it is likely that successful planting mitigation efforts will effectively decrease the duration of EAB infestation–induced increases of soil CO_2 fluxes and buffer the losses of soil C storage within depressional black ash wetlands.

5. Conclusions

Control, girdle, and ash-cut treatments were applied to three sites and used to simulate an EAB disturbance. Water level, soil temperature, and soil CO_2 and CH_4 fluxes were monitored during pre- and posttreatment study periods. Mean soil temperatures were higher in girdle and ash-cut sites compared to the controls. The relative water table position influenced the degree of soil temperature sensitivity to air temperature, where soil temperature more closely paralleled air temperature when water levels were below the wetland soil surface. Consequently, the wetter-than-average posttreatment climatic conditions [17] likely inhibited the magnitude of soil temperature response to disturbance where average or drier-than-average climatic conditions may have yielded even larger differences between treated and untreated sites.

Soil CO_2 fluxes were significantly higher and higher soil CH_4 fluxes occurred more frequently in disturbed sites, but mean soil CH_4 fluxes were similar in disturbed and undisturbed sites. Soil temperature and wetland water levels controlled soil CO_2 and CH_4 fluxes, where warmer and drier conditions yielded higher soil CO_2 fluxes, and warmer and wetter conditions yielded higher soil CH_4 fluxes. The relationship between soil CO_2 and CH_4 fluxes and soil temperature and water level was not influenced by a simulated EAB disturbance. The combination of higher mean soil temperatures, increased herbaceous cover, and the transient large influx of decomposable fine roots following canopy suppression may have contributed to the higher soil CO_2 fluxes detected in disturbed sites. Conversely, the consistent connection with mineral-rich groundwater flow likely constrained methanogenesis, and consequently reduced the relative magnitude of soil CH_4 flux response to a simulated EAB disturbance.

Black ash wetland soil carbon mineralization processes were sensitive, but also showed resilience to EAB-induced impacts. Both CO_2 and CH_4 fluxes were regulated by soil temperature and water level position. The consistent connection with groundwater flow systems and wetter-than-average weather conditions during the posttreatment period likely buffered the relative magnitude of soil CH_4 response to disturbance.

Author Contributions: M.V.G., J.S., J.D., R.K., and T.P. conceived and designed the experiments; M.V.G., J.S., N.B., and J.D. performed the experiments; M.V.G., J.S., and N.B. analyzed the data; and all authors contributed to writing the paper.

Acknowledgments: We appreciate the support of the US Forest Service, in particular Mark Fedora, and the rest of the Ottawa National Forest. Funding for this project was provided by the USDA, the US-EPA Great Lakes Restoration Initiative (DW-12-92429101-0), Michigan Tech's Center for Water and Society, and Michigan Tech's Ecosystem Science Center. Special thanks to Stephen Sebestyen, Alex Mayer, Robert Froese, Nicholas Schreiner, Jarrod Nelson, Leah Harrison, Erica Jones, Jon Bontrager, Dan Hutchinson, John Hrjiblan, Lynette Potvin, and Timothy Veverica for their technical and logistical support.

Conflicts of Interest: The authors declare no conflict of interest.

References

1. McLaughlin, J.W.; Gale, M.R.; Jurgensen, M.F.; Trettin, C.C. Soil organic matter and nitrogen cycling in response to harvesting, mechanical site preparation, and fertilization in a wetland with a mineral substrate. *For. Ecol. Manag.* **2000**, *129*, 7–23. [CrossRef]
2. Sun, G.; McNulty, S.G.; Shepard, J.P.; Amatya, D.M.; Riekerk, H.; Comerford, N.B.; Skaggs, W.; Swift, L. Effects of timber management on the hydrology of wetland forests in the southern united states. *For. Ecol. Manag.* **2001**, *143*, 227–236. [CrossRef]
3. MacFarlane, D.W.; Meyer, S.P. Characteristics and distribution of potential ash tree hosts for emerald ash borer. *For. Ecol. Manag.* **2005**, *213*, 15–24. [CrossRef]
4. Kashian, D.M.; Witter, J.A. Assessing the potential for ash canopy tree replacement via current regeneration following emerald ash borer-caused mortality on southeastern Michigan landscapes. *For. Ecol. Manag.* **2011**, *261*, 480–488. [CrossRef]
5. Flower, C.E.; Knight, K.S.; Gonzalez-Meler, M.A. Impacts of the emerald ash borer (*Agrilus planipennis* fairmaire) induced ash (*Fraxinus* spp.) mortality on forest carbon cycling and successional dynamics in the eastern united states. *Biol. Invasions* **2013**, *15*, 931–944. [CrossRef]
6. Sydnor, T.D.; Bumgardner, M.; Todd, A. The potential economic impacts of emerald ash borer (*Agrilus planipennis*) on Ohio, US, Communities. *Arboricult. Urban For.* **2007**, *33*, 48–54.
7. Kovacs, K.F.; Haight, R.G.; McCullough, D.G.; Mercader, R.J.; Siegert, N.W.; Liebhold, A.M. Cost of potential emerald ash borer damage in U.S. Communities, 2009–2019. *Ecol. Econ.* **2010**, *69*, 569–578. [CrossRef]
8. Gandhi, K.K.; Herms, D. Direct and indirect effects of alien insect herbivores on ecological processes and interactions in forests of eastern north America. *Biol. Invasions* **2010**, *12*, 389–405. [CrossRef]
9. Lovett, G.M.; Canham, C.D.; Arthur, M.A.; Weathers, K.C.; Fitzhugh, R.D. Forest ecosystem responses to exotic pests and pathogens in eastern north America. *BioScience* **2006**, *56*, 395–405. [CrossRef]

10. Kimble, J.M. *The Potential of U.S. Forest Soils to Sequester Carbon and Mitigate the Greenhouse Effect*; CRC Press: Boca Raton, FL, USA, 2003; p. 429.

11. Freeman, C.; Ostle, N.; Kang, H. An enzymic 'latch' on a global carbon store–A shortage of oxygen locks up carbon in peatlands by restraining a single enzyme. *Nature* **2001**, *409*, 149. [CrossRef] [PubMed]

12. Gorham, E. Northern peatlands: Role in the carbon cycle and probable responses to climatic warming. *Ecol. Appl.* **1991**, *1*, 182–195. [CrossRef] [PubMed]

13. Hogberg, P.; Nordgren, A.; Buchmann, N.; Taylor, A.F.S.; Ekblad, A.; Hogberg, M.N.; Nyberg, G.; Ottosson-Lofvenius, M.; Read, D.J. Large-scale forest girdling shows that current photosynthesis drives soil respiration. *Nature* **2001**, *411*, 789–792. [CrossRef] [PubMed]

14. Zerva, A.; Mencuccini, M. Short-term effects of clearfelling on soil CO_2, CH_4, and N_2O fluxes in a sitka spruce plantation. *Soil Biol. Biochem.* **2005**, *37*, 2025–2036. [CrossRef]

15. Moore, T.R.; Dalva, M. The influence of temperature and water-table position on carbon-dioxide and methane emissions from laboratory columns of peatland soils. *J. Soil Sci.* **1993**, *44*, 651–664. [CrossRef]

16. Macdonald, J.A.; Fowler, D.; Hargreaves, K.J.; Skiba, U.; Leith, I.D.; Murray, M.B. Methane emission rates from a northern wetland; response to temperature, water table and transport. *Atmos. Environ.* **1998**, *32*, 3219–3227. [CrossRef]

17. Van Grinsven, M.J.; Shannon, J.P.; Davis, J.C.; Bolton, N.W.; Wagenbrenner, J.W.; Kolka, R.K.; Grant Pypker, T. Source water contributions and hydrologic responses to simulated emerald ash borer infestations in depressional black ash wetlands. *Ecohydrology* **2017**, *10*. [CrossRef]

18. Slesak, R.A.; Lenhart, C.F.; Brooks, K.N.; D'Amato, A.W.; Palik, B.J. Water table response to harvesting and simulated emerald ash borer mortality in black ash wetlands in Minnesota, USA. *Can. J. For. Res.* **2014**, *44*, 961–968. [CrossRef]

19. Blodau, C. Carbon cycling in peatlands—A review of processes and controls. *Environ. Rev.* **2002**, *10*, 111–134. [CrossRef]

20. Nancy, D.; Narasinha, S.; Peter, W.; Shashi, V.; Elon, V.; Eville, G.; Patrick, C.; Robert, H.; Cheryl, K.; Joseph, Y.; et al. Carbon emissions from peatlands. In *Peatland Biogeochemistry and Watershed Hydrology at the Marcell Experimental Forest*; CRC Press: Boca Raton, FL, USA, 2011; pp. 297–347.

21. McFee, W.W.; Kelly, J.M. Soil carbon in northern forested wetlands: Impacts of silvicultural practices. *Carbon* **1995**, *1*, 437–461.

22. Trettin, C.C.; Jurgensen, M.F.; Grigal, D.F.; Gale, M.R.; Jeglum, J.R. *Northern Forested Wetlands Ecology and Management*; CRC Press: Boca Raton, FL, USA, 1996.

23. Eswaran, H.; Van Den Berg, E.; Reich, P. Organic carbon in soils of the world. *Soil Sci. Soc. Am. J.* **1993**, *57*, 192–194. [CrossRef]

24. Bridgham, S.; Megonigal, J.P.; Keller, J.; Bliss, N.; Trettin, C. The carbon balance of north American wetlands. *Wetlands* **2006**, *26*, 889–916. [CrossRef]

25. Freeman, C.; Kim, S.Y.; Lee, S.H.; Kang, H. Effects of elevated atmospheric CO_2 concentrations on soil microorganisms. *J. Microbiol.* **2004**, *42*, 267–277. [PubMed]

26. Blodau, C.; Basiliko, N.; Moore, T.R. Carbon turnover in peatland mesocosms exposed to different water table levels. *Biogeochemistry* **2004**, *67*, 331–351. [CrossRef]

27. Dalva, M.; Moore, T.R.; Arp, P.; Clair, T.A. Methane and soil and plant community respiration from wetlands, Kejimkujik National Park, Nova Scotia: Measurements, predictions, and climatic change. *J. Geophys. Res. Atmos.* **2001**, *106*, 2955–2962. [CrossRef]

28. Segers, R. Methane production and methane consumption: A review of processes underlying wetland methane fluxes. *Biogeochemistry* **1998**, *41*, 23–51. [CrossRef]

29. Aller, R.C. Bioturbation and remineralization of sedimentary organic-matter—Effects of redox oscillation. *Chem. Geol.* **1994**, *114*, 331–345. [CrossRef]

30. van Hulzen, J.B.; Segers, R.; van Bodegom, P.M.; Leffelaar, P.A. Temperature effects on soil methane production: An explanation for observed variability. *Soil Biol. Biochem.* **1999**, *31*, 1919–1929. [CrossRef]

31. Fissore, C.; Giardina, C.P.; Kolka, R.K.; Trettin, C.C. Soil organic carbon quality in forested mineral wetlands at different mean annual temperature. *Soil Biol. Biochem.* **2009**, *41*, 458–466. [CrossRef]

32. Hagerty, S.B.; Van Groenigen, K.J.; Allison, S.D.; Hungate, B.A.; Schwartz, E.; Koch, G.W.; Kolka, R.K.; Dijkstra, P. Accelerated microbial turnover but constant growth efficiency with warming in soil. *Nat. Clim. Chang.* **2014**, *4*, 903–906. [CrossRef]

33. Daulat, W.E.; Clymo, R.S. Effects of temperature and watertable on the efflux of methane from peatland surface cores. *Atmos. Environ.* **1998**, *32*, 3207–3218. [CrossRef]

34. Kettunen, A.; Kaitala, V.; Lehtinen, A.; Lohila, A.; Alm, J.; Silvola, J.; Martikainen, P.J. Methane production and oxidation potentials in relation to water table fluctuations in two boreal mires. *Soil Biol. Biochem.* **1999**, *31*, 1741–1749. [CrossRef]

35. Peters, V.; Conrad, R. Sequential reduction processes and initiation of CH_4 production upon flooding of oxic upland soils. *Soil Biol. Biochem.* **1996**, *28*, 371–382. [CrossRef]

36. Moore, T.R.; Heyes, A.; Roulet, N.T. Methane emissions from wetlands, southern hudson-bay lowland. *J. Geophys. Res. Atmos.* **1994**, *99*, 1455–1467. [CrossRef]

37. Amaral, J.A.; Knowles, R. Methane metabolism in a temperate swamp. *Appl. Environ. Microbiol.* **1994**, *60*, 3945–3951. [PubMed]

38. Shannon, R.D.; White, J.R. A three-year study of controls on methane emissions from two Michigan peatlands. *Biogeochemistry* **1994**, *27*, 35–60. [CrossRef]

39. Ballantyne, D.; Hribljan, J.; Pypker, T.; Chimner, R. Long-term water table manipulations alter peatland gaseous carbon fluxes in northern Michigan. *Wetl. Ecol. Manag.* **2014**, *22*, 35–47. [CrossRef]

40. Johnson, C.P.; Pypker, T.G.; Hribljan, J.A.; Chimner, R.A. Open top chambers and infrared lamps: A comparison of heating efficacy and CO_2/CH_4 dynamics in a northern Michigan peatland. *Ecosystems* **2013**, *16*, 736–748. [CrossRef]

41. Prasad, A.M.; Iverson, L.R. *Little's Range and FIA Importance Value Database for 135 Eastern US Tree Species*; Northeastern Research Station: Delaware, OH, USA, 2003.

42. Davis, J.C.; Shannon, J.P.; Bolton, N.W.; Kolka, R.K.; Pypker, T. Vegetation responses to simulated emerald ash borer infestation in *Fraxinus nigra* dominated wetlands of Upper Michigan, USA. *Can. J. For. Res.* **2017**, *47*, 319–330. [CrossRef]

43. Soil Survey Staff. Web Soil Survey. Natural Resources Conservation Service, United States Department of Agriculture. Available online: http://websoilsurvey.Nrcs.Usda.Gov/ (accessed on 14 November 2014).

44. NOAA-NCEI. National Oceanic and Atomospheric Administration. National Centers for Environmental Information: Asheville, NC, USA. Available online: https://www.ncei.noaa.gov/ (accessed on 15 November 2017).

45. McCullough, D.G.; Schneeberger, N.F.; Katovich, S.A.; Siegert, N.W. Pest Alert: Emerald Ash Borer. USDA For. Serv., Northeastern Area, State & Private Forestry: Newtown Square, PA, USA, 2015. NA-PR-02-04. Available online: http://www.na.fs.fed.us/spfo/pubs/pest_al/eab/eab.pdf (accessed on 5 December 2017).

46. Tuittila, E.-S.; Komulainen, V.-M.; Vasander, H.; Nykänen, H.; Martikainen, P.J.; Laine, J. Methane dynamics of a restored cut-away peatland. *Glob. Chang. Biol.* **2000**, *6*, 569–581. [CrossRef]

47. Carroll, P.; Crill, P. Carbon balance of a temperate poor fen. *Glob. Biogeochem. Cycles* **1997**, *11*, 349–356. [CrossRef]

48. R Core Team. *R: A Language and Environment for Statistical Computing [Computer Software]*; R: Foundation for Statistical Computing: Vienna, Austria, 2016.

49. Bates, D.; Mächler, M.; Bolker, B.; Walker, S. Fitting linear mixed-effects models using LME4. *J. Stat. Softw.* **2015**, *67*, 1–48. [CrossRef]

50. Skoog, D.A.; Holler, J.; Crouch, S.R. Liquid chromatography. In *Principles of Instrumental Analysis*, 6th ed.; Thomson Cengage Learning: Boston, MA, USA, 2006; pp. 816–855.

51. Hewett, P.; Ganser, G.H. A comparison of several methods for analyzing censored data. *Ann. Occup. Hyg.* **2007**, *51*, 611–632. [PubMed]

52. Millard, S.P. *Envstats, an R Package for Environmental Statistics*; Springer Science & Business Media: Berlin, Germany, 2013.

53. Nakagawa, S.; Schielzeth, H. A general and simple method for obtaining R2 from generalized linear mixed-effects models. *Methods Ecol. Evol.* **2013**, *4*, 133–142. [CrossRef]

54. Lenth, R.V. Lsmeans: Least-Squares Means. R Package Version 2013, Volume 2. Available online: https://www.jstatsoft.org/article/view/v069i01/0 (accessed on 18 October 2016).

55. Halekoh, U.; Højsgaard, S.; Højsgaard, M.S.; Matrix, I. Package 'pbkrtest'. 2014. Available online: https://cran.r-project.org/web/packages/pbkrtest/index.html (accessed on 23 September 2016).

56. Gałecki, A.; Burzykowski, T. *Linear Mixed-Effects Models Using R*; Springer: New York, NY, USA, 2013.

57. Fox, J. Effect displays in R for generalised linear models. *J. Stat. Softw.* **2003**, *8*, 1–27. [CrossRef]

58. Wickham, H.; Chang, W. Ggplot2: An Implementation of the Grammar of Graphics. 2015. Available online: http://ftp.auckland.ac.nz/software/CRAN/src/contrib/Descriptions/ggplot.html (accessed on 11 August 2016).

59. Walko, R.L.; Band, L.E.; Baron, J.; Kittel, T.G.; Lammers, R.; Lee, T.J.; Ojima, D.; Pielke, R.A., Sr.; Taylor, C.; Tague, C. Coupled atmosphere–biophysics–hydrology models for environmental modeling. *J. Appl. Meteorol.* **2000**, *39*, 931–944. [CrossRef]

60. Kolka, R.K.; Sebestyen, S.D.; Verry, E.S.; Brooks, K.N. *Peatland Biogeochemistry and Watershed Hydrology at the Marcell Experimental Forest*; CRC Press Taylor & Francis Group: Boca Raton, FL, USA, 2011.

61. Noh, N.J.; Shannon, J.; Bolton, N.; Davis, J.; Van Grinsven, M.; Pypker, T.; Kolka, R.; Wagenbrenner, J. Carbon dioxide fluxes from coarse dead wood in black ash wetlands. *Forests* **2018**, in press.

62. Dinsmore, K.; Skiba, U.; Billett, M.; Rees, R. Effect of water table on greenhouse gas emissions from peatland mesocosms. *Plant Soil* **2009**, *318*, 229–242. [CrossRef]

63. Chimner, R.A.; Cooper, D.J. Influence of water table levels on CO_2 emissions in a colorado subalpine fen: An in situ microcosm study. *Soil Biol. Biochem.* **2003**, *35*, 345–351. [CrossRef]

64. Leifeld, J.; Steffens, M.; Galego-Sala, A. Sensitivity of peatland carbon loss to organic matter quality. *Geophys. Res. Lett.* **2012**, *39*. [CrossRef]

65. Updegraff, K.; Pastor, J.; Bridgham, S.D.; Johnston, C.A. Environmental and substrate controls over carbon and nitrogen mineralization in northern wetlands. *Ecol. Appl.* **1995**, *5*, 151–163. [CrossRef]

66. Yavitt, J.B.; Williams, C.J.; Wieder, R.K. Production of methane and carbon dioxide in peatland ecosystems across north America: Effects of temperature, aeration, and organic chemistry of pent. *Geomicrobiol. J.* **1997**, *14*, 299–316. [CrossRef]

67. Best, E.P.H.; Jacobs, F.H.H. The influence of raised water table levels on carbon dioxide and methane production in ditch-dissected peat grasslands in the netherlands. *Ecol. Eng.* **1997**, *8*, 129–144. [CrossRef]

68. Ewel, K.C.; Cropper, W.P., Jr.; Gholz, H.L. Soil CO_2 evolution in florida slash pine plantations. Ii. Importance of root respiration. *Can. J. For. Res.* **1987**, *17*, 330–333. [CrossRef]

69. Hanson, P.J.; Edwards, N.T.; Garten, C.T.; Andrews, J.A. Separating root and soil microbial contributions to soil respiration: A review of methods and observations. *Biogeochemistry* **2000**, *48*, 115–146. [CrossRef]

70. Edwards, N.T.; Harris, W.F. Carbon cycling in a mixed deciduous forest floor. *Ecology* **1977**, *58*, 431–437. [CrossRef]

71. Burns, R.M.; Honkala, B.H. *Silvics of North America*; United States Department of Agriculture: Washington, DC, USA, 1990; Volume 2.

72. Costilow, K.; Knight, K.; Flower, C. Disturbance severity and canopy position control the radial growth response of maple trees (*Acer* spp.) in forests of northwest Ohio impacted by emerald ash borer (*Agrilus planipennis*). *Ann. For. Sci.* **2017**, *74*, 10. [CrossRef]

73. Thomas, K.L.; Benstead, J.; Davies, K.L.; Lloyd, D. Role of wetland plants in the diurnal control of CH_4 and CO_2 fluxes in peat. *Soil Biol. Biochem.* **1996**, *28*, 17–23. [CrossRef]

74. Nave, L.; Gough, C.; Maurer, K.; Bohrer, G.; Hardiman, B.; Le Moine, J.; Munoz, A.; Nadelhoffer, K.; Sparks, J.; Strahm, B. Disturbance and the resilience of coupled carbon and nitrogen cycling in a north temperate forest. *J. Geophys. Res. Biogeosci.* **2011**, *116*. [CrossRef]

75. Nuckolls, A.E.; Wurzburger, N.; Ford, C.R.; Hendrick, R.L.; Vose, J.M.; Kloeppel, B.D. Hemlock declines rapidly with hemlock woolly adelgid infestation: Impacts on the carbon cycle of southern Appalachian forests. *Ecosystems* **2009**, *12*, 179–190. [CrossRef]

76. Davis, J.C. *Vegetation Dynamics and Nitrogen Cycling Responses to Simulated Emerald Ash Borer Infestation in Fraxinus nigra-Dominated Wetlands of Upper Michigan, USA*; Michigan Technological University: Houghton, MI, USA, 2016.

77. Zehnder, A.; Stumm, W. Geochemistry and biogeochemistry of anaerobic habitats. In *Biology of Anaerobic Microorganisms*; John Wiley & Sons, Inc.: New York, NY, USA, 1988; pp. 1–38.

78. Dunfield, P.; Knowles, R.; Dumont, R.; Moore, T.R. Methane production and consumption in temperate and subarctic peat soils: Response to temperature and PH. *Soil Biol. Biochem.* **1993**, *25*, 321–326. [CrossRef]

79. Moore, T.R.; Dalva, M. Methane and carbon dioxide exchange potentials of peat soils in Aerobic and Anaerobic laboratory incubations. *Soil Biol. Biochem.* **1997**, *29*, 1157–1164. [CrossRef]

80. Gauci, V.; Gowing, D.J.G.; Hornibrook, E.R.C.; Davis, J.M.; Dise, N.B. Woody stem methane emission in mature wetland alder trees. *Atmos. Environ.* **2010**, *44*, 2157–2160. [CrossRef]

81. Erdmann, G.G.; Crow, T.R.; Ralph, M., Jr.; Wilson, C.D. *Managing Black Ash in the Lake States*; General Technical Report NC-115; US Dept. of Agriculture, Forest Service, North Central Forest Experiment Station: St. Paul, MN, USA, 1987; Volume 115.
82. Flower, C.E.; Gonzalez-Meler, M.A. Responses of temperate forest productivity to insect and pathogen disturbances. *Annu. Rev. Plant Biol.* **2015**, *66*, 547–569. [CrossRef] [PubMed]
83. Looney, C.E.; D'Amato, A.W.; Palik, B.J.; Slesak, R.A. Canopy treatment influences growth of replacement tree species in *Fraxinus nigra* forests threatened by the emerald ash borer in Minnesota, USA. *Can. J. For. Res.* **2016**, *47*, 183–192. [CrossRef]
84. Bolton, N.; Shannon, J.; Davis, J.; Grinsven, M.V.; Noh, N.J.; Schooler, S.; Kolka, R.; Pypker, T.; Wagenbrenner, J. Methods to improve survival and growth of planted alternative species seedlings in black ash ecosystems threatened by emerald ash borer. *Forests* **2018**, *9*, 146. [CrossRef]
85. Iverson, L.; Knight, K.S.; Prasad, A.; Herms, D.A.; Matthews, S.; Peters, M.; Smith, A.; Hartzler, D.M.; Long, R.; Almendinger, J. Potential species replacements for black ash (*Fraxinus nigra*) at the confluence of two threats: Emerald ash borer and a changing climate. *Ecosystems* **2016**, *19*, 248–270. [CrossRef]
86. Zehetgruber, B.; Kobler, J.; Dirnböck, T.; Jandl, R.; Seidl, R.; Schindlbacher, A. Intensive ground vegetation growth mitigates the carbon loss after forest disturbance. *Plant Soil* **2017**, *420*, 239–252. [CrossRef] [PubMed]
87. Hayhoe, K.; VanDorn, J.; Croley, T., II; Schlegal, N.; Wuebbles, D. Regional climate change projections for Chicago and the US great lakes. *J. Great Lakes Res.* **2010**, *36*, 7–21. [CrossRef]

forests

MDPI

Article

Modest Effects of Host on the Cold Hardiness of Emerald Ash Borer [†]

Lindsey D. E. Christianson [1] and Robert C. Venette [2,*]

[1] Department of Entomology, University of Minnesota, 1980 Folwell Ave, Saint Paul, MN 55108, USA; chri1203@umn.edu
[2] Northern Research Station, USDA Forest Service, 1561 Lindig St., Saint Paul, MN 55108, USA
* Correspondence: rvenette@fs.fed.us; Tel.: +1-651-649-5028
† Adapted from a Master's Thesis, Department of Entomology, University of Minnesota, Saint Paul, MN 55108, USA.

Received: 31 March 2018; Accepted: 7 June 2018; Published: 12 June 2018

Abstract: The emerald ash borer, *Agrilus planipennis* Fairmaire, is invading North America and Europe but has not yet reached its ultimate distribution. Geographic differences in host availability and winter temperatures might affect where this species will occur. In central North America, black ash (*Fraxinus nigra*) is more abundant than green ash (*F. pennsylvanica*) at northern latitudes, but much of our current understanding of *A. planipennis* cold tolerance is based on observations of overwintering larvae from green ash. The effects of black and green ash on the cold hardiness of *A. planipennis* larvae were measured over three winters. Supercooling point, the temperature at which insect bodily fluids spontaneously begin to freeze, was marginally greater for larvae from artificially-infested black ash than green ash in one trial, but not in three others. Host species also did not consistently affect mortality rates after larval exposure to subzero temperatures, but larvae from black ash were less cold hardy than larvae from green ash when there were differences. Comparisons of mortality rates among chilled (unfrozen) and frozen larvae indicated that overwintering *A. planipennis* larvae are primarily freeze avoidant, and this cold tolerance strategy is unaffected by host. All of our studies suggest that *A. planipennis* larvae from black ash are not more cold hardy that larvae from green ash. Where temperatures annually decline below ~−30 °C, overwintering morality may substantially affect the population dynamics and future impacts from this invasive alien species.

Keywords: *Fraxinus nigra*; *Fraxinus pennsylvanica*; biogeography; ecophysiology; overwintering mortality; freeze avoidance

1. Introduction

The invasions of North America and Eurasia by the emerald ash borer, *Agrilus planipennis* Fairmaire (Coleoptera: Buprestidae), portend significant ecological, economic, and social impacts, many of which have been realized and others theorized. Black ash, *Fraxinus nigra* Marsh., blue ash, *F. quadrangulata* Michx., green ash, *F. pennsylvanica* Marsh., and white ash, *F. americana* L. have been colonized naturally by this insect in eastern North America [1–3], and tens of millions of ash trees have died from infestation [4–6]. The loss of ash could detrimentally affect native arthropods [7] and ecosystem functions [8,9]. Where emerald ash borer has invaded, significant economic costs associated with mitigative treatments, tree removals, and tree replacements and lost home equity have been incurred [10,11]. Further, losses of ash trees may be associated with an increase in cardiopulmonary disease in urban areas [12] and threaten culturally significant uses of the resource by some Native Americans and First Nations [6,13]. The ultimate magnitude of these impacts depends, in part, on the final geographic distribution of the species.

Continued spread of emerald ash borer in North America and Europe indicates that the adventive populations of the species have not yet reached spatial equilibria with their environments [6,14,15]. For many insects, host availability and climate suitability are among the primary determinants of where a species may (not) persist [16]. A worst-case scenario for emerald ash borer assumes that all areas where *Fraxinus* spp. grow will also be climatically suitable for the insect. Inductive species distribution models, which compare the climates where emerald ash borer occurs (in Asia, North America, or both) to the rest of North America, suggest that 35% or more of the geographic range of ash may be climatically sub-optimal for emerald ash borer [17]. Variation in temperature (as opposed to precipitation) contributes to more than 50% of these models, and the mean cold temperature of the coldest quarter alone may contribute up to 35% [17]. Model forecasts of suitability differ considerably along the northern, western, and southern limits of the geographic range of ash [17]. Ecophysiological studies may be useful to resolve these differences, and studies of the cold tolerance of *A. planipennis* may be particularly informative at high elevations or northern latitudes.

Agrilus planipennis typically overwinter as late fourth instars (commonly called the "J-stage") in the outer sapwood but may overwinter as earlier instars in or near phloem when individuals undergo a two-year lifecycle [6]. A number of studies suggest that *A. planipennis* larvae might be freeze avoidant (=freeze intolerant) and capable of surviving some sub-zero temperatures. Insects that avoid freezing by supercooling produce polyols and thermal hysteresis proteins to stave off internal ice formation until low temperatures are reached [18,19]. The supercooling point is the temperature at which the spontaneous freezing of body fluids begins and has been used, with criticism [20,21], as a measure of the lowest temperature at which a freeze-avoidant insect species can survive [22,23]. Wu et al. [24] were the first to report supercooling points for the species between −23 and −26 °C in China. Venette and Abrahamson [25] similarly reported a mean supercooling point of −25 °C for winter-acclimated larvae from green ash. Crosthwaite et al. [26] reported seasonal changes in the supercooling point for larvae from green ash in southern Ontario and measured the lowest mean supercooling points from November through February of approximately −30 °C. They also provided the first direct evidence of freeze avoidance for the species [26]. Sobek-Swant et al. [27] found that once cold acclimated, *A. planipennis* supercooling points increased only marginally after one week when exposed to 10 °C or warmer, and only after two weeks at 0–4 °C. All North American studies of the cold tolerance of *A. planipennis* focused on larvae from green ash.

For some insects, the host species on which immature stages develop can affect supercooling points and overwintering ability [28–33]. The objective of this study was to measure the effect of host species on the cold hardiness of *A. planipennis* larvae. Specifically, we hypothesized that the cold hardiness of larvae from black ash would differ from larvae from green ash. While both ash species are considered quality hosts for larval development, *A. planipennis* adults prefer green ash over black ash [2,3,34], and larvae extracted more essential and non-essential amino acids from green ash than from black ash [35]. Black ash tends to be more abundant than green ash in northern stands in the northeastern United States [36], so any impact of host on cold hardiness would have significant impacts on the potential geographic distribution of the species. In this study, supercooling points were measured for *A. planipennis* larvae from artificially- and naturally-infested hosts, and lower lethal temperatures were measured for larvae from naturally-infested hosts.

2. Materials and Methods

2.1. Supercooling Points of A. planipennis from Artificially-Infested Ash

Because of the low numbers of infested black ash within the distribution of *A. planipennis* in Minnesota, we artificially infested cut logs with laboratory reared larvae. *Agrilus planipennis* were reared by following methods modified from Keena et al. [37]. To obtain adults, multiple naturally-infested green ash were felled in Minneapolis and Saint Paul, MN, USA in late May and early June 2011, cut into 38–56 cm lengths, and placed in 66.8 (length) by 31.75 cm (diameter) cardboard rearing tubes with vented plastic lids in a controlled temperature greenhouse. Rearing tubes were checked daily for newly emerged adults. Adults were transferred to 1.9 L plastic jars (United States Plastic Corporation, Lima, OH, USA) with a mesh lid (aluminum window screen with an opening size of 1.41 × 1.59 mm or stainless steel wire mesh with an opening size of 6.4 × 6.4 mm overlapped for openings approximately one quarter of the size). The bottom of the jar and the lid were lined with grocery store bleached coffee filter paper to provide a substrate for oviposition. Adults were also given a water pick and fresh green ash foliage from a mature tree. Jars were cleaned every other day, at which time the filter paper, water, and foliage were replaced. Eggs on the filter paper were counted and transferred to 37 mL sealed plastic cups (Solo Cup Company, Highland Park, IL, USA). Rearing jars and egg cups were stored in an illuminated growth chamber at 26 °C with a photoperiod of 16:8 (L:D) h. Cups were checked daily for newly emerged neonates, which were inserted into cut logs on the same day.

Green and black ash logs from Grand Rapids, MN, USA, far from any known infestations of *A. planipennis*, were cut in July 2011. Two logs, each from a different tree, from each host species were used in 2011. Logs ranged in size from 12–16 cm in diameter and 90–94 cm in length. A total of 26 neonates, all that were available, were infested from 11–18 July 2011, with 13 neonates in each species of ash (no more than 1 larva per 205 cm^2 of log surface area). Neonates were only placed in the top 60 cm of the log because each log was set in a plastic potting dish with 2.5–5 cm of water. Notches 2–3 cm wide and down to the phloem were made in the bark with a 1.25 cm chisel, and a neonate was placed at the bottom of each notch. Laboratory infested logs were kept in a greenhouse throughout August (mean temperature June–August 2011 = 27.6 ± 2.6 °C SD) before they were moved into an unheated garage in preparation for winter temperature acclimation. The bark was peeled from the logs, and larvae were extracted in November 2011. All recovered larvae had reached the "J"-stage, as characterized by Chamorro et al. [38]. Six of thirteen infested *A. planipennis* larvae had successfully developed in black ash, and five were extracted without injury. Seven of thirteen larvae developed in green ash, and four were successfully recovered.

The rearing process was repeated in 2012. Uninfested logs were cut in July 2012 from Grand Rapids, MN, USA and ranged in size from 25.5–33 cm in diameter and 118–125 cm in length. Two logs, each from a different tree, of each host species were artificially infested from 16 July–19 July 2012, with 28 neonates infested into green ash and 29 infested into black ash and held as before (mean temperature June–August 2012 = 28.4 ± 3.3 °C). Artificially infested logs were peeled, and larvae were extracted in January 2013. Eight of 28 larvae developed to the late fourth instar in green ash, and four were successfully recovered. Seven of 29 larvae developed to the same instar in black ash, and four were extracted.

Supercooling points of *A. planipennis* larvae were measured by following protocols modified from Carrillo et al. [39]. Each larva was weighed to the nearest tenth of a milligram before being placed in a trimmed 9 mm (diameter) × 27 mm (length) gelatin capsule (size 000) (Capsuline, Pompano Beach, FL, USA). Capsules prevented larvae from contacting the high vacuum grease (Dow Corning, Midland, MI, USA) that was used to attach the capsule to the thermocouple. Coiled copper-constantan thermocouples, as described in Hanson and Venette [40], were connected to multi-channel data loggers (USB-TC, Measurement Computing, Norton, MA, USA) so that temperatures for up to 16 insects could be recorded at the same time. Each insect and thermocouple was insulated within a polystyrene cube designed to provide a cooling rate of 1 °C min^{-1} below 0 °C in a −80 °C freezer [39]. Temperatures were recorded once per second using TracerDAQ software (Measurement Computing, Norton, MA, USA). Larvae were cooled until an exotherm was detected. The supercooling point was the lowest temperature reached before the onset of the exotherm.

Data were analyzed in R version 3.1.1 [41] or in SAS v. 9.4 [42]. Linear regression (in the R package "MASS") was used to assess the relationship between the supercooling point and larval mass. Larval supercooling points were compared between hosts in each year and between years for each host by using a Wilcoxon rank sum test (PROC NPAR1WAY in SAS), because not all data were normally distributed (Table 1). Statistical *p*-values were calculated exactly to account for small sample sizes.

Table 1. Median and mean (±SEM) supercooling points for larvae from green and black ash that were artificially infested (lab) or naturally infested (field) during the winters of 2011–2012, 2012–2013, or 2013–2014.

	n	Median (°C)	Mean ± SE (°C)	Shapiro-Wilk (W, p)
Lab 2011–2012				
Black ash	5	−25.5	−24.8 ± 0.97	0.91, 0.48
Green ash	4	−24.0	−22.1 ± 3.32	0.81, 0.13
Lab 2012–2013				
Black ash	4	−29.2 *	−30.4 ± 1.42	0.74, 0.03
Green ash	4	−34.2 *	−33.8 ± 1.59	0.93, 0.61
Field 2012–2013				
Black ash	41	−32.6	−31.2 ± 0.71	0.87, < 0.001
Green ash	30	−32.9	−29.8 ± 5.43	0.84, < 0.001
Field 2013–2014				
Black ash	25	−30.6	−31.3 ± 0.63	0.95, 0.3
Green ash	36	−32.1	−31.4 ± 0.61	0.95, 0.1

Shapiro-Wilk provides a test for normality in the distribution of supercooling points. * A Wilcoxon test indicates a moderately significant effect of host (Wilcoxon S = 24, p = 0.06).

2.2. Cold Tolerance of A. planipennis Larvae from Naturally-Infested Ash

Larvae of *A. planipennis* from naturally-infested green and black ash trees were collected in the winters of 2012–2013 and 2013–2014. In January 2013, six green ash and nine black ash trees were removed from the Fort Snelling Golf Course, in Minneapolis, MN (44.886286° latitude, −93.195189° longitude). *Agrilus planipennis* was confirmed at this location on 13 August 2012 by the Minnesota Department of Agriculture but could have been present for ≥3 years given the extent of infestation in some trees (RCV, personal observation). Logs (5–20 cm in diameter and approx. 90 cm in length) were stored outside our research facility in Saint Paul. Bark was peeled from logs and larvae were extracted between 10 January and 28 February 2013. In the winter of 2013–2014, at least 24 infested trees each of green and black ash were identified at Great River Bluffs State Park (approx. 43.944° latitude, −91.385° longitude) in Winona County, MN. *Agrilus planipennis* was first confirmed in the park on 14 September 2011. Trees were felled in November and December 2013 and January and early February 2014. Logs (5–15 cm in diameter and approx. 120 cm in length) were cut and returned to St. Paul, MN (Permit: Minnesota Department of Agriculture, State Formal Quarantine No. RF-1 036, RF-1076, RF-2074 Emerald Ash Borer, Section VI: Special Exemptions). Logs were again held outdoors and were peeled between 30 December 2013 and 27 March 2014 to extract larvae. All larvae were noted as early (first–third) instar or fourth instar (including the "J-stage"), whether or not a larva was recovered without injury, based on descriptions provided by Chamorro et al. [38]. The proportions of individuals that were fourth instars were compared between hosts in the same year and within the same host between years by using a z-test of proportions.

Extracted larvae from both winters were stored at 0 °C for 24–96 h before measuring larval mass and exposing larvae to cold. This holding time allowed minor injuries from extraction to become visible, but was not long enough for the overwintering larvae to lose cold acclimation [27].

Laboratory cooling experiments of field-collected larvae were performed from 11 January–1 March 2013 and 3 January–20 March 2014. In January 2013, supercooling points were measured for 13 larvae from green ash and 15 larvae from black ash, as previously described. Results informed cold exposure treatments for lower lethal temperature studies. The lower lethal temperature studies generally followed procedures described in Stephens et al. [43] and Hefty et al. [44]. The experimental design was a randomized-complete-block with time as the block. In each block, up to ten larvae from black and green ash (20 larvae total) were randomly assigned to one of five temperature treatments (i.e., two larvae from each host species per temperature). Treatment temperatures in the winter of 2012–2013 were −35 °C, −30 °C, −25 °C, −20 °C, and a room temperature control (approx. 25 °C). The temperatures were selected to detect pre-freeze mortality, if it occurred. The coldest exposure temperature was approximately 2.5 °C lower than the median supercooling point that we measured and was expected to cause 60–80% mortality. This expectation came from the distribution of supercooling points and an assumption that the onset of an exotherm would cause death. Treatment temperatures in the winter of 2013–2014 were −40 °C, −35 °C, −30 °C, −25 °C, and a room temperature control (approx. 25 °C). These treatments provided a more thorough analysis of larval condition before or after freezing as they were balanced around the median supercooling point. A total of 111 larvae from green ash and 100 larvae from black ash were tested in the winter of 2012–2013; 47 larvae from green and black ash (94 larvae total) were tested in the winter of 2013–2014.

Larvae were prepared and cooled as previously described for the measurement of supercooling points. In this case, larvae were removed from the freezer and held at room temperature within ~5 s of when they reached the treatment temperature, whether an exotherm was detected during the course of cooling or not. If an exotherm was detected, the supercooling point was noted. Control larvae (held at room temperature) were placed in gelatin capsules and affixed to thermocouples with high vacuum grease, just as the chilled larvae were.

To determine if larvae had survived, cold-exposed and control larvae were removed from capsules and placed in covered 24-well cell culture plates (Corning, Tewksbury, MA, USA) at room temperature in the dark. Culture plates were kept in a plastic storage container with a loose fitting lid for gas exchange and wet paper towels to keep the humidity high. First observations were taken 72 h after cold exposure and then every other day until pupation. Color and movement were noted. Larvae that were active during the observation period or had pupated (only applicable to fourth instars) were classified as having survived. Larvae that had not molted or pupated at 13 days after cold exposure, did not move, or were discolored, were categorized as dead. For the study in the winter of 2012–2013, we focused our analysis on larval condition at one week post cold exposure and at the time of pupation, while in the winter of 2013–2014, we focused on observations at one week because mold had overtaken larvae thereafter. Larvae <1 mg were not included in lower lethal temperature studies because initial work suggested high mortality from extraction or handling, even among larvae that showed no initial signs of injury.

The relationship between supercooling point and mass, a proxy for development, was analyzed by linear regression in R. Because no relationship between these two measures was found (F = 0.654, df = 1, 129, p = 0.420), supercooling points were not analyzed by size class.

Supercooling point frequency distributions by host were tested for normality with the Shapiro-Wilk test in R (function *shapiro.test*). Because supercooling points of larvae from naturally infested ash in the winter of 2012–2013 were not normally distributed (Table 1), supercooling point data were analyzed by using non-parametric survival analysis in the survival package in R [45]. This statistical approach also allowed us to take advantage of information provided by individuals that were chilled but did not give an exotherm before being removed from the freezer. In our case, temperature (measured as the difference from room temperature) replaced time in the survival analysis, and the binary independent variable was whether each larva had an exotherm or not. When exotherms were observed, the supercooling points provided the exact temperatures at which freezing began, and these data are considered non-censored. A larva that did not give an exotherm before reaching the target temperature was considered a right-censored observation (i.e., we presumed that the supercooling point occurred at an unspecified temperature colder than the removal temperature). To estimate the probability that an individual would begin to freeze, given that it had been cooled to a particular temperature, non-parametric Kaplan-Meier curves were estimated from non-censored and right-censored observations using the *survfit* function in the R survival package [45]. We used the *survdiff* function to test for differences (α = 0.05) in Kaplan-Meier curves between each host in the same year or between years for the same host.

Host influence on larval mortality after cold exposure was analyzed in two ways. The potential effect of host species on the extent of larval mortality at each target exposure temperature was tested by a z-test of proportions. Mixed linear models with a logit-link function were used to relate the extent of mortality to exposure temperature, host, and the interaction of host and exposure temperature (Proc GLIMMIX in SAS). Block and thermocouple within block were included as random effects.

The determination of a species' cold-tolerance strategy, i.e., chill intolerant, freeze avoidant, or freeze tolerant, depends on a comparison of mortality rates among chilled and frozen individuals at a particular temperature [46]. In our studies, "frozen" equated to the onset of freezing, as evident from the detection of an exotherm; the extent of freezing was not measured. We followed the statistical procedures described in Cira et al. [47]. For each host species within a winter, the difference in mortality rates between chilled and frozen individuals was tested by using Fisher's exact test (Chi square) in SAS (Proc TABULATE). The exact test accounted for small sample sizes ($n < 5$) in some cases and converged to the large sample approximation as sample size increased. If the mortality rates of chilled individuals were equal to the mortality rates of frozen individuals and the mortality rates of chilled individuals increased as exposure temperature decreased, chill intolerance would be indicated. If the mortality rates of chilled individuals were less than the mortality rates of frozen individuals and the mortality rates of chilled individuals remained constant as exposure temperature decreased, freeze avoidance would be indicated. If the mortality rates of frozen individuals were less than 1, freeze tolerance (or perhaps partial freeze tolerance) would be suggested. We also tested whether the mortality rate of chilled (unfrozen) individuals differed from the mortality of controls kept at room temperature, again using Fisher's exact test. Because no differences were detected, likely as a joint consequence of relatively high mortality among control larvae and small sample sizes, we pooled mortality results across exposure temperatures (excluding controls) for chilled and frozen individuals, respectively, and tested for differences in mortality rates by using Fisher's exact test.

3. Results

3.1. Supercooling Points of A. planipennis Larvae from Artificially-Infested Ash

In 2011, the supercooling points of larvae from artificially-infested green and black ash were not significantly different (Table 1; Wilcoxon S = 21, $p = 0.45$). Supercooling points of larvae from black ash ranged from -26.7 to -22.0 °C and from green ash ranged from -27.1 to -12.4 °C. The supercooling point of -12.4 °C was atypically high, but removing this potential outlier (adjusted mean $\pm$ SE, -25.4 ± 1.0 °C) still did not reveal an effect of host (Wilcoxon S = 12, $p = 0.39$). In 2013, host had a marginally significant effect on the supercooling points of artificially-infested larvae (Wilcoxon S = 21, $p = 0.06$). Supercooling points of larvae from black ash, ranging from -34.6 to -28.5 °C, were greater than larvae from green ash, ranging from -36.8 to -29.3 °C (Table 1). Supercooling points occurred at warmer temperatures in 2011 (November) than in 2013 (January) for larvae from black ash (difference in medians = 3.7 °C, Wilcoxon S = 10, $p < 0.01$) and green ash (difference in medians = 9.7 °C, Wilcoxon S = 26, $p = 0.01$).

3.2. Cold Tolerance of A. planipennis Larvae from Naturally-Infested Ash

Stage distributions. In the winter of 2012–2013, 165 *A. planipennis* larvae were recovered from green ash, and a significantly greater percentage (99.4% $\pm$ 0.6%; [$\hat{p} \pm$ SE] $\times$ 100) were fourth instars than among the 209 larvae recovered from black ash (92.3% $\pm$ 1.8%; Z = 3.64, $p < 0.001$). In the winter of 2013–2014, 203 and 170 *A. planipennis* larvae were recovered from green ash and black ash, respectively, and the percentages that were fourth instars (46.8% $\pm$ 3.5% and 53.5% $\pm$ 3.8%) did not differ (Z = 1.30, $p = 0.099$). A significantly greater percentage of larvae were fourth instars in the winter of 2012–2013 (Minneapolis) than 2013–2014 (Great River Bluffs State Park) for green ash (Z = 14.80, $p < 0.001$) and black ash (Z = 9.14, $p < 0.001$).

Supercooling points. Supercooling points were not significantly different between host species in the winter of 2012–2013 ($\chi^2 = 1.9$, df = 1, $p = 0.17$; Figure 1) or 2013–2014 ($\chi^2 = 0$, df = 1, $p = 0.90$; Figure 1). In the winter of 2012–2013, the lowest supercooling points of larvae from black ash and green ash were -37.5 °C and -36.2 °C, respectively, and in the winter of 2013–2014, they were -37.3 °C and

−38.0 °C, respectively (Figure 1). Supercooling points of larvae from black ash were not significantly different between winters (χ^2 = 1.2, df = 1, p = 0.28; Figure 1), but were different for larvae from green ash (χ^2 = 6.4, df = 1, p = 0.01; Figure 1).

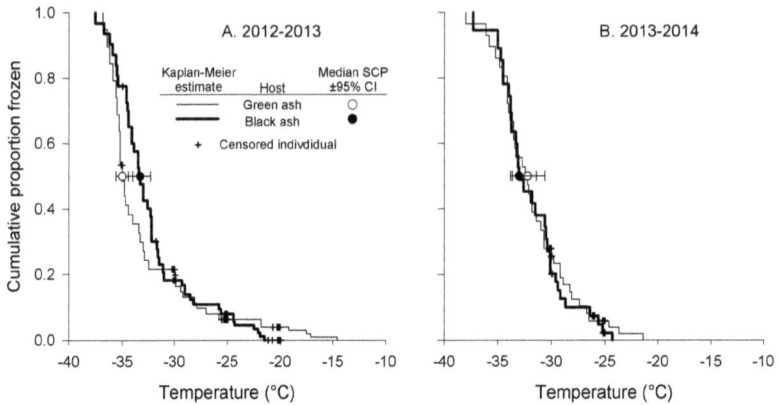

Figure 1. Kaplan-Meier estimates of the likelihood that an *A. planipennis* larva will begin to freeze (i.e., give a supercooling point) after exposure to a particular temperature, by tree species, in (**A**) the winter of 2012-2013 and (**B**) the winter of 2013-2014. Censored individuals did not start to freeze by their lowest exposure temperature.

Mortality from cold exposure. In the winter of 2012–2013, no effect of host on mortality rates of *A. planipennis* larvae could be detected one week after cold exposure (Figure 2A). Mortality rates of control larvae were 5.0% ± 4.9% (SE) from black ash and 20.8% ± 8.3% from green ash at this time. At 0 °C (i.e., the intercept of the statistical model), no effect of host on mortality rate was detectable (F = 1.3, df = 1, 170, p = 0.25). The mortality rate increased significantly as the coldest exposure temperature decreased (F = 9.3, df = 1, 170, p = 0.003). When exposure temperatures reached −35 °C, mortality was 59.1% ± 10.5% in larvae from black ash and 45.4% ± 10.6% in larvae from green ash. Host species did not affect the change in larval mortality rate as the coldest exposure temperatures declined (F = 1.5, df = 1, 170, p = 0.22).

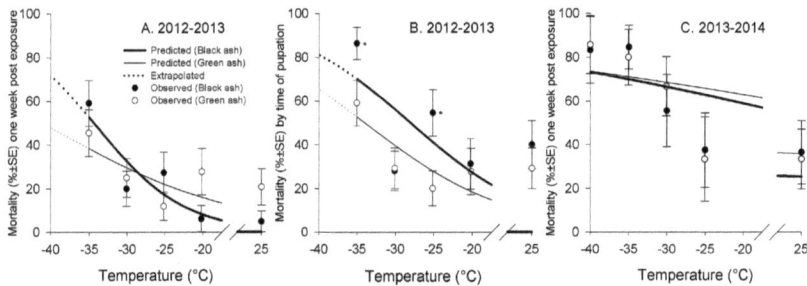

Figure 2. Mortality of *A. planipennis* larvae from black and green ash in the winters of 2012–2013 and 2013–2014 in response to acute exposure to different low temperatures. Mortality was assessed at: (**A**) one week after exposure; (**B**) again 13 days after exposure at pupation; and (**C**) one week after exposure. Asterisks indicate statistically significant differences in morality between larvae from black and green ash at that temperature. Host did not significantly affect the predicted mortality rates at 0 °C or the change in mortality rates as temperatures changed.

Mortality rates by the time of pupation (within two weeks of cold exposure) were typically ~40% greater than at one week after cold exposure (increase ranged from zero- to five-fold; Figure 2A,B). At this time, mortality rates were greater for larvae from black ash than green ash at −35 °C (Z = 2.13, *p* = 0.017) and −25 °C (Z = 2.50, *p* = 0.0062; Figure 2B). No statistically significant differences were detected when the coldest exposure temperature was −30 °C, −20 °C, or room temperature (Z ≤ 0.09, *p* ≥ 0.18). Although larvae from black ash had greater mortality rates than larvae from green ash in four of the five exposure treatments, host species did not affect the projected level of mortality at 0 °C (F = 0.1, df = 1, 170, *p* = 0.92) or the change in mortality rate as the coldest exposure temperatures declined (F = 0.08, df = 1, 170, *p* = 0.77). The mortality rate increased as the coldest exposure temperatures decreased for larvae from both host species (F = 7.2, df = 1, 170, *p* = 0.008; Figure 2B).

In the winter of 2013–2014, no differences were detected in mortality rates between *A. planipennis* larvae from black and green ash at one week after cold exposure (Z ≤ 0.52, *p* ≥ 0.31; Figure 2C). Mortality rates among larvae that were kept at room temperature were ~37%. Although mortality rates increased as the coldest exposure temperatures declined (F = 5.1, df = 1, 87, *p* = 0.03), host species had no discernable effect on the projected level of mortality at 0 °C (F = 0.3, df = 1, 87, *p* = 0.6) or on the change in mortality rate as exposure temperatures declined (F = 0.2, df = 1, 87, *p* = 0.7). Of larvae that were briefly exposed to −40 °C, ~85% were dead 7 days later. In the winter of 2012–2013, the mortality rates of control *A. planipennis* larvae from black and green ash were ~10% at one week after testing (Table 2). Mortality levels among chilled (unfrozen) individuals did not significantly differ from controls at any subzero temperature exposure. No larvae from black or green ash began to freeze (i.e., gave an exotherm) at temperatures ≥−20 °C (Table 2). When chilled (unfrozen) and frozen larvae were present after being exposed to the same temperature, the mortality rate of chilled larvae was less than one third of the mortality rate of frozen larvae; the difference was statistically significant for individuals from black ash exposed to −25 °C and from green ash exposed to −30 °C or −35 °C (Table 2). When observations among exposure temperatures were combined, the proportion of larvae that were dead one week after being chilled was significantly less than after being frozen. The mortality rate among all chilled individuals combined within a host species was not different from controls for larvae from black (χ^2 = 0.29, df = 1, *p* = 0.59) or green ash (χ^2 = 1.14, df = 1, *p* = 0.28). The proportions of larvae that were dead one week after being frozen were significantly less than 1.0 for larvae from black ash (exact upper 95% confidence interval = 0.87) and green ash (exact upper 95% confidence interval = 0.89).

In the winter of 2013–2014, host species, chilling, and freezing had qualitatively similar effects on mortality rates (one week post exposure) of *A. planipennis* larvae as in the winter of 2012–2013. The mortality rates of chilled (unfrozen) individuals did not significantly differ from the mortality rates of control larvae (~44% for both host species) at any exposure temperature (Table 2). No larvae began to freeze at temperatures ≥−30 °C. The mortality rates of chilled (unfrozen) larvae were numerically less for frozen larvae from black and green ash at each exposure temperature, but the differences were not statistically different until observations from all exposure treatments were combined (Table 2). Then, the mortality rates of frozen individuals were twice as great as the mortality of chilled individuals for both host species. The combined mortality rate of chilled individuals was not different from the control for larvae from black (χ^2 = 0.18, df = 1, *p* = 0.67) or green ash (χ^2 = 0.25, df = 1, *p* = 0.62). The combined mortality rate of frozen individuals was less than 1.0 for larvae from black ash (exact upper 95% confidence interval = 0.99) and green ash (exact upper 95% confidence interval = 0.97).

Table 2. Mortality of chilled and frozen *A. planipennis* larvae from green and black ash in the winters of 2012–2013 and 2013–2014 at one week after acute exposure to a subzero temperature.

	Chilled [a]		Frozen		Chilled		Frozen	
Temp. (°C)	n [b]	Mortality %	n	Mortality %	n	Mortality %	n	Mortality %
2012–2013		*Black ash*				*Green ash*		
−35	2	0.0	20	65.0	8	0.0	14	71.4 *
−30	22	13.6	3	66.7	19	5.3	5	100 *
−25	17	5.9	5	100 *	22	9.1	3	33.3
−20	16	6.3	0	-	18	27.8	0	-
25	20	5.0	0	-	24	20.1	0	-
Combined [c]	57	8.8	28	71.4 *,†	67	11.9	22	72.3 *,†
2013–2014								
−40	0	-	6	83.3	0	-	7	85.7
−35	1	0.0	12	91.7	1	0.0	9	88.9
−30	9	55.6	0	-	7	57.1	5	80.0
−25	8	37.5	0	-	6	33.3	0	-
25	11	36.4	0	-	12	33.3	0	-
Combined [c]	18	44.4	18	88.9 *,†	14	42.9	21	85.7 *,†

[a] an exotherm was not observed for chilled larvae but was for frozen larvae; the extent of freezing was not measured. [b] n = number of individuals that were chilled or had started to freeze by the exposure temperature. [c] combined values did not include 25 °C (room temperature controls). * mortality rates from chilling and freezing were significantly different ($p < 0.05$) as determined by Fisher's exact test (Chi square) with one degree of freedom. † combined mortality of frozen individuals was significantly less than one, as determined by the lack of overlap with exact 95% confidence intervals. For each host species in each winter, the mortality of chilled (unfrozen) larvae at each exposure temperature was not statistically different ($p > 0.05$) from larvae held at room temperature, as determined by Fisher's exact test (Chi square) with one degree of freedom.

4. Discussion

Host species had idiosyncratic effects on the cold tolerance of *A. planipennis* larvae. In only one of four trials did host have an effect on the supercooling point, and in this case, the effect was statistically marginal (Table 1). Similarly, in 16 comparisons of larval mortality after exposure to subzero temperatures, statistically-significant effects of host were only detected in two instances, both in the winter of 2012–2013 for larvae whose coldest exposure temperatures were −35 or −25 °C (Figure 2). In all cases where a possible effect was detected, larvae from black ash were less cold hardy than larvae from green ash. While these sporadic differences are intriguing to us, most of the data from this study indicate that the cold tolerance of *A. planipennis* larvae from black ash and green ash does not differ. In some cases, small sample sizes reduced the power of the statistical tests to detect a difference. Thus, we believe that the correct interpretation of these findings, in aggregate, is that overwintering *A. planipennis* larvae from black ash were not more cold tolerant than larvae from green ash.

This interpretation of our findings has important ramifications. Much of the previous knowledge of *A. planipennis* cold tolerance was based on larvae that had developed in green ash, and this information had been used to forecast where *A. planipennis* might (not) be able to overwinter, irrespective of the host species that were present [48]. Our findings suggest that mortality levels of larvae in black ash should be expected to be the same as larvae in green ash when exposed to the same temperature, but occasionally might be greater. Accurate forecasts depend on reliable information about the relationship between temperature and mortality and on measures of the temperature(s) that overwintering larvae experience. Although Vermunt et al. [49] developed models to forecast temperatures beneath the bark of green ash and white ash (*F. americana*), similar relationships have yet to be developed for black ash in mesic or hydric sites (but see Christianson [50]).

Our results are based on acute exposures of larvae to subzero temperatures and need to be interpreted with caution. The mortality measured in this study does not account for the potential effects of prolonged exposure to subzero temperatures. In general, mortality increases within limits as exposure time to a constant, low temperature increases [51]. In this respect, our estimates of mortality are conservative because larvae in the field would be expected to experience some of these temperatures for minutes, hours, or days when they occur, not seconds. Conversely, our estimates of mortality may be liberal because they are based on a cooling rate of 1 °C/min. Although this rate has been widely accepted as a standard laboratory protocol [52,53], it has been challenged for its ecological relevance [54]. With slower cooling rates that are common in temperate climates, supercooling points can be different, as can temperature-mortality relationships, e.g., [55]. The utility of these laboratory-based measures of cold tolerance, irrespective of the cooling rate that was used, can only be evaluated through comparisons with independent observations of mortality from the field. Such comparisons were beyond the scope of this study.

Our study indicates that overwintering *A. planipennis* larvae, whether developing in black or green ash, are primarily freeze avoidant. If larvae stay in an unfrozen state, mortality rates appear to remain unaffected, even as the coldest exposure temperature decreases. Any effects of cold on the mortality of these individuals could not be distinguished from the effects of larvae being extricated from their hosts and handled during the study, despite efforts to exclude injured individuals from the experiments. However, once larvae begin to freeze, mortality rates increase markedly. Freeze avoidant insects often have a low supercooling point, whilst freeze tolerant insects have a greater supercooling point [18]. The results of the present study are consistent with this pattern. The results are also in agreement with the conclusions of Crosthwaite et al. [26]; yet, in the winter of 2012–2013, we were surprised to find four larvae that had frozen, or had at least started to freeze (as evident by the detection of an exotherm), and continued to live for up to 13 days after cold exposure. Because we did not measure the extent of ice formation, it would be presumptuous to conclude that these individuals were freeze tolerant. Such individuals might be classified as partially-freeze tolerant [46,56], but the ecological ramifications of such a cold tolerance strategy are debatable [56,57].

Our studies help to provide a more robust characterization of the cold tolerance of *A. planipennis* by measuring the extent of mortality and the proportion of insects that may have frozen at different exposure temperatures [46]. Because the vast majority of overwintering *A. planipennis* larvae die upon freezing, the supercooling point provides a convenient indicator of the expected level of mortality at different temperatures. In portions of our study, the supercooling points of *A. planipennis* larvae were lower than previously reported. The supercooling points of artificially infested larvae measured in November 2011 (~−25 °C) were similar to previously reported values. However, the mean supercooling points of artificially-infested larvae in January 2013 and field-collected larvae in January-March 2013 and January–March 2014 were nearly 2–6 °C lower than previously reported mean supercooling points and were significantly colder than those from November 2011.

The differences in supercooling points from previous studies and ours could also be related to the extent of cold acclimation that occurs during the fall and early winter. The warmer supercooling points we recorded in November 2011 could have been because testing was performed before the insects had fully acclimatized to winter or the insects were responding to a warmer than average fall, which preceded a warmer than average winter [58]. However, two years of study by Crosthwaite et al. [26] showed that the mean supercooling point declined from October through mid-November, but did not change statistically thereafter for the remainder of the winter. The annual variation in supercooling points and temperature-mortality relationships measured in this study hint that the degree of cold tolerance of *A. planipennis* is a plastic response to as yet undefined cues experienced during autumn or winter. If *A. planipennis* larvae are able to adjust their cold hardiness in response to these cues, our data seem to show a lower limit to *A. planipennis'* capacity to cold harden: the fall and winter of 2013–2014 were colder than average, but supercooling points and the mortality from freezing were not statistically different from the results collected during the more typical Minnesota winter of 2012–2013. Potential genetic differences in populations of *A. planipennis* could also explain different reports of cold tolerance, but genetic testing suggests that the U.S. populations most likely stem from a single introduction [59]. Regional differences in the quality of host species could also be a factor, but the attribute of the host that could drive these differences remains to be identified.

Cold often determines the northern limits of an invading insect's range, e.g., [60,61]. *Agrilus planipennis* has yet to achieve its ultimate distribution in North America. The potential geographic range expansion by *A. planipennis* into northern Minnesota where black ash is abundant in lowland black ash-American elm-red maple forests [62] is a particular concern. Minnesota has more than 900 million ash trees, about 75% of which are black ash [63]. Black ash and green ash comprise almost half of Minnesota's timber resource by volume in lowland forests and about one quarter of Minnesota's total forest resource [63,64]. Black ash is one of the few native tree species that grows in bogs and poorly drained soils [65,66]. Green ash is a popular boulevard tree in communities throughout the state and is a common native species to the central and southern part of Minnesota [64].

Seven years after first being detected in North America, *A. planipennis* was found in Minnesota for the first time in Saint Paul [67]. Eight years after the insect arrived in the state, most of Minnesota's ash remains uninfested. The effect of temperatures <−30 °C on the distribution and impact of *A. planipennis* in this state and elsewhere in North America remain to be determined, but overwintering mortality in these areas could have substantial impacts on the rates of population growth and spread. Areas with regular exposure to temperatures <−35 °C may provide thermal refugia that are vital to the local persistence of native ash stands.

Author Contributions: Conceptualization, L.D.E.C. and R.C.V.; Funding acquisition, R.C.V.; Methodology, L.D.E.C. and R.C.V.; Data collection, L.D.E.C.; Writing-original draft, L.D.E.C. and R.C.V.

Acknowledgments: This project was funded by the Minnesota Environment and Natural Resources Trust Fund as recommended by the Legislative-Citizen Commission on Minnesota Resources. We also thank Doug Kastendick and Mitch Slater from the U.S.D.A. Forest Service Northern Research Station in Grand Rapids, MN and the Chippewa National Forest for providing uninfested ash material. We sincerely appreciate the assistance of Paul Castillo from the U.S.D.A. Forest Service, Northern Research Station in Saint Paul, MN, for his help in the field. Thank you to the Fort Snelling Golf Course, City of Minneapolis, and City of Saint Paul for working with us to find and remove infested material within the Twin Cities Metro Area; to Mark Abrahamson, Monika Chandler, and Jonathan Osthus from the Minnesota Department of Agriculture; and to Rick Samples and Shawn Fritcher of the Minnesota Department of Natural Resources for finding, cutting, and transporting materials in the field. Student workers Jade Thomason, Joseph Pohnan, and Benjamin Davis were indispensable in the lab, helping with larva extraction. Finally, we would like to acknowledge Brian Aukema and Tracy Twine for reviews of an earlier draft of this manuscript.

Conflicts of Interest: The authors declare no conflict of interest.

References

1. Wallander, E. Systematics of *Fraxinus* (Oleaceae) and evolution of dioecy. *Plant Syst. Evol.* **2008**, *273*, 25–49. [CrossRef]

2. Anulewicz, A.C.; McCullough, D.G.; Cappaert, D.L.; Poland, T.M. Host range of the emerald ash borer (*Agrilus planipennis* Fairmaire) (Coleoptera: Buprestidae) in North America: Results of multiple-choice field experiments. *Environ. Entomol.* **2008**, *37*, 230–241. [CrossRef]

3. Anulewicz, A.C.; McCullough, D.G.; Miller, D.L. Oviposition and development of emerald ash borer (*Agrilus planipennis*) (Coleoptera: Buprestidae) on hosts and potential hosts in no-choice bioassays. *Great Lakes Entomol.* **2006**, *39*, 99–112.

4. McCullough, D.G.; Poland, T.M.; Anulewicz, A.C.; Cappaert, D. Emerald ash borer (Coleoptera: Buprestidae) attraction to stressed or baited ash trees. *Environ. Entomol.* **2009**, *38*, 1668–1679. [CrossRef] [PubMed]

5. Marshall, J.M.; Smith, E.L.; Mech, R.; Storer, A.J. Estimates of *Agrilus planipennis* infestation rates and potential survival of ash. *Am. Midl. Nat.* **2013**, *169*, 179–193. [CrossRef]

6. Herms, D.A.; McCullough, D.G. Emerald ash borer invasion of North America: History, biology, ecology, impacts, and management. *Annu. Rev. Entomol.* **2014**, *59*, 13–30. [CrossRef] [PubMed]

7. Gandhi, K.J.K.; Herms, D.A. Direct and indirect effects of alien insect herbivores on ecological processes and interactions in forests of eastern North America. *Biol. Invasions* **2010**, *12*, 389–405. [CrossRef]

8. Slesak, R.A.; Lenhart, C.F.; Brooks, K.N.; D'Amato, A.W.; Palik, B.J. Water table response to harvesting and simulated emerald ash borer mortality in black ash wetlands in Minnesota, USA. *Can. J. For. Res.* **2014**, *44*, 961–968. [CrossRef]

9. Nisbet, D.; Kreutzweiser, D.; Sibley, P.; Scarr, T. Ecological risks posed by emerald ash borer to riparian forest habitats: A review and problem formulation with management implications. *For. Ecol. Manag.* **2015**, *358*, 165–173. [CrossRef]

10. Kovacs, K.F.; Haight, R.G.; McCullough, D.G.; Mercader, R.J.; Siegert, N.W.; Liebhold, A.M. Cost of potential emerald ash borer damage in US communities, 2009–2019. *Ecol. Econ.* **2010**, *69*, 569–578. [CrossRef]

11. Lovett, G.M.; Weiss, M.; Liebhold, A.M.; Holmes, T.P.; Leung, B.; Lambert, K.F.; Orwig, D.A.; Campbell, F.T.; Rosenthal, J.; McCullough, D.G.; et al. Nonnative forest insects and pathogens in the United States: Impacts and policy options. *Ecol. Appl.* **2016**, *26*, 1437–1455. [CrossRef] [PubMed]

12. Donovan, G.H.; Butry, D.T.; Michael, Y.L.; Prestemon, J.P.; Liebhold, A.M.; Gatziolis, D.; Mao, M.Y. The relationship between trees and human health evidence from the spread of the emerald ash borer. *Am. J. Prev. Med.* **2013**, *44*, 139–145. [CrossRef] [PubMed]

13. Costanza, K.K.L.; Livingston, W.H.; Kashian, D.M.; Slesak, R.A.; Tardif, J.C.; Dech, J.P.; Diamond, A.K.; Daigle, J.J.; Ranco, D.J.; Neptune, J.S.; et al. The Precarious State of a Cultural Keystone Species: Tribal and Biological Assessments of the Role and Future of Black Ash. *J. For.* **2017**, *115*, 435–446. [CrossRef]

14. Selikhovkin, A.V.; Popovichev, B.G.; Mandelshtam, M.Y.; Vasaitis, R.; Musolin, D.L. The frontline of invasion: The current northern limit of the invasive range of emerald ash borer, *Agrilus planipennis* Fairmaire (Coleoptera: Buprestidae), in European Russia. *Balt. For.* **2017**, *23*, 309–315.

15. Valenta, V.; Moser, D.; Kuttner, M.; Peterseil, J.; Essl, F. A high-resolution map of emerald ash borer invasion risk for southern Central Europe. *Forests* **2015**, *6*, 3075–3086. [CrossRef]

16. Venette, R.C. The challenge of modelling and mapping the future distribution and impact of invasive alien species. In *Pest Risk Modelling and Mapping for Invasive Alien Species*; Venette, R.C., Ed.; CAB International: Wallingford, UK, 2015; pp. 1–17.

17. Sobek-Swant, S.; Kluza, D.A.; Cuddington, K.; Lyons, D.B. Potential distribution of emerald ash borer: What can we learn from ecological niche models using Maxent and GARP? *For. Ecol. Manag.* **2012**, *281*, 23–31. [CrossRef]

18. Turnock, W.J.; Fields, P.G. Winter climates and cold hardiness in terrestrial insects. *Eur. J. Entomol.* **2005**, *102*, 561–576. [CrossRef]

19. Leather, S.R.; Walters, K.F.A.; Bale, J.S. *The Ecology of Insect Overwintering*; Cambridge University Press: New York, NY, USA, 1993; p. 255.

20. Baust, J.G.; Rojas, R.R. Insect cold hardiness: Facts and fancy. *J. Insect Physiol.* **1985**, *31*, 755–759. [CrossRef]

21. Renault, D.; Salin, C.; Vannier, G.; Vernon, P. Survival at low temperatures in insects: What is the ecological significance of the supercooling point? *Cryo Lett.* **2002**, *23*, 217–228.

22. Bale, J.S. Insect cold hardiness—Freezing and supercooling—An ecophysiological perspective. *J. Insect Physiol.* **1987**, *33*, 899–908. [CrossRef]

23. Sømme, L. Supcooling and winter survival in terrestrial arthropods. *Comp. Biochem. Physiol. A Physiol.* **1982**, *73*, 519–543. [CrossRef]

24. Wu, H.; Li, M.-L.; Yang, Z.-Q.; Wang, X.-Y.; Wang, H.-Y.; Bai, L. Cold hardiness of *Agrilus planipennis* and its two parasitoids, *Spathius agrili* and *Tetrastichus planipennisi*. *Chin. J. Biol. Control* **2007**, *23*, 119–122.

25. Venette, R.C.; Abrahamson, M. *Cold Hardiness of Emerald Ash Borer, Agrilus planipennis: A New Perspective*; Black Ash Symposium: Bemidji, MN, USA; U.S. Department of Agriculture, Forest Service, Chippewa National Forest: Bemidji, MN, USA, 2010.

26. Crosthwaite, J.C.; Sobek, S.; Lyons, D.B.; Bernards, M.A.; Sinclair, B.J. The overwintering physiology of the emerald ash borer, *Agrilus planipennis* Fairmaire (Coleoptera: Buprestidae). *J. Insect Physiol.* **2011**, *57*, 166–173. [CrossRef] [PubMed]

27. Sobek-Swant, S.; Crosthwaite, J.C.; Lyons, D.B.; Sinclair, B.J. Could phenotypic plasticity limit an invasive species? Incomplete reversibility of mid-winter deacclimation in emerald ash borer. *Biol. Invasions* **2012**, *14*, 115–125. [CrossRef]

28. Yuill, J.S. Cold hardiness of two species of bark beetles in California forests. *J. Econ. Entomol.* **1941**, *34*, 702–709. [CrossRef]

29. Liu, Z.D.; Gong, P.Y.; Heckel, D.G.; Wei, W.; Sun, J.H.; Li, D.M. Effects of larval host plants on over-wintering physiological dynamics and survival of the cotton bollworm, *Helicoverpa armigera* (Hubner) (Lepidoptera: Noctuidae). *J. Insect Physiol.* **2009**, *55*, 1–9. [CrossRef] [PubMed]

30. Liu, Z.D.; Gong, P.Y.; Wu, K.J.; Wei, W.; Sun, J.H.; Li, D.M. Effects of larval host plants on over-wintering preparedness and survival of the cotton bollworm, *Helicoverpa armigera* (Hubner) (Lepidoptera: Noctuidae). *J. Insect Physiol.* **2007**, *53*, 1016–1026. [CrossRef] [PubMed]

31. Hiiesaar, K.; Williams, I.; Luik, A.; Metspalu, L.; Muljar, R.; Jogar, K.; Karise, R.; Mand, M.; Svilponis, E.; Ploomi, A. Factors affecting cold hardiness in the small striped flea beetle, *Phyllotreta undulata*. *Entomol. Exp. Appl.* **2009**, *131*, 278–285. [CrossRef]

32. Trudeau, M.; Mauffette, Y.; Rochefort, S.; Han, E.; Bauce, E. Impact of host tree on forest tent caterpillar performance and offspring overwintering mortality. *Environ. Entomol.* **2010**, *39*, 498–504. [CrossRef] [PubMed]

33. Rosenberger, D.W.; Aukema, B.H.; Venette, R.C. Cold tolerance of mountain pine beetle among novel eastern pines: A potential for trade-offs in an invaded range? *For. Ecol. Manag.* **2017**, *400*, 28–37. [CrossRef]

34. Pureswaran, D.S.; Poland, T.M. Host selection and feeding preference of *Agrilus planipennis* (Coleoptera: Buprestidae) on ash (*Fraxinus* spp.). *Environ. Entomol.* **2009**, *38*, 757–765. [CrossRef] [PubMed]

35. Chen, Y.G.; Ulyshen, M.D.; Poland, T.M. Differential utilization of ash phloem by emerald ash borer larvae: Ash species and larval stage effects. *Agric. For. Entomol.* **2012**, *14*, 324–330. [CrossRef]

36. Prasad, A.M.; Iverson, L.R.; Matthews, S.; Peters, M. A Climate Change Atlas for 134 Forest Tree Species of the Eastern United States. Available online: https://www.nrs.fs.fed.us/atlas/tree (accessed on 30 March 2018).

37. Keena, M.A.; Gould, J.R.; Bauer, L.S. *Developing an Effective and Efficient Rearing Method for the Emerald Ash Borer, Emerald Ash Borer Reseach and Technology Development Meeting, Pittsburgh, PA, USA, 21 October 2009*; Lance, D., Buck, J., Binion, D., Reardon, R., Mastro, V., Eds.; USDA Forest Service, Forest Health Technology Enterprise Team: Pittsburgh, PA, USA, 2010; pp. 34–35.

38. Chamorro, M.L.; Volkovitsh, M.G.; Poland, T.M.; Haack, R.A.; Lingafelter, S.W. Preimaginal stages of the emerald ash borer, *Agrilus planipennis* Fairmaire (Coleoptera: Buprestidae): An invasive pest on ash trees (*Fraxinus*). *PLoS ONE* **2012**, *7*, 12. [CrossRef] [PubMed]

39. Carrillo, M.A.; Kaliyan, N.; Cannon, C.A.; Morey, R.V.; Wilcke, W.F. A simple method to adjust cooling rates for supercooling point determination. *Cryo Lett.* **2004**, *25*, 155–160.

40. Hanson, A.A.; Venette, R.C. Thermocouple design for measuring temperatures of small insects. *Cryo Lett.* **2013**, *34*, 261–266.

41. R Core Team. R: A Language and Environment for Statistical Computing. Available online: http://www.R-project.org/ (accessed on 2 May 2017).

42. SAS Institute. *SAS Version 9.4*; SAS Institute Inc.: Cary, NC, USA, 2013.

43. Stephens, A.R.; Asplen, M.K.; Hutchison, W.D.; Venette, R.C. Cold hardiness of winter-acclimated *Drosophila suzukii* (Diptera: Drosophilidae) adults. *Environ. Entomol.* **2015**, *44*, 1619–1626. [CrossRef] [PubMed]

44. Hefty, A.R.; Seybold, S.J.; Aukema, B.H.; Venette, R.C. Cold tolerance of *Pityophthorus juglandis* (Coleoptera: Scolytidae) from northern California. *Environ. Entomol.* **2017**, *46*, 967–977. [CrossRef] [PubMed]

45. Therneau, T.M. A Package for Survival Analysis in S. R Package Version 2.37-7. Available online: http://cran.R-project.org/package=survival (accessed on 14 February 2014).

46. Sinclair, B.J. Insect cold tolerance: How many kinds of frozen? *Eur. J. Entomol.* **1999**, *96*, 157–164.

47. Cira, T.M.; Venette, R.C.; Aigner, J.; Kuhar, T.; Mullins, D.E.; Gabbert, S.E.; Hutchison, W.D. Cold tolerance of *Halyomorpha halys* (Hemiptera: Pentatomidae) across geographic and temporal scales. *Environ. Entomol.* **2016**, *45*, 484–491. [CrossRef] [PubMed]

48. DeSantis, R.D.; Moser, W.K.; Gormanson, D.D.; Bartlett, M.G.; Vermunt, B. Effects of climate on emerald ash borer mortality and the potential for ash survival in North America. *Agric. For. Meteorol.* **2013**, *178*, 120–128. [CrossRef]

49. Vermunt, B.; Cuddington, K.; Sobek-Swant, S.; Crosthwaite, J. Cold temperature and emerald ash borer: Modelling the minimum under-bark temperature of ash trees in Canada. *Ecol. Model.* **2012**, *235*, 19–25. [CrossRef]

50. Christianson, L.D.E. Host Influence on the Cold Hardiness of Agrilus Planipennis. Master's Thesis, University of Minnesota, Saint Paul, MN, USA, 2014.

51. Salt, R.W. Time as a factor in the freezing of under-cooled insects. *Can. J. Res.* **1950**, *28*, 285–291. [CrossRef]

52. Salt, R.W. Effect of cooling rate on freezing temperatures of supercooled insects. *Can. J. Zool.* **1966**, *44*, 655–659. [CrossRef]

53. Andreadis, S.S.; Milonas, P.G.; Savopoulou-Soultani, M. Cold hardiness of diapausing and non-diapausing pupae of the European grapevine moth, *Lobesia botrana*. *Entomol. Exp. Appl.* **2005**, *117*, 113–118. [CrossRef]

54. Terblanche, J.S.; Hoffmann, A.A.; Mitchell, K.A.; Rako, L.; le Roux, P.C.; Chown, S.L. Ecologically relevant measures of tolerance to potentially lethal temperatures. *J. Exp. Biol.* **2011**, *214*, 3713–3725. [CrossRef] [PubMed]

55. Mohammadzadeh, M.; Izadi, H. Cooling rate and starvation affect supercooling point and cold tolerance of the Khapra beetle, *Trogoderma granarium* Everts fourth instar larvae (Coleoptera: Dermestidae). *J. Therm. Biol.* **2018**, *71*, 24–31. [CrossRef] [PubMed]

56. Hawes, T.C.; Wharton, D.A. Tolerance of freezing in caterpillars of the New Zealand Magpie moth (*Nyctemera annulata*). *Physiol. Entomol.* **2010**, *35*, 296–300. [CrossRef]

57. Morey, A.C.; Venette, R.C.; Santacruz, E.C.N.; Mosca, L.A.; Hutchison, W.D. Host-mediated shift in the cold tolerance of an invasive insect. *Ecol. Evol.* **2016**, *6*, 8267–8275. [CrossRef] [PubMed]

58. Ault, T.R.; Henebry, G.M.; de Beurs, K.M.; Schwartz, M.D.; Betancourt, J.L.; Moore, D. The false spring of 2012, earliest in North American record. *Eos* **2013**, *94*, 181–188. [CrossRef]

59. Bray, A.M.; Bauer, L.S.; Poland, T.M.; Haack, R.A.; Cognato, A.I.; Smith, J.J. Genetic analysis of emerald ash borer (*Agrilus planipennis* Fairmaire) populations in Asia and North America. *Biol. Invasions* **2011**, *13*, 2869–2887. [CrossRef]

60. Ungerer, M.J.; Ayres, M.P.; Lombardero, M.J. Climate and the northern distribution limits of *Dendroctonus frontalis* Zimmermann (Coleoptera : Scolytidae). *J. Biogeogr.* **1999**, *26*, 1133–1145. [CrossRef]

61. Crozier, L. Winter warming facilitates range expansion: Cold tolerance of the butterfly *Atalopedes campestris*. *Oecologia* **2003**, *135*, 648–656. [CrossRef] [PubMed]

62. Erdmann, G.G.; Crow, T.R.; Peterson, R.M., Jr.; Wilson, C.D. *Managing Black Ash in the Lake States, General Technical Report NC-115*; USDA Forest Service, North Central Forest Experiment Station: Saint Paul, MN, USA, 1987.

63. Miles, P.D.; Heinzen, D.; Mielke, M.E.; Woodall, C.W.; Butler, B.J.; Piva, R.J.; Meneguzzo, D.M.; Perry, C.H.; Gormanson, D.D.; Barnett, C.J. *Minnesota's Forest 2008: Statistics, Methods and Quality Assurance*; Bull. NRS-50; U.S. Department of Agriculture Forest Service, Northern Research Station: Newtown Square, PA, USA, 2011.

64. VanderSchaaf, C.L. Minnesota's Forest Resources 2011. Available online: http://files.dnr.state.mn.us/forestry/um/forestresourcesreport_11.pdf (accessed on 30 March 2018).

65. Palik, B.J.; Ostry, M.E.; Venette, R.C.; Abdela, E. Tree regeneration in black ash (*Fraxinus nigra*) stands exhibiting crown dieback in Minnesota. *For. Ecol. Manag.* **2012**, *269*, 26–30. [CrossRef]

66. Telander, A.C.; Slesak, R.A.; D'Amato, A.W.; Palik, B.J.; Brooks, K.N.; Lenhart, C.F. Sap flow of black ash in wetland forests of northern Minnesota, USA: Hydrologic implications of tree mortality due to emerald ash borer. *Agric. For. Meteorol.* **2015**, *206*, 4–11. [CrossRef]

67. MN-DNR. Emerald Ash Borer (EAB). Available online: https://www.dnr.state.mn.us/invasives/terrestrialanimals/eab/index.html (accessed on 30 March 2018).

forests

MDPI

Review

Evaluating Adaptive Management Options for Black Ash Forests in the Face of Emerald Ash Borer Invasion

Anthony W. D'Amato [1,*], **Brian J. Palik** [2], **Robert A. Slesak** [3], **Greg Edge** [4], **Colleen Matula** [4] and **Dustin R. Bronson** [4]

1. Rubenstein School of Environment and Natural Resources, University of Vermont, 204e Aiken Center, 81 Carrigan Drive, Burlington, VT 05405, USA
2. USDA Forest Service Northern Research Station, 1831 Hwy 169 East, Grand Rapids, MN 55744, USA; bpalik@fs.fed.us
3. Minnesota Forest Resources Council, 1530 Cleveland Ave. North, St. Paul, MN 55108, USA; raslesak@umn.edu
4. Wisconsin Department of Natural Resources, 101 S. Webster St., Madison, WI 53703, USA; Gregory.Edge@Wisconsin.gov (G.E.); colleen.matula@wisconsin.gov (C.M.); Dustin.Bronson@wisconsin.gov (D.R.B.)
* Correspondence: awdamato@uvm.edu; Tel.: +1-802-656-8030

Received: 18 April 2018; Accepted: 12 June 2018; Published: 13 June 2018

Abstract: The arrival and spread of emerald ash borer (EAB) across the western Great Lakes region has shifted considerable focus towards developing silvicultural strategies that minimize the impacts of this invasive insect on the structure and functioning of black ash (*Fraxinus nigra*) wetlands. Early experience with clearcutting in these forests highlighted the risks of losing ash to EAB from these ecosystems, with stands often retrogressing to marsh-like conditions with limited tree cover. Given these experiences and an urgency for increasing resilience to EAB, research efforts began in north-central Minnesota in 2009 followed by additional studies and trials in Michigan and Wisconsin to evaluate the potential for using regeneration harvests in conjunction with planting of replacement species to sustain forested wetland habitats after EAB infestations. Along with these more formal experiments, a number of field trials and demonstrations have been employed by managers across the region to determine effective ways for reducing the vulnerability of black ash forest types to EAB. This paper reviews the results from these recent experiences with managing black ash for resilience to EAB and describes the insights gained on the ecological functioning of these forests and the unique, foundational role played by black ash.

Keywords: black ash; adaptive silviculture; emerald ash borer; Lake States; hydrology; habitat type

---

1. Introduction

Novel stressors, such as non-native insects and diseases, present a significant challenge to the long-term sustainable management of forest ecosystems around the globe [1]. For example, recent estimates for the United States indicate that almost two thirds of all forestlands are susceptible to damage from non-native insects and diseases already present in the country [2]. In many cases, these organisms are host-specific with the ability to functionally eliminate a given tree species from an ecosystem [3]. As such, there is a significant need for the development of adaptive management strategies that increase forest resilience to the loss of a constituent tree species as a way to maintain key ecosystem functions after infestation by non-native species [4]. This need becomes urgent when the host species is considered foundational to ecosystem functioning, such that its loss cannot readily be absorbed by other species in the ecosystem [5].

Forest management guidelines developed in response to non-native insects and diseases have generally focused on altering host species structure and abundance to minimize spread and impacts of a given organism. These approaches may include removing all potential host species from forested areas within a set distance of known infestations [6,7] or selective removal of larger host trees to reduce availability of brood trees [8,9]. Although these strategies are warranted in areas proximal to introduction points and ongoing infestations, their application in areas not currently threatened may generate greater ecological and economic impacts than the non-native forest pest itself (e.g., [10]). To date, less attention has been devoted to developing appropriate silvicultural strategies for increasing the resilience of these unimpacted areas to maintain long-term management options and ecological functions even after a pest's establishment.

The introduced emerald ash borer (*Agrilis planipennis* Fairmaire [EAB]) is one of the more significant non-native insects threatening North American forests given the importance of ash species across numerous forest types and regions [11], and the ability of this insect to cause widespread mortality within 2–6 years of invasion [12]. Most ash species exist as minor components of mixed species forests [11] and loss of canopy ash to EAB will eliminate the unique ecological functions of ash, but forested conditions are expected to persist in these affected ecosystems. One exception are black ash (*Fraxinus nigra* Marsh.)-dominated lowland forests in the western Great Lakes region, where black ash often forms almost pure stands, with up to 95% of stems composed of this species [5]. As black ash is a foundational species in these forests, expectations regarding EAB impacts include not only a loss of canopy black ash, but also a concomitant shift in ecosystem structure and function to marsh-like conditions with little to no tree component [13,14]. In particular, historic experience with clearcutting and recent field simulations of EAB impacts indicate that the loss of canopy black ash may cause the water table to rise and increase the duration of water at or near the soil surface [13,14]. These changes represent not only a significant threat to organisms and processes associated with current black ash ecosystems [5], but also a significant challenge to managers tasked with sustaining these ecosystems in the face of the threat of EAB invasion.

As with other forest types threatened by non-native insects and diseases, the management guidance for black ash forests was developed based on research and management experience developed prior to the introduction of EAB. This historical guidance largely focused on silvicultural strategies that encouraged the dominance of black ash over time [13], a condition in contemporary contexts that is highly vulnerable to EAB impacts. Recent management recommendations for addressing EAB in these and other ash forest types have primarily focused on minimizing establishment and spread of the insect through reducing host abundance [8,9]. Although these approaches, when applied in conjunction with insecticides, may reduce spread of EAB in localized, urban settings, they represent a reactive strategy to a current outbreak as opposed to a proactive approach that aims to increase resilience to impacts prior to invasion. Moreover, black ash can dominate extensive areas (1–40 ha) presenting significant logistical challenges for widespread application of insecticides to protect all black ash forests. Given the potential for significant ecosystem phase shifts following EAB mortality or pre-emptive liquidation harvests in lowland black ash forests (cf. [13]), there is a great need for silvicultural strategies that reduce vulnerability of these ecosystems to EAB prior to its arrival. In addition, given the continued spread of this insect and the infeasibility of treating all black ash-dominated areas, there may also be a growing need for rehabilitation approaches to reintroduce forest cover to places where the insect or associated reactive management have resulted in a shift to non-forest conditions.

The overall objective of this paper is to review recent research and experience managing black ash ecosystems to provide general guidance for the adaptive management of these ecosystems in the face of EAB invasion. This includes describing the general ecological setting of these forests as a foundation for adaptive silviculture prescriptions, summarizing insights from long-term research trials evaluating the effectiveness of different silvicultural treatments at increasing the non-ash components in these forests, and discussing potential rehabilitation approaches for returning other species to these areas after EAB has eliminated black ash. To date, EAB has spread to 32 states in the US and three Canadian provinces since its discovery in Michigan in 2002, and has killed nearly 100% of ash stems >2.5 cm diameter at breast height (DBH; 1.3 m) in areas proximal to where it was first introduced [15]. In most cases, the majority of ash trees are killed within 2–6 years of invasion; however, individual trees and stump sprouts have persisted for over a decade in some green ash (*Fraxinus pennsylvanica* Marsh.)-dominated wetlands in southern Michigan, indicating areas may not always experience complete ash mortality following invasion [16]. As such, the management recommendations reviewed in this work are not definitive given the great uncertainty around long-term EAB impacts and silvicultural outcomes, nor is it expected that they can be applied across all black ash-dominated areas and ownerships. Instead, they are meant provide a framework for approaching this threat to increase resilience and sustain a forested condition over the long term, particularly given the significant losses of ash observed in all ecosystem types impacted by EAB (>62–100% of basal area; [15,16]). At the time of writing, EAB had not been detected in the areas serving as the focus for this review (northern MI, MN, and WI) and the experiments and recommendations largely assumed EAB arrival within the next 10–15 years based on location of known infestations (southern MN and WI, eastern Upper Peninsula of Michigan) relative to expected rates of spread [17].

2. Ecological Context of Black Ash Wetlands

A central challenge to developing adaptive management approaches for addressing the threat of EAB is the limited amount of information on the ecology and management of black ash forests relative to other forest ecosystems in the Great Lakes region. Black ash has long been a significant component of the cultures and livelihoods of Native Americans and First Nations people across the range of this species [15]. However, its lower historic value to the timber industry in the Great Lakes region resulted in greater historical emphasis of research on more productive, upland forest types, as well as lowland conifers. The introduction of EAB to the region in the early 2000s motivated considerable effort towards generating a greater understanding of the ecology and dynamics of these forests as a foundation for developing appropriate management responses. The following section summarizes our current understanding of the variety in black ash forest habitats as well as several key findings from recent investigations into the hydrology and regeneration dynamics following episodic dieback and their relevance to adaptive management in the context to EAB.

2.1. Black Ash Forest Habitat Types

Black ash wetland forests are complex and exist across a wide range of biophysical settings covering over 1 million ha in Minnesota, Michigan, and Wisconsin [18]. Early work describing the vegetation and silviculture of these forests recognized this complexity and demonstrated the importance of water table depth and nutrient conditions in affecting regeneration dynamics and forest composition [13,19,20]. Nevertheless, historical management guidance and forest classification systems often lumped all black ash-dominated forests into a single "black ash swamp" or "lowland hardwood" category with little consideration for the important variations in soil, nutrient, and vegetation dynamics that occur in these ecosystems [21–23]. The advent of forest habitat type classification systems (FHTCS) for the Lake States region during the 1980s and 1990s resulted in a greater appreciation for the range in black ash-dominated forests that existed across the landscape. In particular, they generated a framework that has been critical for recognizing variations in soil, moisture, and vegetation conditions that exist across sites when developing management strategies for addressing EAB [20,24,25].

The first FHTCS to describe the variation in black ash communities were developed in Minnesota, owing in part to the comparatively high abundance of black ash forests in this state relative to Wisconsin and Michigan [20]. Two main black ash forest types, Wet Ash Forests (WFn55) and Very Wet Ash Forests (WFn64), were described by this FHTCS and both are dominated by >75% black ash in the canopy and understory layers [25]. The variation between these two communities is driven by the influence of flooding duration and organic soil depth in affecting nutrient dynamics. The primary distinguishing features between WFn64 and WFn55 communities are peaty versus mineral soils and a greater duration of standing water during the growing season in WFn64 communities. These differences have important management implications, as even the creation of larger group selection openings (0.2 ha) has led to a rise in local water tables and loss of tree cover in areas classified as WFn64 communities (Peter Bundy, personal communication). The importance of overstory tree cover in regulating water table dynamics during recruitment events has also been reinforced by examinations of the age structure of old-growth examples for these two habitat types, which indicate these systems are strongly multi-aged with historic development influenced primarily by gap-scale disturbances (Figure 1; D'Amato et al. in prep). A similar multi-aged population structure was also documented across numerous old-growth black ash riparian forests in the Lake Duparquet region of Quebec [26].

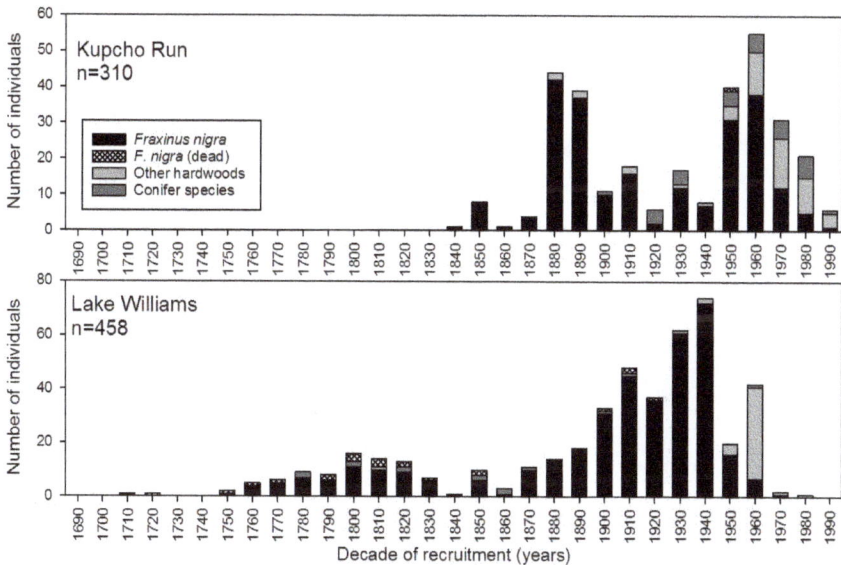

Figure 1. Age structure for two old-growth black ash forests in Minnesota. Sample size corresponds to number of cored individuals for which recruitment ages were obtainable, including several downed, dead black ash. The Kupcho Run site is classified as a "Very Wet Ash Forest (WFn64)", whereas the Lake Williams site is classified as a "Wet Ash Forest (WFn55)" based on Aaseng et al. [25]. Other hardwood species are largely *Ulmus americana* L., *Tilia americana* L., *Betula alleghaniensis* Britt., *Acer rubrum* L., and *Quercus macrocarpa* Michx. Conifer species are *Abies balsamea* (L.) Mill., *Picea glauca* (Moench) Voss, and *Thuja occidentalis* L.

In contrast to the FHTCS developed for Minnesota, early classification systems developed for Wisconsin and Michigan focused primarily on more productive upland forest types leaving key knowledge gaps regarding lowland forest, including black ash ecosystems [24]. The arrival of EAB to the region motivated renewed focus on these ecosystems and led to the development of the Wisconsin Wetland Forest Habitat Type guide using methods and interpretation similar to the forest habitat type guide for upland forests [27]. This guide includes management interpretations provided by the Wisconsin Department of Natural Resources (WDNR; [27]) and has been an important tool for foresters across the state in interpreting complex, wetland forest systems.

The wetland habitat typing system developed for Wisconsin covers five biophysical regions in the northern portion of the state and recognizes 12 lowland *Fraxinus* (black ash) types that vary in understory vegetation, nutrient availability, soils, and hydrology [27]. Consistent with the silvics of the species, black ash-dominated habitat types are generally characterized by medium rich nutrient regimes (e.g., Figure 2). The Black Ash–Balsam Fir–Red Maple/Sensitive Fern (FnAbArOn) type is the most widespread black ash habitat type in Wisconsin and is characterized by organic soils that can vary from 30 to 90 cm in depth overlaying mineral soils [27]. Habitat type field monitoring of these areas identified organic soil depth as a key driving factor of nutrient availability and duration of flooding, with deeper organic soils having greater potential for hydrological changes and rutting associated with timber harvest. As with the Minnesota habitat types, the less rich habitat types tend to regenerate predominately to black ash following harvesting and have become areas where artificial regeneration is being considered to enhance the non-ash components [28].

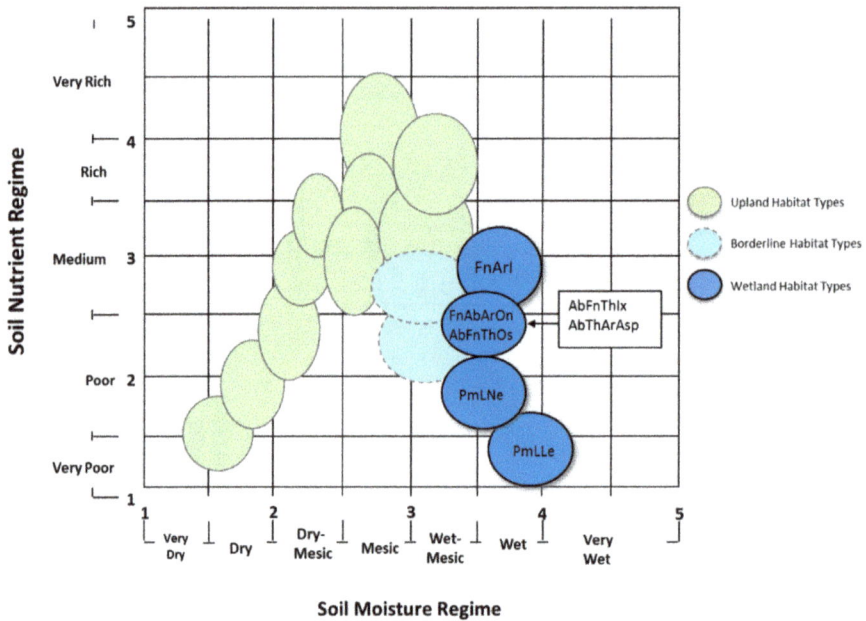

Figure 2. Synecological coordinates of forest habitat types in north-central Wisconsin (Region 3) based on understory plant species communities, soil moisture, and nutrient regimes. Habitat types in blue are recently added wetland habitat types, including several that are dominated by black ash: Fraxinus-Acer/Impatiens (FnArI), Fraxinus-Abies-Acer/Onoclea (FnAbArOn), Abies-Fraxinus-Thuja/Ilex (AbFnThIx), and Abies-Fraxius-Thuja/Osmunda (AbFnThOs). Black ash-dominated habitat types in general are most mesic in nutrient regimes relative to other forested wetland communities in the region. From Kotar and Burger [27].

Ash-dominance is a common feature across abovementioned habitat types developed in Minnesota and Wisconsin; however, the abundance of non-ash species in the overstory can be quite variable within and across habitat types reflecting variations in local site conditions and land-use history. For example, non-ash species, including *Quercus macrocarpa*, *Ulmus americana*, and *Abies balsamea*, made up 16–30% of the basal area across old-growth WFn64 communities in Minnesota [29]. Similarly, black ash abundance ranged from 25% to 85% of basal area in isolated, depressional wetlands in the Upper Peninsula of Michigan, with *Betula alleghaniensis* and *Acer rubrum* often constituting the other primary overstory species [30]. The variation in these community types, particularly in relation to the amount of non-ash trees present in the overstory and regeneration layers, is an important consideration when determining vulnerability to EAB impacts and developing appropriate management responses [31]. Much of the experience with managing black ash forests for resilience to EAB has been in the context of the most vulnerable conditions and habitat types in MN and WI (i.e., >70% black ash) and application of these approaches to other systems and regions will require adaptation to local site conditions and levels of non-ash in the canopy.

2.2. Hydrologic Setting of Black Ash Forests

Black ash wetlands occur in a number of hydrogeomorphic settings that span a range of scales including large expansive complexes [14], floodplains [32], areas along streams and rivers [33], and in topographic low depressions [34,35]. Topography and its influence on drainage is a primary factor influencing wetland hydrology, but soil parent material can also play a large role. In the large expansive wetlands of northern Minnesota, presence of a dense layer of thick clay results in a perched water table that is the primary driver of hydrology. Soils range from deep organic peats (especially in depressional wetlands) to mineral soils overlain by shallow layers of muck, but all are poorly to very-poorly drained. Water source is primarily from precipitation, either as direct rainfall or snow inputs [35] or indirect from localized drainage [34].

Depending on the dominant hydrological process(es), hydroperiod length in black ash wetlands can be quite variable, but most sites are characterized as having water tables at or above the soil surface during a portion of the growing season. Wetlands fed by consistent groundwater sources, whether from deep regional flows or areas occurring at topographic lows in the local landscape, may have very little water level variability throughout the year with prolonged surface water presence [34]. In large expansive wetland complexes, water level variability is typically intermediate, where water tables are generally at or above the surface in spring, followed by a large drawdown in early summer until a stable minimum is reached later in the season (Figure 3).

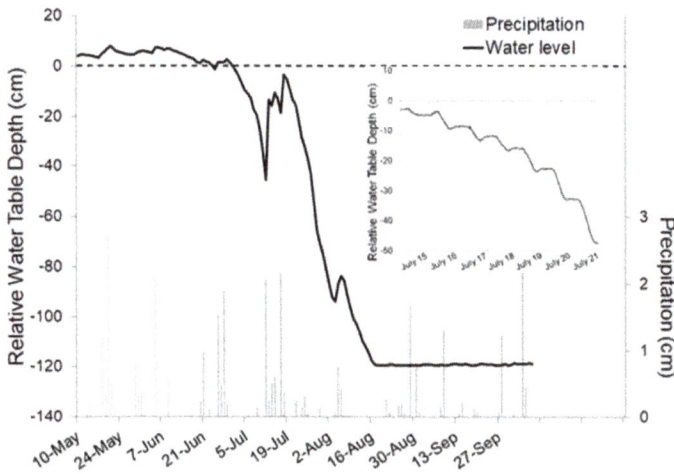

Figure 3. Water table hydrograph for a black ash wetland in northern Minnesota (classified as WFn64 based on Aaseng et al. [25]). Water tables are above the soil surface following spring snowmelt, and then rapidly drawdown once black ash leaf expansion occurs (usually in mid-June). The influence of black ash transpiration on this pattern is readily apparent in the diurnal water table pattern once water tables are below the soil surface (inset).

The dominance of black ash (typically greater than 80% of the canopy; [29]) in wetland settings reflects its unique physiology that allows for transpiration during periods of inundation [36,37]. Because of this, black ash strongly regulates the hydrologic regime with water table drawdown occurring rapidly following black ash leaf expansion in the spring (Figure 3). Evidence of this biological influence is clearly evident in the diurnal water table signature, with pronounced drawdown during the day when trees transpire followed by slight recharge or stability during the night (Figure 3 inset). Black ash transpiration (and its influence on overall evapotranspiration [ET]) is thought to be greatest when water tables are below the surface and at intermediate depths (~50 cm, [38]).

Given the role of ET in site hydrology, a reduction in black ash transpiration is likely to alter water table dynamics, causing the water table to rise and increase the duration of water at or near the soil surface [13,14]. Because of this, black ash wetlands are commonly believed to be highly susceptible to hydrologic alteration following EAB mortality [14,34]. Slesak et al. [14] used experimental girdling to mimic EAB mortality, which caused the water table to be elevated and near the soil surface for longer periods of time relative to control stands. Using a similar approach but in a different hydrogeomorphic setting, Van Grinsven et al. [34] found the effect of black ash loss was more muted, largely because the hydrologic regime was influenced by continuous subsurface inflow associated with topographic position in the landscape. Although there are clear effects of ash loss on water table dynamics [13,14], the magnitude and duration of these effects is not clear. Recent work by Diamond et al. [39] showed that clearcutting caused a distinct change in hydrologic regime that persists for at least five years. Although hydrologic alteration at the site scale is likely following black ash loss, it is not clear how such changes will alter hydrology at larger scales such as the watershed.

2.3. Black Ash Decline and Natural Regeneration Potential for Non-Ash Species

Finding effective adaptive management approaches for black ash ecosystems in the face of EAB is complicated by a long-standing episodic dieback/decline problem that occurs throughout much of the range of black ash, including the Lake States and northeastern United States [40–42]. Most dieback episodes are not associated with a specific disease or insect pest, but include areas of crown dieback and tree mortality, particularly on more hydric sites [42]. In fact, the cause is still in doubt, with speculation that it relates to spring drought [41,43] or, conversely, overly wet conditions [42,44]. Regardless, dieback episodes can affect large areas of black ash with over 9000 ha affected in Minnesota in 2009 alone [41].

Regardless of cause, examining successional potential in declining stands can provide insight into the future of EAB impacted black ash wetlands. Unfortunately, this future may not be bright. While black ash regenerates prolifically from stump-sprouts [13], and research has documented high numbers of (likely) sprout-origin black ash seedlings after harvesting [45–47], densities may decline significantly as trees grow into larger sizes [22,48], potentially limiting self-replacement. Moreover, this regeneration presumably will suffer the same fate as canopy black ash, i.e., death from EAB after a susceptible size is reached (approximately 2.5 cm diameter).

Natural successional replacement of black ash after EAB mortality may be limited as well. In Minnesota wetlands, for instance, canopy species other than black ash occur in very low densities, generally much less than 20% of relative density [23,48,49]. Consequently, there is not a high potential for replacement of EAB killed black ash from the current sapling or tree layers. Moreover, the regeneration layer in these forests is generally dominated by speckled alder (*Alnus incana* [L.] Moench) and beaked hazel (*Corylus cornuta* Marsh.), with only limited abundance of canopy potential species [48]. This regeneration bottleneck may be exacerbated with loss of black ash from EAB if, as discussed above, this results in an upward fluctuating water table, further increasing the potential for regeneration failures.

3. Management Experience with Strategies for Increasing Resilience of Black Ash Wetlands to EAB

3.1. Research Trials Examining Artificial Regeneration Options to Increase Non-Ash Components

The importance of black ash in many lowland habitat types and the hydrological responses observed following the loss of this species to harvesting or decline episodes, led to the establishment of several adaptive silvicultural studies focused on strategies for increasing ecosystem resilience in advance of EAB. A primary component of this work was the evaluation of artificial regeneration of non-ash species as a strategy to maintain tree cover following the loss of overstory ash. Although there is no true cultural or ecological replacement for black ash, the emphasis of these studies was to evaluate approaches for replacing the core ecological functions provided by black ash (e.g., hydrologic regulation, mature tree habitat, live-tree carbon storage) so as to maintain forested wetland conditions following infestation by EAB. The first of these studies was established in north-central Minnesota in a region containing the greatest concentration of lowland black ash forests in the US [14]. This work included canopy treatments intended to emulate EAB (girdling of all ash in 1.4 ha areas), potential management responses to the threat of EAB (group selection and clearcut harvests), and untreated controls, all in conjunction with operational plantings of 12 different tree species [49]. A later, companion study was established based off this work in smaller, depressional black ash wetlands in the Upper Peninsula of Michigan with a specific focus on the effect of planting microsite on seedling survival [50]. Finally, planting trials of non-ash species in riparian black ash systems in Wisconsin were developed to evaluate the effectiveness of underplanting several non-ash species to increase resilience of these forests [50].

The wide range of species evaluated as potential replacements for black ash largely reflects the great uncertainty and limited prior experience with planting trees in these lowland forests. The majority of these species commonly co-occur at low abundance in black ash wetlands, including American elm (*Ulmus americana*), red maple (*Acer rubrum*), northern white cedar (*Thuja occidentalis*), bur oak (*Quercus macrocarpa*), yellow birch (*Betula alleghaniensis*), balsam poplar (*Populus balsamifera*), basswood (*Tilia americana*), white spruce (*Picea glauca*), tamarack (*Larix laricina* [Du Roi] K. Koch), balsam fir (*Abies balsamea*), and trembling aspen (*Populus tremuloides* Michx; Table 1). Other species found in wetland forests in the study regions of interest have also been included in these trials, including black spruce (*Picea mariana* (Mill.) Britton, Sterns & Poggenb.), cottonwood (*Populus deltoides* W. Bartram ex Marshall), and American sycamore (*Platanus occidentalis* L.). Wetland species currently found in the climate zone immediately south of these areas, including swamp white oak (*Quercus bicolor* Willd.), hackberry (*Celtis occidentalis* L.), and silver maple (*Acer saccharinum* L.) have also been evaluated in an attempt to anticipate both EAB and climate change impacts [49,50]. Finally, Manchurian ash (*Fraxinus mandschurica* Rupr.) has been included in Minnesota trials due to its resistance to EAB and potential to serve as a replacement for traditional Native American basket making materials historically provided by black ash (G. Swanson, personal communication).

Table 1 summarizes general recommendations for artificial regeneration of non-black ash species based on early patterns in seedling survival and growth. Survival has been generally quite low for most species examined (e.g., <40%; [49]); however, swamp white oak has consistently performed well in a variety of black ash habitat types (e.g., Figure 4). This greater survival (>75%; [49]) is consistent with findings from planting trials in simulated floodplain conditions that indicate high potential for this species for reforestation efforts in seasonally flooded environments [51]. Swamp white oak early height growth rates have also been similar to those documented in other planting trials and have ranged from 5 to 12 cm year^{-1} [51]. Within species, seedling survival has largely been greatest under an undisturbed or partial black ash canopy, reflecting the importance of ash in regulating water table dynamics during the establishment phase and reinforcing historic recruitment dynamics in these forests (e.g., Figure 1; [49]). In addition, overstory ash may also limit levels of competition experienced by planted seedlings, as loss of ash often leads to an increased abundance of understory species, including lake sedge and speckled alder (Looney et al., 2017). Planting location has also influenced seedling success with planting on hummocks significantly increasing the survival of more mesic species, such as American basswood, given the shorter duration of seedling exposure to flooding on these microsites [50].

Table 1. Species planted as part of adaptive strategies for minimizing the impacts on emerald ash borer on black ash wetlands. General recommendation for species (R = recommended, NR = not recommended) are based on 3–5-year patterns in survival and growth and are presented based on performance in research trials in Minnesota [47,49], Michigan, and Wisconsin [50], as well as operational plantings.

Species	Stock Type	MN	MI	WI	Notes
Acer rubrum	Bare-root (2+0), containerized (90 cm³)	NR	NR	R	Success observed with bare-root stock in riparian black ash forest
Acer saccharinum	Bare-root (4+0)	-	R	-	Success observed with large bare-root stock
Abies balsamea	Bare-root (2+0, 3+0)	-	NR	NR	Very low survival regardless of cultural treatment
Betula alleghaniensis	Containerized (90 cm³)	NR	NR	NR	Very low survival with containerized stock. Marginal survival (36%) for bare-root stock planted in clearcut prepared by fecon mower
Celtis occidentalis	Bare-root (2+0), Containerized (336 cm³)	R	-	R	Greatest success when planted in group selection harvests in depressional wetlands
Fraxinus mandshurica	Bare-root (3+0)	NR	-	-	Limited cold tolerance, despite high survival of sprouts
Larix laricina	Bare-root (2+0), containerized (60 cm³)	NR	NR	R	Success observed with bare-root and containerized stock in clearcut prepared with fecon mower
Picea glauca	Bare-root (3+0)	-	-	R	Success observed with bare-root stock in clearcut prepared with fecon mower
Picea mariana	Containerized (90 cm³)	NR	NR	R	Success observed with bare-root and containerized stock in clearcut prepared with fecon mower
Pinus strobus	Bare-root (3+0)	-	-	R	Success observed with bare-root stock in clearcut prepared with fecon mower
Platanus occidentalis	Bare-root (1+0)	-	-	R	Success observed with bare-root stock in riparian black ash forest
Populus balsamifera	Containerized (164 cm³)	R	-	-	Greatest success when planted in clearcut and group selection harvests in depressional wetlands
Populus deltoides	Bare-root (1+0)	NR	-	-	Very low survival regardless of cultural treatment
Populus tremuloides	Containerized (90 cm³)	NR	-	NR	Marginal success observed with containerized stock in clearcut prepared with fecon mower
Quercus bicolor	Bare-root (1+0)	R	-	R	Greatest success in group selection harvests in depressional wetlands; successful under intact canopy of riparian black ash forests
Quercus macrocarpa	Bare-root (3+0)	-	NR	-	Greatest success when planted on hummocks
Thuja occidentalis	Bare-root (3+0), Containerized (60 cm³)	R	R	R	Greatest success in intact canopies and group selection harvests; high incidence of browse on seedlings
Tilia americana	Bare-root (3+0)	-	R	-	Greatest success when planted on hummocks
Tsuga canadensis	Bare-root (3+0)	-	-	NR	
Ulmus americana	Containerized (1890 cm³)	R	R	-	Greatest success when planted in group selection harvests and on hummocks using DED[1] tolerant cultivars

[1] Dutch elm disease (*Ophiostoma novo-ulmi*).

Figure 4. (a) Swamp white oak (*Quercus bicolor*) planted in riparian black ash forest two years prior to photo in southern Wisconsin as part of a strategy to underplant non-ash species to increase ecosystem resilience to emerald ash borer (EAB); (b) Balsam poplar (*Populus balsamifera*) planted five years prior to photo in 0.04 ha group selection harvest as part of an adaptive approach for addressing the threat of EAB to the extensive black ash swamps in north-central Minnesota. Arrows in photos identify terminal shoot of planted individuals, which was at 0.8 and 1.6 m height in (a) and (b), respectively.

Given that few of the species selected for the adaptive approaches being tested are widely planted in operational settings, there has been limited opportunity to evaluate how stock type and size affects species performance. General patterns suggest larger, bare-root or containerized stock have generally had greater survival and growth when planted in black ash wetlands (Table 1). This includes large, containerized Dutch elm disease (*Ophiostoma novo-ulmi* Brasier)-tolerant American elm, which has also shown promise in other adaptive plantings to address EAB threats to riparian green ash (*Fraxinus pennsylvanica*) forests in Ohio [52]. Black ash wetlands often have dense understory layers of shrubs and herbaceous species [47] and larger stock sizes may provide a competitive advantage and increase survival on these sites. This dynamic was reflected in a black ash silviculture field trial in Wisconsin in which competition was controlled at the time of planting. In this context, there was no appreciable difference in survival between containerized and bare-root tamarack and black spruce seedlings (>60% survival after three years), whereas both species have fared poorly when planted as containerized stock in other trials lacking competition control (<20% survival; [43]).

3.2. Encouraging Natural Regeneration of Non-Ash Species

There is great uncertainty regarding how effective and operationally feasible planting non-ash species is for increasing the resilience across the large expanse of black ash forests in the Great Lakes region. As a result, there is also considerable interest in the potential for using different regeneration methods to encourage the natural regeneration and recruitment of non-ash species to diversify overall composition and structure. A key source of information on the response of black ash wetlands to different silviculture practices has been the Swamp Hardwood Silviculture Trials established by the Wisconsin DNR beginning in 1977 (Figure 5; [53]). These trials were established prior to the introduction of EAB and correspondingly were focused on developing the most effective strategies for managing and maintaining black ash-dominated forests [28]. Since the introduction of EAB to the region, these trials have been revisited to evaluate the relative effectiveness of different strategies at increasing the non-ash component in these forests, while also maintaining hydrological function.

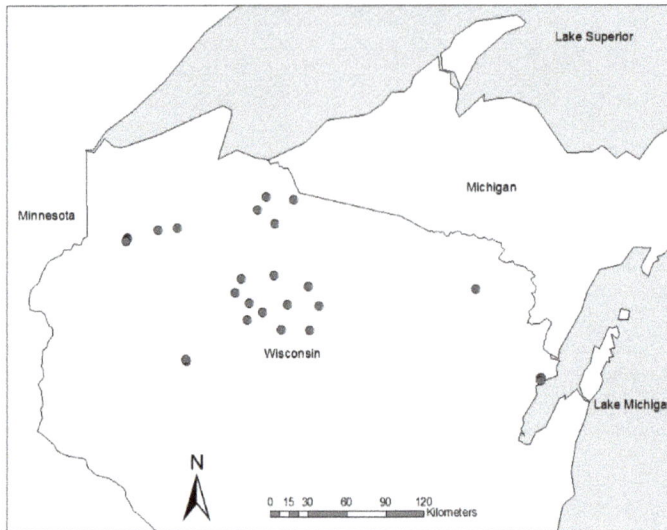

Figure 5. Location of Swamp Hardwood Silviculture Trials maintained by the Wisconsin Department of Natural Resources. Trials are concentrated in north-central Wisconsin, which contains the greatest concentration of black ash wetlands in the state.

The silviculture methods implemented in the Swamp Hardwood Silviculture Trials included clearcut, coppice with standards, diameter limit, strip shelterwood, and single-tree selection with the majority of trials established in FnAbArOn habitat types given the greater abundance of this community type relative others in the state [53]. Within this habitat type, strip shelterwood harvests consisting of 9 m harvested strips alternating with 18 m unharvested strips generally resulted in the highest proportion of non-ash regeneration, particularly when non-ash seed trees were present in unharvested strips (Figure 6). Non-ash species commonly regenerating following these harvests included red maple, yellow birch, and American elm; however, black ash remained the most abundant species in the regeneration layer in this and all other regeneration methods (Figure 6). In addition, all treatments served to increase the abundance of shrub species in these forests indicating follow-up release work may be necessary to maintain non-ash components in these stands (Figure 6). Single-tree selection was the least effective regeneration method at stimulating non-ash regeneration given the smaller canopy openings in this approach largely served to release black ash advance regeneration and shrub species, such as speckled alder. Beyond positive regeneration responses, there was little evidence of elevated water tables in strip shelterwood harvests, whereas swamping was noted for clearcut and diameter-limit methods, although sample size was limited for the latter method. Strip shelterwoods had the lowest amount of harvesting machine traffic and disturbance across sites given equipment is restricted to harvested strips versus across the entire stand in other methods.

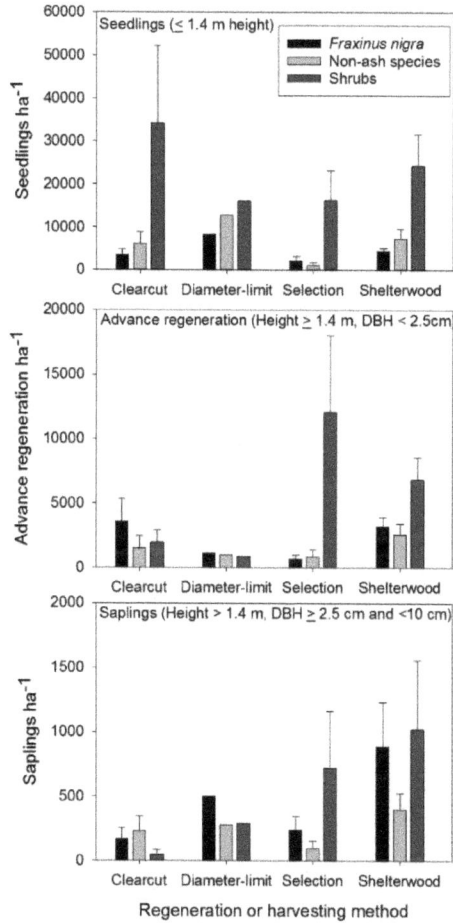

Figure 6. Natural regeneration response to different regeneration methods and diameter-limit cutting as part of the Wisconsin Department of Natural Resources silviculture trials in black ash lowland forests in north-central Wisconsin. Data in figures are based on field measurements of these trials in 2015 and values represent means ± 1 standard error. Sites were harvested between 2003 and 2013 and all exist on Black Ash–Balsam Fir–Red Maple/Sensitive Fern (FnAbArOn) habitat types. Sample sizes for the different regeneration methods and harvesting approaches are: strip shelterwood (*n* = 6), clearcut (*n* = 4), diameter limit (*n* = 1), and single-tree selection (*n* = 4). Non-ash species include *Acer rubrum*, *Betula alleghaniensis*, *Abies balsamea*, *Ulmus americana*, *Betula papyrifera* Marsh., and *Thuja occidentalis*. Shrub species are predominantly *Alnus incana*, *Salix spp.*, and *Corylus cornuta*.

3.3. Decision Support Tools for Guiding Adaptive Management in Black Ash: The Wisconsin DNR Checklist for Evaluating Lowland Ash Stands

Despite the recent focus on managing black ash forests and other ash-dominated lowlands in the context of EAB, current silvicultural guidelines remain tentative and incomplete, due in large part to our limited knowledge and experience managing these ecosystems for species other than ash. Foresters and other natural resources managers have struggled with this knowledge gap while at the same time attempting to develop silvicultural prescriptions that maintain forest productivity and improve resilience to EAB. The *WDNR Checklist for Evaluating Lowland Ash Stands* was developed in an attempt to create a decision support tool that synthesizes our current state of knowledge, drawing upon lessons learned from the research studies and field trials described above [54]. This tool is complemented by other general guidance provided in the region by state agencies and University Extension (e.g., [55]).

The first component of this checklist is a stand assessment designed to track and evaluate site factors recognized as affecting silvicultural options and outcomes in black ash wetland forests based on research discussed above. These factors include an evaluation of site quality (e.g., Wetland Forest Habitat Type Classification System), timber sale operability, potential EAB impact to stand condition, regeneration potential, and hydrological risk. Potential EAB impact is a key determining factor in how managers may treat a given site and is based on the amount of non-ash acceptable growing stock (AGS) that would remain following the loss of black ash. Stands containing ≥40 non-ash AGS per acre or >45% relative density of non-ash AGS are considered less vulnerable to forest conversion following EAB and can be managed according to other forest cover type guidance. Similarly, stand regeneration potential is determined by evaluating non-ash advance regeneration, alternate seed sources, interfering vegetation, and browse pressure. Hydrologic function and the general risk of "swamping" (i.e., water table rise following timber harvesting) is evaluated using a set of observable site factors, such as seasonal inundation period, depth to water table, soil drainage class, depth to mineral soil, and the presence of impeded drainage.

Based on the stand assessment, users are guided to a decision tool to evaluate management options most appropriate for a specific set of stand conditions. Regeneration methods are recommended in vulnerable stands to naturally and artificially establish non-ash species. For example, partial harvest regeneration methods such as strip shelterwood and group selection are recommended for stands with high hydrological risk factors to limit the potential for swamping. The evaluation of stand regeneration potential is central to determining the need for artificial regeneration, as well as if site preparation and browse protection should be applied. It is important to emphasize that the decision tool is not a "cookbook" for every situation, but meant simply to serve as an aid in the prescription writing process when addressing the threat of EAB. To this end, the tool includes additional narratives that synthesizes lessons learned from the Lake States' research studies and field trials, in addition to other research and field experience relevant to managing wetland forests.

3.4. Restoring Forest Cover to Areas Impacted by EAB

Much of the research evaluating adaptive silvicultural strategies for EAB has been developed for areas where EAB is not currently present or is not causing significant impact. Given the challenges with treating all areas prior to its arrival, new research in Wisconsin is now being established to assist foresters with restoring forest cover to black ash wetland sites post-EAB. As with other adaptive strategies, the primary goal is to ultimately maintain forested cover in these wetland systems. Work examining vegetation development following black ash mortality has demonstrated the rapidity with which shrubs, native sedges, and invasive species like reed canary grass (*Phalaris arundinacea* L.) increase on these wetland sites (e.g., [47]) creating a significant challenge to reforestation efforts. There is limited experience with site preparation techniques suitable for addressing these competition conditions on wetland sites and the WDNR has established a new study to better understand the cost and management efforts needed to establish new tree species following complete ash mortality. This study includes three replicates of clearcut harvests of black ash stands >25 hectares to purposely increase flooding duration and allow shrubs and sedges to increase. Study treatments will include brush mowing, herbicide application, and their combination on nested treatments of varying planted tree species that have shown promise as adaptive plantings in previous work (Table 1). In particular, evaluation of these treatments in an earlier silvicultural trial indicate this approach may be affective for establishing planted tamarack and black spruce on these sites [53]. Over the next ten years the WDNR hopes this experiment will generate needed recommendations to foresters on cost-suitable measures for site prep and tree seedling establishment following the loss of black ash to EAB.

4. Conclusions

The invasive EAB has motivated considerable interest and research in understanding the ecological functioning of black ash forests, as well as suitable adaptive silviculture strategies for maintaining these ecological functions even after the arrival of EAB. This work has reinforced the complexity of ecological conditions characterizing black ash-dominated wetlands and the unique, foundational role black ash plays in affecting hydrology, patterns in biodiversity, and vegetation dynamics. Given its unique role, there is truly no replacement for black ash in these forests. Despite this, recent experience with adaptive planting approaches in conjunction with regeneration harvests that maintain canopy cover during the establishment phase (i.e., group selection and shelterwood approaches) suggest there may be opportunities to increase non-ash components to maintain some of the forested wetland functions historically provided by black ash. Results from these studies are still quite short-term; however, they provide a framework for addressing the threat of an introduced pest by focusing on building resilience as opposed to solely focusing on eliminating host tree populations.

Despite the progress made over the past two decades in understanding black ash forests and in developing strategies for increasing their resilience to EAB, there are several key knowledge gaps that remain in relation to our ability to sustain these areas as forested wetlands into the future. One key knowledge gap is the long-term suitability of potential replacement species that are already being evaluated, and a related gap surrounds the suitability of other potential replacements species in these forests, as well as how differences in stock type might affect within species performance. Moreover, general availability of many of the species that have proven successful in early trials is quite limited and a better understanding of nursery capacity for providing planting material necessary for reforesting these areas is crucial. Similarly, there has been little evaluation of other reforestation techniques, particularly aerial seeding of replacement species, which may be more cost-effective and logistically feasible relative to hand planting black ash swamps. The impacts and effectiveness of site preparation treatments in establishing non-ash species and controlling invasive species common in these ecosystems, particularly reed canary grass, also have not been evaluated. Finally, the new field trials in Wisconsin looking at restoration treatments after EAB invasion and tree mortality will provide critical information on approaches for restoring tree cover; however, there is a need for like evaluations across the great variety of habitat types on which black ash currently dominates and associated feedbacks with local site hydrology. Regardless of treatment or site evaluated, all adaptive management efforts should include retention of mature, seed-bearing black ash to maintain its unique ecological functions prior to EAB arrival and provide opportunities for natural resistance and reestablishment after invasion [54].

Author Contributions: A.W.D., B.J.P., and R.A.S. conceived and designed the Minnesota experiments and conceptualized the paper; A.W.D., G.E., C.M., and D.R.B. summarized and integrated results from Wisconsin Lowland Hardwood Field Trials; and all authors contributed to writing the paper.

Acknowledgments: Special thanks to Laura Reuling, Kyle Gill, and Justin Pszwaro for leading field data collections and summaries from the Wisconsin DNR Swamp Hardwood Trials. Christopher Looney and Mitch Slater led field data collections and summaries for adaptive silviculture studies in black ash communities in north-central Minnesota. Nicholas Bolton provided insights on performance and survival of planted seedlings in Michigan and Wisconsin ash forests. Funding was provided by the Minnesota Environment and Natural Resources Trust Fund, the Upper Midwest and Great Lakes Landscape Conservation Cooperative, Department of Interior Northeast Climate Adaptation Science Center, USDA Forest Service Northern Research Station, USDA Forest Service Northeastern Forest Health Protection Program, MN Forest Resources Council, and Wisconsin DNR. Gary Swanson of the Chippewa National forest provided the initial inspiration for the Minnesota study. Comments from three anonymous reviewers and Dr. Marla Emery helped in improving an earlier version of this paper.

Conflicts of Interest: The authors declare no conflict of interest.

References

1. Ayres, M.P.; Lombardero, M.J. Forest pests and their management in the Anthropocene. *Can. J. For. Res.* **2017**, *48*, 292–301.

2. Lovett, G.M.; Weiss, M.; Liebhold, A.M.; Holmes, T.P.; Leung, B.; Lambert, K.F.; Orwig, D.A.; Campbell, F.T.; Rosenthal, J.; McCullough, D.G.; et al. Nonnative forest insects and pathogens in the United States: Impacts and policy options. *Ecol. Appl.* **2016**, *26*, 1437–1455. [CrossRef] [PubMed]

3. Ellison, A.M.; Bank, M.S.; Clinton, B.D.; Colburn, E.A.; Elliott, K.; Ford, C.R.; Foster, D.R.; Kloeppel, B.D.; Knoepp, J.D.; Lovett, G.M.; et al. Loss of foundation species: Consequences for the structure and Dynamics of forested ecosystems. *Front. Ecol. Environ.* **2005**, *3*, 479–486.

4. Waring, K.M.; O'Hara, K.L. Silvicultural strategies in forest ecosystems affected by introduced pests. *For. Ecol. Manag.* **2005**, *209*, 27–41. [CrossRef]

5. Youngquist, M.B.; Eggert, S.L.; D'Amato, A.W.; Palik, B.J.; Slesak, R.A. Potential effects of foundation species loss on Wetland Communities: A case study of black ash wetlands threatened by emerald ash borer. *Wetlands* **2017**, *37*, 787–799.

6. Valachovic, Y.; Lee, C.; Marshall, J.; Scanlon, H. Wildland management of *Phytophthora ramorum* in northern California forests. In Proceedings of the Sudden Oak Death Science Symposium, Santa Rosa, CA, USA, 5–9 March 2007; Forest Service: Albany, CA, USA, 2008; pp. 305–312.

7. Smith, M.T.; Turgeon, J.J.; de Groot, P.; Gasman, B. Asian longhorned beetle Anoplophora glabripennis (Motschulsky): Lessons learned and opportunities to improve the process of eradication and management. *Am. Entomol.* **2009**, *55*, 21–25. [CrossRef]

8. McCullough, D.G.; Siegert, N.W. Estimating potential emerald ash borer (Coleoptera: Buprestidae) populations using ash inventory data. *J. Econ. Entomol.* **2007**, *100*, 1577–1586. [PubMed]

9. McCullough, D.G.; Siegert, N.W.; Bedford, J. Slowing ash mortality: A potential strategy to slam emerald ash borer in outlier sites. In Proceedings of the 20th U.S. Department of Agriculture Interagency Research Forum on Invasive Species, Annapolis, MD, USA, 13–16 January 2009; McManus, K.A., Gottschalk, K.W., Eds.; General Technical Report NRS-P-51. U.S. Department of Agriculture, Forest Service, Northern Research Station: Newtown Square, PA, USA, 2009; pp. 44–46.

10. Kizlinski, M.L.; Orwig, D.A.; Cobb, R.C.; Foster, D.R. Direct and indirect ecosystem consequences of an invasive pest on forests dominated by eastern hemlock. *J. Biogeogr.* **2002**, *29*, 1489–1503. [CrossRef]

11. MacFarlane, D.W.; Meyer, S.P. Characteristics and distribution of potential ash tree hosts for emerald ash borer. *For. Ecol. Manag.* **2005**, *213*, 15–24.

12. Knight, K.S.; Brown, J.P.; Long, R.P. Factors affecting the survival of ash (*Fraxinus* spp.) trees infested by emerald ash borer (*Agrilus planipennis*). *Biol. Invasions* **2013**, *15*, 371–383.

13. Erdmann, G.G.; Crow, T.R.; Peterson, R.M., Jr.; Wilson, C.D. *Managing Black Ash in the Lake States*; General Technical Report, NC-115; USDA Forest Service: Washington, DC, USA, 1987; 12p.

14. Slesak, R.A.; Lenhart, C.F.; Brooks, K.N.; D'Amato, A.W.; Palik, B.J. Water table response to harvesting and simulated emerald ash borer mortality in black ash wetlands in Minnesota, USA. *Can. J. For. Res.* **2014**, *44*, 961–968. [CrossRef]

15. Herms, D.A.; McCullough, D.G. Emerald Ash Borer Invasion of North America: History, Biology, Ecology, Impacts, and Management. *Ann. Rev. Entomol.* **2014**, *59*, 13–30. [CrossRef] [PubMed]

16. Kashian, D.M. Sprouting and seed production may promote persistence of Green ash in the presence of the emerald ash borer. *Ecosphere* **2016**, *7*, e01332. [CrossRef]

17. Iverson, L.; Knight, K.S.; Prasad, A.; Herms, D.A.; Matthews, S.; Peters, M.; Smith, A.; Hartzler, D.M.; Long, R.; Almendinger, J. Potential Species Replacements for Black Ash (*Fraxinus nigra*) at the Confluence of Two Threats: Emerald Ash Borer and a Changing Climate. *Ecosystems* **2016**, *19*, 248–270. [CrossRef]

18. USDA Forest Service Forest Inventory and Analysis (USDA FIA). *Forest Inventory and Analysis Database*; U.S. Department of Agriculture, Forest Service, Northern Research Station: St. Paul, MN, USA, 2018. Available online: http://apps.fs.fed.us/fiadb-downloads/datamart.html (accessed on 6 April 2018).

19. Weber, M. *Factors Affecting Natural Tree Reproduction in Black Ash Communities in Northern Minnesota*; MS Plan B Paper; College of Forestry, University of Minnesota: St. Paul, MN, USA, 1985; 58p.

20. Kurmis, V.; Kim, J.H. *Black Ash Stand Composition and Structure in Carlton County, Minnesota*; Staff Paper Series Number 69; University of Minnesota Agricultural Experiment Station: St. Paul, MN, USA, 1989; 25p.

21. Sterrett, W.D. *The Ashes: Their Characteristics and Management*; USDA Bulletin Number 299; US Department of Agriculture: Washington, DC, USA, 1915; 88p.

22. Eyre, F.H. *Forest Cover Types of the United States and Canada*; Society of American Foresters: Washington, DC, USA, 1980; 148p.

23. Michigan Department of Natural Resources (MIDNR). *Forest Management Guide for Lowland Hardwoods Cover Type*; Department of Natural Resources, Forest Management Division: Lansing, MI, USA, 1993; 4p.

24. Kotar, J.; Kovach, J.A.; Burger, T.L. *A Guide to Forest Communities and Habitat Types of Northern Wisconsin*; Department of Forest Ecology and Management, University of Wisconsin-Madison: Madison, WI, USA, 2002.

25. Aaseng, N.E. *Field Guide to the Native Plant Communities of Minnesota: The Laurentian Mixed Forest Province*; Minnesota Department of Natural Resources: St. Paul, MN, USA, 2003.

26. Tardif, J.; Bergeron, Y. Population dynamics of *Fraxinus nigra* in response to flood-level variations, in Northwestern Quebec. *Ecol. Monogr.* **1999**, *69*, 107–125. [CrossRef]

27. Kotar, J.; Burger, T.L. *Wetland Forest Habitat Type Classification System for Northern Wisconsin*; PUB-FR-627; Wisconsin Department of Natural Resources: Madison, WI, USA, 2017; 251p.

28. Wisconsin Department of Natural Resources (WDNR). *Silviculture Handbook: Chapter 46-Swamp Hardwood Cover Type*; State of Wisconsin, Department of Natural Resources: Madison, WI, USA, 2013.

29. Looney, C.E.; D'Amato, A.W.; Fraver, S.; Palik, B.J.; Reinikainen, M.R. Examining the influences of tree-to-tree competition and climate on size-growth relationships in hydric, multi-aged *Fraxinus nigra* stands. *For. Ecol. Manag.* **2016**, *375*, 238–248. [CrossRef]

30. Davis, J.C.; Shannon, J.P.; Bolton, N.W.; Kolka, R.K.; Pypker, T.G. Vegetation responses to simulated emerald ash borer infestation in *Fraxinus nigra* dominated wetlands of Upper Michigan, USA. *Can. J. For. Res.* **2016**, *47*, 319–330. [CrossRef]

31. Bowen, A.K.M.; Stevens, M.H.H. Predicting the effects of emerald ash borer (*Agrilus planipennis*, Buprestidae) on hardwood swamp forest structure and composition in southern Michigan. *J. Torrey Bot. Soc.* **2018**, *145*, 41–54. [CrossRef]

32. Wright, J.W.; Rauscher, H.M. Black ash. In *Silvics of North America, Volume 2, Hardwoods*; Burns, R.M., Honkala, B.G., Eds.; Agricultural Handbook 654; United States Department of Agriculture (USDA): Washington, DC, USA, 1990; pp. 344–347.

33. Palik, B.J.; Batzer, D.P.; Kern, C. Upland forest linkages to seasonal wetlands: Litter flux, processing, and food quality. *Ecosystems* **2006**, *9*, 142–151. [CrossRef]

34. Van Grinsven, M.J.; Shannon, J.P.; Davis, J.C.; Bolton, N.W.; Wagenbrenner, J.W.; Kolka, R.K.; Pypker, T.G. Source water contributions and hydrologic responses to simulated emerald ash borer infestations in depressional black ash wetlands. *Ecohydrology* **2017**, *10*, e1862. [CrossRef]

35. Lenhart, C.; Brooks, K.; Davidson, M.; Slesak, R.; D'Amato, A. Hydrologic source characterization of black ash wetlands: Implications for EAB response. In Proceedings of the American Water Resources Association Summer Specialty Conference Riparian Ecosystems IV: Advancing Science, Economics and Policy, Denver, CO, USA, 27–29 June 2012.

36. Telander, A.C.; Slesak, R.A.; D'Amato, A.W.; Palik, B.J.; Brooks, K.N.; Lenhart, C.F. Sap flow of black ash in wetland forests of northern Minnesota, USA: Hydrologic implications of tree mortality due to emerald ash borer. *Agric. For. Meteorol.* **2015**, *206*, 4–11. [CrossRef]

37. Shannon, J.; van Grinsven, M.; Davis, J.; Bolton, N.; Noh, N.; Pypker, T.; Kolka, R. Water Level Controls on Sap Flux of Canopy Species in Black Ash Wetlands. *Forests* **2018**, *9*, 147. [CrossRef]

38. Kolka, R.; D'Amato, A.W.; Wagenbrenner, J.; Slesak, R.; Pypker, T.; Youngquist, M.; Grinde, A.; Palik, B. Review of ecosystem level impacts of emerald ash borer on black ash wetlands: What does the future hold? *Forests* **2018**, *9*, 179. [CrossRef]

39. Diamond, J.S.; McLaughlin, D.; Slesak, R.A.; D'Amato, A.W.; Palik, B.J. Forested vs. Herbaceous Wetlands: Can management mitigate ecohydrologic regime shifts from invasive EAB? *J. Environ. Manag.* **2018**, in press.

40. Croxton, R.J. *Detection and Classification of Ash Dieback on Large-Scale Color Aerial Photographs*; Research Paper PSW-RP-35; USDA Forest Service: Washington, DC, USA, 1966.

41. Livingston, W.H.; Hager, A.; White, A.S.; Hobbins, D. Drought associated with brown ash dieback in Maine. *Phytopathology* **1995**, *85*, 1554–1561.

42. Palik, B.J.; Ostry, M.E.; Venette, R.C.; Abdela, E. *Fraxinus nigra* (black ash) dieback in Minnesota: Regional variation and potential contributing factors. *For. Ecol. Manag.* **2011**, *261*, 128–135. [CrossRef]

43. Livingston, W.H.; White, A.S. May drought confirmed as likely cause of brown ash dieback in Maine. *Phytopathology* **1997**, *87*, S59.

44. Trial, H., Jr.; Devine, M.E. *Forest Health Monitoring Evaluation: Brown Ash (Fraxinus nigra) in Maine. A Survey of Occurrence and Health*; Technical Report No. 33; Insect and Disease Management Division: August, ME, USA, 1994.

45. Peterson, C.E. Natural Regeneration after Logging of Black Ash Stands in Central Minnesota. Master's Thesis, College of Natural Resources, University of Minnesota, St. Paul, MN, USA, 1989.

46. Kashian, D.M.; Witter, J.A. Assessing the potential for ash canopy tree replacement via current regeneration following emerald ash borer-caused mortality on southeastern Michigan landscapes. *For. Ecol. Manag.* **2011**, *261*, 480–488. [CrossRef]

47. Looney, C.E.; D'Amato, A.W.; Palik, B.J.; Slesak, R.A.; Slater, M.A. The response of *Fraxinus nigra* forest ground-layer vegetation to emulated emerald ash borer mortality and management strategies in Northern Minnesota, USA. *For. Ecol. Manag.* **2017**, *389*, 352–363. [CrossRef]

48. Palik, B.J.; Ostry, M.E.; Venette, R.C.; Abdela, E. Tree regeneration in black ash (*Fraxinus nigra*) stands exhibiting crown dieback in Minnesota. *For. Ecol. Manag.* **2012**, *269*, 26–30. [CrossRef]

49. Looney, C.E.; D'Amato, A.W.; Palik, B.J.; Slesak, R.A. Overstory treatment and planting season affect survival of replacement tree species in emerald ash borer threatened *Fraxinus nigra* forests in Minnesota, USA. *Can. J. For. Res.* **2015**, *45*, 1728–1738. [CrossRef]

50. Bolton, N.; Shannon, J.; Davis, J.; Grinsven, M.; Noh, N.; Schooler, S.; Kolka, R.; Pypker, T.; Wagenbrenner, J. Methods to improve survival and growth of planted alternative species seedlings in black ash ecosystems threatened by emerald ash borer. *Forests* **2018**, *9*, 146. [CrossRef]

51. Kabrick, J.M.; Dey, D.C.; van Sambeek, J.W.; Coggeshall, M.V.; Jacobs, D.F. Quantifying flooding effects on hardwood seedling survival and growth for bottomland restoration. *New For.* **2012**, *43*, 695–710. [CrossRef]

52. Knight, K.S.; Slavicek, J.M.; Kappler, R.; Pisarczyk, E.; Wiggin, B.; Menard, K. Using Dutch elm disease-tolerant elm to restore floodplains impacted by emerald ash borer. In Proceedings of the 4th International Workshop on Genetics of Host–Parasite Interactions in Forestry: Disease and Insect Resistance in Forest Trees, Eugene, OR, USA, 31 July–5 August 2011; General Technical Report PSW-GTR-240. USDA Forest Service, Pacific Southwest Research Station: Albany, CA, USA, 2012; pp. 317–323.

53. WDNR. *Silviculture Trials Database: Swamp Hardwood*; State of Wisconsin, Department of Natural Resources: Madison, WI, USA, 2015. Available online: https://dnr.wi.gov/topic/forestmanagement/silviculturetrials.html (accessed on 18 April 2018).

54. WDNR. *Checklist for Evaluating Lowland Ash Stands*; State of Wisconsin, Department of Natural Resources: Madison, WI, USA, 2017; 8p.

55. University of Minnesota (UMN) Extension. *Ash Management Guidelines for Private Forest Landowners*; University of Minnesota: St. Paul, MN, USA, 2011; p. 76.

forests

MDPI

Article

Forest Regeneration Following Emerald Ash Borer (*Agrilus planipennis* Fairemaire) Enhances Mesophication in Eastern Hardwood Forests

Benjamin Dolan [1,*] and Jason Kilgore [2]

1   Department of Biology, University of Findlay, 1000 North Main Street, Findlay, OH 45840 USA
2   Biology Department, Washington & Jefferson College, 60 South Lincoln Street, Washington, PA 15301 USA;
    jkilgore@washjeff.edu
*   Correspondence: dolan@findlay.edu; Tel.: +1-419-434-5530

Received: 9 March 2018; Accepted: 10 May 2018; Published: 14 June 2018

Abstract: Emerald ash borer (EAB, *Agrilus planipennis* Fairemaire) is a phloem-feeding beetle that was introduced into North America in the late 20th century and is causing widespread mortality of native ash (*Fraxinus*) species. The loss of an entire genus from the forest flora is a substantial disturbance, but effects vary because of differences in *Fraxinus* dominance and remaining vegetation. At three sites near the center of the North American EAB range, we investigated the impacts of *Fraxinus* mortality on recruitment of woody and non-native vegetation in 14 permanent plots from 2012 to 2017. We used the change in relative *Fraxinus* basal area to determine the impact of EAB on density of woody species and non-native vegetation less than 2.5 cm diameter at breast height (dbh). Changes in canopy cover were not correlated with loss of *Fraxinus* from the overstory, and only the density of shade-tolerant shrubs and saplings increased with *Fraxinus* mortality. Both native and non-native shrub species increased in density at sites where they were present before EAB, but no new invasions were detected following *Fraxinus* mortality. These shifts in understory vegetation indicate that *Fraxinus* mortality enhances the rate of succession to shade-tolerant species.

Keywords: *Fraxinus*; *Agrilus planipennis* Fairemaire; mesophication; forest regeneration; disturbance

---

1. Introduction

In the nearly two decades since emerald ash borer (EAB, *Agrilus planipennis* Fairemaire) was first detected in North America [1], the response has evolved from preventing or mitigating damage to *Fraxinus* [2–4] to understanding how the introduced insect might impact other species and ecosystem function in forest communities [5–7]. Forest plant communities of eastern North America, like any system, are dynamic and subject to change, and EAB's impact on *Fraxinus* is not exceptional. By inducing *Fraxinus* mortality, EAB immediately reduces forest stand basal area and releases resources, which subsequently become available to other species [8–10].

As a component of mesophytic forests of eastern North America, particularly those within Eastern US Lowland Forest Region—Upland Forest Types [11], *Fraxinus* species might well be considered a minor component of the system, represented primarily by *F. americana* L. and *F. pennsylvanica* Marshall [12]. But these species have a wide distribution, and despite their relatively low abundance in this forest type [12], the loss of *Fraxinus* by EAB has the potential to drive changes in plant communities throughout the region, similar in respect to previous introductions that changed forest composition on a broad scale. Chestnut blight (*Cryphonectria parasitica* (Murrill) Barr) led to the widespread loss of *Castanea dentata* (Marshall) Borkh. throughout its native North American range, particularly in the Appalachian region [13], and Dutch elm disease (*Ophiostoma ulmi* Buisman (Melin and Nannf.), *O. novo-ulmi* Brasier) reduced *Ulmus* species to minor understory tree species in mesophytic forests across the continent [14]. More recently, the exotic insects hemlock woolly adelgid (*Adelges tsugae* Annand) and beech scale (*Cryptococcus fagisuga* Lindinger) are influencing the structure and composition of eastern North American forests [15]. In all instances, the change in canopy structure associated with the loss of individual trees has been marked by a shift to increasing dominance of shade-tolerant species in the understory and overstory layers [15–20].

Because the loss of *Fraxinus* from the canopy is similar to that of these species, we might expect similar responses in the understory. Indeed, in regard to changes already described in Ohio and Michigan, where EAB was initially detected and where *Fraxinus* is a relatively important component of the forest ecosystem, the loss of *Fraxinus* from the canopy induced growth of intermediate canopy trees, especially shade-tolerant *Acer* and *Ulmus* [8]. Trees from these genera existed in the canopy prior to EAB and were better able to utilize resources in EAB-induced gaps than trees from other genera, including *Quercus* and *Carya* [8,21–23]. Even in wetter sites, preinvasion surveys indicate that co-dominant shade-tolerant trees are predicted to replace *F. nigra* (Marshall) in Ohio, Michigan, and Minnesota [24–26].

This pattern of change after pest-induced canopy loss generally follows predicted plant community succession from shade-intolerant and intermediate-tolerant species to dominance by shade-tolerant species. This continued process of transition in eastern North American forests has been termed "mesophication" [27], and these changes in canopy composition come with cascade effects and positive feedback on ecosystems, including changes to flammability, soil moisture, nutrient dynamics, decomposition, and forest regeneration [28–30]. EAB-induced *Fraxinus* mortality is likely hastening the process.

Since *Fraxinus* received scant attention prior to EAB, little is known about the long-term response of woody species at the ground level to the release of resources following the loss of *Fraxinus* trees. Initial descriptions of the seedling layer in EAB-impacted forests primarily show shade-tolerant species, including *Acer rubrum* L., *A. saccharum* Marshall., *Ulmus americana* L., *Prunus serotina* Ehrh., and *Quercus rubra* L. [9,23,24,31], but evidence of their persistence over time is lacking, and interactions between native woody seedlings and exotic invasive plants may further influence the composition of forests. Recent evidence suggests that existing invasive plants will increase in size following the release of resources by EAB-induced *Fraxinus* mortality [9].

We seek to understand the changes in richness and abundance of woody species, including their invasive competitors, in the years following EAB invasion. Specifically, we examine forest regeneration at the seedling and sapling layers, including the recruitment of seedlings into the sapling layer over a 5-year period following infestation. In addition, we are investigating shifts in the abundance of woody invasive species, especially those that compete with woody seedlings and saplings and have the potential to alter successional trajectories. We hypothesize that EAB indirectly increases the prevalence of shade-tolerant species in the understory, thus enhancing the process of mesophication.

2. Materials and Methods

2.1. Study Sites

Our 4 study sites are located in the deciduous forests of eastern North America and are associated with 3 colleges and universities: Baldwin-Wallace University (BW; Ohio), University of Findlay (UF; Ohio), and Washington & Jefferson College (WJ; Pennsylvania; Figure 1). All sites contain *Fraxinus* spp. but are generally dominated by other deciduous tree species (Table 1). Total basal area by plot ranges from 18 to 54 $m^2 \cdot ha^{-1}$, including substantial differences in relative contributions of *Fraxinus* (Figure 2). Year of EAB detection was determined by decline of the *Fraxinus* canopy and confirmed by the presence of EAB exit holes. Only 1 site at UF is located in a floodplain; the rest are upland sites. Across these 4 sites, mean growing season length varies from 159 to 173 days, mean precipitation varies from 847 to 1132 mm, and mean annual temperature varies from 9.6 to 11.8 °C (Table 1). Soils are predominantly finer textured, from silty loam to silty clay loam [32].

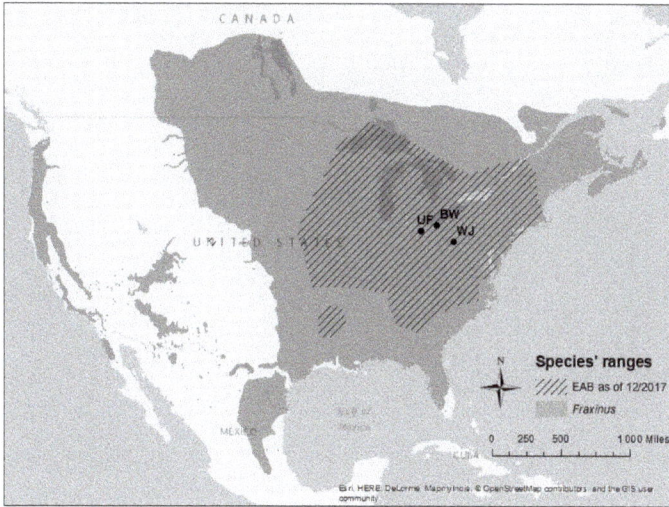

Figure 1. Current distribution of *Fraxinus* spp. (shaded region) and emerald ash borer (EAB, *Agrilus planipennis*, hatched region) in North America. Study sites are indicated on the map: UF, University of Findlay, Findlay, Ohio; BW, Baldwin-Wallace University, Berea, Ohio; and WJ, Washington & Jefferson College, Washington, Pennsylvania. Known EAB distribution as of 1 December 2017 [33].

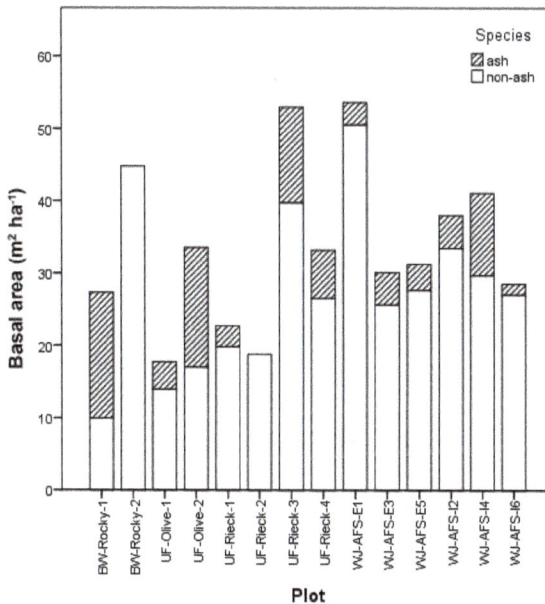

Figure 2. Cumulative basal area of trees by *Fraxinus* spp. (hatched) and non-*Fraxinus* (open) across plots at year of EAB detection at Baldwin-Wallace University (BW, 2015), University of Findlay (UF, 2010), and Washington & Jefferson College (WJ, 2012) sites.

Table 1. Physical characteristics, dominant vegetation, and year of detected emerald ash borer (EAB) infestation by site. Mean growing season, annual precipitation, temperature, elevation, and predominant soil texture were determined following standard Permanent Forest Plot Project (PFPP) methods [34]. MGS: mean growing season length; MAP: mean annual precipitation; MAT: mean annual temperature.

Site	Institution	County	State	MGS (days)	MAP (mm)	MAT (°C)	Mean Elevation (m·asl)	Predominant Soil Texture	Dominant Tree Species (in Order of Dominance)	Mean Plot Basal Area (m²·ha⁻¹)	Year of Detected EAB
Rocky River 41.411217° N, 81.881817° W	Baldwin-Wallace University (BW)	Cuyahoga	OH	173	1132	11.67	233	Silty loam	Fraxinus americana L., Ulmus rubra Muhl., Quercus rubra L., Carya glabra Miller, Acer saccharum Marshall	36.0	2012
Olive Street 41.000292° N, 83.642219° W	University of Findlay (UF)	Hancock	OH	159	847	10.81	240	Silty clay loam	Celtis occidentalis L., Acer saccharum, Fraxinus pennsylvanica Marshall, Carya cordiformis (Wangenh.) K.Koch, Juglans nigra L., Ulmus americana L.	25.7	2010
Rieck Center 40.950716° N, 83.549870° W	University of Findlay (UF)	Hancock	OH	159	857	10.81	250	Silty loam	Acer saccharum, Fraxinus pennsylvanica, Fagus grandifolia Ehrh., Ulmus americana	31.9	2010
Abernathy Field Station (AFS) 40.134110° N, 80.183625° W	Washington & Jefferson College (WJ)	Washington	PA	169	970	9.6	367	Silty loam	Prunus serotina Ehrh., Acer saccharum, Fraxinus americana, Carya ovata (Mill.) K.Koch	37.1	2014

2.2. Data Collection

We used the existing infrastructure of the Permanent Forest Plot Project (PFPP), one of the projects associated with the Ecological Research as Education Network (EREN; http://erenweb.org/). This project was launched in 2012 to establish a set of permanent research plots to address questions related to tree biomass, carbon accumulation, invasive species, and disturbance patterns across a range of sites and ecoregions [34]. Participants establish at least 1 permanent 400-m^2 plot in a forested area, identify and measure the diameter of all trees (>2.5 cm diameter at breast height (dbh), or 1.37 m above the ground), identify and tally all saplings (0–2.5 cm dbh, but at least 1.37 m tall) in 3 randomly selected subplots (25 m^2), and characterize their site and plot using a standard protocol [34]. PFPP data are entered in a secure online database accessible to all PFPP participants.

We also scaffolded 4 variables onto PFPP to evaluate the impact of EAB on *Fraxinus* trees. Tree damage indicates whether the tree is broken, uprooted, or leaned on by another tree (including *Fraxinus*); this variable was incorporated into the PFPP protocol [34]. Ash rating [35] and ash tree breakup [36] indicate the level of decline, mortality, and breaking down of *Fraxinus* trees. We also counted the number of EAB exit holes observed between 1.25 and 1.75 m above the ground as a physical indicator that EAB is associated with the *Fraxinus* decline. These 3 variables are specific to the EAB Impacts Study [37], another EREN project.

To address questions related to the successional dynamics of forests, we built variables to characterize the understory plant community in the Complementary Vegetation (cVeg) Survey [38]. Canopy cover was measured using a concave spherical densiometer (Model C, Robert E. Lemmon, Rapid City, SD, USA), while all woody plants between 0.3 and 1.37 m tall (i.e., shrubs and saplings) were identified and tallied within each of the three 25-m^2 subplots used in the PFPP. In addition, all woody seedlings (<0.3 m tall) and non-woody plants were identified and tallied within 1-m^2 miniplots in each of the subplots. Woody species were classified according to their shade tolerance [39,40].

For our study, we used 2 plots from Rocky River (BW; 2016–2017), 2 plots from the Olive Street site (UF; 2012, 2014–2017), 4 plots from the Rieck Center (UF; 2012, 2014–2017), and 6 plots from Abernathy Field Station (AFS, WJ; 2012, 2014–2017). All of the PFPP, EAB, and cVeg variables were measured at each of the sampling years by site.

2.3. Analyses

Given our hypothesis that the loss of *Fraxinus* is driving a shift in community composition toward more mesophytic, or shade-tolerant, species, as well as increasing the prevalence of invasive plant species, we used the preinvasion relative basal area of *Fraxinus* as the independent variable. This variable was calculated as the proportion of cumulative basal area (BA = $\pi \times (0.5 \times dbh)^2$) of 3 *Fraxinus* species (*F. americana*, *F. pennsylvanica*, *F. quadrangulata* Michx.) relative to all other trees prior to (BW and WJ sites) or at (UF sites) the time of EAB detection. The variables of interest, such as annual changes in woody seedling density, invasive plant density, and sapling and shrub density, partitioned by shade tolerance [39,40], were calculated as the annual change in density from the first to last sampling year and divided by the number of years of sampling, whether we had 2 (BW) or 5 (UF, WJ) years of data. Data from miniplots (woody seedling and forb) and subplots (shrubs and saplings) were pooled by plot, thus plot became the observational unit. Each of these dependent variables was then linearly regressed onto the relative basal area of *Fraxinus*. All analyses were conducted using IBM SPSS Statistics v.24 (IBM Corp., North Castle, NY, USA).

3. Results

To determine whether we could detect changes in light resources at the forest floor, we regressed change in canopy cover on relative *Fraxinus* basal area prior to EAB invasion. Across all four sites, we found no significant relationship between canopy cover and relative *Fraxinus* dominance ($p = 0.46$), thus no evidence that light resources increased more in sites with greater dominance of *Fraxinus* (Figure 3).

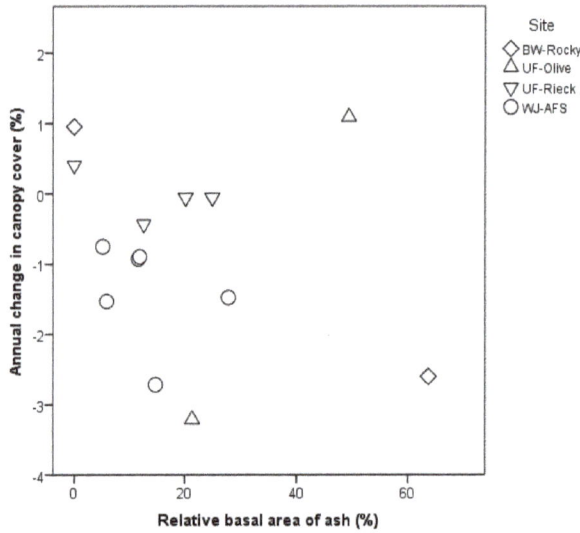

Figure 3. Mean annual change in canopy cover as a function of relative basal area of *Fraxinus* spp. prior to EAB invasion across four sites ($r^2 = 0.048$, $p = 0.450$). AFS: Abernathy Field Station.

The relative basal area of *Fraxinus* has not significantly affected the density of woody seedlings across sites ($p > 0.18$), whether with combined species or when partitioned by shade tolerance (Figure 4), yet some patterns are emerging. For example, shade-tolerant species increased in density at a higher rate in sites with more *Fraxinus* trees. Seedling density for *F. americana* generally declined across all sites, yet *F. pennsylvanica* seedlings generally increased in density at both UF sites (in supplementary file Table S1). Across UF and WJ sites, the density of *Parthenocissus quinquefolia* (L.) Planch. seedlings increased since EAB invasion. Other species, including the non-native *Rosa multiflora* Thunb. experienced both increased and decreased density within sites (Table S1).

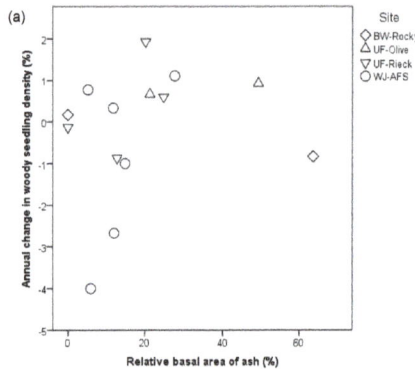

Figure 4. *Cont.*

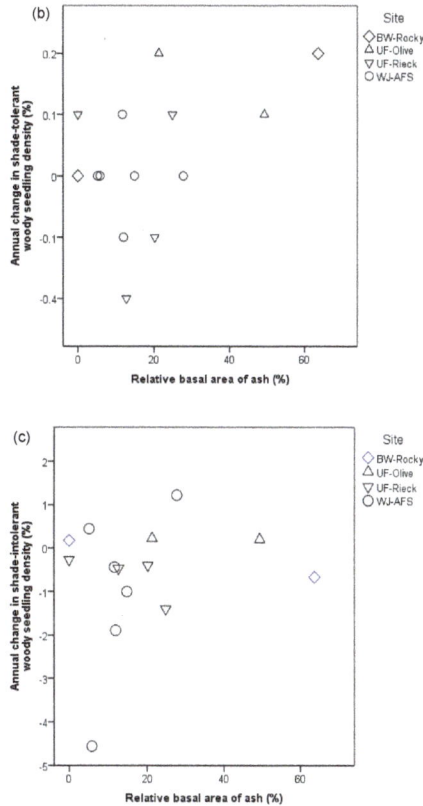

Figure 4. Effects of relative *Fraxinus* basal area on change in density of (**a**) woody seedlings of all species ($r^2 = 0.04$. $p = 0.482$), (**b**) shade-tolerant species ($r^2 = 0.14$, $p = 0.184$), and (**c**) shade-intolerant species ($r^2 = 0.03$, $p = 0.526$).

While total shrub density (Figure 5a), including shade-intolerant species (Figure 5c), did not respond to changes in relative basal area of *Fraxinus*, shade-tolerant shrubs significantly ($p = 0.005$) increased in density (Figure 5b). This response appears to be driven by individual plots at the BW (Rocky-1) and UF (Olive-2) sites, both of which had higher losses of *Fraxinus* than other plots (Figure 2); two native shrubs, *Lindera benzoin* L. and *Asimina triloba* (L.) Dunal, respectively, substantially increased in these plots, but the non-native *Rosa multiflora* also increased in density over this same period at UF (Olive-2) (in supplementary file Table S2). Saplings of *Fraxinus* spp. generally increased at UF (Rieck) and WJ (AFS), as did the native *Celtis occidentalis* at UF (Rieck) and non-native *Celastrus orbiculatus* Thunb. and native *Rubus* spp. at WJ (AFS). However, *Lindera benzoin* generally decreased in density at WJ (AFS), while the non-native *Lonicera* spp. and *Rosa multiflora* Thunb. increased in the UF (Olive-1) plot that experienced widespread canopy tree loss from the June 2012 North American derecho (Table S2).

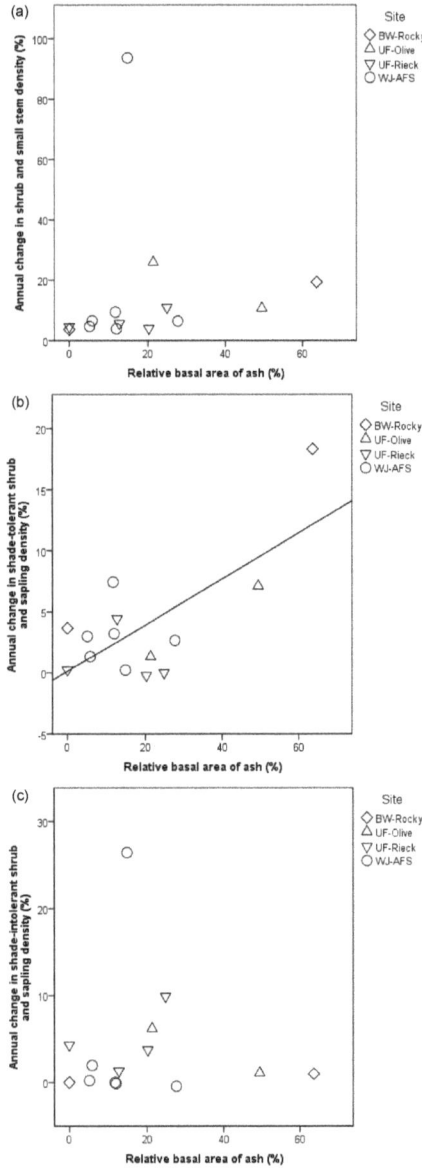

Figure 5. Effects of loss of *Fraxinus* on change in density of (**a**) shrubs and saplings of all species ($r^2 = 0.01$, $p = 0.764$), (**b**) shade-tolerant species ($r^2 = 0.50$, $p = 0.005$), and (**c**) shade-intolerant species ($r^2 = 0.003$, $p = 0.853$).

Relative basal area of *Fraxinus* was not a significant predictor of density of non-native vegetation in the ground layer or the shrub and sapling layer in 2017 (Figure 6). Only two plots, one each at WJ (AFS) and UF (Olive), saw above-average increases in non-native species density, yet both sites had relatively low relative amounts of *Fraxinus* basal area. And though three non-native species, *Celastrus orbiculatus* Thunb., *Lonicera maackii* (Rupr.) Maxim. and *Rosa multiflora*, increased in density in the shrub layer within sites during the period of study (Table S2), the overall change in non-native species density does not show a relationship to the decrease in relative *Fraxinus* basal area.

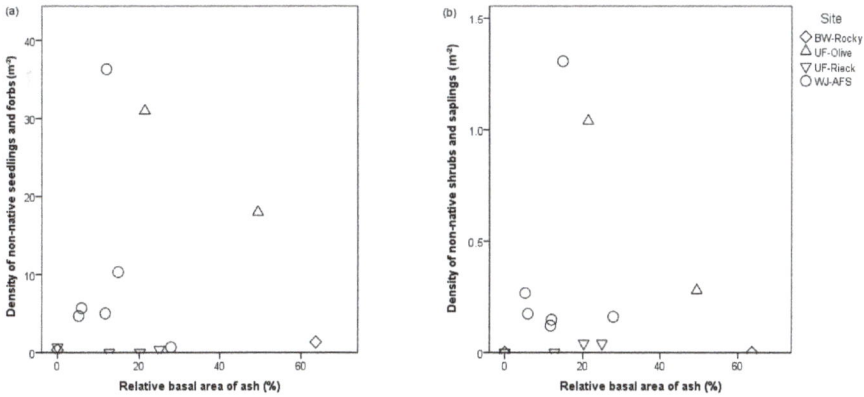

Figure 6. Effects of relative *Fraxinus* basal area on density of non-native vegetation, including (**a**) woody seedlings and forbs in 2017 ($r^2 = 0.003$, $p = 0.843$) and (**b**) shrubs and saplings ($r^2 = 0.001$, $p = 0.955$) in 2017.

4. Discussion

The loss of *Fraxinus* from mesophytic North American hardwood forests and the subsequent release of resources appear to have little influence on the distribution and abundance of forest regeneration at the seedling and sapling level. Furthermore, we found no evidence that this loss alters the process of mesophication in the same manner as acute, severe disturbances that typically impact canopy cover.

4.1. Canopy Cover

Light is the primary resource expected to change at the ground layer as a result of the death of large overstory *Fraxinus* trees [41], yet our evidence suggests that the protracted decline and ultimate death of *Fraxinus* individuals does not create conditions that substantially alter the light regime on the forest floor, as measured by our canopy cover index. Although canopy cover decreased at a rate of 0%–3% annually, there was no detectable relationship to *Fraxinus* mortality (Figure 3). In order to meet the physiological demands of shade-intolerant species, like *Quercus* and *Carya*, light levels need to reach 25%–50% full sunlight [42], which is often reached with more severe disturbances, like shelterwood harvesting [43].

Research on canopy trees in response to EAB-induced *Fraxinus* mortality shows that existing canopy trees respond positively. Gaps concurrently fill with non-*Fraxinus* species [23], likely influencing the lack of substantial change in annual canopy cover over time. Recent evidence using LIDAR (Light Detection and Ranging) indicates that single-tree death in northern hardwood forests is not sufficient to generate canopy gaps where the subcanopy is well developed [44], suggesting that the gradual canopy opening that occurs with EAB-induced *Fraxinus* mortality is less likely to enrich the ground layer with light than it is to promote the growth of existing subcanopy trees.

Indeed, in northwest Ohio, the relative growth rates of the shade-tolerant species *Acer saccharum* and *Ulmus americana* were faster than other species in the forest canopy following *Fraxinus* mortality [8], and smaller trees exhibited faster growth rates than larger ones, suggesting that subcanopy individuals benefit more than the co-dominant overstory trees. Trees remaining in the canopy of our forest sites include mesophytic species, typically from *Acer*, *Prunus*, *Ulmus*, and *Celtis* (Table 1), which supports our hypothesis that *Fraxinus* loss will increase the speed at which forests are dominated by more shade-tolerant species.

4.2. Understory Composition and Density

The composition of woody species in the seedling layer is relatively stable following the loss of *Fraxinus* to EAB. We detected no statistically significant relationship between the loss of *Fraxinus* and the change in seedling density (Figure 4). In regard to individual species, there is a large amount of variability among sites and plots. Species with the largest, most consistent increases over time include the native vine *Parthenocissus quinquefolia*, while the non-native shrub *Rosa multiflora* increased or decreased across sites, indicating a lack of relationship with *Fraxinus* loss. Interestingly, the number of *Fraxinus* seedlings increased at both UF (Olive) plots, where overstory *Fraxinus* trees had succumbed to EAB beginning in 2012 (Table S1). Recent evidence suggests that *Fraxinus* seedbanks do not persist beyond the life of the tree [45], indicating that an off-site seed source persists for this site.

The shrub and sapling layer was more responsive to *Fraxinus* mortality than the seedling layer. As the amount of *Fraxinus* basal area increased in plots, the density of shade-tolerant species increased ($r^2 = 0.50$, $p = 0.005$). This was in contrast to the lack of association detected between *Fraxinus* basal area and the change in shade-intolerant species density ($r^2 = 0.003$, $p = 0.853$). The increase in shade-tolerant species density coincides with evidence of established trees in the mid-canopy responding to the resources made available through the mortality of *Fraxinus* individuals. The response of individual species varied across sites, but those with increases at more than one site include shade-tolerant *Lindera benzoin* and *Asimina triloba*, both native understory species.

Our findings are not unusual, as shrub and sapling density has been shown to increase in response to species-specific canopy dieback elsewhere. In Europe, fungal-induced *Fraxinus excelsior* L. dieback promoted the growth of widely distributed shade-tolerant shrubs and understory trees, including *Corylus avellana* L., *Prunus padus* L., and *Lonicera xylosterum* L. [46]. In North America, after Dutch elm disease induced *Ulmus* decline, the intermediate- and shade-tolerant shrubs *Alnus rugosa* (L.) Moench, *Viburnum recognitum* Fern., and *Cornus stolonifera* L. in wetland habitats responded more positively in gaps created by the death of *Ulmus* stems in the overstory than seedlings of other canopy trees [47]. In southeastern Wisconsin, species composition and density of woody seedlings were not associated with density of dead *Ulmus*, yet the density of shrubs, especially *Ribes* spp., *Rubus* spp., and *Cornus racemosa* Lam., was associated with *Ulmus* mortality [48].

4.3. Invasive Species

No new invasive species were detected in the research plots. For plots with existing invasive plants, some became more numerous, including *Lonicera maackii*, *Rosa multiflora*, and *Celastrus orbiculatus*, but because of variability among plots, we were unable to identify *Fraxinus* mortality as the primary cause of the increase at our sites. Prior to EAB invasion, one plot at the UF (Olive) site had large numbers of *L. maackii* and *R. multiflora*; these species increased in density over time. This particular site, however, was also strongly influenced by a storm that caused significant loss of the overstory in 2012. The confounding disturbances of storm damage and EAB-induced *Fraxinus* mortality cannot be distinguished. Recent evidence suggests that the amount of *Fraxinus*, and thus the amount of light increasing beneath the canopy, influences the response of existing *L. maackii*; in a comprehensive study of population-level responses to EAB, existing *L. maackii* individuals had increased stem diameter following *Fraxinus* mortality, while the number of individual plants, amount of fruiting, and height of the shrubs were not impacted [9].

Additionally, in three plots in Pennsylvania, *Celastrus orbiculatus* had higher annual increases in density than other species following EAB, but because of losses to this species in other plots at the same site and an overall lack of relationship between *Fraxinus* basal area and shrub and sapling density (Figure 6b; $r^2 = 0.001$, $p = 0.955$), the increase in this species was not likely caused by the loss of *Fraxinus*.

5. Conclusions

Our evidence shows that the resources released through *Fraxinus* mortality are insufficient to modify the process of mesophication in mesophytic forests, but instead generate changes that resemble those observed with small gaps. The primary beneficiaries of released resources include individuals already existing in the understory canopy, among them small shade-tolerant trees, saplings, and shrubs typical of mature mesophytic forests, conforming to what is observed in forest gap dynamics [8,49–52]. This enhanced mesophication resulting from the loss of *Fraxinus* has allowed shade-tolerant species, including those restricted to understory canopy positions, like *Asimina triloba* (L.) Dunal and *Lindera benzoin* L., to increase in the shrub and sapling layer at faster rates than shade-intolerant species (Figure 5). As mesophication has been noted throughout much of the eastern North American forest complex [28,53–55], we would expect to see similar patterns throughout the region where *F. pennsylvanica* and *F. americana* are the primary *Fraxinus* species represented in the canopy.

Importantly, some of our sites are in regions with relatively high amounts of *Fraxinus* compared to other mesophytic forest regions of eastern North America [11]. Our research primarily includes sites with relative *Fraxinus* basal area within the range of 5–25%, and we find no compelling evidence that the loss of trees at our sites impacts regeneration and recruitment. Where *Fraxinus* is a smaller component of the forest tree community, we expect that, like our sites, these areas will not see dramatic shifts in seedling and shrub composition resulting from the loss of *Fraxinus* in the overstory.

With the exception of *Fraxinus*, EAB is not a disturbance that influences the abundance of woody species regeneration and subsequent recruitment in sites where *Fraxinus* is a minor component of the canopy vegetation. The five species of *Fraxinus* most common to eastern hardwood forests, *F. americana*, *F. pennsylvanica*, *F. nigra*, *F. profunda* Bush, and *F. quadrangulata*, have been classified as critically endangered and on the brink of extinction by the International Union for Conservation of Nature (IUCN) [56–60]. Conservation of the genus is important, but the consequence of losing the species has little influence on immediate future successional pathways of woody vegetation in the forests where *Fraxinus* is a minor component.

Supplementary Materials: The following are available online at http://www.mdpi.com/1999-4907/9/5/353/s1: Table S1. Annual change in density (ha-1) of woody seedlings (<0.30 m tall) by plot after EAB invasion. Empty cells indicate species absence. Table S2. Annual change in density (ha^{-1}) of shrubs and saplings (woody plants, 0.30–1.37 m tall) by plot after EAB invasion. Empty cells indicate species absence.

Author Contributions: B.J.D. and J.S.K. developed the research questions, conceived the design for the EAB Impacts Study and Complementary Vegetation (cVeg) Survey, collected the data, performed the analyses, and wrote the manuscript. Both B.J.D. and J.S.K. provided edits and comments during the drafting of the paper.

Funding: This research received no external funding.

Acknowledgments: The Ecological Research as Education Network (EREN) was funded with the support of a Research Coordination Grant from the Research Coordination Network—Undergraduate Biology Education Program of the U.S. National Science Foundation (DEB 0955344, 2010–2016). We are grateful to our home institutions for funding and for supporting travel to work together, and we are thankful for the contributions of data, discussions, and analyses of changes in the invasive plant species by Kathryn Flinn (Baldwin-Wallace University). For access to field sites, we thank Cleveland Metroparks (Rocky River), University of Findlay (Olive Street, Rieck Center), and the Abernathy family (Abernathy Field Station). We appreciate the online site data gained from the PFPP database. We also thank the myriad undergraduate students and two high school students who contributed to data collection, and the reviewers who improved this manuscript with insightful comments and suggestions.

Conflicts of Interest: The authors declare no conflict of interest.

References

1. Haack, R.A.; Jendak, E.; Houping, L.; Marchant, K.R.; Petrice, T.R.; Poland, T.M.; Ye, H. The emerald ash borer: A new exotic pest in North America. *Newsl. Mich. Entomological Soc.* **2002**, *47*, 1–5.
2. Poland, T.M.; McCullough, D.G. Emerald Ash Borer: Invasion of the Urban Forest and the Threat to North America's Ash Resource. *J. For.* **2006**, *104*, 118–124. [CrossRef]

3. Animal and Plant Health Inspection Service, USDA. Emerald Ash Borer: Quarantine and Regulations. Available online: https://www.federalregister.gov/d/03-25881 (accessed on 15 April 2018).
4. Herms, D.A.; McCullough, D.G. Emerald Ash Borer Invasion of North America: History, Biology, Ecology, Impacts, and Management. *Annu. Rev. Entomol.* **2014**, *59*, 13–30. [CrossRef] [PubMed]
5. Perry, K.I.; Herms, D.A.; Klooster, W.S.; Smith, A.; Hartzler, D.M.; Coyle, D.R.; Gandhi, K.J.K. Downed Coarse Woody Debris Dynamics in Ash (*Fraxinus* spp.) Stands Invaded by Emerald Ash Borer (*Agrilus planipennis* Fairmaire). *Forests* **2018**, *9*, 191. [CrossRef]
6. Ricketts, M.P.; Flower, C.E.; Knight, K.S.; Gonzalez-Meler, M.A. Evidence of Ash Tree (*Fraxinus* spp.) Specific Associations with Soil Bacterial Community Structure and Functional Capacity. *Forests* **2018**, *9*, 187. [CrossRef]
7. Marché, J.D., II. *The Green Menace: Emerald Ash Borer and the Invasive Species Problem*; Oxford University Press: Oxford, UK, 2017; ISBN 978-0-19-066892-1.
8. Flower, C.E.; Knight, K.S.; Gonzalez-Meler, M.A. Impacts of the emerald ash borer (*Agrilus planipennis* Fairmaire) induced ash (*Fraxinus* spp.) mortality on forest carbon cycling and successional dynamics in the eastern United States. *Biol. Invasions* **2013**, *15*, 931–944. [CrossRef]
9. Hoven, B.M.; Gorchov, D.L.; Knight, K.S.; Peters, V.E. The effect of emerald ash borer-caused tree mortality on the invasive shrub Amur honeysuckle and their combined effects on tree and shrub seedlings. *Biol. Invasions* **2017**, *19*, 2813–2836. [CrossRef]
10. McCullough, D.G.; Siegert, N.W. Estimating potential emerald ash borer (Coleoptera: Buprestidae) populations using ash inventory data. *J. Econ. Entomol.* **2007**, *100*, 1577–1586. [CrossRef] [PubMed]
11. Vankat, J.L. A classification of the forest types of North America. *Vegetatio* **1990**, *88*, 53–66. [CrossRef]
12. MacFarlane, D.W.; Meyer, S.P. Characteristics and distribution of potential ash tree hosts for emerald ash borer. *For. Ecol. Manag.* **2005**, *213*, 15–24. [CrossRef]
13. Anagnostakis, S.L. Chestnut blight: The classical problem of an introduced pathogen. *Mycologia* **1987**, *79*, 23–37. [CrossRef]
14. Karnosky, D.F. Dutch elm disease: A review of the history, environmental implications, control, and research needs. *Environ. Conserv.* **1979**, *6*, 311–322. [CrossRef]
15. Morin, R.S.; Liebhold, A.M. Invasions by two non-native insects alter regional forest species composition and successional trajectories. *For. Ecol. Manag.* **2015**, *341*, 67–74. [CrossRef]
16. Elliott, K.J.; Swank, W.T. Long-term changes in forest composition and diversity following early logging (1919–1923) and the decline of American chestnut (*Castanea dentata*). *Plant Ecol.* **2007**, *197*, 155–172. [CrossRef]
17. Stephenson, S.L. Changes in a former chestnut-dominated forest after a half century of succession. *Am. Midl. Nat.* **1986**, *116*, 173–179. [CrossRef]
18. Myers, B.R.; Walck, J.L.; Blum, K.E. Vegetation change in a former chestnut stand on the Cumberland Plateau of Tennessee during an 80-year period (1921–2000). *Castanea* **2004**, *69*, 81–91. [CrossRef]
19. Barnes, B.V. Succession in deciduous swamp communities of southeastern Michigan formerly dominated by American elm. *Can. J. Bot.* **1976**, *54*, 19–24. [CrossRef]
20. Parker, G.R.; Leopold, D.J. Replacement of *Ulmus americana* L. in a mature east-central Indiana woods. *Bull. Torrey Bot. Club* **1983**, *110*, 482. [CrossRef]
21. Costilow, K.C.; Knight, K.S.; Flower, C.E. Disturbance severity and canopy position control the radial growth response of maple trees (*Acer* spp.) in forests of northwest Ohio impacted by emerald ash borer (*Agrilus planipennis*). *Ann. For. Sci.* **2017**, *74*, 10. [CrossRef]
22. Burr, S.J.; McCullough, D.G. Condition of green ash (*Fraxinus pennsylvanica*) overstory and regeneration at three stages of the emerald ash borer invasion wave. *Can. J. For. Res.* **2014**, *44*, 768–776. [CrossRef]
23. Kolka, R.K.; D'Amato, A.W.; Wagenbrenner, J.W.; Slesak, R.A.; Pypker, T.G.; Youngquist, M.B.; Grinde, A.R.; Palik, B.J. Review of ecosystem level impacts of emerald ash borer on black ash wetlands: What does the future hold? *Forests* **2018**, *9*, 179. [CrossRef]
24. Iverson, L.; Knight, K.S.; Prasad, A.; Herms, D.A.; Matthews, S.; Peters, M.; Smith, A.; Hartzler, D.M.; Long, R.; Almendinger, J. Potential species replacements for black ash (*Fraxinus nigra*) at the confluence of two threats: Emerald ash borer and a changing climate. *Ecosystems* **2015**, *19*, 248–270. [CrossRef]
25. Davis, J.C.; Shannon, J.P.; Bolton, N.W.; Kolka, R.K.; Pypker, T.G. Vegetation responses to simulated emerald ash borer infestation in *Fraxinus nigra* dominated wetlands of Upper Michigan, USA. *Can. J. For. Res.* **2017**, *47*, 319–330. [CrossRef]

26. Looney, C.E.; D'Amato, A.W.; Palik, B.J.; Slesak, R.A. Canopy treatment influences growth of replacement tree species in *Fraxinus nigra* forests threatened by the emerald ash borer in Minnesota, USA. *Can. J. For. Res.* **2017**, *47*, 183–192. [CrossRef]

27. Nowacki, G.J.; Abrams, M.D. The demise of fire and "mesophication" of forests in the eastern United States. *Bioscience* **2008**, *58*, 123–138. [CrossRef]

28. Alexander, H.D.; Arthur, M.A. Implications of a predicted shift from upland oaks to red maple on forest hydrology and nutrient availability. *Can. J. For. Res.* **2010**, *40*, 716–726. [CrossRef]

29. Alexander, H.D.; Arthur, M.A. Increasing red maple leaf litter alters decomposition rates and nitrogen cycling in historically oak-dominated forests of the eastern U.S. *Ecosystems* **2014**, *17*, 1371–1383. [CrossRef]

30. Kreye, J.K.; Varner, J.M.; Hiers, J.K.; Mola, J. Toward a mechanism for eastern North American forest mesophication: Differential litter drying across 17 species. *Ecol. Appl.* **2013**, *23*, 1976–1986. [CrossRef] [PubMed]

31. Tatina, R. Changes in *Fagus grandifolia* and *Acer saccharum* abundance in an old-growth, beech-maple forest at Warren Woods State Park, Berrien County, Michigan, USA. *Castanea* **2015**, *80*, 95–102. [CrossRef]

32. Natural Resources Conservation Service, United States Department of Agriculture. Web Soil Survey. Available online: https://websoilsurvey.sc.egov.usda.gov/ (accessed on 7 July 2017).

33. Initial County EAB Detections in North America 2017. Available online: https://www.aphis.usda.gov/plant_health/plant_pest_info/emerald_ash_b/downloads/MultiState.pdf (accessed on 11 February 2018).

34. Kuers, K.; Lindquist, L. PFPP Protocols and Datasheets. Available online: http://erenweb.org/new-page/carbon-storage-project/permanent-plot-protocol/pfpp-protocol-files/ (accessed on 9 March 2018).

35. Smith, A. Effects of community structure on forest susceptibility and response to the emerald ash borer invasion of the Huron River watershed in southeast Michigan. Master's Thesis, The Ohio State University, Columbus, OH, USA, 2006.

36. Knight, K.S.; Flash, B.P.; Kappler, R.H.; Throckmorton, J.A.; Grafton, B.; Flower, C.E. *Monitoring Ash (Fraxinus spp.) Decline and Emerald Ash Borer (Agrilus planipennis) Symptoms in Infested Areas*; General Technical Report NRS-139; U.S. Department of Agriculture, Forest Service, Northern Research Station: Newtown Square, PA, USA, 2014; p. 18.

37. Dolan, B.J.; Kilgore, J.S. Emerald Ash Borer Project. Available online: http://erenweb.org/new-page/eab/ (accessed on 11 February 2018).

38. Kilgore, J.S.; Dolan, B.J. Complementary Vegetation Survey (cVeg). Available online: http://erenweb.org/new-page/cveg/ (accessed on 11 February 2018).

39. Baker, F.S. A revised tolerance table. *J. For.* **1949**, *47*, 179–181. [CrossRef]

40. Niinemets, Ü.; Valladares, F. Tolerance to shade, drought, and waterlogging of temperate northern hemisphere trees and shrubs. *Ecol. Monogr.* **2006**, *76*, 521–547. [CrossRef]

41. Canham, C.D.; Denslow, J.S.; Platt, W.J.; Runkle, J.R.; Spies, T.A.; White, P.S. Light regimes beneath closed canopies and tree-fall gaps in temperate and tropical forests. *Can. J. For. Res.* **1990**, *20*, 620–631. [CrossRef]

42. Gottschalk, K.W. Shade, leaf growth and crown development of *Quercus rubra, Quercus velutina, Prunus serotina* and *Acer rubrum* seedlings. *Tree Physiol.* **1994**, *14*, 735–749. [CrossRef] [PubMed]

43. Parker, W.C.; Dey, D.C. Influence of overstory density on ecophysiology of red oak (*Quercus rubra*) and sugar maple (*Acer saccharum*) seedlings in central Ontario shelterwoods. *Tree Physiol.* **2008**, *28*, 797–804. [CrossRef] [PubMed]

44. Senécal, J.-F.; Doyon, F.; Messier, C. Tree death not resulting in gap creation: An investigation of canopy dynamics of northern temperate deciduous forests. *Remote Sens.* **2018**, *10*, 17. [CrossRef]

45. Klooster, W.S.; Herms, D.A.; Knight, K.S.; Herms, C.P.; McCullough, D.G.; Smith, A.; Gandhi, K.J.K.; Cardina, J. Ash (*Fraxinus* spp.) mortality, regeneration, and seed bank dynamics in mixed hardwood forests following invasion by emerald ash borer (*Agrilus planipennis*). *Biol. Invasions* **2014**, *16*, 859–873. [CrossRef]

46. Pušpure, I.; Laiviņš, M.; Matisons, R.; Gaitnieks, T. Understory changes in *Fraxinus excelsior* stands in response to dieback in Latvia. *Proc. Latv. Acad. Sci.* **2016**, *70*, 131–137. [CrossRef]

47. Huenneke, L.F. Understory response to gaps caused by the death of *Ulmus americana* in central New York. *Bull. Torrey Bot. Club* **1983**, *110*, 170–175. [CrossRef]

48. Dunn, C.P. Shrub layer response to death of *Ulmus americana* in southeastern Wisconsin lowland forests. *Bull. Torrey Bot. Club* **1986**, *113*, 142. [CrossRef]

49. Poulson, T.L.; Platt, W.J. Replacement patterns of beech and sugar maple in Warren Woods, Michigan. *Ecology* **1996**, *77*, 1234–1253. [CrossRef]
50. Yamamoto, S.-I. The gap theory in forest dynamics. *Bot. Mag. Tokyo* **1992**, *105*, 375–383. [CrossRef]
51. Hart, J.L.; Grissino-Mayer, H.D. Gap-scale disturbance processes in secondary hardwood stands on the Cumberland Plateau, Tennessee, USA. *Plant Ecol.* **2008**, *201*, 131–146. [CrossRef]
52. Cowell, C.M.; Mark Cowell, C.; Hoalst-Pullen, N.; Jackson, M.T. The limited role of canopy gaps in the successional dynamics of a mature mixed *Quercus* forest remnant. *J. Veg. Sci.* **2010**, *21*, 201–212. [CrossRef]
53. Flatley, W.T.; Lafon, C.W.; Grissino-Mayer, H.D.; LaForest, L.B. Changing fire regimes and old-growth forest succession along a topographic gradient in the Great Smoky Mountains. *For. Ecol. Manage.* **2015**, *350*, 96–106. [CrossRef]
54. Chapman, J.I.; McEwan, R.W. Thirty years of compositional change in an old-growth temperate forest: The role of topographic gradients in oak-maple dynamics. *PLoS ONE* **2016**, *11*, e0160238. [CrossRef] [PubMed]
55. Knopp, P.D. The distribution of *Quercus rubra* in the Maumee Lake Plain of southeastern Michigan. *Am. Midl. Nat.* **2012**, *168*, 70–92. [CrossRef]
56. Jerome, D.; Westwood, M.; Oldfield, S.; Romero-Severson, J. Fraxinus Americana. Available online: http://www.iucnredlist.org/details/61918430/0 (accessed on 11 February 2018).
57. Westwood, M.; Oldfield, S.; Jerome, D.; Romero-Severson, J. *Fraxinus quadrangulata*. Available online: http://www.iucnredlist.org/details/61919112/0 (accessed on 11 February 2018).
58. Jerome, D.; Westwood, M.; Oldfield, S.; Romero-Severson, J. *Fraxinus nigra*. Available online: http://www.iucnredlist.org/details/61918683/0 (accessed on 11 February 2018).
59. Westwood, M.; Oldfield, S.; Jerome, D.; Romero-Severson, J. *Fraxinus pennsylvanica*. Available online: http://www.iucnredlist.org/details/61918934/0 (accessed on 11 February 2018).
60. Westwood, M.; Jerome, D.; Oldfield, S.; Romero-Severson, J. *Fraxinus profunda*. Available online: http://www.iucnredlist.org/details/61919022/0 (accessed on 11 February 2018).

MDPI
St. Alban-Anlage 66
4052 Basel
Switzerland
Tel. +41 61 683 77 34
Fax +41 61 302 89 18
www.mdpi.com

Forests Editorial Office
E-mail: forests@mdpi.com
www.mdpi.com/journal/forests

www.ingramcontent.com/pod-product-compliance
Lightning Source LLC
Chambersburg PA
CBHW051718210326
41597CB00032B/5529